FLA. SOLAR ENERGY CENTER LIBRARY

**Catalysis for
Sustainable Energy Production**

*Edited by
Pierluigi Barbaro and
Claudio Bianchini*

Related Titles

Crabtree, R. H. (ed.)

Handbook of Green Chemistry - Green Catalysis

2009

ISBN: 978-3-527-31577-2

Parmon, V., Vorontsov, A., Kozlov, D., Smirniotis, P.

Photocatalysis

Catalysts, Kinetics and Reactors

2008

ISBN: 978-3-527-31784-4

Züttel, A., Borgschulte, A., Schlapbach, L. (eds.)

Hydrogen as a Future Energy Carrier

2008

ISBN: 978-3-527-30817-0

Endres, F., MacFarlane, D., Abbott, A. (eds.)

Electrodeposition from Ionic Liquids

2008

ISBN: 978-3-527-31565-9

Wasserscheid, P., Welton, T. (eds.)

Ionic Liquids in Synthesis

2008

ISBN: 978-3-527-31239-9

Roberts, S. M.

Catalysts for Fine Chemical Synthesis

Volumes 1-5. Set

2007

ISBN: 978-0-470-51605-8

Sheldon, R. A., Arends, I., Hanefeld, U.

Green Chemistry and Catalysis

2007

ISBN: 978-3-527-30715-9

Catalysis for Sustainable Energy Production

Edited by
Pierluigi Barbaro and Claudio Bianchini

WILEY-VCH Verlag GmbH & Co. KGaA

The Editors

Dr. Pierluigi Barbaro
ICCOM/CNR
Via Madonna del Piano 10
50019 Sesto Fiorentino
Italy

Dr. Claudio Bianchini
ICCOM/CNR
Area della Ricerca
Via Madonna Del Piano 10
50019 Sesto Fiorentino
Italy

All books published by Wiley-VCH are carefully produced. Nevertheless, authors, editors, and publisher do not warrant the information contained in these books, including this book, to be free of errors. Readers are advised to keep in mind that statements, data, illustrations, procedural details or other items may inadvertently be inaccurate.

Library of Congress Card No.: applied for

British Library Cataloguing-in-Publication Data
A catalogue record for this book is available from the British Library.

Bibliographic information published by the Deutsche Nationalbibliothek
The Deutsche Nationalbibliothek lists this publication in the Deutsche Nationalbibliografie; detailed bibliographic data are available on the Internet at http://dnb.d-nb.de.

© 2009 WILEY-VCH Verlag GmbH & Co. KGaA, Weinheim

All rights reserved (including those of translation into other languages). No part of this book may be reproduced in any form – by photoprinting, microfilm, or any other means – nor transmitted or translated into a machine language without written permission from the publishers. Registered names, trademarks, etc. used in this book, even when not specifically marked as such, are not to be considered unprotected by law.

Composition Thomson Digital, Noida, India
Printing Betz-Druck GmbH, Darmstadt
Bookbinding Litges & Dopf GmbH, Heppenheim
Cover Design Schulz Grafik-Design, Fußgönheim

Printed in the Federal Republic of Germany
Printed on acid-free paper

ISBN: 978-3-527-32095-0

Contents

Foreword *XIII*
Epilogue *XVII*
List of Contributors *XIX*

Part One Fuel Cells *1*

1 The Direct Ethanol Fuel Cell: a Challenge to Convert Bioethanol Cleanly into Electric Energy *3*
Claude Lamy, Christophe Coutanceau, and Jean-Michel Leger
1.1   Introduction *3*
1.2   Principles and Different Kinds of Fuel Cells *4*
1.2.1   Working Principles of a Fuel Cell *4*
1.2.1.1 The Thermodynamics of Fuel Cells *5*
1.2.1.2 The Kinetics of Fuel Cells *6*
1.2.1.3 Catalysis of Fuel Cell Reactions *9*
1.2.2   Different Types of Fuel Cells *14*
1.2.2.1 Fuels for Fuel Cells *14*
1.2.2.2 Hydrogen-fed Fuel Cells *16*
1.2.2.3 Methanol- and Ethanol-fed Fuel Cells *16*
1.3   Low-temperature Fuel Cells (PEMFCs and DAFCs) *17*
1.3.1   Proton Exchange Membrane Fuel Cell (PEMFC) *17*
1.3.1.1 Principle of a PEMFC *17*
1.3.1.2 The Proton Exchange Membrane *18*
1.3.1.3 The Electrode Catalysts *19*
1.3.1.4 The Membrane–Electrode Assembly *19*
1.3.1.5 The Bipolar Plates *19*
1.3.1.6 Auxiliary and Control Equipment *19*
1.3.2   Direct Ethanol Fuel Cell (DEFC) *21*
1.3.2.1 Principle of the Direct Ethanol Fuel Cell *21*
1.3.2.2 Reaction Mechanisms of Ethanol Oxidation *22*
1.3.2.3 DEFC Tests *26*
1.4   Solid Alkaline Membrane Fuel Cell (SAMFC) *29*

Catalysis for Sustainable Energy Production. Edited by P. Barbaro and C. Bianchini
Copyright © 2009 WILEY-VCH Verlag GmbH & Co. KGaA, Weinheim
ISBN: 978-3-527-32095-0

1.4.1	Development of a Solid Alkaline Membrane for Fuel Cell Application *29*
1.4.2	Anodic Catalysts in Alkaline Medium *32*
1.4.3	Cathodic Catalysts in Alkaline Medium *38*
1.5	Conclusion *42*
	References *42*

2 Performance of Direct Methanol Fuel Cells for Portable Power Applications *47*
Xiaoming Ren

2.1	Introduction *47*
2.2	Experimental *49*
2.3	Results and Discussion *51*
2.3.1	Water Balance, Maximum Air Feed Rate and Implications for Cathode Performance *51*
2.3.2	Stack Performance *57*
2.3.3	Thermal Balance and Waste Heat Rejection *64*
2.3.4	Stack Life Test Results *65*
2.4	Conclusions *67*
	References *68*

3 Selective Synthesis of Carbon Nanofibers as Better Catalyst Supports for Low-temperature Fuel Cells *71*
Seong-Hwa Hong, Mun-Suk Jun, Isao Mochida, and Seong-Ho Yoon

3.1	Introduction *71*
3.2	Preparation and Characterization of CNFs and Fuel Cell Catalysts *73*
3.2.1	Preparation of Typical CNFs *73*
3.2.2	Preparation of Nanotunneled Mesoporous H-CNF *73*
3.2.3	Preparation of Fuel Cell Catalysts *74*
3.2.4	Performance Characterization of Fuel Cell Catalysts *74*
3.3	Results *74*
3.3.1	Structural Effects of CNFs *74*
3.3.2	Catalytic Performance of CNFs in Half and Single Cells *76*
3.3.3	Structure of Nanotunneled Mesoporous Thick H-CNF *78*
3.3.4	Catalytic Performance of Nanotunneled Mesoporous Thick H-CNF *78*
3.3.5	Effect of the Dispersion of Thin and Very Thin H-CNFs on the Catalyst Activity *81*
3.4	Discussion *84*
	References *86*

4 Towards Full Electric Mobility: Energy and Power Systems *89*
Pietro Perlo, Marco Ottella, Nicola Corino, Francesco Pitzalis, Mauro Brignone, Daniele Zanello, Gianfranco Innocenti, Luca Belforte, and Alessandro Ziggiotti

| 4.1 | Introduction *89* |

4.2	The Current Grand Challenges	89
4.3	Power–Energy Needed in Vehicles	90
4.3.1	Basic Formulation	90
4.3.2	Well to Wheel Evaluations	92
4.3.3	Specific Calculations for Ideal Electric Powertrains	92
4.3.4	A Roadmap of Feasibility with Batteries and Supercapacitors	95
4.3.5	The Need for Range Extenders	96
4.3.5.1	Direct Thermoelectric Generators	98
4.4	A Great New Opportunity for True Zero Emissions	101
4.5	Advanced Systems Integration	102
4.6	Conclusion and Perspectives	103
	References	104

Part Two Hydrogen Storage 107

5 Materials for Hydrogen Storage 109
Andreas Züttel

5.1	The Primitive Phase Diagram of Hydrogen	109
5.2	Hydrogen Storage Methods	109
5.3	Pressurized Hydrogen	111
5.3.1	Properties of Compressed Hydrogen	112
5.3.2	Pressure Vessel	113
5.3.3	Volumetric and Gravimetric Hydrogen Density	114
5.3.4	Microspheres	115
5.4	Liquid Hydrogen	117
5.4.1	Liquefaction Process	118
5.4.2	Storage Vessel	119
5.4.3	Gravimetric and Volumetric Hydrogen Density	120
5.5	Physisorption	121
5.5.1	Van der Waals Interaction	121
5.5.2	Adsorption Isotherm	122
5.5.3	Hydrogen and Carbon Nanotubes	123
5.6	Metal Hydrides	128
5.6.1	Interstitial Hydrides	128
5.6.2	Hydrogen Absorption	130
5.6.3	Empirical Models	133
5.6.4	Lattice Gas Model	137
5.7	Complex Hydrides	141
5.7.1	Tetrahydroalanates	143
5.7.2	Tetrahydroborates	148
5.8	Chemical Hydrides (Hydrolysis)	154
5.8.1	Zinc Cycle	154
5.8.2	Borohydride	156
5.9	New Hydrogen Storage Materials	157
5.9.1	Amides and Imides ($-NH_2$, $=NH$)	158

5.9.2	bcc Alloys	160
5.9.3	AlH$_3$	160
5.9.4	Metal Hydrides with Short H-H-Distance	161
5.9.5	MgH$_2$ with a New Structure	162
5.9.6	Destabilization of MgH$_2$ by Alloy Formation	162
5.9.7	Ammonia Storage	163
5.9.8	Borazane	163
	References	165

Part Three H$_2$ and Hydrogen Vectors Production 171

6 Catalyst Design for Reforming of Oxygenates 173
Loredana De Rogatis and Paolo Fornasiero

6.1	Introduction	173
6.2	Catalyst Design	179
6.2.1	Impregnated Catalysts: the Role of Metal, Support and Promoters	181
6.2.2	Emerging Strategies: Embedded Catalysts	183
6.3	Reforming Reactions: Process Principles	185
6.3.1	Catalytic Steam Reforming	185
6.3.2	Catalytic Partial Oxidation	188
6.3.3	Autothermal Reforming	189
6.3.4	Aqueous Phase Reforming	190
6.4	Key Examples of Oxygenate Reforming Reactions	193
6.4.1	Methanol	193
6.4.2	Ethanol	197
6.4.3	Dimethyl Ether	203
6.4.4	Acetic Acid	207
6.4.5	Sugars	210
6.4.6	Ethylene Glycol	214
6.4.7	Glycerol	219
6.5	Conclusions	222
6.6	List of Abbreviations	223
	References	224

7 Electrocatalysis in Water Electrolysis 235
Edoardo Guerrini and Sergio Trasatti

7.1	Introduction	235
7.2	Thermodynamic Considerations	237
7.3	Kinetic Considerations	239
7.3.1	Equilibrium Term (ΔE)	240
7.3.2	Ohmic Dissipation Term (IR)	240
7.3.2.1	Cell Design	241
7.3.3	Stability Term (ΔV_t)	242
7.3.4	Overpotential Dissipation Term ($\Sigma\eta$)	243
7.3.5	Electrocatalysis	244

7.3.5.1	Theory of Electrocatalysis	245
7.4	The Hydrogen Evolution Reaction	248
7.4.1	Reaction Mechanisms	248
7.4.2	Electrocatalysis	249
7.4.3	Materials for Cathodes	251
7.4.4	Factors of Electrocatalysis	252
7.5	The Oxygen Evolution Reaction	255
7.5.1	Reaction Mechanisms	255
7.5.2	Anodic Oxides	256
7.5.3	Thermal Oxides (DSA)	257
7.5.4	Electrocatalysis	259
7.5.5	Factors of Electrocatalysis	260
7.5.6	Intermittent Electrolysis	263
7.6	Electrocatalysts: State-of-the-Art	264
7.7	Water Electrolysis: State-of-the-Art	265
7.8	Beyond Oxygen Evolution	265
	References	267

8 Energy from Organic Waste: Influence of the Process Parameters on the Production of Methane and Hydrogen 271
Michele Aresta and Angela Dibenedetto

8.1	Introduction	271
8.2	Experimental	273
8.2.1	Methanation of Residual Biomass	273
8.2.2	Bioconversion of Glycerol	274
8.2.2.1	Characterization of Strains K1–K4	274
8.2.2.2	Use of Strains K1, K2 and K3	274
8.2.2.3	Use of Strain K4	274
8.2.2.4	Tests Under Aerobic Conditions	275
8.2.2.5	Tests Under Microaerobic or Anaerobic Conditions	275
8.3	Results and Discussion	275
8.3.1	Biogas from Waste	275
8.3.2	Dihydrogen from Bioglycerol	279
	References	284

9 Natural Gas Autothermal Reforming: an Effective Option for a Sustainable Distributed Production of Hydrogen 287
Paolo Ciambelli, Vincenzo Palma, Emma Palo, and Gaetano Iaquaniello

9.1	Introduction	287
9.2	Autothermal Reforming: from Chemistry to Engineering	294
9.2.1	The Catalyst	294
9.2.2	Kilowatt-scale ATR Fuel Processors	298
9.3	Thermodynamic Analysis	299
9.3.1	Effect of Preheating the Reactants	300

9.3.2	Effect of $O_2:CH_4$ and $H_2O:CH_4$ Molar Feed Ratios	300
9.4	A Case Study	303
9.4.1	Laboratory Apparatus and ATR Reactor	303
9.4.2	ATR Reactor Setup: Operating Conditions	306
9.4.3	ATR Reactor Setup: Start-up Phase	306
9.4.4	ATR Reactor Setup: Influence of Preheating the Reactants	307
9.4.5	Catalytic Activity Test Results	309
9.5	Economic Aspects	313
9.6	Conclusions and Perspectives	316
	References	317

Part Four Industrial Catalysis for Sustainable Energy 321

10 The Use of Catalysis in the Production of High-quality Biodiesel 323
Nicoletta Ravasio, Federica Zaccheria, and Rinaldo Psaro

10.1	Introduction	323
10.2	Heterogeneous Transesterification and Esterification Catalysts	328
10.2.1	Heterogeneous Basic Catalysts	328
10.2.2	Heterogeneous Acid Catalysts	330
10.3	Selective Hydrogenation in Biodiesel Production	336
10.4	Conclusions and Perspectives	341
	References	342

11 Photovoltaics – Current Trends and Vision for the Future 345
Francesca Ferrazza

11.1	Introduction	345
11.2	Market: Present Situation and Challenges Ahead	346
11.3	Crystalline Silicon Technology	348
11.3.1	From Feedstock to Wafers	348
11.3.2	From Wafers to Cells and Modules	349
11.3.3	Where to Cut Costs	351
11.4	Thin Films	353
11.4.1	Technology and Improvement Requirements	354
11.5	Other Technology-related Aspects	355
11.6	Advanced and Emerging Technologies	357
11.7	System Aspects	359
11.8	Conclusions	361
	References	362

12 Catalytic Combustion for the Production of Energy 363
Gianpiero Groppi, Cinzia Cristiani, Alessandra Beretta, and Pio Forzatti

12.1	Introduction	363
12.2	Lean Catalytic Combustion for Gas Turbines	364
12.2.1	Principles and System Requirements	364
12.2.2	Design Concepts and Performance	366

12.2.2.1	Fully Catalytic Combustor	366
12.2.2.2	Fuel Staging	367
12.2.2.3	Partial Catalytic Hybrid Combustor	367
12.3	Fuel-rich Catalytic Combustion	370
12.4	Oxy-fuel Combustion	372
12.5	Microcombustors	373
12.6	Catalytic Materials	375
12.6.1	Structured Substrate	376
12.6.2	Active Catalyst Layer	376
12.6.2.1	PdO-based Catalysts	377
12.6.2.2	Metal-substituted Hexaaluminate Catalysts	381
12.6.2.3	Rich Combustion Catalysts	382
12.7	Conclusions	387
	References	388

13 Catalytic Removal of NO_x Under Lean Conditions from Stationary and Mobile Sources 393

Pio Forzatti, Luca Lietti, and Enrico Tronconi

13.1	Introduction	393
13.2	Selective Catalytic Reduction	395
13.2.1	Standard SCR Process	395
13.2.2	SCR Applications: Past and Future	399
13.2.3	Modeling of the SCR Reactor	400
13.2.3.1	Steady-state Modeling of the SCR Reactor	400
13.2.3.2	Unsteady-state Kinetics of the Standard SCR Reaction	401
13.2.3.3	Unsteady-state Models of the Monolith SCR Reactor	406
13.2.4	Fast SCR	409
13.2.4.1	Mechanism of Fast SCR	409
13.2.4.2	Unsteady-state Models of the Monolith $NO/NO_2/NH_3$ SCR Reactor	412
13.3	NO_x Storage Reduction	414
13.3.1	NSR Technology	414
13.3.2	Storage of NO_x	415
13.3.2.1	Mechanistic Features	415
13.3.2.2	Kinetics	421
13.3.2.3	Effect of CO_2	422
13.3.3	Reduction of Stored NO_x	424
13.3.3.1	Mechanism of the Reduction by H_2 of Stored NO_x	425
13.3.3.2	Identification of the Reaction Network During Reduction of Stored NO_x by H_2	428
13.4	Open Issues and Future Opportunities	432
	References	433

Index 439

Foreword
Gabriele Centi

Catalysis is a major science behind sustainable energy. No matter what the energy source is – oil, natural gas, coal, biomass, solar – a clean, sustainable energy future will involve catalysis. However, at the same time we are facing a rapidly evolving energetic scenario, due to the faster rise in the cost of oil and natural gas with respect to the levels predicted only few years ago, the changing geo-political strategies and the increasing awareness of society regarding sustainability. Challenges and priority areas for catalysis in sustainable energy, across the energy distribution system and for all major energy sources are changing rapidly. The next 20 years will be dominated by major changes in the uses of catalysis for energy, due to the following main drivers:

- high fossil fuel costs
- the geo-political and social (more than economic) need to increase the use of biomasses, but not in competition with food
- introduction and/or extension of the use of more eco-friendly energy vectors (electron, hydrogen)
- the need not to postpone any longer a major step forward in the use of solar energy
- progressive introduction of delocalized energy production
- social pressure for sustainability, in terms of impact on the environment and life quality of energy production and use and efficient use of energy
- the need to mitigate climate changes and reduce greenhouse gas emissions.

These drivers push to accelerate the transition towards more cost-effective renewable energy technologies and new lower energy demand technologies. In the transition to a new smart-energy world, there is the need to find more efficient ways of producing, refining and using fossil fuels, maximizing the use of actual resources and introducing new technologies to use low-value fossil fuels (heavy residues, coal, oil shale and similar materials). Improved catalysts are needed to increase process efficiency, reduce energy and produce cleaner products such as ultralow sulfur diesel.

Catalysis for Sustainable Energy Production. Edited by P. Barbaro and C. Bianchini
Copyright © 2009 WILEY-VCH Verlag GmbH & Co. KGaA, Weinheim
ISBN: 978-3-527-32095-0

For a sustainable hydrogen economy, hydrogen will need to be produced using renewable energy and a sustainable feedstock rather than from petrochemical sources. A successful transition to the hydrogen economy will require a reliable and efficient means of creating, transporting and storing hydrogen. Significant research is still required before hydrogen fuel cells will become competitive, such as for the development of better catalysts for fuel cells, the investigation of new, reliable solutions and catalysts for direct alcohol fuel cells and the development of new ideas to combine chemical and electrical energy production, preferably using waste as raw materials.

Reducing the energy demand of the chemical industry and improving its sustainability require the development of novel approaches and consideration of alternative sources of energy for chemical reactions such as photochemical, microwave and ultrasonic. These may offer significant energy saving over more conventional energy sources and also allow progress towards a relevant strategic objective to implement sustainable chemistry: process intensification to develop novel modular and energy-efficient chemical processes. Catalysis is the core technology to achieve these objectives.

Catalysis is thus an enabling factor for sustainable energy, but a new approach to catalysis is needed in order to address the major changes that we will face in the near future. For this reason, IDECAT (a European Network of Excellence on Catalysis) organized in 2006 two thematic (brainstorm-like) workshops dedicated to the identification of a roadmap and priorities for a catalytic approach for sustainable energy production and for the direct catalytic conversion of renewable feedstocks into energy.

The first workshop, 'Catalysis for Renewables', was held in Rolduc (Kerkrade, The Netherlands) on 16–18 May 2006 and was dedicated mainly to the discussion of the catalytic process options for the conversion of renewable feedstocks into energy and chemicals. A book was published as result of this effort: Gabriele Centi and Rutger A. van Santen (eds), *Catalysis for Renewables: from Feedstock to Energy Production*, Wiley VCH Verlag GmbH, Weinheim, 2007 (ISBN 978-3-527-31788-2). The different chapters in this book cover various aspects mainly related to biomass conversion, but with some chapters also discussing the relationship between catalysis and solar energy and H_2 as the bridge to a sustainable energy system. The concluding chapter reported a perspective critical synthesis of topics discussed in the book and a concise presentation of the Research Strategic Agenda (SRA) of catalysis for renewables.

A second workshop, 'Catalysis for Sustainable Energy Production', was held in Sesto Fiorentino (Florence, Italy) from 29 November to 1 December 2006. The structure and approach of this workshop were similar to those of the first, but the focus was on (i) fuel cells, (ii) hydrogen and methane storage and (iii) H_2 production from old to new processes, including those using renewable energy sources. The present book is based on this second workshop and reports a series of invited contributions which provide both the 'state-of-the-art' and frontier research in the field. Many contributions are from industry, but authors were also asked to focus their description on the identification of priority topics and problems. The active discussions during the workshop are reflected in the various chapters of this book.

Although a specific chapter dedicated to the role of catalysis in energy production is not included, because most of the aspects were already covered in the final chapter of the previous book, the reader can easily use the final sections of each chapter to identify priorities for research on catalysis for sustainable energy.

The workshop was organized by the CNR (National Research Council of Italy) and INSTM (Italian Interuniversity Consortium for the Science and Technology of Materials) in the frame of the activities of the EU Network of Excellence (NoE) IDECAT (Integrated Design of Catalytic Nanomaterials for Sustainable Production).

The objective of this NoE is to strengthen research in catalysis by the creation of a coherent framework of research, know-how and training between the various disciplinary catalysis communities (heterogeneous, homogeneous and bio-catalysis) with the objective of achieving a lasting integration between the main European institutions in this area. IDECAT will create (starting from Spring 2008) the virtual Institute 'European Research Institute on Catalysis' (ERIC), which is intended to be the main reference point for catalysis in Europe and a reference center for catalysis and sustainable energy.

IDECAT focuses its research actions on (i) the synthesis and mastering of nano-objects, the materials of the future for catalysis, integrating the concepts common also to other nanotechnologies, (ii) bridging the gap between theory and modeling, surface science and kinetic/applied catalysis and also between heterogeneous, homogeneous and biocatalytic approaches and (iii) developing an integrated design of catalytic nanomaterials.

The objectives of IDECAT are to:

1. Create a critical mass of expertise going beyond collaboration.
2. Create a strong cultural thematic identity on nanotech-based catalysts.
3. Increase cost-effectiveness of European research.
4. Establish a frontier research portfolio able to promote innovation in catalysis use especially at the SMEs level.
5. Increase potential for training and education in multidisciplinary approaches to nanotech-based catalysis.
6. Spread excellence beyond the NoE to both the scientific community and to the citizen.

Next-generation catalysts should achieve zero waste emissions and use selectively the energy in chemical reactions. In addition, they will permit the development of new bio-mimicking catalytic transformations, new clean energy sources and chemical storage methods, the utilization of new and/or renewable raw materials and reuse of the waste, solving global issues (greenhouse gas emissions, water and air quality) and realizing smart devices. These challenging objectives can be reached only through a synergic interaction between the best catalytic research centers and permanent and strong interaction with companies and public institutions. This is the scope of IDECAT.

In conclusion, this book constitutes a further step in IDECAT's aim to develop a coherent framework of activities to create sustainable energy and society through catalysis. This book is at the same time an updated overview of the state-of-the-art and a roadmap, which defines new directions, opportunities and needs for R&D. Finally, warm thanks are due to Dr Pierluigi Barbaro of CNR (ICCOM, Florence) and Dr ssa Serena Orsi (INSTM and CNR, Florence), whose continued support enabled both the workshop cited to take place and this book on *Catalysis for Sustainable Energy Production* to be produced.

Gabriele Centi

INSTM and University Messina, Italy
IDECAT Coordinator

Epilogue
Claudio Bianchini

This book collects many of the most outstanding contributions presented at the workshop 'Catalysis for Sustainable Energy Production' held in Sesto Fiorentino (Florence, Italy) from 29 November to 1 December 2006. Unlike the previous book in the IDECAT series, *Catalysis for Renewables: from Feedstock to Energy Production*, edited by Gabriele Centi and Rutger A. van Santen, the focus of this book is on the technological innovation related to sustainable energy production rather than on specific catalytic processes. Indeed, it was our principal aim to provide the reader with a updated view of the technology involved in relevant energy production processes, with particular emphasis on critical issues in the fields of low-temperature fuel cells, hydrogen and methane storage, hydrogen and biodiesel production by conventional and bio-processes, photovoltaics and catalytic combustion.

The authors were asked to highlight the deficiencies in current technologies and, at the same time, to offer a perspective to devise possible solutions. My feeling is that the mission has been fully accomplished. The series of contributions dealing with direct alcohol fuel cells are paradigmatic in this sense: the complexity of the chemical processes involved at either electrode and of the devices so far developed has received more attention than traditional electrochemical results with appealing power density curves. Likewise, the sustainable production of hydrogen and its unconventional storage have been faced up to with the intention of offering not only a view of the 'state-of-the art' but also mainly to warn the reader that the way to success is still long and hard.

Finally, we are confident that this book may be a good companion as well as a guide to the very many senior and young researchers who have decided to use their talents to offer new perspectives to Sustainable Energy production.

List of Contributors

Michele Aresta
University of Bari
Department of Chemistry and CIRCC
Campus Universitario
70126 Bari
Italy

Luca Belforte
Centro Ricerche Fiat
Strada Torino 50
10043 Orbassano
Italy

Alessandra Beretta
Politecnico di Milano
Dipartimento di Energia
Laboratory of Catalysis and Catalytic Processes
Piazza Leonardo da Vinci 32
20133 Milan
Italy

Mauro Brignone
Centro Ricerche Fiat
Strada Torino 50
10043 Orbassano
Italy

Paolo Ciambelli
University of Salerno
Department of Chemical and Food Engineering
Via Ponte Don Melillo
84084 Fisciano
Italy

Nicola Corino
Centro Ricerche Fiat
Strada Torino 50
10043 Orbassano
Italy

Christophe Coutanceau
CNRS – Université de Poitiers
UMR 6503
Laboratory of Electrocatalysis
40 avenue du Recteur Pineau
86022 Poitiers
France

Cinzia Cristiani
Politecnico di Milano
Dipartimento di Chimica
Materiali e Ingegneria Chimica 'G. Natta'
Piazza Leonardo da Vinci 32
20133 Milan
Italy

Angela Dibenedetto
University of Bari
Department of Chemistry and CIRCC
Campus Universitario
70126 Bari
Italy

Francesca Ferrazza
Eni SpA
Direzione Strategie e Sviluppo
P. le Mattei 1
00144 Rome
Italy

Paolo Fornasiero
University of Trieste
Chemistry Department
Center of Excellence for Nanostructured
Materials (CENMAT) and INSTM–
Trieste Research Unit
Via L. Giorgieri 1
34127 Trieste
Italy

and

ICCOM–CNR Associate Research Unit

Pio Forzatti
Politecnico di Milano
Dipartimento di Energia
Laboratory of Catalysis and Catalytic
Processes
Piazza Leonardo da Vinci 32
20133 Milan
Italy

Gianpiero Groppi
Politecnico di Milano
Dipartimento di Energia
Laboratory of Catalysis and Catalytic
Processes
Piazza Leonardo da Vinci 32
20133 Milan
Italy

Edoardo Guerrini
University of Milan
Department of Physical Chemistry and
Electrochemistry
Via Venezian 21
20133 Milan
Italy

Seong-Hwa Hong
Kyushu University
Institute for Materials Chemistry and
Engineering
Kasuga Koen 6-1
Kasuga
Fukuoka 816-8580
Japan

Gaetano Iaquaniello
Technip KTI SpA
Viale Castello della Magliana 75
00148 Rome
Italy

Gianfranco Innocenti
Centro Ricerche Fiat
Strada Torino 50
10043 Orbassano
Italy

Mun-Suk Jun
Kyushu University
Institute for Materials Chemistry and
Engineering
Kasuga Koen 6-1
Kasuga
Fukuoka 816-8580
Japan

Claude Lamy
CNRS – Université de Poitiers
UMR 6503
Laboratory of Electrocatalysis
40 avenue du Recteur Pineau
86022 Poitiers
France

List of Contributors

Jean-Michel Leger
CNRS – Université de Poitiers
UMR 6503
Laboratory of Electrocatalysis
40 avenue du Recteur Pineau
86022 Poitiers
France

Luca Lietti
Politecnico di Milano
Dipartimento di Energia
Laboratory of catalysis and catalytic processes
Pza Leonardo da Vinci 32
20133 Milan
Italy

Isao Mochida
Kyushu University
Institute for Materials Chemistry and Engineering
Kasuga Koen 6-1
Kasuga
Fukuoka 816-8580
Japan

Marco Ottella
Centro Ricerche Fiat
Strada Torino 50
10043 Orbassano
Italy

Vincenzo Palma
University of Salerno
Department of Chemical and Food Engineering
Via Ponte Don Melillo
84084 Fisciano
Italy

Emma Palo
University of Salerno
Department of Chemical and Food Engineering
Via Ponte Don Melillo
84084 Fisciano
Italy

Francesco Pitzalis
Centro Ricerche Fiat
Strada Torino 50
10043 Orbassano
Italy

Pietro Perlo
Centro Ricerche Fiat
Strada Torino 50
10043 Orbassano
Italy

Rinaldo Psaro
Consiglio Nazionale delle Ricerche
Istituto di Scienze e Tecnologie Molecolari
Dipartimento di Chimica Inorganica Metallorganica e Analitica
Via G. Venezian 21
20133 Milan
Italy

Nicoletta Ravasio
Consiglio Nazionale delle Ricerche
Istituto di Scienze e Tecnologie Molecolari
Dipartimento di Chimica Inorganica Metallorganica e Analitica
Via G. Venezian 21
20133 Milan
Italy

Xiaoming Ren
ACTA Spa
Via di Lavoria 56/G
56040 Crespina
Italy

Loredana De Rogatis
University of Trieste
Chemistry Department
Center of Excellence for Nanostructured
Materials (CENMAT) and INSTM–
Trieste Research Unit
Via L. Giorgieri 1
34127 Trieste
Italy

Sergio Trasatti
University of Milan
Department of Physical Chemistry and
Electrochemistry
Via Venezian 21
20133 Milan
Italy

Enrico Tronconi
Politecnico di Milano
Dipartimento di Energia
Laboratory of catalysis and catalytic
processes
Pza Leonardo da Vinci 32
20133 Milan
Italy

Seong-Ho Yoon
Kyushu University
Institute for Materials Chemistry and
Engineering
Kasuga Koen 6-1
Kasuga
Fukuoka 816-8580
Japan

Federica Zaccheria
Consiglio Nazionale delle Ricerche
Istituto di Scienze e Tecnologie
Molecolari
Dipartimento di Chimica Inorganica
Metallorganica e Analitica
Via G. Venezian 21
20133 Milan
Italy

Daniele Zanello
Centro Ricerche Fiat
Strada Torino 50
10043 Orbassano
Italy

Alessandro Ziggiotti
Centro Ricerche Fiat
Strada Torino 50
10043 Orbassano
Italy

Andreas Züttel
Empa Materials Science and Technology
Department of Energy
Environment and Mobility
Section Hydrogen and Energy
8600 Dübendorf
Switzerland

Part One
Fuel Cells

1
The Direct Ethanol Fuel Cell: a Challenge to Convert Bioethanol Cleanly into Electric Energy
Claude Lamy, Christophe Coutanceau, and Jean-Michel Leger

1.1
Introduction

Discovered in England in 1839 by Sir William Grove, a fuel cell (FC) is an electrochemical device which transforms directly the heat of combustion of a fuel (hydrogen, natural gas, methanol, ethanol, hydrocarbons, etc.) into electricity [1]. The fuel is electrochemically oxidized at the anode, without producing any pollutants (only water and/or carbon dioxide are released into the atmosphere), whereas the oxidant (oxygen from the air) is reduced at the cathode. This process does not follow Carnot's theorem, so that higher energy efficiencies are expected: 40–50% in electrical energy, 80–85% in total energy (electricity + heat production).

There is now a great interest in developing different kinds of fuel cells with several applications (in addition to the first and most developed application in space programs) depending on their nominal power: stationary electric power plants (100 kW–10 MW), power train sources (20–200 kW) for the electrical vehicle (bus, truck and individual car), electricity and heat co-generation for buildings and houses (5–20 kW), auxiliary power units (1–100 kW) for different uses (automobiles, aircraft, space launchers, space stations, uninterruptible power supply, remote power, etc.) and portable electronic devices (1–100 W), for example, cell phones, computers, camcorders [2, 3].

For many applications, hydrogen is the most convenient fuel, but it is not a primary fuel, so that it has to be produced from different sources: water, fossil fuels (natural gas, hydrocarbons, etc.), biomass resources and so on. Moreover, the clean production of hydrogen (including the limitation of carbon dioxide production) and the difficulties with its storage and large-scale distribution are still strong limitations for the development of such techniques [2, 3]. In this context, other fuels, particularly those, like alcohols, which are liquid at ambient temperature and pressure, are more convenient due to the ease of their handling and distribution.

Therefore, alcohols have begun to be considered as valuable alternative fuels, because they have a high energy density (6–9 kW h kg^{-1}, compared with 33 kW h kg^{-1}

Catalysis for Sustainable Energy Production. Edited by P. Barbaro and C. Bianchini
Copyright © 2009 WILEY-VCH Verlag GmbH & Co. KGaA, Weinheim
ISBN: 978-3-527-32095-0

for pure hydrogen without a storage tank and about 11 kW h kg^{-1} for gasoline) and they can be obtained from renewable sources (e.g. bioethanol from biomass feedstock). Ethanol is an attractive fuel for electric vehicles, since it can be easily produced in great quantities by the fermentation of cellulose-containing raw materials from agriculture (corn, wheat, sugar beet, sugar cane, etc.) or from different wastes (e.g. agricultural wastes containing lignocellulosic residues). In addition, in some countries such as Brazil and the USA and more recently in France (with the E85 fuel containing 85% of ethanol), ethanol is already distributed through the fuel station network for use in conventional automobiles with internal combustion engines (flex-fuel vehicles). Moreover, for portable electronics, particularly cell phones, ethanol can advantageously replace methanol, which is used in the direct methanol fuel cell (DMFC), since it is less toxic and has better energy density and similar kinetics at low temperature.

DMFCs and direct ethanol fuel cells (DEFCs) are based on the proton exchange membrane fuel cell (PEMFC), where hydrogen is replaced by the alcohol, so that both the principles of the PEMFC and the direct alcohol fuel cell (DAFC), in which the alcohol reacts directly at the fuel cell anode without any reforming process, will be discussed in this chapter. Then, because of the low operating temperatures of these fuel cells working in an acidic environment (due to the protonic membrane), the activation of the alcohol oxidation by convenient catalysts (usually containing platinum) is still a severe problem, which will be discussed in the context of electrocatalysis. One way to overcome this problem is to use an alkaline membrane (conducting, e.g., by the hydroxyl anion, OH^-), in which medium the kinetics of the electrochemical reactions involved are faster than in an acidic medium, and then to develop the solid alkaline membrane fuel cell (SAMFC).

After rehearsing the working principles and presenting the different kinds of fuel cells, the proton exchange membrane fuel cell (PEMFC), which can operate from ambient temperature to 70–80 °C, and the direct ethanol fuel cell (DEFC), which has to work at higher temperatures (up to 120–150 °C) to improve its electric performance, will be particularly discussed. Finally, the solid alkaline membrane fuel cell (SAMFC) will be presented in more detail, including the electrochemical reactions involved.

1.2
Principles and Different Kinds of Fuel Cells

1.2.1
Working Principles of a Fuel Cell

The principles of the fuel cell are illustrated in Figure 1.1. The electrochemical cell consists of two electrodes, an anode and a cathode, which are electron conductors, separated by an electrolyte [e.g. a proton exchange membrane (PEM) in a PEMFC or in a DAFC], which is an ion conductor (as the result of proton migration and diffusion inside the PEM). An elementary electrochemical cell converts directly the chemical

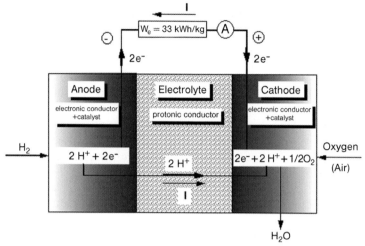

Figure 1.1 Schematic diagram of a hydrogen/oxygen fuel cell.

energy of combustion in oxygen (i.e. the Gibbs free energy change, $-\Delta G$) of a given fuel (hydrogen, natural gas, hydrocarbons, kerosene, alcohols, biomass resources and wastes) into electricity [4–8].

Electrons liberated at the anode (negative pole of the cell) by the electro-oxidation of the fuel pass through the external circuit (producing electric energy equal to $-\Delta G$) and arrive at the cathode (positive pole), where they reduce oxygen (from air). Inside the fuel cell, the electric current is transported by migration and diffusion of the electrolyte ions (H^+, OH^-, O^{2-}, CO_3^{2-}), for example, H^+ in a PEMFC.

1.2.1.1 The Thermodynamics of Fuel Cells

At the anode, the electro-oxidation of hydrogen takes place as follows:

$$H_2 \rightarrow 2H^+ + 2e^- \qquad E_1^\circ = 0.000 \, \text{V/SHE} \tag{1.1}$$

whereas the cathode undergoes the electro-reduction of oxygen:

$$O_2 + 4H^+ + 4e^- \rightarrow 2H_2O \qquad E_2^\circ = 1.229 \, \text{V/SHE} \tag{1.2}$$

where E_i° are the standard electrode potentials versus the standard hydrogen (reference) electrode (SHE). The standard cell voltage is thus $E_{eq}^\circ = E_2^\circ - E_1^\circ = 1.229 \, \text{V} \approx 1.23 \, \text{V}$. This corresponds to the overall combustion reaction of hydrogen in oxygen:

$$H_2 + \frac{1}{2}O_2 \rightarrow H_2O \tag{1.3}$$

with the following thermodynamic data, under standard conditions: $\Delta G^\circ = -237 \, \text{kJ mol}^{-1}$; $\Delta H^\circ = -286 \, \text{kJ mol}^{-1}$ of H_2.

The protons produced at the anode cross over the membrane, ensuring the electrical conductivity inside the electrolyte, whereas the electrons liberated at the anode reach the cathode (where they reduce oxygen) through the external circuit. This process produces an electric energy, $W_{el} = nFE_{eq}^{\circ} = -\Delta G^{\circ}$, corresponding to an energy mass density of the fuel $W_e = -\Delta G^{\circ}/(3600M) = 32.9 \text{ kW h kg}^{-1} \approx 33 \text{ kW h kg}^{-1}$, where $M = 0.002$ kg is the molecular weight of hydrogen, $F = 96\,485$ C the Faraday constant (i.e. the absolute value of the electric charge of 1 mol of electrons) and $n = 2$ the number of electrons involved in the oxidation of one hydrogen molecule.

The standard electromotive force (emf), E_{eq}°, at equilibrium (no current flowing) under standard conditions is then calculated as follows:

$$E_{eq}^{\circ} = -\frac{\Delta G^{\circ}}{nF} = \frac{237 \times 10^3}{2 \times 96485} = E_2^{\circ} - E_1^{\circ} = 1.229 \text{ V} \approx 1.23 \text{ V} \quad (1.4)$$

The working of the cell under reversible thermodynamic conditions does not follow Carnot's theorem, so that the theoretical energy efficiency, ε_{cell}^{rev}, defined as the ratio between the electrical energy produced $(-\Delta G^{\circ})$ and the heat of combustion $(-\Delta H^{\circ})$ at constant pressure, is

$$\varepsilon_{cell}^{rev} = \frac{W_e}{-\Delta H^{\circ}} = \frac{nFE_{eq}^{\circ}}{-\Delta H^{\circ}} = \frac{\Delta G^{\circ}}{\Delta H^{\circ}} = 1 - \frac{T\Delta S^{\circ}}{\Delta H^{\circ}} = \frac{237}{286} = 0.83 \quad (1.5)$$

for the hydrogen/oxygen fuel cell at 25 °C.

This theoretical efficiency is much greater (by a factor of about 2) than that of a thermal combustion engine, producing the reversible work, W_r, according to Carnot's theorem:

$$\varepsilon_{therm}^{rev} = \frac{W_r}{-\Delta H^{\circ}} = \frac{Q_1 - Q_2}{Q_1} = 1 - \frac{Q_2}{Q_1} = 1 - \frac{T_2}{T_1} = 0.43 \quad (1.6)$$

for, for example, $T_1 = 350$ °C and $T_2 = 80$ °C, where Q_1 and Q_2 are the heat exchanged with the hot source and cold source, respectively.

1.2.1.2 The Kinetics of Fuel Cells

However, under working conditions, with a current density j, the cell voltage $E(j)$ becomes smaller than the equilibrium cell voltage E_{eq}, as the result of three limiting factors: (i) the overvoltages η_a and η_c at both electrodes due to a rather low reaction rate of the electrochemical reactions involved (η is defined as the difference between the working electrode potential E_i and the equilibrium potential E_i^{eq}, so that $E_i = E_i^{eq} + \eta$), (ii) the ohmic drop $R_e j$ both in the electrolyte and interface resistances R_e and (iii) mass transfer limitations for reactants and products (Figure 1.2).

The cell voltage, $E(j)$, defined as the difference between the cathode potential E_2 and the anode potential E_1, can thus be expressed as

$$\begin{aligned} E(|j|) &= E_2(|j|) - E_1(|j|) = E_2^{eq} + \eta_c - (E_1^{eq} + \eta_a) - R_e|j| \\ &= E_{eq} - (|\eta_a| + |\eta_c| + R_e|j|) \end{aligned} \quad (1.7)$$

where the overvoltages η_a (anodic reaction, i.e. oxidation of the fuel with $\eta_a > 0$) and η_c (cathodic reaction, i.e. reduction of the oxidant with $\eta_c < 0$) take into account both the

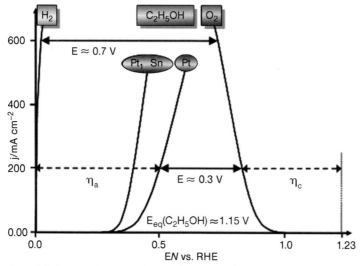

Figure 1.2 Comparative current density (j) vs potential (E) curves for H_2 and EtOH electro-oxidation at different Pt-based catalytic anodes and oxygen electro-reduction at a Pt cathode.

low kinetics of the electrochemical reactions involved (charge transfer overvoltage or activation polarization) and the limiting rate of mass transfer (mass transfer overvoltage or concentration polarization) – see Figure 1.3.

In the $E(j)$ characteristics one may distinguish three zones associated with these energy losses:

- *Zone I:* the E vs j linear curve corresponds to ohmic losses $R_e|j|$ in the electrolyte and interface resistances; a decrease of the specific electric resistance R_e from 0.3 to 0.15 $\Omega\,cm^2$ gives an increase in the current density j (at 0.7 V) from 0.25 to 0.4 A cm^{-2}, that is, an increase in the energy efficiency and in the power density of 1.6-fold.

- *Zone II:* the E vs $\ln(|j|/j_0)$ logarithmic curve corresponds to the charge transfer polarization, that is, to the activation overvoltages due to a relatively low electron exchange rate at the electrode–electrolyte interface, particularly for the oxygen reduction reaction whose exchange current density j_0 is much smaller than that of the hydrogen oxidation; an increase in j_0 from 10^{-8} to 10^{-6} A cm^{-2} leads to an increase in the current density j (at 0.7 V) from 0.4 to 0.9 A cm^{-2}, that is, an increase in the energy efficiency and in the power density of 3.6-fold compared with the initial data.

- *Zone III:* the E vs $\ln(1-|j/j_l|)$ logarithmic curve corresponds to concentration polarization, which results from the limiting value j_l of the mass transfer limiting current density for the reactive species and reaction products to and/or from the electrode active sites; an increase in j_l from 1.4 to 2.2 A cm^{-2} leads to a further

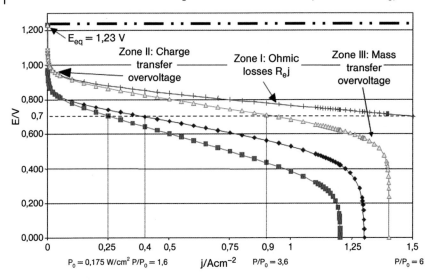

Figure 1.3 Theoretical $E(j)$ electric characteristics of an H_2/O_2 fuel cell: (■) $j_0 = 10^{-8}$ A cm^{-2}, $R_e = 0.30\,\Omega\,\text{cm}^2$, $j_l = 1.2$ A cm^{-2}; (♦) $j_0 = 10^{-8}$, $R_e = 0.15$, $j_l = 1.3$; (△) $j_0 = 10^{-6}$, $R_e = 0.15$, $j_l = 1.4$; (+) $j_0 = 10^{-6}$, $R_e = 0.10$, $j_l = 2.2$.

increase in the current density j (at 0.7 V) from 0.9 to 1.5 A cm^{-2}, that is, an increase in the energy efficiency and in the power density of 6-fold compared with the initial curve.

Hence these three key points will determine the energy efficiency and the specific power of the elementary fuel cell: an improvement in each component of the cell will increase the power density from 0.175 to 1.05 W cm^{-2}, that is, an increase by a factor of 6. As a consequence, for the fuel cell systems the weight and volume will be decreased by a similar factor, for a given power of the system, and presumably the overall cost will be diminished. The improvement in the components of the elementary fuel cell thus has a direct effect on the system technology and therefore on the overall cost.

For a PEMFC, fed with reformate hydrogen and air, the working cell voltage is typically 0.8 V at 500 mA cm^{-2}, which leads to a voltage efficiency ε_E given by

$$\varepsilon_E = \frac{E(j)}{E_{eq}^\circ} = 1 - \frac{|\eta_a(j)| + |\eta_c(j)| + R_e|j|}{E_{eq}^\circ} = \frac{0.8}{1.23} = 0.65 \quad (1.8)$$

The overall energy efficiency (ε_{cell}) thus becomes

$$\varepsilon_{cell} = \frac{W_{el}}{-\Delta H^\circ} = \frac{n_{exp} F E(|j|)}{-\Delta H^\circ} = \frac{n_{exp}}{n_{th}} \frac{E(|j|)}{E_{eq}^\circ} \frac{n_{th} F E_{eq}^\circ}{-\Delta H^\circ} = \varepsilon_F \varepsilon_E \varepsilon_{cell}^{rev} \quad (1.9)$$

where the faradaic efficiency $\varepsilon_F = n_{exp}/n_{th}$ is the ratio between the number of electrons n_{exp} effectively exchanged in the overall reaction and the theoretical numbers of electrons n_{th} for complete oxidation of the fuel (to H_2O and CO_2). $\varepsilon_F = 1$ for hydrogen oxidation, but $\varepsilon_F = 4/12 = 0.33$ for ethanol oxidation stopping at the acetic acid stage (four electrons exchanged instead of 12 electrons for complete oxidation to CO_2) – see Section 1.3.2.1.

As an example, the overall energy efficiency of an H_2/O_2 fuel cell, working at 0.8 V under 500 mA cm^{-2}, is

$$\varepsilon_{cell}^{H_2/O_2} = 1 \times 0.65 \times 0.83 = 0.54 \tag{1.10}$$

whereas that of a DEFC working at 0.5 V under 100 mA cm^{-2} (assuming complete oxidation to CO_2) would be

$$\varepsilon_{cell}^{C_2H_5OH/O_2} = 1 \times 0.44 \times 0.97 = 0.43 \text{ since } \varepsilon_F = 1,$$
$$\varepsilon_E = \frac{0.5}{1.14} = 0.44 \text{ and } \varepsilon_{cell}^{rev} = \frac{1325}{1366} = 0.97 \tag{1.11}$$

with $\Delta G° = -1325$ kJ mol^{-1}, $E_{eq}° = (1325 \times 10^3)/(12 \times 96\,485) = 1.144$ V and $\Delta H° = -1366$ kJ mol^{-1} of ethanol (see Section 1.3.2.1).

These energy efficiencies are better than those of the best thermal engines (diesel engines), which have energy efficiency of the order of 0.40.

Therefore, it appears that the only way to increase significantly the overall energy efficiency is to increase ε_E and ε_F, that is, to decrease the overvoltages η and the ohmic drop $R_e j$ and to increase the faradaic efficiency for complete oxidation, since ε_{cell}^{rev} is given by the thermodynamics (one can increase it slightly by changing the pressure and temperature operating conditions). For the hydrogen/oxygen fuel cell, usually $\varepsilon_F \approx 1$, but it can be much lower in the case of incomplete combustion of the fuel (see, e.g., the case of the DEFC in Section 1.3.2.1). The decrease in $|\eta|$ is directly related to the increase in the rate of the electrochemical reactions occurring at both electrodes. This is typically the field of electrocatalysis, where the action of both the electrode potential and the catalytic electrode material will synergistically increase the reaction rate v [9–11].

1.2.1.3 Catalysis of Fuel Cell Reactions

Electrocatalysis and the Rate of Electrochemical Reactions For a given electrochemical reaction $A + ne^- \rightleftharpoons B$, which involves the transfer of n electrons at the electrode/electrolyte interface, the equilibrium potential, called the electrode potential, is given by the Nernst law:

$$E_{eq}^{A/B} = E_0^{A/B} + \frac{RT}{nF} \ln \frac{a_A}{a_B} \tag{1.12}$$

where $E_0^{A/B}$ is the standard electrode potential, as measured versus the standard hydrogen electrode (SHE), the potential of which is zero at 25 °C by definition, and a_i is the activity of reactant i.

As soon as the electrode potential takes a value $E^{A/B}$ different from the equilibrium potential $E_{eq}^{A/B}$, an electrical current of intensity I passes through the interface, the magnitude of which depends on the deviation $\eta = E^{A/B} - E_{eq}^{A/B}$ from the equilibrium

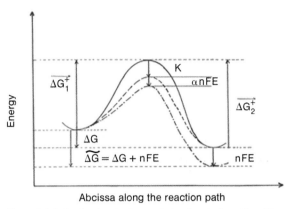

Figure 1.4 Activation barrier for an electrochemical reaction. K is the decrease in activation energy due to the electrode catalyst and αnFE is that due to the electrode potential E.

potential. η, which is called the overvoltage, is positive for an oxidation reaction (anodic reaction B → A + ne^-) and negative for a reduction reaction (cathodic reaction A + ne^- → B). The current intensity I is proportional to the rate of reaction v, that is, $I = nFv$. For heterogeneous reaction v is proportional to the surface area S of the interface, so that the kinetics of electrochemical reactions are better defined by the intrinsic rate $v_i = v/S$ and the current density $j = I/S = nFv_i$.

The electrical characteristics $j(E)$ can then be obtained by introducing the exponential behavior of the rate constant with the electrochemical activation energy, $\widehat{\Delta G}^{\ddagger} = \Delta G_0^{\ddagger} - \alpha nFE$, which comprises two terms: the first ($\Delta G_0^{\ddagger}$) is the chemical activation energy and the second (αnFE) is the electrical part of the activation energy. This latter is a fraction α ($0 \leq \alpha \leq 1$) of the total electric energy, nFE, coming from the applied electrode potential E, where α is called the charge transfer coefficient (Figure 1.4). In the theory of absolute reaction rate, one obtains, for a first-order electrochemical reaction (the rate of which is proportional to the reactant concentration c_i):

$$j = nFvi = nFk(T, E)c_i = nFk_0 c_i e^{-(\widehat{\Delta G}^{\ddagger}/RT)}$$

$$= nFk_0 c_i e^{-(\Delta G_0^{\ddagger}/RT)} e^{(\alpha nFE/RT)} = j_0 e^{(\alpha nFE/RT)} \quad (1.13)$$

This equation contains the two essential activation terms met in electrocatalysis: (i) an exponential function of the electrode potential E and (ii) an exponential function of the chemical activation energy $\Delta G_0^{\ddagger}$. By modifying the nature and structure of the electrode material, one may decrease $\Delta G_0^{\ddagger}$, by a given amount K (see Figure 1.4), thus increasing j_0, as the result of the catalytic properties of the electrode. This leads to an increase of the reaction rate v_i, that is, the current density j.

Electrocatalytic Oxidation of Hydrogen The rate constant of the hydrogen oxidation reaction (HOR), as measured by the exchange current density j_0 (i.e. the current

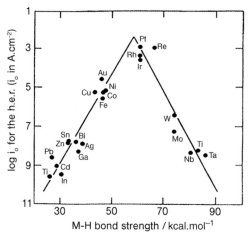

Figure 1.5 Exchange current density of the hydrogen reaction as a function of the metal–hydrogen bond energy [12].

density at the equilibrium potential $E_{eq}^{A/B}$, where the anodic current is exactly compensated by the cathodic current, so that the overall current is zero) depends greatly on the electrode material (Figure 1.5).

These values can be correlated with the heat of adsorption of hydrogen on the catalytic metal since the oxidation mechanism, apart from diffusion and mass transport limitations, is controlled by an adsorption step in a two consecutive step mechanism:

$$H_2 \rightarrow 2H_{ads} \qquad \text{dissociative adsorption of hydrogen} \qquad (1.14)$$

$$2(H_{ads} \rightarrow H_{aq}^+ + e^-) \qquad \text{electron transfer reaction} \qquad (1.15)$$

$$H_2 \rightarrow 2H_{aq}^+ + 2e^- \qquad \text{overall oxidation reaction} \qquad (1.16)$$

where H_{aq}^+ is a proton solvated by the electrolytic medium (usually an aqueous electrolyte). This leads to a Volcano plot (Figure 1.5) with a maximum of activity for the transition noble metals (Pt, Pd, Rh).

Oxidation of Alcohols in a Direct Alcohol Fuel Cell The electrocatalytic oxidation of an alcohol (methanol, ethanol, etc.) in a direct alcohol fuel cell (DAFC) will avoid the presence of a heavy and bulky reformer, which is particularly convenient for applications to transportation and portable electronics. However, the reaction mechanism of alcohol oxidation is much more complicated, involving multi-electron transfer with many steps and reaction intermediates. As an example, the complete oxidation of methanol to carbon dioxide:

$$CH_3OH + H_2O \rightarrow CO_2 + 6H^+ + 6e^- \qquad (1.17)$$

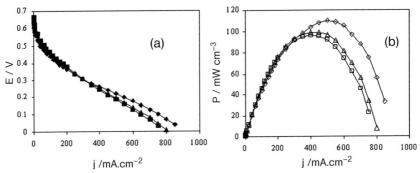

Figure 1.6 (a) $E(j)$ and (b) $P(j)$ curves at 110 °C of a DMFC with different Pt/Ru atomic ratios in the anode catalyst (■, 50:50; ▲, 70:30; ◆, 80:20).

involves the transfer of six electrons and the formation of many adsorbed species and reaction intermediates, among them adsorbed CO, which blocks the active sites of platinum catalysts. As a result of such a complex reaction, the oxidation overvoltage η_a is relatively high (0.3–0.5 V), so that more effective electrocatalysts are needed in order to increase the reaction rate and thus to decrease η_a. Therefore, several bi- and trimetallic catalysts were developed, among which Pt/Ru-based electrocatalysts lead to the best performance.

Pt/Ru electrocatalysts are currently used in DMFC stacks of a few watts to a few kilowatts. The atomic ratio between Pt and Ru, the particle size and the metal loading of carbon-supported anodes play a key role in their electrocatalytic behavior. Commercial electrocatalysts (e.g. from E-Tek) consist of 1:1 Pt/Ru catalysts dispersed on an electron-conducting substrate, for example carbon powder such as Vulcan XC72 (specific surface area of 200–250 m^2 g^{-1}). However, fundamental studies carried out in our laboratory [13] showed that a 4:1 Pt/Ru ratio gives higher current and power densities (Figure 1.6).

This may be explained by the bifunctional theory of electrocatalysis developed by Watanabe and Motoo [14], according to which Pt activates the dissociative chemisorption of methanol to CO, whereas Ru activates and dissociates water molecules, leading to adsorbed hydroxyl species, OH. A surface oxidation reaction between adsorbed CO and adsorbed OH becomes the rate-determining step. The reaction mechanism can be written as follows [15]:

$$Pt + CH_3OH \rightarrow Pt-CH_3OH_{ads} \quad (1.18)$$

$$Pt-CH_3OH_{ads} \rightarrow Pt-CHO_{ads} + 3H^+ + 3e^- \quad (1.19)$$

$$Pt-CHO_{ads} \rightarrow Pt-CO_{ads} + H^+ + e^- \quad (1.20)$$

$$Ru + H_2O \rightarrow Ru-OH_{ads} + H^+ + e^- \quad (1.21)$$

$$Pt\text{-}CO_{ads} + Ru\text{-}OH_{ads} \rightarrow Pt + Ru + CO_2 + H^+ + e^- \qquad (1.22)$$

Overall reaction:
$$CH_3OH + H_2O \rightarrow CO_2 + 6H^+ + 6e^- \qquad (1.23)$$

In this reaction mechanism, three or four Pt sites are involved in methanol dissociation, whereas only one Ru site is involved in water activation, so that the best Pt/Ru atomic ratio is between 3:1 and 4:1 [15].

Some Pt/Ru-based trimetallic electrocatalysts, such as Pt/Ru/Mo, give enhanced catalytic activity leading to a power density, in an elementary single DMFC, at least twice that of Pt/Ru catalyst.

Electrocatalytic Reduction of Dioxygen The electrocatalytic reduction of oxygen is another multi-electron transfer reaction (four electrons are involved) with several steps and intermediate species [16]. A four-electron mechanism, leading to water, is in competition with a two-electron mechanism, giving hydrogen peroxide. The four-electron mechanism on a Pt electrode can be written as follows:

$$Pt + O_2 \rightarrow Pt\text{-}O_{2ads} \qquad (1.24)$$

$$Pt\text{-}O_{2ads} + H^+ + e^- \rightarrow Pt\text{-}O_2H_{ads} \qquad (1.25)$$

$$Pt + Pt\text{-}O_2H_{ads} \rightarrow Pt\text{-}O_{ads} + Pt\text{-}OH_{ads} \qquad (1.26)$$

$$Pt\text{-}O_{ads} + H^+ + e^- \rightarrow Pt\text{-}OH_{ads} \qquad (1.27)$$

$$2Pt\text{-}OH_{ads} + 2H^+ + 2e^- \rightarrow 2Pt + 2H_2O \qquad (1.28)$$

Overall reaction:
$$O_2 + 4H^+ + 4e^- \rightarrow 2H_2O \qquad (1.29)$$

This complex reduction reaction leads to a relatively high overvoltage – at least 0.3 V – thus decreasing the cell voltage of the fuel cell by the same quantity. Pt–X binary catalysts (with X = Cr, Ni, Fe, …) give some improvements in the electrocatalytic properties compared with pure Pt dispersed on Vulcan XC72 [17].

In addition, in a DAFC, the proton exchange membrane is not completely alcohol tight, so that some alcohol leakage to the cathodic compartment will lead to a mixed potential with the oxygen electrode. This mixed potential will decrease further the cell voltage by about 0.1–0.2 V. It turns out that new electrocatalysts insensitive to the presence of alcohols are needed for the DAFC.

Transition metal compounds, such as organic macrocycles, are known to be good electrocatalysts for oxygen reduction. Furthermore, they are inactive for alcohol oxidation. Different phthalocyanines and porphyrins of iron and cobalt were thus dispersed in an electron-conducting polymer (polyaniline, polypyrrole) acting as a conducting matrix, either in the form of a tetrasulfonated counter anion or linked to

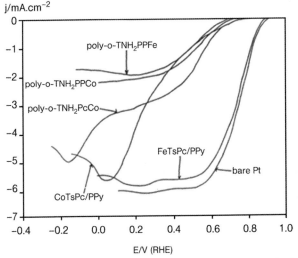

Figure 1.7 Activity of different electrodes containing macrocycles for the oxygen reduction reaction.

the monomer before its electro-polymerization Among the different macrocyclic compounds investigated in our laboratory, tetrasulfonated iron phthalocyanine (FeTsPc), incorporated in polypyrrole, has a catalytic activity close to that of pure platinum (Figure 1.7).

However, the stability with time and/or temperature of these modified electrodes needs to be greatly improved for practical applications (except maybe for power sources in portable electronics working at room temperature).

Further, Pt-based binary catalysts, such as Pt/Cr and Pt/Ni [18], are less sensitive to the presence of alcohol than pure Pt, giving hope for the development of better catalysts for oxygen reduction.

1.2.2
Different Types of Fuel Cells

Different types of fuel cells have been developed and are classified mainly according to (i) the type of fuel, (ii) the operating temperature range and/or electrolyte or (iii) the direct or indirect utilization of fuel [19] . The state-of-the-art and the different applications of fuel cells are summarized in Table 1.1.

1.2.2.1 Fuels for Fuel Cells
For fuel cells operating at low (<100 °C) and intermediate temperatures (up to 200 °C), H_2 and H_2–CO_2 (with minimal amounts of CO) are the ideal fuels. The H_2–CO_2 gas mixture is produced by steam reforming/water gas shift conversion or partial oxidation/water gas shift conversion of primary or secondary organic fuels. On a large scale, hydrogen is produced from the primary fuels, that is, natural gas, oil or coal gasification, but it can be generated by water electrolysis using nuclear power plants, avoiding the direct production of CO_2 and CO.

Table 1.1 Status of fuel cell technologies.

Type of fuel cell[a]; operating fuel and temperature	Power rating (kW)	Fuel efficiency (%)	Power density (mW cm^{-2})	Lifetime (h)	Capital cost ($ kW^{-1})	Applications
SOFC; CH$_4$, coal?, 800–1000 °C	25–5000	50–60	200–400	8000–40 000	1500	Base-load and Intermediate load, power generation, co-generation
MCFC; CH$_4$, coal?, 650 °C	100–5000	50–55	150–300	10 000–40 000	1250	Base-load and intermediate load, power generation – co-generation
PAFC; CH$_4$, CH$_3$OH, oil, 200 °C	200–10 000	40–45	200–300	30 000–40 000	200–3000	On-site integrated energy systems, transportation (fleet vehicles), load-leveling
AFC; H$_2$, 80 °C	20–100	65	250–400	3000–10 000	1000	Space flights, space stations, transportation, APU
PEMFC; H$_2$, CH$_3$OH, 25–100 °C	0.01–250	40–50	500–1000	10 000–100 000	50–2000	Transportation, stand-by power, portable power, space stations
DMFC; CH$_3$OH, 25–150 °C	0.001–10	30–45	50–200	1000–10 000	1000	Portable power, stand-by power, transportation (?), APU
SAMFC; H$_2$, CH$_3$OH, C$_2$H$_5$OH, NaBH$_4$, 25–80 °C	0.001–0.1	30–60	10–100	?	?	Portable power

[a] SOFC = solid oxide fuel cell; MCFC = molten carbonate fuel cell; PAFC = phosphoric acid fuel cell; AFC = alkaline fuel cell; PEMFC = proton exchange membrane fuel cell; DMFC = direct methanol fuel cell; SAMFC = Solid alkaline membrane fuel cell.

These fuels (pure H_2, H_2–CO_2, H_2–CO–CO_2) may also be produced from renewable energy sources, such as biomass, solar, windmill and hydroelectric power.

Hydrogen is a secondary fuel and, like electricity, is an energy carrier. It is the most electroactive fuel for fuel cells operating at low and intermediate temperatures. Methanol and ethanol are the most electroactive alcohol fuels, and, when they are electro-oxidized directly at the fuel cell anode (instead of being transformed in a hydrogen-rich gas in a fuel processor), the fuel cell is called a DAFC: either a DMFC (with methanol) or a DEFC (with ethanol).

1.2.2.2 Hydrogen-fed Fuel Cells

Hydrogen/oxygen (air) fuel cells are classified according to the type of electrolyte used and the working temperature:

- Solid oxide fuel cell (SOFC) working between 700 and 1000 °C with a solid oxide electrolyte, such as yttria-stabilized zirconia (ZrO_2–8% Y_2O_3), conducting by the O^{2-} anion.

- Molten carbonate fuel cell (MCFC) working at about 650 °C with a mixture of molten carbonates (Li_2CO_3–K_2CO_3) as electrolyte, conducting by the CO_3^{2-} anion.

Both of the above are high-temperature fuel cells.

- Phosphoric acid fuel cell (PAFC) working at 180–200 °C with a porous matrix of PTFE-bonded silicon carbide impregnated with phosphoric acid as electrolyte, conducting by the H^+ cation. This medium-temperature fuel cell is now commercialized by ONSI (USA), mainly for stationary applications.

- Alkaline fuel cell (AFC) working at 80 °C with concentrated potassium hydroxide as electrolyte, conducting by the OH^- anion. This kind of fuel cell, developed by IFC (USA), is now used in space shuttles.

- Proton exchange membrane fuel cell (PEMFC) working at around 70 °C with a polymer membrane electrolyte, such as Nafion, which is a solid proton conductor (conducting by the H^+ cation).

- Direct methanol fuel cell (DMFC) working between 30 and 110 °C with a proton exchange membrane (such as Nafion) as electrolyte, which realizes the direct oxidation of methanol at the anode.

- Solid alkaline membrane fuel cell (SAMFC) working at moderate temperatures (20–80 °C) for which an anion-exchange membrane (AEM) is the electrolyte, electrically conducting by, for example, hydroxyl ions (OH^-).

1.2.2.3 Methanol- and Ethanol-fed Fuel Cells

In addition to hydrogen as a fuel, methanol or ethanol can be directly converted into electricity in a DAFC, the great progress of which resulted from the use of a proton exchange membrane acting both as an electrolyte (instead of the aqueous electrolytes previously used) and as a separator preventing the mixing of fuel and oxidant. A DAFC can work at moderate temperatures (30–50 °C) for portable applications, but now the tendency is to look for new membranes that are less permeable to alcohol and work at higher temperatures (80–120 °C) to increase the rate of the electrochemical reactions involved (oxidation of alcohol and reduction of oxygen) and to manage better the heat produced, either to use it in a co-generation system or to evacuate it in transportation applications.

The last three fuel cells (PEMFC, DAFC and SAMFC) are low-temperature fuel cells. In this chapter, the discussion will be focused on these fuel cells, particularly the PEMFC and the DAFC, since they can accommodate biomass fuels, either after fuel processing to obtain reformate hydrogen or directly with bioethanol.

For these low-temperature fuel cells, the development of catalytic materials is essential to activate the electrochemical reactions involved. This concerns the electro-oxidation of the fuel (reformate hydrogen containing some traces of CO, which acts as a poisoning species for the anode catalyst; methanol and ethanol, which have a relatively low reactivity at low temperatures) and the electro-reduction of the oxidant (oxygen), which is still a source of high energy losses (up to 30–40%) due to the low reactivity of oxygen at the best platinum-based electrocatalysts.

1.3
Low-temperature Fuel Cells (PEMFCs and DAFCs)

1.3.1
Proton Exchange Membrane Fuel Cell (PEMFC)

The PEMFC is nowadays the most advanced low-temperature fuel cell technology [19, 20], because it can be used in several applications (space programs, electric vehicles, stationary power plants, auxiliary power units, portable electronics). The progress made in one application is greatly beneficial to the others.

1.3.1.1 Principle of a PEMFC
An elementary PEMFC consists of a thin film (10–200 μm) of a solid polymer electrolyte (a protonic membrane, such as Nafion), on both sides of which the electrode structures (fuel anode and oxygen cathode) are pasted, giving a membrane–electrode assembly (MEA) (Figure 1.8). A single cell delivers a cell voltage of 0.5–0.9 V (instead of the theoretical emf of 1.23 V under standard equilibrium conditions), depending on the working current density. Many elementary cells, electrically connected by bipolar plates, are assembled together (in series and/or in parallel) to reach the nominal voltage (such as 48 V for electric vehicles) and the nominal power of the fuel cell stack.

In PEMFCs working at low temperatures (20–90 °C), several problems need to be solved before the technological development of fuel cell stacks for different applications. This concerns the properties of the components of the elementary cell, that is, the proton exchange membrane, the electrode (anode and cathode) catalysts, the membrane–electrode assemblies and the bipolar plates [19, 20]. This also concerns the overall system with its control and management equipment (circulation of reactants and water, heat exhaust, membrane humidification, etc.).

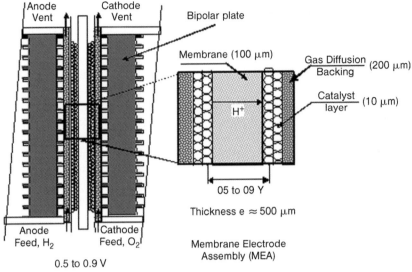

Figure 1.8 Schematic representation of a PEMFC elementary cell (and its MEA).

1.3.1.2 The Proton Exchange Membrane

The proton exchange membrane, which is a solid polymer electrolyte, plays a key role in the PEMFC. It allows the electrical current to pass through it thanks to its H^+ ionic conductivity and prevents any electron current through it so that electrons are obliged to circulate in the external electric circuit to produce the electric energy corresponding to the combustion reaction of the fuel. Moreover, it must avoid any gas leakage between the anode and cathode compartments, so that no chemical combination between hydrogen and oxygen is directly allowed. It must also be mechanically, thermally (up to 150 °C in order to increase the working temperature of the cell) and chemically stable. Finally, its lifetime must be sufficient for practical applications (4 000 h for automotive to 40 000 h for stationary applications).

The actually developed PEMFCs have a Nafion membrane, which partially fulfills these requirements, since its thermal stability is limited to 100 °C and its proton conductivity decreases strongly at higher temperatures because of its dehydration. On the other hand, it is not completely tight to liquid fuels (such as alcohols). This becomes more important as the membrane is thin (a few tens of micrometers). Furthermore, its actual cost is too high (more than $500 € m^{-2}$), so that its use in a PEMFC for an electric car is not cost competitive.

Therefore, new membranes are being investigated with improved stability and conductivity at higher temperatures (up to 150 °C) [21]. For power FCs, the increase in temperature will increase the rate of the electrochemical reactions occurring at both electrodes, that is, the current density at a given cell voltage and the specific power. Furthermore, thermal management and heat utilization will be improved, particularly for residential applications with heat co-generation and for mobile applications to exhaust excess heat.

1.3.1.3 The Electrode Catalysts

One of the main problems with low-temperature (20–80 °C) PEMFCs is the relatively slow kinetics of the electrochemical reactions involved, such as oxygen reduction at the cathode and fuel (hydrogen from a reformate gas or alcohols) oxidation at the anode. The reaction rates can only be increased by the simultaneous action of the electrode potential and electrode material (electrocatalytic activation). Moreover, increasing the working temperature from 80 to 150 °C would strongly increase (by a factor of at least 100–1000) the rates of the electrochemical reactions (thermal activation). All these combined effects would increase the cell voltage by about 0.1–0.2 V, since at room temperature the anode and cathode overvoltages are close to 0.2–0.4 V, which decreases the cell voltage by 0.4–0.7 V, leading to values close to 0.5–0.7 V instead of the theoretical cell voltage of 1.23 V.

The investigation of new electrocatalysts, particularly Pt-based catalysts, that are more active for oxygen reduction and fuel oxidation (hydrogen from reformate gas or alcohols) is thus an important point for the development of PEMFCs [16, 17, 22, 23].

1.3.1.4 The Membrane–Electrode Assembly

The realization of the MEA is a crucial point for constructing a good fuel cell stack. The method currently used consists in hot-pressing (at 130 °C and 35 kg cm^{-2}) the electrode structures on the polymer membrane (Nafion). This gives non-reproducible results (in terms of interface resistance) and this is difficult to industrialize. New concepts must be elaborated, such as the continuous assembly of the three elements in a rolling tape process (as in the magnetic tape industry) or successive deposition of the component layers (microelectronic process) and so on.

1.3.1.5 The Bipolar Plates

The bipolar plates, which separate both electrodes of neighboring cells (one anode of a cell and one cathode of the other), have a triple role:

- to ensure the electron conductivity between two neighboring cells;
- to allow the distribution of reactants (gases and liquids in the case of alcohols) to the electrode catalytic sites and to evacuate the reaction products (H_2O and CO_2 in the case of alcohols);
- thermal management inside the elementary cell by evacuating the excess heat.

The bipolar plates are usually fabricated with non-porous machined graphite or corrosion-resistant metal plates. Distribution channels are engraved in these plates. Metallic foams can also be used for distributing the reactants. One key point is to ensure a low ohmic resistance inside the bipolar plate and at the contact with the MEA. Another point is to use materials with high corrosion resistance in the oxidative environment of the oxygen cathode.

1.3.1.6 Auxiliary and Control Equipment

A more detailed picture of a PEMFC system, including the auxiliary and control equipments, is shown in Figure 1.9.

20 | *1 The Direct Ethanol Fuel Cell: a Challenge to Convert Bioethanol Cleanly into Electric Energy*

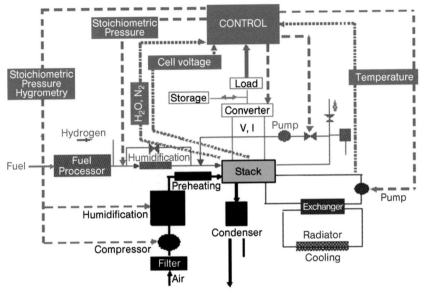

Figure 1.9 Detailed scheme of a PEMFC system with its auxiliary and control equipments.

Fuel supply is usually from liquid hydrogen or pressurized gaseous hydrogen. For other fuels, a fuel processor is needed, which includes a reformer, water gas shift reactors and purification reactors, in order to decrease the amount of CO to an acceptable level (below a few tens of ppm), which would otherwise poison the platinum-based catalysts. This equipment is still heavy and bulky and limits the dynamic response of the fuel cell stack, particularly for the electric vehicle in some urban driving cycles.

On the other hand, the other auxiliary equipment depends greatly on the stack characteristics:

- air compressor, the characteristics of which are related to the pressures supported by the proton exchange membrane;
- humidifiers for the reacting gases with controlled humidification conditions;
- preheating of gases to avoid condensation phenomena;
- hydrogen recirculation and purging systems of the anode compartment;
- cooling system for the MEAs;
- control of pressure valves and/or of gas flows;
- DC/DC or DC/AC electric converters.

The system control must ensure correct working of the system, not only under steady-state conditions, but also during power transients. All the elementary cells must be controlled (the cell voltage of each elementary cell, if possible) and the purging system must be activated in the case of a technical hitch.

According to different applications (stationary power plants, power sources for electric vehicle, auxiliary power units, etc.), different specifications and system design are required.

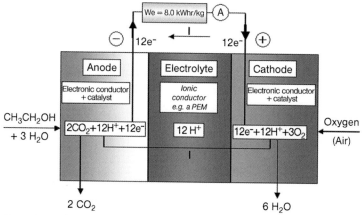

Figure 1.10 Schematic diagram of a DEFC.

1.3.2
Direct Ethanol Fuel Cell (DEFC)

1.3.2.1 Principle of the Direct Ethanol Fuel Cell
A schematic diagram of a DEFC is shown in Figure 1.10.

The DEFC transforms directly the Gibbs energy of combustion of ethanol into electricity, without a fuel processor. This greatly simplifies the system, reducing its volume and cost [22, 23]. The important development of DEFCs is due to the use of a proton exchange membrane as electrolyte, instead of a liquid acid electrolyte, as done previously.

At the anode, the electro-oxidation of ethanol takes place as follows, leading, in the case of complete oxidation, to CO_2:

$$CH_3CH_2OH + 3H_2O \rightarrow 2CO_2 + 12H^+ + 12e^- \quad E_1^\circ = 0.085 \text{ V vs SHE}$$
(1.30)

whereas the electro-reduction of oxygen occurs at the cathode:

$$O_2 + 4H^+ + 4e^- \rightarrow 2H_2O \quad E_2^\circ = 1.229 \text{ V vs SHE}$$
(1.31)

where E_i° are the electrode potentials versus SHE. This corresponds to the overall combustion reaction of ethanol into oxygen:

$$CH_3CH_2OH + 3O_2 \rightarrow 2CO_2 + 3H_2O \quad (1.32)$$

with the thermodynamic data, under standard conditions:

$$\Delta G^\circ = -1325 \text{ kJ mol}^{-1}; \Delta H^\circ = -1366 \text{ kJ mol}^{-1} \text{ of ethanol} \quad (1.33)$$

This gives a standard emf at equilibrium:

$$E_{eq}^\circ = -\frac{\Delta G^\circ}{nF} = \frac{1325 \times 10^3}{12 \times 96485} = E_2^\circ - E_1^\circ = 1.144 \text{ V} \quad (1.34)$$

with $n = 12$ the number of electrons exchanged per molecule for complete oxidation to CO_2. The corresponding electrical energy, $W_{el} = nFE°_{eq} = -\Delta G°$, leads to a mass energy density $W_e = -\Delta G°/(3600\,M) = 8\,kWh\,kg^{-1}$, where $M = 0.046\,kg$ is the molecular weight of ethanol.

The theoretical energy efficiency, under reversible standard conditions, defined as the ratio between the electrical energy produced $(-\Delta G°)$ and the heat of combustion $(-\Delta H°)$ at constant pressure, is – see Equation 1.5

$$\varepsilon_{cell}^{rev} = \frac{\Delta G°}{\Delta H°} = \frac{1325}{1366} = 0.97 \tag{1.35}$$

However, under working conditions, with a current density j, the cell voltage $E(j)$ is lower than E_{eq} – see Equation 1.7 – so that the practical energy efficiency, for a DEFC working at 0.5 V and 100 mA cm^{-2} with complete oxidation to CO_2, would be – see Equation 1.9

$$\varepsilon_{cell}^{C_2H_5OH/O_2} = \varepsilon_F \times \varepsilon_E \times \varepsilon_{cell}^{rev} = 1 \times 0.437 \times 0.97 = 0.424 \tag{1.36}$$

since the potential efficiency $\varepsilon_E = E(j)/E_{eq} = 0.5/1.14 = 0.437$ and the faradaic efficiency $\varepsilon_F = n_{exp}/n_{th} = 1$ for complete oxidation to CO_2. This is similar to that of the best thermal engine (diesel engine). However, if the reaction process stops at the acetic acid stage, which involves the transfer of four electrons (instead of 12 for complete oxidation), the efficiency will be reduced by two-thirds, reaching only 0.15.

An additional problem arises from ethanol crossover through the proton exchange membrane. It results that the platinum cathode experiences a mixed potential, since both the oxygen reduction and ethanol oxidation take place at the same electrode. The cathode potential is therefore lower, leading to a decrease in the cell voltage and a further decrease in the voltage efficiency.

1.3.2.2 Reaction Mechanisms of Ethanol Oxidation

The electrochemical oxidation of ethanol has been extensively studied at platinum electrodes [22–34]. The first step is the dissociative adsorption of ethanol, either via an O-adsorption or a C-adsorption process [25, 26], to form acetaldehyde (AAL) according to the following reaction equations. Indeed, it was shown by Hitmi et al. [34] that AAL was formed at potentials lower than 0.6 V vs RHE. Thus:

$$Pt + CH_3CH_2OH \rightarrow Pt-(OCH_2CH_3)_{ads} + H^+ + e^- \tag{1.37}$$

or

$$Pt + CH_3CH_2OH \rightarrow Pt-(CHOHCH_3)_{ads} + H^+ + e^- \tag{1.38}$$

followed (at $E < 0.6$ V vs RHE) by

$$Pt-(OCH_2CH_3)_{ads} \rightarrow Pt + CHOCH_3 + H^+ + e^- \tag{1.39}$$

or

$$Pt-(CHOHCH_3)_{ads} \rightarrow Pt + CHOCH_3 + H^+ + e^- \tag{1.40}$$

AAL has to be readsorbed to oxidize further either into acetic acid or carbon dioxide. To complete the oxidation reaction leading to both of these species, an extra oxygen atom is needed, which has to be brought by activated (adsorbed) water molecules at the platinum surface.

Thus, as soon as AAL is formed, it can adsorb on platinum sites leading to a Pt–COCH$_3$ species at $E < 0.6$ V vs RHE:

$$Pt + CHOCH_3 \rightarrow Pt-(COCH_3)_{ads} + H^+ + e^- \qquad (1.41)$$

Further oxidation, without breaking of the –C–C– bond, may occur at potentials >0.6 V vs RHE, through the activation of water molecules at platinum sites:

$$Pt + H_2O \rightarrow Pt-(OH)_{ads} + H^+ + e^- \quad \text{at } E > 0.6 \text{ V vs RHE} \qquad (1.42)$$

$$Pt-(COCH_3)_{ads} + Pt-(OH)_{ads} \rightarrow 2Pt + CH_3COOH \qquad (1.43)$$

On the other hand, SNIFTIRS measurements have clearly shown that Pt is able to break the –C–C– bond, leading to adsorbed CO species at relatively low anode potentials (from 0.3 V vs RHE) [35]. However, Iwasita and Pastor [26] found some traces of CH$_4$ at potentials lower than 0.4 V vs RHE. Thus:

$$Pt-(COCH_3)_{ads} + Pt \rightarrow Pt-(CO)_{ads} + Pt-(CH_3)_{ads} \quad \text{at } E > 0.3 \text{ V vs RHE}$$
$$(1.44)$$

and

$$Pt-(CH_3)_{ads} + Pt-(H)_{ads} \rightarrow 2Pt + CH_4 \quad \text{at } E < 0.4 \text{ V vs RHE}$$
$$(1.45)$$

At potentials higher than 0.6 V vs RHE, the dissociative adsorption of water occurs on platinum, providing –OH adsorbed species, able to oxidize further the adsorption residues of ethanol. Then, oxidation of adsorbed CO species occurs as was shown by FTIR reflectance spectroscopy and CO stripping experiments [36]:

$$Pt-(CO)_{ads} + Pt-(OH)_{ads} \rightarrow 2Pt + CO_2 + H^+ + e^- \qquad (1.46)$$

AAL can also be oxidized, leading to acetic acid (AA), as follows:

$$Pt-(CHOCH_3)_{ads} + Pt-(OH)_{ads} \rightarrow 2Pt + CH_3COOH + H^+ + e^- \qquad (1.47)$$

In a recent study, the analysis of the reaction products at the outlet of the anode compartment of a DEFC fitted with a Pt/C anode showed that only AA, AAL and CO$_2$ could be detected by HPLC [24]. Depending on the electrode potential, AAL, AA, CO$_2$ and traces of CH$_4$ are observed, the main products being AAL and AA (Table 1.2), AA being considered as a final product because it is not oxidized under smooth conditions. Long-term electrolysis experiments on a Pt catalyst show that AAL is detected at potentials as low as 0.35 V vs RHE, whereas no AA was detected in this potential range.

Table 1.2 Chemical yields in acetaldehyde, acetic acid and CO_2 for the electro-oxidation of ethanol at Pt/C, Pt–Sn (90:10)/C and Pt–Sn-Ru (86:10:4)/C catalysts under DEFC operating conditions at 80 °C for 4 h.

Parameter	Electrocatalyst		
	Pt/C	$Pt_{0.9}Sn_{0.1}$/C	$Pt_{0.86}Sn_{0.1}Ru_{0.04}$/C
Metal loading (wt%)	60	60	60
Current density (mA cm^{-2})	8	32	32
Cell voltage (V)	0.3–0.35	0.45–0.49	0.5–0.55
AAL/Σ products (%)	47.5	15.4	15.2
AA/Σ products (%)	32.5	76.9	75
CO_2/Σ products (%)	20	7.7	9.8

From the results given in Table 1.2, it appears that the addition of tin to platinum greatly favors the formation of AA compared with AAL, as explained by the bifunctional mechanism [14].

Several added metals were investigated to improve the kinetics of ethanol oxidation at platinum-based electrodes, including ruthenium [27, 28], lead [29] and tin [22, 30]. Of these, tin appeared to be very promising. Figure 1.11 shows the polarization curves of ethanol electro-oxidation recorded at a slow sweep rate (5 mV s^{-1}) on different platinum-based electrodes. Pt–Sn(0.9:0.1)/C displays the

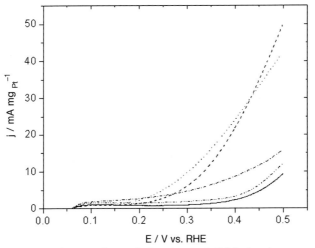

Figure 1.11 Electro-oxidation of ethanol on Pt/C (full line) and different Pt-based (dashed and dotted lines) catalysts with 0.1 mg Pt cm^{-2} loading. 0.1 M $HClO_4$ + 1 M C_2H_5OH; 5 mV s^{-1}; 3000 rpm; 20 °C. (———), Pt/XC72; (- - - - -), Pt–Sn (90:10)/XC72; (·········), Pt–Sn (80:20)/XC72; (·—·—·—), Pt–Ru (80:20)/XC72; (-·-·-·-·-), Pt-Mo (80:20)/XC72.

best activity in the potential range 0.15–0.5 V vs RHE, as it gives a higher oxidation current density than the other catalysts. The role of tin can also be established by analyzing the distribution of reaction products at the anode outlet of a fuel cell fitted with a Pt–Sn(0.9 : 0.1)/C anode and from *in situ* IR reflectance spectroscopic measurements [24, 37].

Table 1.2 indicates that alloying platinum with tin led to important changes in the product distribution: an increase in the AA chemical yield and a decrease in the AAL and CO_2 chemical yields. The presence of tin seems to allow, at lower potentials, the activation of water molecules and the oxidation of AAL species into AA. In the same manner, the amount of CO_2 decreased, which can be explained by the need for several adjacent platinum atoms (three or four) to realize the dissociative adsorption of ethanol into CO species, via breaking the C–C bond. In the presence of tin, 'dilution' of platinum atoms can limit this reaction. The effect of tin, in addition to the activation of water molecules, may be related to some electronic effects (ligand effects) on the CO oxidation reaction [38].

On Pt–Sn, assuming that ethanol adsorbs only on platinum sites, the first step can be the same as for platinum alone. However, as was shown by SNIFTIRS experiments [37], the dissociative adsorption of ethanol on a PtSn catalyst to form adsorbed CO species takes place at lower potentials than on a Pt catalyst, between 0.1 and 0.3 V vs RHE, whereas on a Pt catalyst the dissociative adsorption of ethanol takes place at potentials between 0.3 and 0.4 V vs RHE. Hence it can be stated that the same reactions occur at lower potentials and with relatively rapid kinetics. Once intermediate species such as Pt–$(COCH_3)_{ads}$ and Pt–$(CO)_{ads}$ are formed, they can be oxidized at potentials close to 0.3 V vs RHE, as confirmed by CO stripping experiments, because OH species are formed on tin at lower potentials [39, 40]:

$$Sn + H_2O \rightarrow Sn-(OH)_{ads} + H^+ + e^- \quad (1.48)$$

then adsorbed acetyl species can react with adsorbed OH species to produce AA according to

$$Pt-(COCH_3)_{ads} + Sn(OH)_{ads} \rightarrow Pt + Sn + CH_3COOH \quad (1.49)$$

Similarly, Pt–$(CO)_{ads}$ species are oxidized as follows:

$$Pt-(CO)_{ads} + Sn-(OH)_{ads} \rightarrow Pt + Sn + CO_2 + H^+ + e^- \quad (1.50)$$

This mechanism explains also the higher efficiency of Pt–Sn in forming AA compared with Pt at low potentials ($E < 0.35$ V vs RHE), as was shown by electrolysis experiments. Indeed, adsorbed OH species on Sn atoms can be used to oxidize adsorbed CO species to CO_2 or to oxidize adsorbed –$COCH_3$ species to CH_3COOH, according to the bifunctional mechanism [14].

On the other hand, the yield of CO_2 with a Pt/C catalyst is double that with a Pt–Sn/C catalyst (see Table 1.2). This can be explained by the need to have several adjacent platinum sites to adsorb dissociatively the ethanol molecule and to break the C–C bond. As soon as some tin atoms are introduced between platinum atoms, this latter reaction is disadvantaged.

The spectro-electrochemical study of the adsorption and oxidation of ethanol and the HPLC analyses of reaction products underline the necessity to activate water molecules at lower potentials in order to increase the activity of the catalyst and the selectivity towards either AA or CO_2 formation, which means the improvement of the potential ε_E and faradaic ε_F efficiencies. To perform this, modification of platinum by another metal is necessary.

1.3.2.3 DEFC Tests

The preparation of electrodes and MEAs have been described elsewhere [15]. FC tests (determination of the cell voltage E and power density P versus the current density j) were carried out in a single DEFC with 5 or 25 cm^2 geometric surface area electrodes using a Globe Tech test bench, purchased from ElectroChem (USA) (Figure 1.12).

The cell voltage E and current density j were recorded using a high-power potentiostat (Wenking Model HP 88) interfaced with a variable resistance in order to fix the current applied to the cell and with a PC to apply constant current sequences and to store the data.

The reaction products at the outlet of the anode side of the DEFC were analyzed quantitatively by HPLC, as described previously [24]. Large surface area electrodes (25 cm^2) were used in order to have a sufficient amount of products. The current density was kept constant and the voltage of the cell was simultaneously measured as a function of time. The ethanol flow rate was chosen close to 2 mL min^{-1} in order to perform long-term experiments (for at least 4 h) to obtain enough reaction products suitable for chemical analysis.

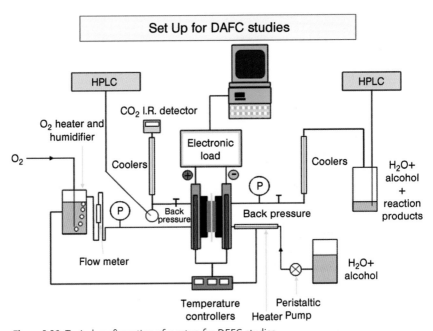

Figure 1.12 Typical configuration of a setup for DEFC studies.

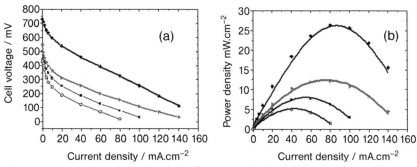

Figure 1.13 Fuel cell characteristics of a 5 cm^2 DEFC recorded at 110 °C. Influence of the nature of the bimetallic catalysts (80:20 atomic ratio with 30% metal loading). Anode catalyst, 1.5 mg cm^{-2}; cathode catalyst, 2 mg cm^{-2} (40% Pt/XC72 from E-TEK); membrane, Nafion 117; ethanol concentration, 1 M. (■) Pt/XC72; (◆) Pt–Sn (80:20)/XC72; (▶) Pt–Ru (80:20)/XC72; (◇) Pt-Mo (80:20)/XC72.

The experiments were carried out using Pt/C, Pt–Sn/C and Pt–Sn–Ru/C catalysts and in each case no other reaction products than AAL, AA and CO_2 were detected. The addition of tin to platinum not only increases the activity of the catalyst towards the oxidation of ethanol and therefore the electrical performance of the DEFC, but also changes greatly the product distribution: the formation of CO_2 and AAL is lowered, whereas that of AA is greatly increased (Table 1.2).

Typical electrical performances obtained with several Pt-based electrocatalysts are shown in Figure 1.13. The use of platinum alone as anode catalyst leads to poor electrical performance, the open circuit voltage (OCV) being lower than 0.5 V and the maximum current density reaches only 100 mA cm^{-2}, leading to a maximum power density lower than 7 mW cm^{-2} at 110 °C. The addition of Ru and especially of Sn to platinum in the anode catalyst greatly enhances the electrical performance of the DEFC by increasing the OCV to 0.75 V, which indicates that the modified catalysts are less poisoned by adsorbed species coming from ethanol chemisorption than the Pt/C catalyst. For Pt–Sn (80:20), current densities up to 150 mA cm^{-2} are reached at 110 °C, giving a maximum power density greater than 25 mW cm^{-2}, which means an electrical performances four times higher than those obtained with Pt/C. The increase in the electrical performance indicates that the bimetallic catalyst is more active for ethanol electro-oxidation than the Pt/C catalyst.

With the best electrocatalyst, that is, Pt–Sn (90:10)/XC72, the effect of temperature on the cell voltage E and power density P versus current density j characteristics is shown in Figure 1.14. It appears clearly that increasing the temperature greatly increases the performance of the cell, from a maximum power density close to 5 mW cm^{-2} at 50 °C to 25 mW cm^{-2} at 110 °C, that is, five times higher.

This confirms the difficulty of oxidizing ethanol at low temperatures and the necessity to work at temperatures higher than 100 °C to enhance the electrode kinetics and, thus, the performance of the DEFC.

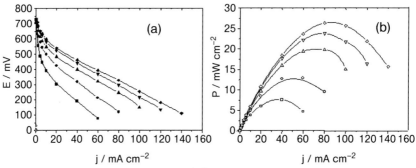

Figure 1.14 Fuel cell characteristics of a 25 cm² DEFC recorded with a 30% Pt–Sn (90:10) catalyst. Influence of the working temperature. Anode catalyst, 1.5 mg cm^{-2} [30% Pt–Sn (90:10)/XC72]; cathode catalyst, 2 mg cm^{-2} (40% Pt/XC72 from E-TEK); membrane, Nafion 117; ethanol concentration, 1 M. (■) 50 °C; (●) 70 °C; (▲) 90 °C; (▼) 100 °C; (◆) 110 °C.

Finally the performance of a DEFC with an anode containing a higher amount of platinum (i.e. 60 wt% instead of 30 wt% for the previous experiments) were determined with three Nafion membranes, N117, N115 and N112, of different thickness (180, 125 and 50 μm, respectively) to investigate the effect of ethanol crossover through the membrane. Figure 1.15 shows the ethanol crossover as measured by following the ethanol concentration in a second cell compartment, initially containing no ethanol, separated by the Nafion membrane from the first one containing 1 M ethanol. The crossover rate through N117 is about half that through N112. The better behavior of N117 is confirmed in Figure 1.16, showing the comparative electrical characteristics of a DEFC having as electrolyte one of the three Nafion membranes and an anode with a higher platinum loading (60% Pt–Sn (90:10)/XC72): the DEFC with N117 displays the highest OCV (0.8 V) and leads to a power density of 52 mW cm^{-2} at 90 °C.

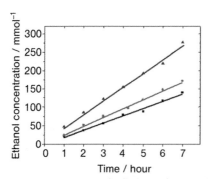

Figure 1.15 Behavior of different Nafion membranes. Ethanol permeability measurements of different Nafion membranes. $T = 25$ °C. (■) Nafion 117; (●) Nafion 115; (▲) Nafion 112.

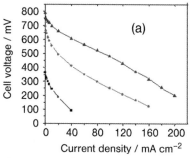 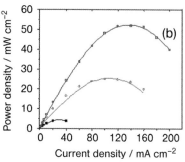

Figure 1.16 Fuel cell characteristics of a DEFC recorded at 90 °C with a 60% Pt–Sn (90 : 10)/XC72 catalyst for different Nafion membranes. (▲) Nafion 117; (●) Nafion 115; (■) Nafion 112.

1.4
Solid Alkaline Membrane Fuel Cell (SAMFC)

The rate of oxidation of alcohols [41] and reduction of oxygen [41, 42] is higher in alkaline than acidic media, so that the use of an AEM instead of a PEM brings new opportunities to develop DEFC with the concept of a SAMFC. For that purpose, in addition to the choice of an AEM with sufficiently good conductivity and stability, the investigation of electrode reaction catalysts, particularly non-noble metals, is challenging.

1.4.1
Development of a Solid Alkaline Membrane for Fuel Cell Application

The first key component of a membrane fuel cell is the membrane electrolyte. Its central role lies in the separation of the two electrodes and the transport of ionic species (e.g. hydroxyl ion, OH^-, in an AEM), between them. In general, quaternary ammonium groups are used as anion-exchange groups in these materials. However, due to their low stability in highly alkaline media [43, 44], only a few membranes have been evaluated for use as solid polymer electrolytes in alkaline fuel cells.

The first methanol-fed PEMFC working with an AEM was conceived by Hunger in 1960 [15, 45]. This system contained an AEM with porous catalytic electrodes pressed on both sides and led to relatively poor electrical performance (1 mA cm^{-2} at 0.25 V at room temperature with methanol and air as the reactants). Since this first attempt, many studies have been carried out to develop alkaline membranes.

Chloromethylated aromatic polymers of the polycondensation type are generally used to produce strongly alkaline AEMs [46]. The backbone positive charge of AEMs is generally provided by quaternary ammonium groups. For instance, Fang and Shen developed a quaternized poly(phthalazinone ether sulfone ketone) membrane (PESK) for alkaline fuel cell applications [47]. They obtained a maximum conductivity ranging from 5.2×10^{-3} to 0.14 S cm^{-1} depending on the concentration of the KOH solution and good thermal stability of the polymer up to 150 °C. Quaternized polyether sulfone cardo polymers (QPES-C) were also studied by Li et al. [48, 49]. They determined their conductivity and permeability to methanol. The conductivity of the membrane in 1.0 M NaOH solution increased from 4.1×10^{-2} to 9.2×10^{-2} S cm^{-1} with increase in temperature from room temperature to 70 °C, whereas the methanol diffusion coefficient inside the membrane increased from 5.7×10^{-8} to 1.2×10^{-7} cm^2 s^{-1} over this temperature range, which is at least 40 times lower than those for a Nafion membrane [49]. Although these membranes display high ionic conductivity, high thermal stability and low methanol permeability and therefore they may be suitable for use in direct alcohol alkaline fuel cells, no fuel cell data are available so far to compare with the current performance of DEFCs.

Another important class of membranes studied for SAMFC application is prepared from perfluorinated backbone polymers. Yu and Scott studied the electrochemical performance of an alkaline DMFC working with an AEM [50]. The commercial ADP-Morgane membrane provided by Solvay consisted of a cross-linked fluorinated polymer carrying quaternary ammonium as exchange groups. Its thickness in the fully humidified state is 150–160 μm and its specific resistance in 1 M NaOH solution is close to 0.5 Ω cm^2. The fuel cell performance gave a maximum power density close to 10 mW cm^{-2} with a Pt/C anode (60 wt% Pt, 2.19 mg cm^{-2} on non-Teflonized Toray 90 carbon paper) and a Pt/C cathode (60 wt% Pt, 2.07 mg cm^{-2} on 20% wet-proofed Toray 90 carbon paper), at a cell operating temperature of 60 °C and air pressure of 2 bar in 2 M methanol and 1 M NaOH. This membrane was also used by Demarconnay and co-workers to compare the performance of a direct methanol fuel cell with that of a direct ethylene glycol fuel cell [51, 52]. In both cases, they obtained a maximum power density close to 20 mW cm^{-2} at 20 °C, with similar electrodes mechanically pressed against the membrane (Pt 40 wt%/C, 2.0 mg Pt cm^{-2}), using pure oxygen at a pressure of 1 bar and 2 M methanol in 1 M NaOH. Varcoe and co-workers developed and characterized quaternary ammonium radiation grafted alkaline anion-exchange membranes (AAEMs), such as poly(ethylene-co-tetrafluoroethylene) (ETFE) [53, 54] and poly(hexafluoropropylene-co-tetrafluoroethylene) (FEP) with ion-exchange capacity as high as 1.35 meq OH$^-$ g^{-1} and conductivity (as determined from electrochemical impedance spectroscopy) up to 0.023 ± 0.001 S cm^{-1} at 50 °C, which are between 20 and 50% of the values for the commercial acid-form Nafion115 membrane [55, 56]. In order to avoid carbonation of the electrode (carbonate precipitation) and to improve the long-term operation stability, they worked in fuel solutions without the undesirable addition of M$^+$OH$^-$. A peak power density of 130 mW cm^{-2} for an H$_2$/O$_2$ fuel cell was obtained, while a maximum power density of 8.5 mW cm^{-2} was obtained in a metal cation-free methanol/O$_2$ fuel cell working at 2–2.5 bar back-pressure and 80 °C.

However, fluorine-containing polymers are expensive and the use of hydrocarbon-only membranes (C–H backbone) could be interesting considering their manufacturing and availability. According to Hübner and Roduner [57], at high *in situ* pH, the oxidative radical mechanism for polymer degradation is suppressed. Moreover according to Matsuoka et al. [58], an alcohol penetrating the AEM may protect it from attack by peroxide. They used a polyolefin backbone chain from Tokuyama (Japan), on which was fixed tetraalkylammonium groups giving a membrane with a thickness of 240 μm. The conductivity of this membrane was close to 14 mS cm^{-1}. Alkaline direct alcohol fuel cell tests were performed with 1.0 M of different alcohols (methanol, ethylene glycol, *meso*-erythritol, xylitol) in 1.0 M KOH, with a 1.0 mg Pt cm^{-2} Pt/C catalyst as cathode and 4.5 mg Pt–Ru cm^{-2} Pt–Ru/C catalyst as anode. Ethylene glycol led to the highest power density (9.8 mW cm^{-2}) at 50 °C.

Agel *et al.* [59] and Stoica *et al.* [60] also used non-fluorine-containing polymers, such as polyepichlorhydrin homo- and copolymers. The ionic conduction of these polymers, which can be assimilated to polyether polymers, is insured by the presence of quaternary ammonium groups as exchange groups. An ion-exchange capacity of 1.3 mol kg^{-1} and a conductivity of 1.3×10^{-2} S cm^{-1} at 60 °C and RH = 98% were found with ionomers based on polyepichlorhydrin copolymers using allyl glycidyl ether as cross-linking agent, but no fuel cell results were given. However, using a polyepichlorhydrin homopolymer, H$_2$/O$_2$ fuel tests were carried out without and with an interfacial solution made of poly(acrylic acid) and 1 M KOH between the electrodes and the membrane, giving maximum power densities close to 20 and 40 mW cm^{-2}, respectively, at 25 °C with $P_{O_2} = P_{H_2} = 1$ bar and electrodes containing 0.13 mg Pt cm^{-2}.

The plasma route for the synthesis of conductive polymers was developed recently [61]. Starting from a monomer containing triethylamine or triallylamine groups, the membrane is synthesized by plasma polymerization of the monomer vapor in a glow discharge to form a thin film partially composed of amine functions. Then, the amine functions are quaternized through methylation by immersion of the plasma film in methyl iodide solution. The structure of the plasma membranes consists of a polyethylene-type matrix containing a mixture of primary, secondary, tertiary and quaternary amine groups. The polymers are amorphous, dense materials, highly cross-linked, and show a disorganized structure. Electrochemical impedance spectroscopic measurements revealed a rather low ion-exchange capacity and low conductivity, but this could be counterbalanced by their low thickness (a few microns).

All the membranes considered were based on quaternary ammonium groups as anion-exchange groups. However, these conductive groups may decompose in concentrated alkaline solution following the Hofmann degradation reaction [43]. It is crucial how these groups are attached to the polymer backbone, especially if one considers that, during fuel cell operation, the pH may increase up to 14. Under these conditions, the presence of β-hydrogen atoms with respect to the nitrogen atom may represent a serious drawback for the stability in alkaline media [62]. The decomposition of the backbone polymer matrix represents another factor to be taken into account when designing an effective AEM. Basically, so far there does not exist any stable AEM in concentrated sodium hydroxide solution at high temperature. Although perfluorinated cation-exchange membranes show excellent stability in the presence of oxidizing agents and concentrated hydroxide solution, a stable perfluorinated amine is difficult to obtain because of the specific properties of fluorine compounds, which readily decompose as follows:

$$R_f CF_2 NH_2 \rightarrow R_f CN + 2HF \tag{1.51}$$

Thus, some workers have investigated a completely different approach by using polymer films and introducing OH^- conductivity by treating the films with concentrated KOH. For example, poly(vinyl alcohol) (PVA) cross-linked with sulfosuccinic acid [63] or glutaraldehyde [64], doped with TiO_2 and so on, [64], were developed by Yang et al. The DMFC, consisting of an MnO_2/C air cathode, a Pt–Ru black/C anode and a PVA/TiO_2 composite polymer membrane, allowed a maximum peak power density of about 7.5 mW cm^{-2} to be reached at 60 °C and 1 bar in 2 M KOH + 2 M CH_3OH solution. Xing and Savadogo [65] developed a 40 µm thick alkali-doped polybenzimidazole (PBI) membrane, which exhibited ionic conductivity as high as 9×10^{-2} S cm^{-1} with 6 M KOH-doped PBI at 90 °C. The performance of an H_2/O_2 SAMFC based on the KOH-doped PBI was similar to that of an H_2/O_2 PEMFC with Nafion 117, that is, close to 0.7–0.8 W cm^{-2}.

1.4.2
Anodic Catalysts in Alkaline Medium

Platinum-based catalysts are widely used in low-temperature fuel cells, so that up to 40% of the elementary fuel cell cost may come from platinum, making fuel cells expensive. The most electroreactive fuel is, of course, hydrogen, as in an acidic medium. Nickel-based compounds were used as catalysts in order to replace platinum for the electrochemical oxidation of hydrogen [66, 67]. Raney Ni catalysts appeared among the most active non-noble metals for the anode reaction in gas diffusion electrodes. However, the catalytic activity and stability of Raney Ni alone as a base metal for this reaction are limited. Indeed, Kiros and Schwartz [67] carried out durability tests with Ni and Pt–Pd gas diffusion electrodes in 6 M KOH medium and showed increased stability for the Pt–Pd-based catalysts compared with Raney Ni at a constant load of 100 mA cm^{-2} and at temperatures close to 60 °C. Moreover, higher activity and stability could be achieved by doping Ni–Al alloys with a few percent of transition metals, such as Ti, Cr, Fe and Mo [68–70].

In the case of an SAMFC, working with a Pt/C anode and fuelled with either methanol or ethylene glycol, very similar power densities are achieved, that is, both are close to 20 mW cm^{-2} at 20 °C, as shown in Figure 1.17 [51].

However, the most extensively investigated anode catalysts for DEFCs, in either acidic or alkaline media, contain binary and ternary combinations based almost exclusively on Pt–Ru and Pt–Sn [71, 72]. The superior performance of these binary and ternary electrocatalysts for the oxidation of ethanol, compared with Pt alone, is attributed to the bifunctional mechanism [14] and to the electronic interaction between Pt and the alloying metals [73]. Matsuoka et al. [74] studied the electro-oxidation of different alcohols and polyols for direct alkaline fuel cell applications using an AEM (from Tokumaya, Japan) functionalized with tetraalkylammonium groups as cationic groups (thickness = 240 µm). They obtained the best performance with ethylene glycol, achieving a maximum power density close to 10 mW cm^{-2} with a Pt–Ru catalyst at the anode. The relatively good reactivity of ethylene glycol (EG) makes it a good candidate for SAMFC applications. It is less toxic than methanol, its specific energy is close to that of alcohols and both carbons carry alcoholic groups. It can then be

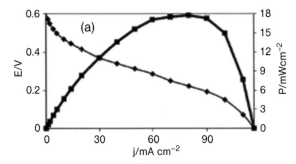

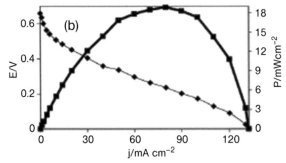

Figure 1.17 Cell voltage and power density vs current density curves for (a) 2 M methanol in 4 M NaOH solution; (b) 2 M ethylene glycol in 4 M NaOH solution. Anode and cathode catalysts, laboratory-made Pt (40 wt%)/C prepared via the Bönnemann method, 2 mg Pt cm^{-2}; commercial anionic membrane, Morgane ADP from Solvay; $T = 20\,°C$.

assumed that its oxidation into oxalate species ($^-$COO–COO$^-$) can be achieved, as proposed by several authors [75–79]. Thus, 8 mol of electrons are exchanged per mole of EG, instead of 10 for complete oxidation into CO_2, which means a faradaic efficiency of 80%. Metals other than Pt, such as Pd [80, 81], Au [82], Sn, Cd and Pb [83], display or enhance the catalytic activity towards EG electro-oxidation. Kadirgan *et al.* studied the effect of bismuth adatoms deposited on platinum surface by underpotential deposition (UPD) [84, 85]. They observed that the activity towards EG electro-oxidation is decreased in acidic medium, whereas it is highly enhanced in alkaline medium. Cnobloch *et al.* used catalysts containing Pt–PdBi as anode for an AFC working with 6 M KOH electrolyte [86]. A power of 225 W with a 52-cell stack was achieved. Coutanceau *et al.* studied the electrocatalytic behavior of nanostructured $Pt_{1-x}Pd_x$/C towards EG electro-oxidation [51]. The highest electroactivity was found with a Pt:Pd atomic ratio of ca 1 : 1 for a metal loading on carbon of 20%, as shown in Figure 1.18. The synergetic effect of palladium when added to platinum was confirmed in an SAMFC by comparing the cell performance obtained at 20 °C using a Pt (20 wt%)/C anode with that obtained using a $Pt_{0.5}Pd_{0.5}$ (20 wt%)/C anode (Figure 1.19).

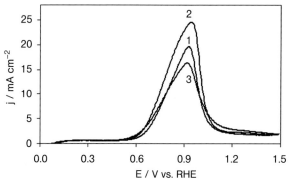

Figure 1.18 Polarization curves for the oxidation of 0.1 M ethylene glycol in 0.2 M NaOH solution recorded on (1) $Pt_{0.25}Pd_{0.75}$ (20 wt%)/C, (2) $Pt_{0.5}Pd_{0.5}$ (20 wt%)/C and (3) $Pt_{0.75}Pd_{0.25}$ (20 wt%)/C; $v = 50$ mV s^{-1}; $T = 20\,°C$. Catalysts were prepared according to the Bönnemann method.

$Pt_{1-x}Bi_x/C$ and $Pt_{1-(x+y)}Pd_xBi_y/C$ catalysts were also studied for EG electrooxidation in alkaline medium [52]. The highest electrocatalytic activity was found for the atomic compositions $Pt_{0.9}Bi_{0.1}$ and $Pt_{0.45}Pd_{0.45}Bi_{0.1}$. On the basis of electrochemical, *in situ* IR reflectance spectroscopy and high-performance liquid chromatography (HPLC) measurements, it was shown that the modification of platinum and platinum-palladium catalysts with bismuth led to the decrease in the CO_2 and formic acid yields (e.g. in the ability to break the C–C bond), whereas the yield in oxalic acid was increased, as shown in Figure 1.20a and b in the case of Pt/C and $Pt_{0.45}Pd_{0.45}Bi_{0.1}$, respectively. The different Bi-containing catalysts were used in a 5 cm^2 SAMFC anode and compared with respect to the electro-oxidation of EG (Figure 1.21). Finally, the activity increases in the following order: Pd/C < Pt/C < Pt–Pd/C < Pt–Bi < C < Pt–Pd–Bi/C.

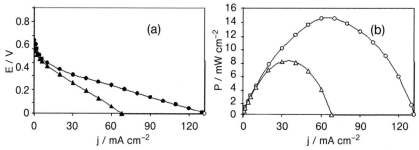

Figure 1.19 (a) Cell voltage and (b) power density vs current density curves recorded at 20 °C for the oxidation of 2 M ethylene glycol in 4 M NaOH. Anodes: (▲, △) Pt (20 wt%)/C and (●, ○) $Pt_{0.5}Pd_{0.5}$ (20 wt%)/C. Laboratory-made cathode Pt (40 wt%)/C and anode catalysts prepared according to the Bönnemann method. All electrodes were loaded with 2 mg cm^{-2} of metal. Anionic membrane, Morgane ADP from Solvay.

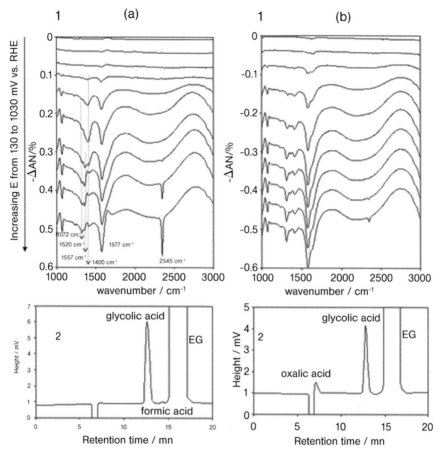

Figure 1.20 (a) SPAIR spectra recorded during the electro-oxidation of ethylene glycol on Pt/C in 0.2 M NaOH + 0.1 M EG, at various potentials (each 100 mV) from 130 to 1030 mV/RHE (1) and chromatograms of the anodic outlet of a SAMFC working with a Pt/C anode without fuel recirculation (2). (b) SPAIR spectra recorded during the electro-oxidation of ethylene glycol on $Pt_{0.45}Pd_{0.45}Bi_{0.1}$/C in 0.2 M NaOH + 0.1 M EG, at various potentials (each 100 mV) from 130 to 1030 mV/RHE (1) and chromatograms of the anodic outlet of a SAMFC working with a $Pt_{0.45}Pd_{0.45}Bi_{0.1}$/C anode without fuel recirculation (2). Electrocatalysts were prepared according to the 'water-in-oil' microemulsion method.

In the case of ethanol, Pd-based electrocatalysts seem to be slightly superior to Pt-based catalysts for electro-oxidation in alkaline medium [87], whereas methanol oxidation is less activated. Shen and Xu studied the activity of Pd/C promoted with nanocrystalline oxide electrocatalysts (CeO_2, Co_3O_4, Mn_3O_4 and nickel oxides) in the electro-oxidation of methanol, ethanol, glycerol and EG in alkaline media [88]. They found that such electrocatalysts were superior to Pt-based electrocatalysts in terms of activity and poison tolerance, particularly a Pd–NiO/C electrocatalyst, which led to a negative shift of the onset potential of the oxidation of ethanol by ca 300 mV compared

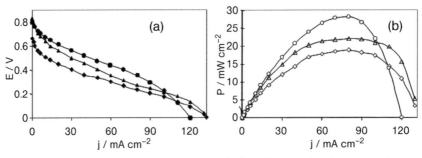

Figure 1.21 (a) Cell voltage and (b) power density vs current density curves recorded at 20 °C for the oxidation of 2 M ethylene glycol in 4 M NaOH. Anodes: (◆, ◇) Pt (40 wt%)/C; (▲, △) $Pt_{0.9}Bi_{0.1}$ (50 wt%)/C; (●, ○) $Pt_{0.45}Pd_{0.45}Bi_{0.1}$ (50 wt%)/C. Laboratory-made cathode catalyst Pt (40 wt%)/C prepared according to the Bönnemann method. Anode catalysts prepared according to the 'water-in-oil' microemulsion method. All electrodes were loaded with 2 mg cm^{-2} of metal. Anionic membrane was Morgane-ADP from Solvay.

with a Pt/C electrocatalyst under the same experimental conditions. Recently, nanostructured electrocatalysts based on Fe–Co–Ni alloys, known with the acronym HYPERMEC [89], were synthesized. Electrodes containing these catalysts have proved to be efficient in DEFCs with anion-exchange membranes, delivering power densities as high as 30–40 mW cm^{-2} at room temperature in self-breathing cells and up to 60 mW cm^{-2} at 80 °C in active systems. However, the catalyst durability is relatively low due to the slow formation of a metal oxide layer on the metal particle surface [90]. But preliminary data from CNR and ACTA show unequivocally that the catalyst stability with time under working conditions can be remarkably improved, up to several thousand hours, by decorating non-noble metal catalysts with tiny amounts of palladium using original electroless or spontaneous deposition techniques [91].

However, the oxidation reaction of alcohol is very difficult to activate at low temperature, even in alkaline medium. This leads to poor cell performance. Other hydrogen 'reservoirs' were then studied for fuelling the SAMFC. Among them, $NaBH_4$ seems to be the most promising one [in that case the direct fuel cell is called the direct borohydride fuel cell (DBFC)]. Indeed, the specific energy density of $NaBH_4$ is about 9.3 kW h kg^{-1} [92, 93], that is, higher than that of ethanol (8.0 kW h kg^{-1}). However, several oxidation paths are possible:

- the direct oxidation path:

$$BH_4^- + 8OH^- \rightarrow BO_2^- + 6H_2O + 8e^- \quad E^\circ_{NaBH_4} = -1.24 \text{ V vs SHE} \tag{1.52}$$

- the indirect oxidation path:

$$BH_4^- + 2H_2O \rightarrow BO_2^- + 4H_2 \tag{1.53}$$

$$H_2 + 2OH^- \rightarrow 2H_2O + 2e^- \quad E^\circ_{H_2} = -0.83 \text{ V vs SHE} \tag{1.54}$$

Assuming an electrode potential of 0.40 V vs SHE for oxygen reduction in a 1 M alkaline medium (pH = 14), the direct path gives a cell voltage of 1.64 V, whereas the indirect path leads to a cell voltage of 1.23 V. Therefore, catalysts which favor the direct oxidation path are preferable in order to achieve higher cell performance and higher energy efficiency. First, this implies working in alkaline solution with pH higher than 12 in order to limit the borohydride hydrolysis reaction [94]. According to Elder and Hickling [95], only Pt, Pd and Ni, which display good activity for ionizing hydrogen, are able to oxidize borohydride at low overvoltages. Liu and co-workers [96, 97] achieved a power density of 35 mW cm^{-2} at room temperature with a DBFC working with a nickel-based anode and a silver-based cathode and of 290 mW cm^{-2} at 60 °C with a mixture of surface-treated Zr–Ni Laves phase alloys, AB$_2$ (Zr$_{0.9}$Ti$_{0.1}$Mn$_{0.6}$V$_{0.2}$ Co$_{0.1}$Ni$_{1.1}$), and Pd deposited on carbon as anode catalyst and platinum as cathode catalyst. However, these catalysts also activate greatly the hydrolysis reaction of borohydride and therefore favor the indirect oxidation path. Gold is known to be an active metal for the electro-oxidation of borohydride and inactive for that of hydrogen. According to Chatenet et al. [98], gold allows an almost complete utilization of the reducer since around 7.5 electrons per borohydride molecule are consumed with this material versus 4 with platinum. However, the onset potential of BH$_4^-$ oxidation is shifted from about 0 V vs RHE with a Pt/C catalyst to about 0.35 V vs RHE with a Au/C catalyst, as shown in Figure 1.22.

Subsequently, to obtain a powerful DBFC system with high efficiency, Latour and co-workers proposed to manage the hydrolysis of borohydride to release an accurate quantity of molecular hydrogen such that the anode is able to oxidize at a given functioning point [99]. For this purpose, they developed Pt–Ni catalysts and claimed that the preliminary results were very promising [100].

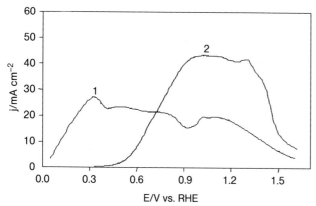

Figure 1.22 Polarization curves for the oxidation of 0.01 M NaBH$_4$ in a 1 M NaOH solution recorded on (1) Pt (50 wt%)/C and (2) Au (50 wt%)/C; $v = 20$ mV s^{-1}, $\Omega = 1000$ rpm, $T = 20$ °C. Catalysts were prepared according to the 'water in oil' microemulsion method.

1.4.3
Cathodic Catalysts in Alkaline Medium

Some electrocatalysts as alternatives to platinum have been studied as active catalysts for the oxygen reduction reaction (ORR) in alkaline media, such as palladium-based catalysts [101], ruthenium-based catalysts [102], iron-porphyrin or phthalocyanine catalysts [103, 104], nickel and cobalt catalysts and nickel–cobalt–spinel catalysts [105–107], and especially manganese oxide-based catalysts [108–113]. In the case of palladium catalysts, the onset of the reduction wave is shifted by 50 mV towards more negative potentials in comparison with platinum electrocatalysts. At the beginning of the reduction wave, water is the main reaction product, whereas at higher cathodic overpotentials, hydrogen peroxide becomes the main reaction product [101]. Ruthenium-based catalysts have shown poor performance in the oxygen reduction reaction because of the oxide film that is formed at the catalyst surface; as a result, the electrode becomes inactive towards oxygen reduction [102]. To achieve good activity and stability towards ORR, iron porphyrins have to be pyrolyzed at about 700 °C. Then, the onset of the reduction wave is close to that obtained with a platinum catalyst. Moreover, the main reaction product remains water, even though some hydrogen peroxide is produced particularly at high overvoltages [103, 104]. The nickel- or cobalt-based electrocatalysts showed poor activity for ORR in alkaline media. Oxygen reduction on these materials occurs only via a two-electron mechanism, producing hydrogen peroxide as the main product. On the other hand, spinel of cobalt and nickel oxides, $Ni_xCo_{3-x}O_4$ ($0 < x < 1$), showed better activity towards ORR, but the reaction mechanism via four electrons is not the main one and a large amount of hydrogen peroxide is formed [105–107]. Manganese oxides have been extensively studied and it was shown that the activity of these catalysts depends on the kind of oxides. The most active were the oxides α-MnO_2, γ-MnO_2, Mn_2O_3 and MnOOH [108, 110, 111]. The reaction mechanism on these catalysts involves the formation of the HO_2^- intermediate via the transfer of two electrons followed by a dismutation reaction of this species into OH^- and O_2. The kinetics of the dismutation reaction is so fast, particularly on the MnOOH catalyst, that apparently four electrons are exchanged during the ORR [109, 113]. Manganese oxides doped with Ni and/or Mg have also been studied in terms of activity towards ORR [41, 112, 113]. It appeared that the use of nickel as doping agent allowed the best activity to be obtained. Indeed, the reduction wave was shifted by 100 mV towards higher potentials and started only 50 mV lower than the reduction wave obtained at a platinum catalyst. However, the mechanism of reaction remains the same as for manganese oxides alone, that is, via the HO_2^- dismutation reaction. Ag-based cathodes have an extensive history with alkaline fuel cells, for example Raney Ag-based cathodes were used by Siemens to develop AFCs for military applications [114]. Moreover, Ag nanoparticles deposited on carbon are very easy to prepare and are known to be active towards the ORR [115].

Demarconnay *et al.* used a method derived from that of the so-called 'water-in-oil microemulsion' method to prepare well dispersed Ag/C catalysts [116]. The onset of the oxygen reduction wave is only shifted by 50 mV towards lower potentials on an Ag/C catalyst compared with that obtained on a Pt/C catalyst and the limiting current

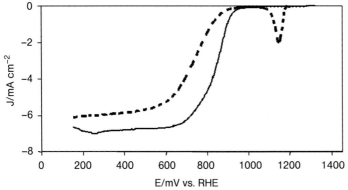

Figure 1.23 j(E) polarization curves recorded at a rotation rate Ω = 2500 rpm in an O_2-saturated 0.1 M NaOH electrolyte (T = 20 °C, ν = 5 mV s^{-1}) for (dashed line) 20 wt% Ag/C and (solid line) 20 wt% Pt/C. Ag/C catalyst was prepared according to the 'water-in-oil' microemulsion method, whereas Pt/C catalyst was prepared using the Bönnemann method.

densities are very close for both catalysts (Figure 1.23), indicating that a similar mechanism is involved at least at high overvoltage. On the basis of rotating disk electrode and rotating ring disk electrode experiments, it was shown that nanodispersed Ag particles on carbon powder led to the reduction of molecular oxygen mainly to water, which is similar to the result obtained with platinum (see Table 1.3).

According to the results obtained from rotating ring disk experiments and the determination of Tafel slopes, it was concluded that the mechanism of oxygen electroreduction on an Ag/C catalyst is similar to that on a Pt/C catalyst at high potentials, that is, interesting potentials for fuel cell applications. Lee *et al.* also showed that a 30 wt% Ag electrode displayed the same electroactivity towards the ORR as a 10 wt% electrode and that the activity could be enhanced by doping Ag with Mg [115].

Table 1.3 Total number of exchanged electrons n_t as determined by RDE^b and RRDE^d experiments.

E (V vs RHE)	KL slopes^a at Pt/C	KL slopes^a at Ag/C	20 wt% Ag/C				
			n_{RDE}^b	j_D^c (mA cm^{-2})	$(j_{R,1}^0 - j_{R,1})^c$ (μA cm^{-2})	n_{RRDE}^d	$p(H_2O)$
0.87	—	—	—	0.121	0	4	100
0.77	6.06	—	—	1.161	0	4	100
0.67	5.91	6.70	3.6	2.735	10.1	3.9	96.4
0.57	6.17	6.76	3.5	3.605	21.6	3.9	94.2
0.47	6.22	6.79	3.5	3.967	34.2	3.8	91.7
0.37	6.07	6.71	3.6	4.236	40.2	3.8	90.9
0.27	6.10	6.60	3.7	4.448	49.2	3.8	89.5

Slopes^a determined from the Koutecky–Levich (KL) plots; j_D and j_R are the disk and ring current densities, respectively^c. $p(H_2O)$ is the fraction of water produced.

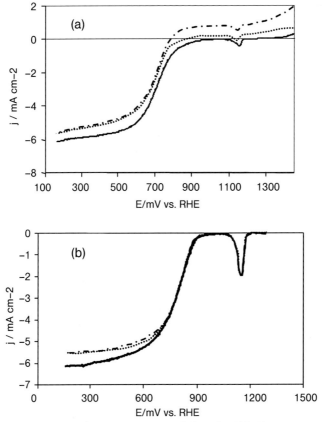

Figure 1.24 $j(E)$ polarization curves recorded on 20 wt% Ag/C at a rotation rate $\Omega = 2500$ rpm in an O_2-saturated 0.1 M NaOH electrolyte in the presence of alcohol ($T = 20\,°C$, $v = 5$ mVs).
(a) (—) Methanol-free electrolyte; (········) 0.1 M methanol; (········) 0.5 M methanol. (b) (—) EG-free electrolyte' (········) 0.1 M EG; (········) 0.5 M EG. Catalysts were prepared according to the 'water-in-oil' microemulsion method.

Moreover, for methanol concentrations higher than 0.1 M, the Ag/C cathode catalyst displayed higher tolerance towards methanol than Pt/C (Figure 1.24a) and almost total insensitivity towards the presence of EG (Figure 1.24b).

Other platinum-free materials, such as cobalt and iron macrocycles, display total insensitivity to the presence of methanol [117–119], whereas manganese oxides show very good tolerance towards the presence of $NaBH_4$ [120]. However, for portable applications, the fuel concentration has to be very high in order to increase the stored energy density and hence the autonomy of the system. Nevertheless, the intrinsic electroactivity of platinum free catalysts being lower than that of platinum, it could be interesting to modify platinum catalysts with foreign metal elements, in order to improve their tolerance towards the presence of the fuel and to increase their electrocatalytic activity towards the ORR. Numerous studies have shown that platinum-based binary alloyed electrocatalysts such as Pt–Fe, Pt–Co, Pt–Ni and Pt–Cr exhibit in acid electrolytes a higher catalytic activity for the ORR than pure platinum [17, 121–124] and a higher tolerance to the presence of alcohol [18, 125–127]. The observed electrocatalytic enhancement was interpreted either by an electronic factor, that is, the change in the d-band vacancy of Pt upon alloying, and/or geometric effects (Pt coordination number and Pt–Pt distance). Both effects should enhance the reaction rate of oxygen adsorption and the breaking of the O–O bond during the reduction reaction. For example, the lattice parameter a_0 in the case of a cubic Pt–X catalyst (X = Fe, Co, Ni) decreases with increasing content of the alloying component X, leading to changes in the catalytic behavior. In the case of Pt–Ni alloys, the maximum electrochemical activity for the ORR is obtained with 30 at.% of Ni [128]. Lemire et al. also showed that the presence of highly uncoordinated atoms is very important for CO adsorption at Au nanoparticles [129]. The same effect can also be true for oxygen adsorption at platinum nanoparticles. In addition to electronic, geometric, coordination and size effects, it was also proposed by Shukla et al. [130], that foreign metals, such as Cr, can protect platinum from oxidation and hence enhance the electrocatalytic activity of platinum towards the ORR. In alkaline medium, Adžič et al. pointed out that the addition of bismuth adatoms to an Au electrode led to greatly enhanced electrocatalytic activity towards the ORR [131].

Demarconnay et al. prepared nanodispersed Pt–Bi catalysts with different atomic compositions for the ORR [132]. They showed that a $Pt_{0.8}Bi_{0.2}/C$ catalyst led to a shift of the onset potential of the ORR towards higher potentials together with a higher tolerance towards the presence of EG (Figure 1.25) compared with a pure platinum catalyst. They explained this enhancement in activity and tolerance by the protection of the platinum surface by bismuth oxides. In the potential range where the ORR starts to occur, Bi_2O_5 surface species start to undergo a reduction reaction in Bi_2O_4, liberating oxide-free sites of platinum by decreasing the coverage by higher oxides of

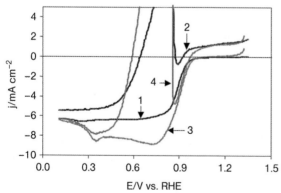

Figure 1.25 $j(E)$ polarization curves for the oxygen reduction reaction recorded in oxygen-saturated 0.2 M NaOH: (1) at Pt/C; (2) at Pt/C in the presence of 0.1 M EG; (3) at $Pt_{0.8}Bi_{0.2}/C$; (4) at $Pt_{0.8}Bi_{0.2}/C$ in the presence of 0.1 M EG ($v = 5\,mV\,s^{-1}$, $\Omega = 2500$ rpm; $T = 20\,°C$). Catalysts were prepared according to the 'water-in-oil' microemulsion method.

bismuth. The newly accessible oxide-free platinum sites become active for oxygen reduction. Moreover, the liberation of oxide-free platinum sites in a potential range where EG is not expected to adsorb at platinum leads to a higher tolerance of Pt–Bi towards the presence of EG.

1.5
Conclusion

Ethanol is one of the most popular fuels produced from biomass (by fermentation of sugar-containing crops or of numerous agricultural and municipal wastes) and has a relatively high energy density (about two-thirds of that of gasoline and one-quarter of that of pure hydrogen). Its clean and efficient direct conversion into energy then becomes a challenge, particularly its conversion into electricity inside a fuel cell. The development of PEMFCs and more recently of SAMFCs brings new opportunities to the use of ethanol in a direct fuel cell, working either in acidic (PEM) or alkaline (AEM) medium.

In an acidic medium, a PEMFC fed with ethanol allows power densities up to $60\,\text{mW}\,\text{cm}^{-2}$ to be reached at high temperatures (80–120 °C), but this needs platinum-based catalysts, which may prevent wider applications for portable electronic devices. On the other hand, in an alkaline medium, the activity of non-noble catalysts for ethanol or ethylene glycol oxidation and oxygen reduction is sufficient to reach power densities of the order of $20\,\text{mW}\,\text{cm}^{-2}$ at room temperature. This opens up the hope of developing SAMFCs that are particularly efficient for large-scale portable applications.

References

1 Grove, W.R. (1839) *Philosophical Magazine*, **14**, 127–130. Grove, W.R; (1843) *Journal of the Franklin Institute*, **35**, 277–280.
2 Gosselink, J.W. (2002) *International Journal of Hydrogen Energy*, **27**, 1125–1129.
3 Ströbel, R., Oszcipok, M., Fasil, M., Rohland, B., Jörissen, L. and Garche, J. (2002) *Journal of Power Sources*, **105**, 208–215.
4 Tilak, B.V., Yeo, R.S. and Srinivasan, S. (1981) *Comprehensive Treatise of Electrochemistry*, Vol. 3 (eds J.O'M. Bockris, B.E. Conway, E. Yeager and R.E. White), Plenum Press, New York, pp. 39–122.
5 Appleby, A.J. and Foulkes, F.R. (1989) *Fuel Cell Handbook*, Van Nostrand Reinhold, New York, pp. 203–500.
6 Lamy, C. and Léger, J.-M. (1994) *Journal de Physique IV*, **4**, C1, 253–281.
7 Kordesch, K. and Simader, G. (1996) *Fuel Cells and Their Applications*, VCH Verlag GmbH, Weinheim.
8 Vielstich, W., Lamm, A. and Gasteiger, H. (2003) *Handbook of Fuel Cells: Fundamental, Technology and Applications*, Vol. 1, John Wiley & Sons, Ltd, Chichester.
9 Bockris, J.O'M. and Reddy, A.K.N. (1972) *Modern Electrochemistry*, Vol. 2, Plenum Press, New York, p. 1141.
10 Sakellaropoulos, G.P. (1981) *Advances in Catalysis*, Vol. 30 (eds D.D. Eley, H. Pines and P.B. Weisz), Academic Press, New York, p. 218.
11 Appleby, A.J., Conway, B.E. and Bockrisv, J.O'M. (1983) *Comprehensive Treatise of*

Electrochemistry, Vol. 7 (eds E. Yeager, S.U.M. Khan, R.E. White), Plenum Press, New York, pp. 173–239.

12 Krishtalik, L.I. (1970) *Advances in Electrochemistry and Electrochemical Engineering*, Vol. 7 (ed. P. Delahay), Interscience, New York.Trasatti, S. (1972) *Journal of Electroanalytical Chemistry*, **39**, 163–184.

13 Lamy, C., Léger, J.-M., Srinivasan, S. and Bockris, J.O'M. (2001) in *Modern Aspects of Electrochemistry*, Vol. 34 (eds B.E. Conway and R.E. White), Kluwer Academic/Plenum Publishers, New York, pp. 53–118.

14 Watanabe, M. and Motoo, S. (1975) *Journal of Electroanalytical Chemistry*, **60**, 267–273; 275–283.

15 Dubau, L., Coutanceau, C., Garnier, E., Léger, J.-M. and Lamy, C. (2003) *Journal of Applied Electrochemistry*, **33**, 419–429.

16 Tarasevich, M.R., Sadkowski, A. and Yeager, E. (1983) *Comprehensive Treatise of Electrochemistry*, Vol. 7 (eds B.E. Conway, J.O'.M. Bockris, E. Yeager, S.U.M. Khan and R.E. White), Plenum Press, New York, pp. 301–398.

17 Ralph, T.R. and Hogarth, M.P., (2002) *Platinum Metals Review*, **46**, 3–14.

18 Yang, H., Alonso-Vante, N., Léger, J.-M. and Lamy, C. (2004) *Journal of Physical Chemistry. B*, **108**, 1938–1947.

19 Stevens, P., Novel-Catin, F., Hammou, A., Lamy, C. and Cassir, M. (2000) *Traité Génie Electrique, D3 340* (ed R. Bonnefille), Techniques de l'Ingénieur, Paris, pp. 1–28.

20 Lamy, C., Kauffmann, J.-M. and Grenier, J.-C. (2003) *Livre Blanc 'Programme ENERGIE du CNRS*, Ministère Délégué à la Recherche et aux Nouvelles Technologies, Paris, pp. 21–41.

21 Jones, D.J., Rozière, J., Marrony, M., Lamy, C., Coutanceau, C., Léger, J.-M., Hutchinson, H. and Dupont, M. (2005) *Fuel Cells Bulletin*, **10**, 12–15.

22 Lamy, C., Belgsir, E.M. and Léger, J.-M. (2001) *Journal of Applied Electrochemistry*, **31**, 799–809.

23 Lamy, C., Lima, A., Le Rhun, V., Delime, F., Coutanceau, C. and Léger, J.-M. (2002) *Journal of Power Sources*, **105**, 283–296.

24 Rousseau, S., Coutanceau, C., Lamy, C. and Léger, J.-M. (2006) *Journal of Power Sources*, **158**, 18–24.

25 Rightmire, R.A., Rowland, R.L., Boos, D.L. and Beals, D.L. (1964) *Journal of the Electrochemical Society*, **111**, 242.

26 Iwasita, T. and Pastor, E. (1994) *Electrochimica Acta*, **39**, 531.Iwasita, T. and Pastor, E. (1994) *Electrochimica Acta*, **39**, 547.

27 Wang, J., Wasmus, S. and Savinell, R.F. (1995) *Journal of the Electrochemical Society*, **142**, 4218.

28 Souza, J., Rabelo, F.J.B., de Moraes, I.R. and Nart, F.C. (1997) *Journal of Electroanalytical Chemistry*, **420**, 17.

29 Hitmi, H. (1992) PhD Thesis, University of Poitiers.

30 Delime, F., Léger, J.-M. and Lamy, C. (1999) *Journal of Applied Electrochemistry*, **29**, 1249.

31 Rezzouk, A. (1994) PhD Thesis, University of Poitiers.

32 Bonarowska, M., Malinowski, A. and Karpinski, Z. (1999) *Applied Catalysis A: General*, **188**, 145.

33 Aboul-Gheit, A.K., Menoufy, M.F. and El-Morsi, A.K. (1990) *Applied Catalysis*, **61**, 283.

34 Hitmi, H., Belgsir, E.-M., Léger, J.-M., Lamy, C. and Lezna, R.O. (1994) *Electrochimica Acta*, **39**, 407.

35 Perez, J.M., Beden, B., Hahn, F., Aldaz, A. and Lamy, C. (1989) *Journal of Electroanalytical Chemistry*, **262**, 251.

36 Lamy, C., Rousseau, S., Belgsir, E.-M., Coutanceau, C. and Léger, J.-M. (2004) *Electrochimica Acta*, **49**, 3901–3908.

37 Vigier, F., Coutanceau, C., Hahn, F., Belgsir, E.M. and Lamy, C. (2004) *Journal of Electroanalytical Chemistry*, **563**, 81–89.

38 Liu, P., Logadottir, A. and Nørskov, J.K. (2003) *Electrochimica Acta*, **48**, 3731.

39 Morimoto, Y. and Yeager, E.B. (1998) *Journal of Electroanalytical Chemistry*, **441**, 77.

40 Shubina, T.E. and Koper, M.T.M. (2002) *Electrochimica Acta*, **47**, 3621.
41 Yang, C.-C. (2004) *International Journal of Hydrogen Energy*, **29**, 135–143.
42 Wang, Y., Li, L., Hu, L., Zhuang, L., Lu, J. and Xu, B. (2003) *Electrochemistry Communications*, **5**, 662–666.
43 Iwamoto, T., Uetake, M. and Umeda, A. (1994) Hofman degradation of strong base anion exchange resin, Proceedings of the 68th Autumn Annual Meeting of the Chemical Society of Japan, Nagoya October 1994, p.472.
44 Matsui, K., Tobita, E., Sugimoto, K., Kondo, K., Seita, T. and Akimoto, A. (1986) *Journal of Applied Polymer Science*, **32**, 4137–4143.
45 Hunger, H.F. (1960) Proceedings of the Annual Power Sources Conference, Vol. 14, p.55.
46 Sata, T., Tsujimoto, M., Yamaguchi, T. and Matsusaki, K. (1996) *Journal of Membrane Science*, **112**, 161–170.
47 Fang, J. and Shen, P.K. (2006) *Journal of Membrane Science*, **285**, 317–322.
48 Li, L., Zhang, J. and Wang, Y.X. (2003) *Journal of Membrane Science*, **226**, 159–167.
49 Li, L. and Wang, Y.X. (2005) *Journal of Membrane Science*, **262**, 1–4.
50 Yu, E.H. and Scott, K. (2004) *Journal of Power Sources*, **137**, 248–256.
51 Coutanceau, C., Demarconnay, L., Léger, J.-M. and Lamy, C. (2006) *Journal of Power Sources*, **156**, 14–19.
52 Demarconnay, L., Brimaud, S., Coutanceau, C. and Léger, J.-M. (2007) *Journal of Electroanalytical Chemistry*, **601**, 169–180.
53 Varcoe, J.R., Slade, R.C.T., Yee, E.L.H., Poynton, S.D. and Driscoll, D.J. (2007) *Journal of Power Sources*, **173**, 194–199.
54 Varcoe, J.R. and Slade, R.C.T. (2006) *Electrochemistry Communications*, **8**, 839–843.
55 Slade, R.C.T. and Varcoe, J.R. (2005) *Solid State Ionics*, **176**, 585–597.
56 Herman, H. Slade, R.C.T. and Varcoe, J.R. (2003) *Journal of Membrane Science*, **218**, 147–163.
57 Hübner, G. and Roduner, E. (1999) *Journal of Materials Chemistry*, **9**, 409–418.
58 Matsuoka, K., Iriyama, Y., Abe, T., Matsuoka, M. and Ogumi, Z. (2005) *Journal of Power Sources*, **150**, 27–31.
59 Agel, E., Bouet, J. and Fauvarque, J.F. (2001) *Journal of Power Sources*, **101**, 267–274.
60 Stoica, D. Ogier, L. Akrour, L. and Alloin, F. Fauvarque, J.-F. (2007) *Electrochimica Acta*, **53**, 1596–1603.
61 Schiedaa, M., Roualdès, S., Durand, J., Martinent, A. and Marsacq, D. (2006) *Desalination*, **199**, 286–288.
62 Varcoe, J.R. and Slade, R.C.T. (2005) *Fuel Cells*, **5**, 187–200.
63 Yang, C.-C., Chiu, S.-J. and Chien, W.-C. (2006) *Journal of Power Sources*, **162**, 21–29.
64 Yang, C.-C. (2007) *Journal of Membrane Science*, **288**, 51–60.
65 Xing, B. and Savadogo, O. (2000) *Electrochemistry Communications*, **2**, 697–702.
66 Kordesch, K.V. (1978) *Journal of the Electrochemical Society*, **125**, 77C–88C.
67 Kiros, Y. and Schwartz, S. (2000) *Journal of Power Sources*, **87**, 101–105.
68 Ewe, H., Justi, E. and Schmitt, A. (1974) *Electrochimica Acta*, **19**, 799–808.
69 Mund, K., Richter, G. and Von Sturm, F. (1977) *Journal of the Electrochemical Society*, **125**, 1–6.
70 Kenjo, T. (1985) *Journal of the Electrochemical Society*, **132**, 383–386.
71 Vigier, F., Rousseau, S., Coutanceau, C., Léger, J.-M. and Lamy, C. (2006) *Topics in Catalysis*, **40**, 111–121.
72 Song, S. and Tsiakara, P. (2006) *Applied Catalysis B: Environmental*, **63**, 187–193.
73 Frelink, T., Visscher, W. and Van Veen, J.A.R. (1996) *Langmuir*, **12**, 3702–3708.
74 Matsuoka, K., Iriyama, Y., Abe, T., Matsuoka, M. and Ogumi, Z. (2005) *Electrochimica Acta*, **51**, 1085–1090.

75 Ewe, H., Justi, E. and Pesditschek, M. (1975) *Energy Conversion*, **15**, 9–14.
76 Hauffe, W. and Heitbaum, J. (1978) *Electrochimica Acta*, **23**, 299–304.
77 Santos, E. and Giordano, M.C. (1985) *Electrochimica Acta*, **30**, 871–878.
78 Kadirgan, F., Bouhier-Charbonnier, E., Lamy, C., Léger, J.-M. and Beden, B. (1990) *Journal of Electroanalytical Chemistry*, **286**, 41–61.
79 de Lima, R.B., Paganin, V., Iwasita, T. and Vielstich, W. (2003) *Electrochimica Acta*, **49**, 85–91.
80 Dalbay, N. and Kadirgan, F. (1990) *Journal of Electroanalytical Chemistry*, **296**, 559–569.
81 Pattabiraman, R. (1997) *Applied Catalysis A: General*, **153**, 9–20.
82 Beden, B., Kadirgan, F., Kahyaoglu, A. and Lamy, C. (1982) *Journal of Electroanalytical Chemistry*, **135**, 329–334.
83 Smirnova, N.W. Petrii, O.A. and Grzejdziak, A. (1988) *Journal of Electroanalytical Chemistry*, **251**, 73–87.
84 Kadirgan, F., Beden, B. and Lamy, C. (1982) *Journal of Electroanalytical Chemistry*, **136**, 119–138.
85 Kadirgan, F., Beden, B. and Lamy, C. (1983) *Journal of Electroanalytical Chemistry*, **143**, 135–152.
86 Cnobloch, H., Gröppel, D., Kohlmüller, H., Kühl, D. and Siemsen, G. (1979) in Power Sources 7. Proceedings of the 11th Symposium, Brighton, 1978, Vol. 24 (ed. J. Thomson), Academic Press, London, p. 389.
87 Xu, C., Cheng, L., Shen, P.K. and Liu, Y. (2007) *Electrochemistry Communications*, **9**, 997–1001.
88 Shen, P.K. and Xu, C. (2006) *Electrochemistry Communications*, **8**, 184–188.
89 Bert, P. and Bianchini, C. (2006) Platinum-free electrocatalysts materials, European Patent EP 1 556 916 B1.
90 Bert, P., Bianchini, C., Giambastiani, G., Miller, H., Santiccioli, S., Tampucci, A. and Vizza, F. (2006) Direct fuel cells comprising a nitrogen compounds and their use, Republica Italiana Dom. It. FI2006A000160.
91 Bert, P., Bianchini, C., Emiliani, C., Giambastiani, G., Santiccioli, S., Tampucci, A. and Vizza, F. (2006) Anode catalysts made of noble metal spontaneously deposited onto nanostructured catalysts based on transition metals, their preparation and use and fuel cells containing them, Republica Italiana Dom. It. FI2006A000180.
92 Amendola, S.C., Onnerud, P., Kelly, M.T., Petillo, P.J., Sharp-Goldman, S.L. and Binder, M. (1999) *Journal of Power Sources*, **84**, 130–133.
93 Li, Z.P., Liu, B.H., Arai, K., Asaba, K. and Suda, S. (2004) *Journal of Power Sources*, **126**, 28–33.
94 Mirkin, M.V. Yang, H. and Bard, A.J. (1992) *Journal of the Electrochemical Society*, **139**, 2212–2217.
95 Elder, J.P. and Hickling, A. (1962) *Transactions of the Faraday Society*, **58**, 1852–1864.
96 Liu, B.H., Li, Z.P., Arai, K. and Suda, S. (2005) *Electrochimica Acta*, **50**, 3719–3725.
97 Li, Z.P., Liu, B.H., Arai, K. and Suda, S. (2005) *Journal of Alloys and Compounds*, **404–406**, 648–652.
98 Chatenet, M., Micoud, F., Roche, I. and Chainet, E. (2006) *Electrochimica Acta*, **51**, 5459–5467.
99 Jamard, R., Latour, A., Salomon, J., Capron, P. and Martinent-Beaumont, A. (2008) *Journal of Power Sources*, **176**, 287–292.
100 Latour, A. (2006) PhD Thesis, Institut National Polytechnique de Grenoble.
101 Yang, Y.-F., Zhou, Y.-H. and Cha, C.-S. (1995) *Electrochimica Acta*, **40**, 2579–2586.
102 Prakash, J. and Joachin, H. (2000) *Electrochimica Acta*, **45**, 2289–2296.
103 Gojkovic, S.L., Gupta, S. and Savinell, R.F. (1999) *Journal of Electroanalytical Chemistry*, **462**, 63–72.
104 Gojkovic, S.L., Gupta, S. and Savinell, R.F. (2000) *Electrochimica Acta*, **45**, 889–897.

105 Heller-Ling, N. Prestat, M. Gautier, J.-L. and Koenig, J.-F. (1997) *Electrochimica Acta*, **42**, 197–202.

106 Hu, Y., Tolmachev, Y.V. and Scherson, D.A. (1999) *Journal of Electroanalytical Chemistry*, **468**, 64–69.

107 Rashkova, V., Kitova, S., Konstantinov, I. and Vitanov, T. (2002) *Electrochimica Acta*, **47**, 1555–1560.

108 Ponce, J., Rehspringer, J.-L., Poillerat, G. and Gautier, J.-L. (2001) *Electrochimica Acta*, **46**, 3373–3380.

109 Matsuki, K. and Kamada, H. (1986) *Electrochimica Acta*, **31**, 13–18.

110 Mao, L., Sotomura, T., Nakatsu, K., Koshiba, N., Zhang, D. and Ohsaka, T. (2002) *Journal of the Electrochemical Society*, **149**, A504–A507.

111 Klapste, B., Vondrak, J. and Velicka, J. (2002) *Electrochimica Acta*, **47**, 2365–2369.

112 Yang, J. and Xu, J.J. (2003) *Electrochemistry Communications*, **5**, 306–311.

113 Mao, L., Zhang, D., Sotomura, T., Nakatsu, K., Koshiba, N. and Ohsaka, T. (2003) *Electrochimica Acta*, **48**, 1015–1021.

114 Strasser, K. (2003) in *Handbook of Fuel Cells – Fundamentals, Technology and Applications*, Vol. 4 (eds W. Vielstisch, H.A. Gasteiger and A. Lamm), John Wiley & Sons, Ltd, Chichester, p. 775.

115 Lee, H.-K., Shim, J.-P., Shim, M.-J., Kim, S.-W. and Lee, J.-S. (1996) *Materials Chemistry and Physics*, **45**, 238–242.

116 Demarconnay, L., Coutanceau, C. and Léger, J.-M. (2004) *Electrochimica Acta*, **49**, 4513–4521.

117 Convert, P., Coutanceau, C., Crouigneau, P., Gloaguen, F. and Lamy, C. (2001) *Journal of Applied Electrochemistry*, **31**, 945–952.

118 Baranton, S., Coutanceau, C., Roux, C., Hahn, F. and Léger, J.-M. (2005) *Journal of Electroanalytical Chemistry*, **577**, 223–234.

119 Zagal, J.H. (2003) in *Handbook of Fuel Cells – Fundamental, Technology and Applications*, Vol. 1 (eds W. Vielstich, A. Lamm and H.A. Gasteiger), John Wiley and Sons, Ltd, Chichester, pp. 544–554.

120 Chatenet, M., Micoud, F., Roche, I., Chainet, E. and Vondrak, J. (2006) *Electrochimica Acta*, **51**, 5452–5458.

121 Paffet, M.T., Beery, J.G. and Gottesfeld, S. (1988) *Journal of the Electrochemical Society*, **135**, 1431–1436.

122 Beard, B.C. and Ross, P.N. (1990) *Journal of the Electrochemical Society*, **137**, 3368–3374.

123 Toda, T., Igarashi, H. and Watanabe, M. (1999) *Journal of Electroanalytical Chemistry*, **460**, 258–262.

124 Neergat, N., Shukla, A. and Gandhi, K.S. (2001) *Journal of Applied Electrochemistry*, **31**, 373–378.

125 Ralph, T.R. and Hogarth, M.P. (2002) *Platinum Metals Review*, **46**, 146–164.

126 Yang, H., Coutanceau, C., Léger, J.-M., Alonso-Vante, N. and Lamy, C. (2005) *Journal of Electroanalytical Chemistry*, **576**, 305–313.

127 Koffi, R.C., Coutanceau, C., Garnier, E., Léger, J.-M. and Lamy, C. (2005) *Electrochimica Acta*, **50**, 4117–4127.

128 Toda, T., Igarashi, H., Uchida, H. and Watanabe, M. (1999) *Journal of the Electrochemical Society*, **146**, 3750–3756.

129 Lemire, C. Meyer, R. Shaikhutdinov, S. and Freund, H.-J. (2003) *Angewandte Chemie International Edition*, **43**, 118–121.

130 Shukla, A., Neergat, M., Parthasarathi, B., Jayaram, V. and Hegde, M.S. (2001) *Journal of Electroanalytical Chemistry*, **504**, 111–119.

131 Adžič, R.R. (1984) *Advances in Electrochemistry and Electrochemical Engineering*, Vol. 13 (eds H. Gerisher and C.W. Tobia) Wiley-Interscience, New York, p 159.

132 Demarconnay, L., Coutanceau, C. and Léger, J.-M. (2008) *Electrochimica Acta*, **53**, 3232–3241.

2
Performance of Direct Methanol Fuel Cells for Portable Power Applications
Xiaoming Ren

2.1
Introduction

Recent reports on direct methanol fuel cells (DMFCs) that use polymer electrolyte membranes and dilute aqueous methanol solutions fed at the anode have demonstrated appreciable performance improvement, both for potential portable power applications operated at or near ambient temperatures and for potential transport applications operated at higher temperatures [1–5]. There are still two major obstacles that hinder the introduction of current DMFC te chnology to widespread commercial applications: the low electro-activity of known catalysts for methanol electro-oxidation and the high permeation rate of methanol through commercially available polymer electrolyte membranes, such as Nafion membranes. The low catalyst activity would require a higher Pt loading, and thus higher cost, in order to achieve appreciable anode performance; however, this cost barrier may be surmountable for portable power applications, as will be discussed further below. As for the lack of suitable methanol-impermeable membranes, methanol crossover through the membrane from the cell anode to the cathode will cause a decrease in fuel efficiency and even adversely affect the cathode performance [6–9]. With commercially available polymer electrolyte membranes, the phenomenon itself and problems caused by methanol crossover can be minimized to a certain extent even in a multi-cell stack operated at relatively constant load levels by proper design of flow fields and electrode structures and by controlling the reactant feed rate, as will be shown in this chapter.

The earliest possible commercial applications of DMFCs may be in the portable power market, where a relatively high cost per unit electric power and energy is set by current battery technology. As the number of mobile electronic devices in the consumer market increases rapidly, there will be an increasingly strong demand for cost-effective and long-lasting portable power sources for these devices. With further development, especially addressing the two above-mentioned technological barriers, DMFC power systems could compete successfully against the

Catalysis for Sustainable Energy Production. Edited by P. Barbaro and C. Bianchini
Copyright © 2009 WILEY-VCH Verlag GmbH & Co. KGaA, Weinheim
ISBN: 978-3-527-32095-0

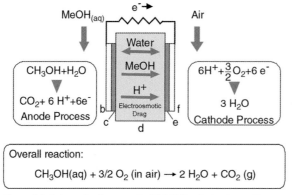

Figure 2.1 Schematic diagram of a DMFC, its electrode reactions and material transport involved, where (b) is the anode backing, (f) the cathode backing, (c) the Pt–Ru anode catalyst layer, (d) the Nafion 117 membrane and (e) the Pt cathode catalyst layer.

advanced batteries available today, providing a longer period of operation without frequent recharging and overcoming limited recharging rates by fast refueling.

Most literature reports have addressed DMFC performance at the single cell level. More relevant for evaluating DMFCs as practical power sources is the performance obtained at the stack level, achieved under operating conditions appropriate for the complete power system to achieve acceptable energy conversion efficiency and with complete thermal and water balances.

Figure 2.1 shows a schematic diagram of the electrode reactions and transport processes involved in an operating DMFC using a polymer electrolyte membrane and fed with an aqueous methanol solution at the anode and ambient dry air at the cathode. Detailed discussions of the methanol crossover, water transport and proton conduction within the polymer electrolyte membrane have been presented previously [8, 10]. It is important to realize that the DMFC cathode is working constantly at or near 'flooding' conditions under the normal operating conditions. This is because, in addition to the amount of water produced at the cathode, there is a substantial flux of water carried across the membrane from the anode side by electro-osmotic drag to the cathode side [10–12]. In principle, by increasing the air feed rate, one can alleviate cathode flooding and thus secure good cathode (and thus the cell) performance. However, as will be shown below, maintaining water balance, achieving high efficiency of the balance of plant and even reaching a desirable stack operating temperature at or over 60 °C become difficult at air feed rates exceeding three times faradaic stoichiometry. Consequently, the cathode structure and flow field used for DMFCs may differ substantially in design from those used for H_2/air fuel cells in order to secure a high cathode performance under flooding conditions with a low air feed rate.

To avoid cathode flooding, a hydrophobic cathode backing and an efficient means to remove water droplets in the cathode flow field are required. We report here measurements of water flux in both liquid and vapor forms in the cathode exhaust as a function of air feed stoichiometry and stack operating temperatures. The results show that, at temperatures below 60 °C and an air feed stoichiometry below three, the cathode exhaust is fully saturated (nearly fully saturated at 60 °C) with water vapor and the exhaust remains saturated after passing through a condenser at a lower temperature. In order to maintain water balance, all of the liquid water and part of the water vapor in the cathode exhaust have to be recovered and returned to the anode side before the cathode exhaust is released to the atmosphere. Because of the low efficiency of a condenser operated with a small temperature gradient between the stack and the environment, a DMFC stack for portable power applications is preferably operated at a low air feed stoichiometry in order to maximize the efficiency of the balance of plant and thus the energy conversion efficiency for the complete DMFC power system. Thermal balance under given operating conditions was calculated here based on the demonstrated stack performance, mass balance and the amount of waste heat to be rejected. The amount of waste heat to be rejected from an operating stack constrains in a significant way the optimal DMFC stack operating conditions and therefore should be considered when designing the complete power system in order to minimize the size and weight and maximize the power and total energy conversion efficiency.

As will be reported, as result of continuous efforts to optimize cell components and cell structure in developing DMFCs for portable power applications, we were able to demonstrate 30-cell DMFC stacks operated at 60 °C and fed with ambient air at 2–3 times faradaic stoichiometry, generating a power density of 320 W L^{-1} (active stack volume) during at least the initial week of testing in our laboratory. An overall DMFC stack energy conversion efficiency of 35% was achieved over a range of stack operating conditions of 0.46–0.57 V per cell. An extended life test over 1100 h on a five-cell stack made of identical cell components and stack configurations was also performed.

2.2
Experimental

Each of the membrane electrode assemblies used for the DMFC stack was prepared by attaching an anode catalyst layer and a cathode catalyst layer on opposite faces of a proton-form Nafion 117 membrane (Du Pont). The anode catalyst layer contained 85 wt% PtRu alloy catalyst (unsupported with a Pt:Ru atomic ratio of 1:1, 70 m^2 g^{-1} BET area; Johnson Matthey) and 15 wt% Nafion recast ionomer; the cathode catalyst layer contained 90 wt% Pt black (30 m^2 g^{-1} BET area; Johnson Matthey) and 10 wt% Nafion recast ionomer. The PtRu catalyst loading on the anode was 8 mg cm^{-2} and the Pt catalyst loading on the cathode was 6 mg cm^{-2}. The active electrode area of the DMFC stack was 45 cm^2, which accounts for only 32% of the stack footprint in the current stack design. Further optimizations in stack design that would reduce the stack weight and size by 50% are currently under way. As appropriate for portable power applications, these DMFC stacks were tested with dry ambient air feed at 0.76 atm realized at Los Alamos altitude. The stack anode was fed in single-pass mode with a 0.5 M aqueous methanol solution preheated by passing through a stainless-steel coil placed together with the stack inside a convection oven. To determine the steady-state stack performance and the flow rate and concentration of methanol of the stack anode exhaust stream, the stack was operated for a duration of 30 min for a given stack voltage and the anode exhaust was collected with a conical beaker cooled in an ice–water bath to minimize methanol loss through evaporation. The concentration of the methanol solution at the anode exhaust was determined from its density measured at 20 °C using a DMA 4500 density meter (Anton Paar).

The fuel cell test station and associated equipment have been described previously [8, 13]. The $V\text{–}I$ curves for the 30-cell stacks reported here were obtained at 0.5 V intervala, stepping down from the open-circuit stack voltage, with a 5 min waiting period before taking the data. The change rate of the stack voltage corresponds to a voltage scan rate of $0.056\,\text{mV}\,\text{s}^{-1}$ per cell. A much longer waiting period (over 30 min) at a given stack voltage was used in the stack mass balance experiments and the resulting stack performance was nearly identical with the $V\text{–}I$ curve reported. This verifies that the $V\text{–}I$ curves reported here closely represent the steady-state stack performance in term of voltage scan rate, without the effects such as hysteresis introduced by the scan direction and a too fast scan rate. Additional experimental conditions are provided in the following sections, together with the related results.

The newly assembled 30-cell stacks (one is shown in Figure 2.2) were immediately run in the DMFC mode at a fixed stack voltage at 60 °C with a 0.5 M methanol solution fed at the anode manifold and dry ambient air fed at the cathode manifold, without subjecting the stacks to any H_2/air break-in conditions. The stacks gradually reached the reported levels of performance within a few hours and remained stable for at least the initial week of testing in our laboratory before they were sent to our partner for system integration. An extended life test over 1000 h on a five-cell stack built identically revealed stable stack performance. During the initial run, the

Figure 2.2 A photograph of the 30-cell DMFC stack shown together with a US Quarter for comparison of dimensions. More detailed stack properties and performance parameters for this current stack design are listed in Table 2.3.

stacks may have gradually reached a stable membrane hydration state and other beneficial attributes, such as improved bonding at the interfaces between the catalyst-coated membrane and backings.

2.3 Results and Discussion

2.3.1 Water Balance, Maximum Air Feed Rate and Implications for Cathode Performance

The experimental setup shown in Figure 2.3 was used to measure the amount of liquid water and water vapor in the cathode exhaust stream at the stack exit under various operating conditions.

A five-cell stack (with a structure identical with that of the 30-cell stack) and a column of Drierite (anhydrous $CaSO_4$) (W.A. Hammond Dryerite) used for trapping water vapor in the stack cathode exhaust stream were placed inside a convection oven set at the stack operating temperature. A T-valve was used to separate the water vapor and liquid water by gravity. The liquid water was directed to the outside of the oven with a peristaltic pump set at a flow rate of $1\,mL\,min^{-1}$, which was slightly higher than the flux of liquid water from the stack cathode exit but significantly lower than the cathode feed rate. The liquid water thus separated from the stack cathode exhaust stream was collected with a conical beaker placed

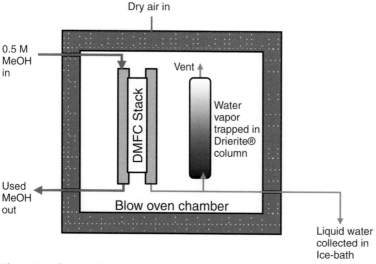

Figure 2.3 Schematic diagram of the experimental setup for measuring amounts of liquid water and water vapor in the air cathode exhaust at a close point of cathode exit of an operating DMFC stack.

Table 2.1 Test results for a five-cell stack fed with a 0.5 M methanol solution at 25 mL min^{-1} and dry oxygen at 0.76 atm at 0.69–1.58 SLPM.

Cell temperature (°C)	J_{cell} (mA cm^{-2})	V_{cell} (V)	R_{HF} (Ω cm^2)	J_x (mA cm^{-2})	W_{total} (g A^{-1} h^{-1} cell^{-1})	Drag coefficient
30	106	0.380	0.250	6	2.24	2.81
40	116	0.415	0.235	15	2.33	2.93
50	145	0.430	0.221	26	2.44	3.07
60	145	0.460	0.196	49	2.57	3.21

in an ice–water bath. The length of tubing connecting stack cathode exit and water liquid–vapor separating valve was kept to a minimum (at ∼3 cm) for accurate measurement of the cathode exhaust saturation level at the stack exit. For each measurement, the stack was fed with a 0.5 M methanol solution at 5 mL min^{-1} per cell at the anode and oxygen at the cathode. For each data point, the stack was operated in a constant-current mode for 2 h and the stack current density tested covers a wide range from 105 mA cm^{-2} at 30 °C to 145 mA cm^{-2} at 60 °C (Table 2.1). Pure oxygen feed at the cathode was used to sustain good cathode performance at the test current density that may not be achievable with air feed at a low flow rate. For the purpose of calculating the water vapor saturation level of the cathode exhaust, the oxygen flow was treated as if an air flow were used. These results are shown in Figures [4–6] for the flux of water in vapor form, the flux of water in liquid form and total flux of water at the stack cathode exit per ampere hour of charge passed, as a function of actual air feed stoichiometry and stack operating temperature. The actual stoichiometry of a reactant is defined as the ratio of the reactant mass flux fed to a DMFC stack to its total consumption rate within the stack. The total consumption of the reactant within a DMFC stack consists of its consumption at the electrode to generate the stack current as defined by Faraday's law and its consumption involved in the direct combination of oxygen with methanol permeating through the membrane at the cathode electrode. If there is no consumption of reactants incurred by methanol crossover in a DMFC stack, the actual stoichiometry becomes the faradaic stoichiometry.

The vapor water flux increases with air feed stoichiometry (Figure 2.4) at the expense of liquid water flux (Figure 2.5) while the total water flux at the stack cathode exit remains nearly constant at a given stack operating temperature (Figure 2.6). By increasing the stack operating temperature, the vapor water flux increases whereas the liquid water flux remains nearly unchanged. The increase in the vapor water flux is due to the increased water partial pressure with temperature. For the liquid water flux, the increase in liquid water flux because of an increase in the electro-osmotic drag of water by the proton current through the membrane at a higher temperature is largely compensated by a higher rate of water evaporation at a higher temperature. The fact the total water flux per ampere hour at the cathode exit is

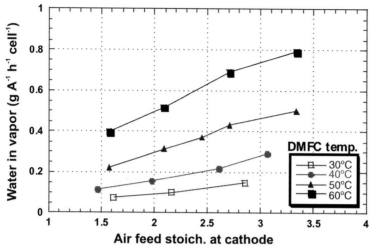

Figure 2.4 Amount of water vapor in the air cathode exhaust at cathode exit point of an operating DMFC stack as a function of stack operating temperature and air feed actual stoichiometry.

invariant with air feed stoichiometry (Figure 2.6) indicates that the cathode exhaust stream is saturated with water vapor for the range of air feed stoichiometry tested. This is because the water flux across the polymer electrolyte membrane is completely controlled by the electro-osmotic drag of water from anode to cath-

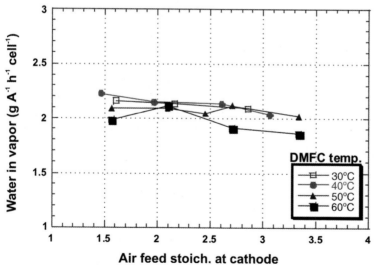

Figure 2.5 Amount of liquid water in the air cathode exhaust at cathode exit point of an operating DMFC stack as a function of stack operating temperature and air feed actual stoichiometry.

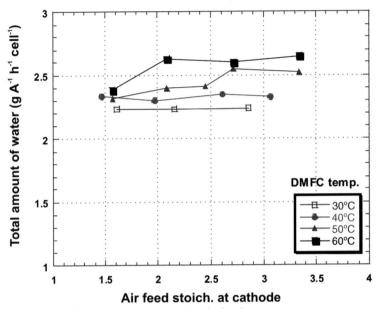

Figure 2.6 Total amount of water in the air cathode exhaust at cathode exit point of an operating DMFC stack as a function of stack operating temperature and air feed actual stoichiometry.

ode [10, 12]. The air exhaust saturation levels shown in Figure 2.7 were calculated from the ratio of the flux of water in vapor form measured to the flux of water needed to saturate the cathode exhaust.

The flux of water vapor in the cathode exhaust at full saturation is

$$W_{vapor}[gA^{-1}h^{-1}] = \frac{p_w}{p_{total} - p_w} \frac{(4.77n - 1) \times 18 \times 3600}{4 \times 96485} \quad (2.1)$$

where P_w is the water vapor partial pressure corresponding to the temperature of the stack cathode exit, P_{total} the total cathode pressure and n the actual air stoichiometry. The oxygen content in the air gives an air to oxygen mole ratio of 4.77, a constant that appears in Equation 2.1. As shown in Figure 2.7, the air cathode exhaust from the DMFC stack is nearly fully saturated at operating temperatures below 60 °C and at an air feed actual stoichiometry up to 3. With the air cathode saturated with water vapor, the cell cathodes are in contact with liquid water that accumulates from the cathode reaction, the oxidation of methanol permeated through the membrane and the electro-osmotic drag of water across the membrane. The total water flux in the DMFC cathode is

$$W_{total}[gA^{-1}h^{-1}] = \left(0.5 + 0.33 \cdot \frac{J_x}{J_{cell}} + \xi\right) \frac{18 \times 3600}{96485} \times J_{cell} \quad (2.2)$$

where J_x is the equivalent current (A) of methanol crossover, J_{cell} the DMFC current (A) and ξ the water electro-osmotic drag coefficient. In this study, the methanol

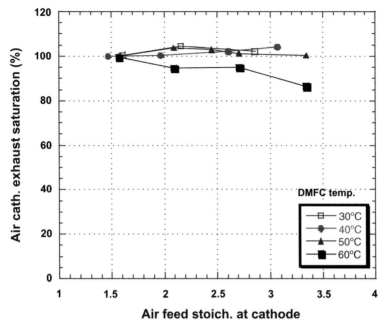

Figure 2.7 Water vapor saturation level of the air cathode exhaust at a close point of cathode exit of an operating DMFC stack as a function of stack operating temperature and air feed actual stoichiometry.

crossover current was determined from methanol mass balance according to:

$$\text{methanol consumption rate} = \text{DMFC current} + \text{methanol crossover current} \quad (2.3)$$

where the overall methanol consumption rate was determined by (methanol inflow rate × methanol inflow concentration) − (methanol outflow rate × methanol outflow concentration).

Finally, the liquid water flux found in the DMFC cathode exhaust is

$$W_{\text{liquid}}[gA^{-1}h^{-1}] = W_{\text{total}}[gA^{-1}h^{-1}] - W_{\text{vapor}}[gA^{-1}h^{-1}] \quad (2.4)$$

Table 2.1 lists the measured value of the DMFC current density, the equivalent current density of methanol crossover, the total water flux at a DMFC cathode and the calculated water electro-osmotic drag coefficient from Equation 2.2 at various DMFC operating temperatures.

Similar values of electro-osmotic drag coefficients for a Nafion 117 membrane in contact with liquid water at both sides and their increase with temperature have been reported previously [10, 12].

In order to maintain water balance during the normal operation of a DMFC stack, part of the water vapor has to be recovered with a condenser, which cools the air cathode exhaust to a lower temperature before it is released to the atmosphere. Based on the finding of water vapor saturation at the stack cathode exit (Figure 2.7), the cathode exhaust stream is expected to remain fully saturated with water vapor after passing through the condenser. Water in vapor form in the exhaust after the condenser will be released and lost to the atmosphere. To maintain water balance, the water loss in vapor form must be less than the amount of water generated from DMFC reactions, which include the cathodic electrode reaction and the oxidation of the methanol permeated through the membrane to the cell cathode. Consequently, the maximum air feed actual stoichiometry has to be less than a limiting value, given by

$$n \leq 2 \times \frac{p_{total} - p_w}{4.77 p_w} + \frac{1}{4.77} \tag{2.5}$$

Hence the maximum air feed actual stoichiometry is a function of water vapor partial pressure corresponding to the air exhaust release temperature (p_w) and the total pressure on the air cathode exhaust (p_{total}). Figure 2.8 shows the maximum air feed actual stoichiometry calculated from Equation 2.5, using water vapor partial pressure from the *CRC Handbook of Chemistry and Physics* [D.R. Lide (ed.), 72nd edn, 1991–92], as a function of air cathode exhaust release temperature.

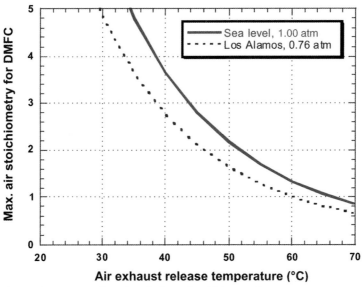

Figure 2.8 Maximum allowed air feed actual stoichiometry for a DMFC power system as a function of air cathode exhaust release temperature.

It shows that, for a condenser outlet temperature of 40 °C, which can be reasonably achieved at commonly encountered environmental temperatures, in order to maintain the water balance for a DMFC power system operated with ambient air feed, the air feed actual stoichiometry is limited to 2.7 at Los Alamos altitude (7200 feet above sea level) and to 3.6 at sea level.

Although under normal stack operating conditions maintaining the water balance requires capturing the necessary amount of water at the cathode and returning it to the anode, such water recovery design may complicate the DMFC power system, in addition to the added parasitic power consumption. For a portable DMFC power pack at a power level of $\sim$80 W, with which we are concerned here, water balance is not considered, in our opinion, as a major obstacle for successful commercial applications. This is because a relatively low level of parasitic power is required to accomplish the water recovery if the air feed at the cathode is not excessively high. A high feed rate of air at the cathode creates a very difficult situation of recovering a large amount of water from the vapor phase at low condenser efficiency. The low condenser efficiency is the result of a relatively low stack temperature realized at a high air flow.

From the foregoing discussion, it is clear that, in a DMFC, the air cathode has to be operated under rather challenging conditions, that is, with a low air feed rate at nearly full water saturation. This type of operating conditions can easily lead to cathode flooding and thus poor and unstable air cathode performance. To secure better air cathode performance, we have made great efforts to improve the ell cathode structure and cathode flow field design to facilitate uniform air distribution and easy water removal. The performance of our 30-cell DMFC stacks operated with dry air feed at low stoichiometry is reported in the following section.

2.3.2
Stack Performance

Three 30-cell DMFC stacks with 45 cm^2 active electrode area were tested under various conditions to evaluate their performance for portable power applications. All three stacks showed nearly identical performance and the results from one stack that had gone through the most complete set of tests are presented here.

Figure 2.9 shows the steady-state V–I curves obtained at various temperatures, ranging from 30 to 70 °C, with a 0.5 M MeOH solution fed to the anode manifold at 120 mL min^{-1} and 0.76 atm dry air fed to the cathode manifold at 7.35 standard liters per minute (SLPM). Figure 2.10 shows the corresponding stack power output, calculated from the V–I curves shown in Figure 2.9. At 30 °C, the stack generated almost half of the power obtained at 60–70 °C. At the design point of operation at 0.46 V per cell, the 30-cell stack generated 80 W gross power at 60–70 °C with an air cathode feed faradaic stoichiometry as low as 2.5. Table 2.2 lists some of the measurements on stack performance and methanol crossover rates at a few selected operation points, each obtained for a period of 30 min of steady-state stack operation in order to obtain good measurements of methanol concentrations

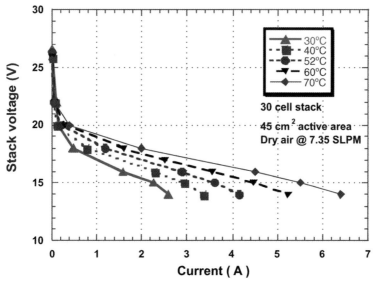

Figure 2.9 Steady-state V–I curves for a 30-cell DMFC stack operated at various temperatures with a 0.5 M methanol solution feed at 125 mL min^{-1} at the anode and with 0.76 atm dry air feed at 7.35 SLPM at the cathode. The steady state of the stack performance was verified by comparing the V–I curves with that of stack performance over 30 min for each given stack voltage listed in Table 2.2.

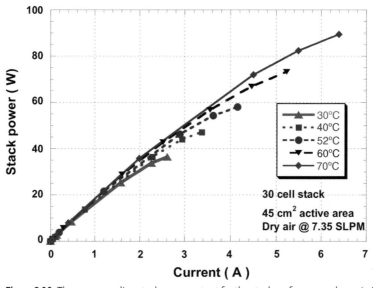

Figure 2.10 The corresponding stack power output for the stack performance shown in Figure 2.9.

Table 2.2 Test results for a 30-cell stack at 60 °C fed with a 0.484 M methanol solution and dry air at 0.76 atm at 7.35 SLPM.

Stack voltage (V)	Stack current (A)	MeOH flow (mL min^{-1})	MeOH concentration out (M)	MeOH crossover (mA cm^{-2})	Fuel efficiency^a (%)	MeOH stoichiometry (faradaic)	MeOH stoichiometry (actual)	Air stoichiometry (faradaic)	Air stoichiometry (actual)
24.42	0.00	14.2	0.309	18	0	—	2.8	—	17
—	0.00	42.3	0.347	41	0	—	3.5	—	7.3
—	0.00	81.7	0.376	63	0	—	4.5	—	4.8
—	0.00	121.0	0.409	65	0	—	6.4	—	4.7
20	0.37	14.2	0.239	17	32	6.0	1.9	37	12
20	0.20	42.3	0.337	40	10	33	3.3	68	6.7
18	1.37	42.3	0.282	33	48	4.8	2.3	9.9	4.7
16	3.29	42.3	0.249	4.0	95	2.0	1.9	4.1	3.9
16	4.00	121.0	0.355	32	74	4.7	3.5	3.4	2.5
15	5.24	121.0	0.337	22	84	3.6	3.0	2.6	2.2
14	6.10	121.0	0.324	16	89	3.1	2.8	2.2	2.0

^{a} Fuel utilization efficiency, calculated by $\dfrac{\text{stack current}}{\text{stack current} + \text{crossover current}} \times 100\%$.

and methanol solution flow rates. The equivalent methanol crossover currents were obtained based on the methanol mass balance described in Equation 2.3. The faradaic stoichiometry was calculated based on stack current only, while the actual stoichiometry was based on total reactant consumption from stack current and methanol crossover current.

Entries in Table 2.2 for a given stack operating voltage show a strong dependence of methanol crossover rate on the methanol solution feed rate and stack current. Under open-circuit conditions, all methanol consumption within the stack was due to methanol crossover through the membrane. With an increase in the methanol solution feed rate, the average concentration of the methanol solution in contact with the membrane increases, resulting an increased methanol crossover rate as observed. With the stack under current, the methanol concentration decreases along the anode flow field from the inlet to outlet because of methanol consumption along the flow path by the anodic electrode reaction and from methanol crossover. Better fuel utilization efficiency can be achieved by adjusting the feed rate of methanol solution at the anode. For example, at a fixed stack voltage of 16 V, it was observed that on increasing the feed rate of methanol solution from 42.3 to 121.0 mL min^{-1}, the stack current (and power) increased from 3.29 to 4.00 A (and the power from 52.6 to 64 W), while the fuel utilization decreases from 95.3 to 73.6%. The decrease in the fuel utilization again corresponded to an increase in the average concentration of the methanol solution in contact with membrane in the anode compartment from 0.367 to 0.420 M with an increase in methanol solution feed rate. At a lower stack voltage, a higher stack power can be achieved at a much higher stack current with a higher methanol feed rate. It can be summarized from the data in Table 2.2 that a fuel utilization efficiency above 80% can be achieved if the methanol feed stoichiometry is limited to <3 under our test conditions. At a very high fuel utilization, such as achieved by a lean-feed scheme (with a very dilute methanol solution fed at the anode at a sufficiently low feed rate), the power output of the stack will be limited and special care is needed to ensure an acceptable uniformity of fuel distribution. To achieve this, one can use a sufficiently dilute methanol solution fed at a high flow rate. Since there is a limitation on the methanol flow rate in a practical device, a compromise should be reached based on the expected normal power demand and load variation for a particular application. A slow change in the stack current and also stack power output can be reasonably followed by simply varying the methanol feed rate in the range of 2–3 times faradaic stoichiometry, which could minimize the methanol crossover rate in a dynamic load situation. Further studies are needed to explore the limitations of this approach. At the designed stack operating voltage of 14 V, the 30-cell stack demonstrated over 90% fuel utilization efficiency and, correspondingly, 35% stack energy conversion efficiency.

Figure 2.11 shows the stack energy conversion efficiency as a function of stack voltage and stack power output. The stack energy conversion efficiency is calculated by the voltage efficiency (= average cell voltage of a stack/1.21 V) times its

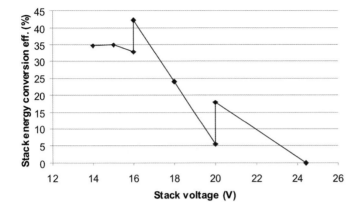

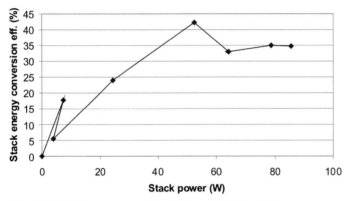

Figure 2.11 Stack energy conversion efficiency plotted against stack current (a) and stack power output (b) from the results of steady-state stack performance obtained at selected operating conditions listed in Table 2.2.

fuel efficiency. For a reasonably wide range of operating conditions (stack voltage from 14 to 17 V or stack power from 40 to 80 W), the stack energy conversion efficiency was maintained at 35%. Results from single-cell tests further indicated that additional stack fuel efficiency and stack performance improvements are possible by operating the stack with a feed of a methanol solution of lower than 0.5 M.

Figure 2.12 shows the cell voltage of the individual cell within the stack at the designed point of operation (stack voltage 14 V). The stack showed a uniform cell voltage distribution, except that the cells near the endplates had a slightly lower cell voltage, which may be due to a lower local temperature for these end cells. The total high-frequency resistance of the stack showed little change with stack current, indicating the full hydration state of the membranes within the stack at all

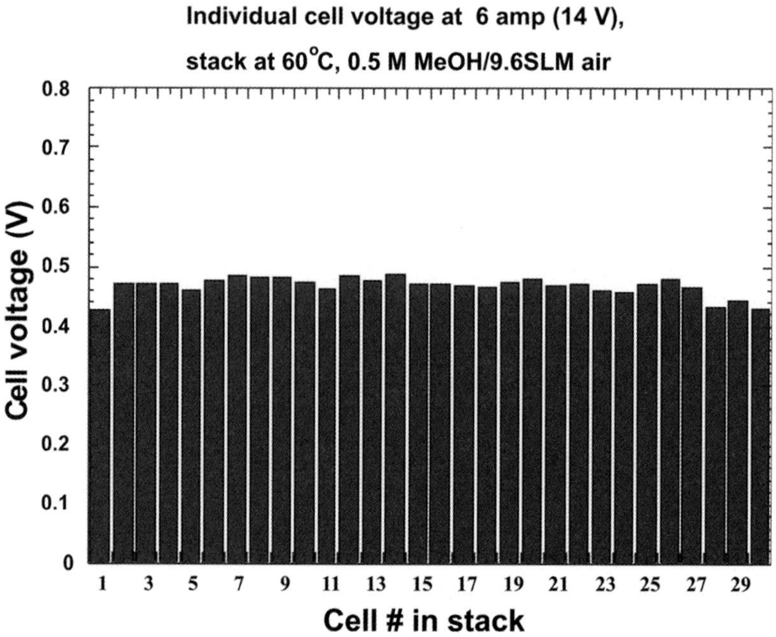

Figure 2.12 Voltages of the individual cells in a 30-cell DMFC stack at 14 V operated at 60 °C with a 0.5 M methanol solution feed at 125 mL min^{-1} at the anode and with 0.76 atm dry air feed at 7.35 SLPM at the cathode. At a stack current of 6.1 A, the corresponding methanol and air feed actual stoichiometries are 2.8 and 2.0, respectively.

operating current. Since no severe dehydration conditions were encountered either in our preparation of the membrane electrode assemblies or in the stack operation and storage at room temperature, the membranes are expected to remain in a full hydration state. To obtain the high-frequency resistance of an individual cell, the stack was fed at both anode and cathode manifolds with oversaturated H_2 prehumidified at 90 °C. A small external voltage of 0.3 V was applied to the stack to drive H_2 oxidation and H_2 evolution at the two cell electrodes of each cell within the stack. Since the resulting current density was very small (<40 mA cm^{-2}) and increased linearly with the applied voltage, we can assume that the measured individual cell voltage is proportional to its cell resistance, with the sum of the individual cell resistances being equal to the measured total stack high-frequency resistance. Figure 2.13 shows a uniform distribution of the high-frequency Press return to continue without the messageresistances of the individual cells within the stack. These values were very similar to those obtained for a single cell under identical operating conditions.

Despite a high cell packing density of 13 cells per inch, the flow resistances for passing both methanol solution and air through the stack were very low, owing to the

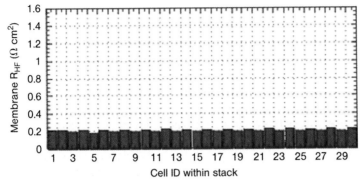

Figure 2.13 High-frequency cell resistance of the individual cells in a 30-cell DMFC stack operated at 60 °C with a 0.5 M methanol solution feed at 125 mL min^{-1} at the anode and with 0.76 atm dry air feed at 7.35 SLPM at the cathode.

unique open flow field structure used. Figure 2.14 shows the air pressure drop across the whole stack under various operating conditions.

A significant fraction of pressure drop for the airflow was due to the tubing connected at the stack cathode exit. The pressure drop also increased with increase in stack operating temperature because of the increased water vapor pressure at higher temperatures. By reducing the pressure drop across fuel cell stack, the

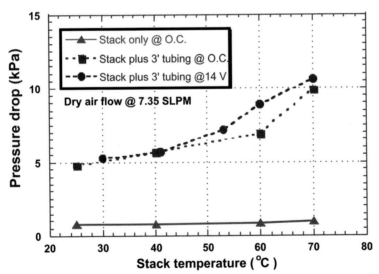

Figure 2.14 Air cathode pressure drop across the 30-cell stack operated at 60 °C with a 0.5 M methanol solution feed at 125 mL min^{-1} at the anode and with 0.76 atm dry air feed at 7.35 SLPM at the cathode.

parasitic power loss of moving reactants through the stack is obviously reduced. With each additional increase of 1 psig pressure drop across the cathode flow field, the $P\Delta V$ work required to push the air flow through amounts to 0.9 W parasitic power loss, which will be further augmented by the inefficiency loss of a given air pump.

2.3.3
Thermal Balance and Waste Heat Rejection

Compared with an H_2/air fuel cell, there is substantial voltage loss associated with both the methanol oxidation electrode and the air reduction electrode. For a typical DMFC operated at a cell voltage of 0.45 V, the anode potential is typically at 0.35 V versus. a dynamic hydrogen electrode (DHE) and the cathode potential is at 0.8 V vs DHE [8, 14]. As a result, for every one part of electric energy generated, the DMFC generates two parts of waste heat. In the following discussion, the DMFC stack is assumed to be adiabatic and all waste heat has to be carried by air through a condenser and/or by the anode-circulating loop of methanol solution through a radiator. Depending on the operating temperature of the DMFC stack, there are basically three configurations for a DMFC system in terms of waste heat rejection: (1) DMFC stack + radiator for a stack operated at a low temperature (<40 °C); (2) DMFC stack + radiator + condenser for a stack operated at an intermediate temperature (>40 °C but <~80 °C); and (3) DMFC stack + condenser for a stack operated at a high temperature (>~80 °C) . In the first case, because of the low stack operating temperature, there is no need to use a condenser to recover water vapor in the cathode exhaust if the air feed stoichiometry is limited to <3 (Figure 2.8). Consequently, the waste heat has to be rejected from the anode-circulating loop of methanol solution by using a radiator. In the second case, the thermal burden of the condenser is determined by the amount of water that needs to be recovered from the vapor phase in the cathode exhaust to maintain water balance and the thermal burden of the radiator is to reject the remaining waste heat. In the third case, all the waste heat is rejected by the condenser in the process of recovering water from the vapor phase in the cathode exhaust in order to maintain water balance. As the stack operating temperature increases, the thermal burden of waste heat rejection shifts from the anode side to the cathode side. However, as the water vapor pressure increases and moves close to atmospheric pressure with an increasing stack-operating temperature, the air feed at cathode will eventually be pressurized to secure acceptable cathode performance. Here, we will only examine the portable DMFC power systems operated with an ambient air feed, that is, a stack operated at low and intermediate temperatures.

Table 2.3 lists the demonstrated DMFC stack properties and performance parameters at 40 and 60 °C. The stack weight and volume had not been the focus of this study and in our next generation of DMFC stack design, reducing the gasket area and redesigning manifolds can achieve over 50% reductions in both stack weight and volume. The stack efficiencies for both DMFC operating temperatures are similar. However, for a stack operated at 40 °C, the fan power

Table 2.3 DMFC stack properties and performance parameters when operated with dry air at 0.76 atm.

Parameter	DMFC at 40 °C without condenser[a]	DMFC at 60 °C with condenser
Stack current (A)	3.15	6.10
Stack voltage (V)	27.0	14
Gross power (W)	85.0	85.4
Number of cells	60	30
Active electrode area (cm^2)	45	45
Unit cell voltage (V)	0.45	0.466
Fuel cell weight (kg)	1.2	0.6
End plates and screws weight (kg)[a]	1.1	1.1
Total stack weight (dry) (kg)	2.3	1.7
Stack volume (L)	1.8	1.0
Air flow at 3 times stoichiometry (L min^{-1})	10.2	9.8
0.5 M MeOH flow rate (mL min^{-1})	180–240	110–140
Water (l) from cathode (g min^{-1})	6.5	6.4
Water (v) from cathode (g min^{-1})	0.0	2.0
Total water recovered from cathode (g min^{-1})	6.5	8.4
Fuel efficiency (%)	88	92
Overall stack efficiency (%)	33	35
Heat generated (W)	172.3	158.6
Heat removed by cathode water evaporation (W)	42.1	40.6
Condenser burden (W)	0	39.0
Radiator burden (W)	130.6	79.0

[a] The 60-cell stack performance at 40 °C is extrapolated from that of a 30-cell stack operated at the same temperature.

and size of the radiator are expected to be considerably larger than those of a stack operated at 60 °C. This is because of the need to reject waste heat at a smaller temperature gradient between the stack and the environment. It seems that a DMFC stack operating at a higher temperature is more likely to attain a higher total system efficiency, partly because of the ease of waste heat rejection at a larger temperature gradient between the stack and the environment and partly because of the enhanced efficiency of the electrochemical processes at higher stack operating temperatures. The optimal stack and system configuration for a stack operated at temperatures higher than 60 °C, even with a moderately pressurized air feed, awaits further investigation regarding issues such as stack performance, water balance, thermal balance, stack efficiency and balance of plant efficiency.

2.3.4
Stack Life Test Results

An extended stack life test was performed on a five-cell stack built with cell components and stack configuration identical with those of the 30-cell stack. The

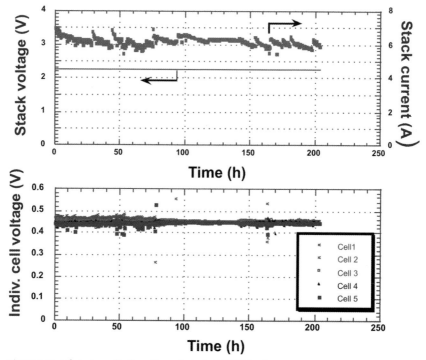

Figure 2.15 Life test results for a five-cell stack at 60 °C for the initial 200 h. The stack was fed with a 0.5 M methanol solution feed at 20 mL min^{-1} at the anode and with 0.76 atm dry air at 1.5 SLPM at the cathode.

life test of a five-cell stack was a considerably easier task to manage than that of a 30-cell stack without an automated control system. The life test of a five-cell stack was conducted at a stack voltage of 2.25 V at a stack temperature of 60 °C with the stack fed with a 0.5 M methanol solution at the anode at 20 mL min^{-1} in a single-pass mode and with ambient dry air at the cathode. Figure 2.15 shows the plot of stack current and individual cell voltage obtained with air feed at three times stoichiometry during the initial 200 h test.

From 200 to 1150 h, the air feed stoichiometry was reduced to 2.2. The stack current during the entire life test period is plotted in Figure 2.16. These test results shows that a stable DMFC stack performance can be achieved over an extended period with a low air feed stoichiometry with our current stack design.

However, we should point out that for real-world application, the environmental conditions and the stack operation mode of recirculating the anode fluid may pose significant challenges to the stack performance stability because of greater possibilities of exposing the stack to contamination and accumulation of contaminants within the stack. Certain precautionary means must be exercised to mitigate degradation of the stack performance caused by contamination.

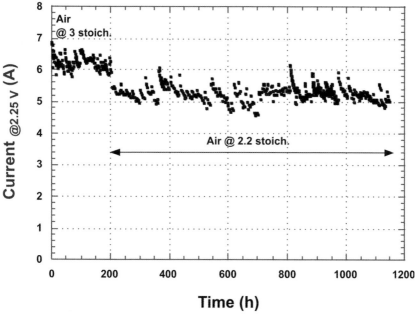

Figure 2.16 Life test results for a five-cell stack at 60 °C for the entire 1150 h.

2.4
Conclusions

Measurements of the liquid water and water vapor fluxes in the air cathode exhaust stream of a DMFC stack show that the exhaust air is nearly fully saturated with water vapor when the stack is operated with ambient air feed at an actual stoichiometry up to 3 at a stack temperature below 60 °C. Under such conditions, water crossover through the membrane by electro-osmotic drag accounts for most of the water in the cathode exhaust. In order to maintain water balance during stack operation, the water loss in vapor form in the released cathode exhaust has to be less than the amount of water produced by the methanol oxidation reaction. With a DMFC stack operated below 40 °C or operated at a higher temperature but with the cathode exhaust cooled below 40 °C using a condenser, the maximum air feed actual stoichiometry is limited to about 2.7 at 7200 feet above sea level or about 3.6 at sea level. The low air feed rate required to maintain water balance and the high flux of water from the anode to the cathode by electro-osmotic drag create challenging conditions for the DMFC cathode. Substantial redesigns and modifications of air cathode H_2/air fuel cell are required in order to secure good air cathode performance in a DMFC. Through such optimizations, we have demonstrated good performance in a 30-cell DMFC stack with a high cell packing density of 13 cells per inch. With an ambient air (at 0.76 atm) feed at an actual stoichiometry of <3, the 30-cell DMFC stack operated at 60–70 °C demonstrated a power density of 320 W L^{-1} active stack volume at 0.46 V per cell. The fuel efficiency achieved was close to 90% and the total stack energy conversion efficiency from methanol to electricity was 35% at the designed operating conditions. Individual cell performance and cell high frequency resistance were uniform across the stack. A comparison study on two DMFC stacks, one operated at 40 °C and the other at 60 °C, showed that, by increasing the stack operating temperature, both the stack efficiency and the efficiency of balance of the plant can be improved. The improvement in stack efficiency is due to improved electrode kinetics, mass transport rates and proton conductivity at higher operating temperatures and the improvement in the efficiency of balance of the plant is due to improved efficiency of waste heat rejection at a larger temperature gradient between the stack and the environment. Within the limitations of stack material stability, it seems worthwhile to explore DMFC stack performance at higher temperatures (>70 °C), even with a moderately pressurized air cathode feed, in order to determine the optimal system efficiency, performance, size and weight for the complete DMFC power system. A stack life test conducted on a five-cell stack built with identical stack components and configuration showed stable stack performance up to 1150 h. The current DMFC fuel cell stack performance demonstrated here should enable DMFC power systems to be used for certain practical applications in selected niche markets in spite of difficulties in using a polymer electrolyte membrane with a high methanol crossover rate and a high precious metal loading and the challenging conditions for the cell cathode in order to maintain the water balance. Polymer electrolyte membranes with low methanol and water crossover rates and more active methanol electro-oxidation catalysts should remain the focal areas of future research for the commercialization of DMFC power systems.

Acknowledgments

This work was supported by the Defense Advanced Research Projects Agency through the Defense Sciences Office and by the US Department of Energy through the Office of Transportation Technology.

References

1 Ren, X., Wilson, M.S. and Gottesfeld, S. (1995) in *Proton Conducting Membrane Fuel Cells I* (eds S. Gottesfeld, G. Halpert and A. Landgrebe), Electrochemical Society, Pennington, NJ, USA, PV95-23, pp. 252–260.
2 Ren, X., Zelenay, P., Thomas, S., Davey, J. and Gottesfeld, S. (2000) *Journal of Power Sources*, **86**, 11.
3 Scott, K. (2000) IEE Seminar on Electris, Hybrid and Fuel Cell Vehicles, April 2000, London, IEE, pp. 1–3.
4 Baldauf, M. and Preidel, W. (1999) *Journal of Power Sources*, **84**, 161.
5 Narayanan, S.R., Valdez, T., Rohatgi, N., Chun, W., Hoover, G. and Halpert, G. (1999) in Fourteenth Annual Battery Conference on Applications and Advances, January 1999, Piscataway, NJ, IEEE, pp. 73–77.
6 Chu, D. and Gilman, S. (1994) *Journal of the Electrochemical Society*, **141**, 1770.
7 Ren, X., Zawodzinski, T.A., Uribe, F., Hongli Dai and Gottesfeld, S. (1995) in *Proton Conducting Membrane Fuel*

Cells I (eds S. Gottesfeld, G. Halpert and A. Landgrebe), Electrochemical Society, Pennington, NJ, PV 95-23, pp. 284–298.

8 Ren, X., Springer, T.E. and Gottesfeld, S. (2000) *Journal of the Electrochemical Society*, **147**, 92.

9 Ren, X., Springer, T.E. and Gottesfeld, S. (1999) in *Proton Conducting Membrane Fuel Cells II* (eds S. Gottesfeld and T.F. Fuller), Electrochemical Society, Pennington, NJ, PV98-27, pp. 341–357.

10 Ren, X., Henderson, W. and Gottesfeld, S. (1997) *Journal of the Electrochemical Society*, **144**, L267–L270.

11 Ren, X., Springer, T.E., Zawodzinski, T.A. and Gottesfeld, S. (2000) *Journal of the Electrochemical Society*, **147**, 466.

12 Ren, X. and Gottesfeld, S. (2001) *Journal of the Electrochemical Society*, **148**, A87–A93.

13 Ren, X., Wilson, M.S. and Gottesfeld, S. (1996) *Journal of the Electrochemical Society*, **143**, L12–L15.

14 Thomas, S.C., Ren, X., Zelenay, P. and Gottesfeld, S. (1999) in *Proton Conducting Membrane Fuel Cells II* (eds S. Gottesfeld and T.F. Fuller), Electrochemical Society, Pennington, NJ, PV98-27, pp. 327–340.

3
Selective Synthesis of Carbon Nanofibers as Better Catalyst Supports for Low-temperature Fuel Cells

Seong-Hwa Hong, Mun-Suk Jun, Isao Mochida, and Seong-Ho Yoon

3.1
Introduction

Low-temperature operational fuel cells such as the direct methanol fuel cell (DMFC) and polymer electrolyte membrane fuel cell (PEMFC) have attracted attention as promising power sources for many applications such as small portable electric devices and automotive and stationary domestic applications [1–8]. Although a number of studies have been performed over the last two decades, further development is still essential to reduce the cost markedly by decreasing the amount of noble metal in the catalyst through improvements in catalytic activity. A breakthrough to enhance the catalytic activity is absolutely required for the wider diffusion of DMFCs and PEMFCs as power sources. PEMFCs show relatively higher power efficiency than DMFCs; however, the preparation, transportation and storage of hydrogen raise severe problems regarding costs, safety and efficiency. In contrast, the easy handling and safety of the storage of methanol are strong advantages for application in mobile power sources.

Common issues with current DMFC anode catalysts are the too high contents of noble metals (Pt and Ru) and the relatively short lifetime. Nanocarbons are expected to contribute to the alleviation of these issues via proper selection and modification of their various forms. Noble metals for DMFC anodes are commonly supported on carbon blacks with large surface areas. Therefore, the dispersion and activation in addition to the stability of the metals on the carbon surface are key factors in addressing the above issues. Carbon black, as a kind of spherical carbon nanoparticles gathered together to form aggregates, has been examined extensively. Its limitations for improving catalyst activity are well established. In this respect, carbon nanofibers (CNFs) with similar size, much larger aspect ratio, distinct graphene alignment and high graphitic properties can replace carbon black to improve catalytic performance due to the much higher active surface area. Such CNFs have been prepared on a large scale in our laboratory [9–17].

Catalysis for Sustainable Energy Production. Edited by P. Barbaro and C. Bianchini
Copyright © 2009 WILEY-VCH Verlag GmbH & Co. KGaA, Weinheim
ISBN: 978-3-527-32095-0

Over the last decade, novel carbonaceous and graphitic support materials for low-temperature fuel cell catalysts have been extensively explored. Recently, fibrous nanocarbon materials such as carbon nanotubes (CNTs) and CNFs have been examined as support materials for anodes and cathodes of fuel cells [18–31]. Mesoporous carbons have also attracted considerable attention for enhancing the activity of metal catalysts in low-temperature DMFC and PEMFC anodes [32–44]. Notwithstanding the many studies, carbon blacks are still the most common supports in industrial practice.

In this chapter, we reviewed the structure-controlled syntheses of CNFs in an attempt to offer better catalyst supports for fuel cell applications. Also, selected carbon nanofibers are used as supports for anode metal catalysts in DMFCs. The catalytic activity and the efficiency of transferring protons to ion-exchange membranes have been examined in half cells and single cells. The effects of the fiber diameter, graphene alignment and porosity on the activity of the CNF-supported catalysts have been examined in detail.

Generally, CNFs show surface areas of 20–300 $m^2 g^{-1}$ according to ordinary evaluation with the N_2 BET method [9–17]. Such a surface area basically originates from the free surface of the CNF because the CNF usually has no porosity. However, the effective surface area for the best dispersion of noble metals on the surface of CNF is still very small. To increase the effective surface area of mesoporosity, nanotunneled mesoporous CNFs were obtained through a two-step gasification procedure: careful preparation of CNFs having graphitic layers substantially angled (herringbone) with respect to the fiber axis, followed by a selective drilling of nanosized tunnels of 10–40 nm diameter along with the aligned graphitic layers through the gasification process. Catalytic gasification from the outside of the nanofiber into its central part was achieved in a substantially transverse fashion using a careful selection and dispersion of nanosized irons under specific reaction conditions [17]. The parent and nanotunneled H-CNFs were characterized by means of scanning electron microscopy (SEM), transmission electron microscopy (TEM), scanning tunneling microscopy (STM) and Raman techniques, as reported previously [17]. Such mesoporous CNFs may have a very high effective surface area and excellent electric conductivity and electrochemical stability. The supporting conditions of Pt–Ru noble metals on mesoporous CNFs are also important for controlling the activity of the resultant catalysts.

As mentioned above, CNFs have virtually no microporosity; the smaller the CNF diameter, the greater is their outer surface area. Very thin CNFs, having an average diameter of 40 nm, were also examined to enhance the catalyst activity through the enlargement of the effective outer surface area. The enhanced dispersion of thin CNFs using the nanodispersion developed is expected to increase the surface availability for supporting noble metals. The better dispersion of thin CNF must be a key technology for the preparation of DMFC catalysts.

The catalytic performance of CNFs was compared with that of commercial catalysts from Johnson Matthey and E-TEK, which are recognized as standard materials.

3.2
Preparation and Characterization of CNFs and Fuel Cell Catalysts

3.2.1
Preparation of Typical CNFs

Five main types of CNFs, platelet (P-CNF), tubular (T-CNF), thick herringbone (thick H-CNF), thin herringbone (thin H-CNF) and very thin herringbone (very thin H-CNF) were selectively prepared and examined as supports of anode catalysts for DMFCs. P-CNF was synthesized from carbon monoxide over a pure iron catalyst at 600 °C, whereas thick H-CNF was obtained from ethylene over a copper–nickel catalyst [Cu–Ni (2:8 w/w)]. An Fe–Ni alloy (6:4 w/w) was used for the selective synthesis of T-CNFs from carbon monoxide gas at 650 °C [15, 16].

Thin CNFs were synthesized using Ni–Fe/MgO as a catalyst from ethylene gas at 500 °C. The Ni–Fe/MgO catalyst (4:1:5 metal mole ratio) was prepared from nickel (II) nitrate hexahydrate [$Ni(NO_3)_2 \cdot 6H_2O$], iron(III) nitrate hydrate [$Fe(NO_3)_2 \cdot 9H_2O$] and magnesium nitrate hexahydrate [$Mg(NO_3)_2 \cdot 6H_2O$)]. The very thin CNFs were synthesized over a Ni/MgO catalyst from ethylene at 500 °C. The Ni/MgO catalyst (1:1 metal mole ratio) was prepared using nickel(II) nitrate hexahydrate and magnesium nitrate hexahydrate. For the preparation of the Ni–Fe/MgO and Ni/MgO catalysts, each aqueous solution was mixed at room temperature. The mixed solutions were dried slowly at 120 °C until the gel state and subsequently heated to 180 °C for oxidation at a heating rate of $1 \,°C\,min^{-1}$ in air and maintained there for 1 h. The catalyst powder was used for CNF synthesis after grinding. After the synthesis of the CNFs, the catalysts were removed by treatment with 10% HCl for 1 week.

3.2.2
Preparation of Nanotunneled Mesoporous H-CNF

The preparation procedure for mesoporous CNF was described in detail previously [17]. The Fe precursor as gasification catalyst was dispersed on the surface of H-CNF by the incipient impregnation method using an aqueous solution of iron nitrate (Wako Pure Chemical Industries, Japan) at an Fe:nanofiber ratio of 5:100 (w/w); then the impregnated CNF was dried at 150 °C in a vacuum oven for 2 h. The Fe–CNF mixture, which was located on a flat quartz boat in an ordinary horizontal furnace of 50 mm diameter was heated at a $10\,°C\,min^{-1}$ to 850 °C and maintained there for 3 h under different flow rates of hydrogen and a fixed amount of $1000\,mL\,min^{-1}$ of helium. The carbon yield by gasification was controlled with the flow rate of hydrogen, which was 750, 1000, 1500 or $2000\,mL\,min^{-1}$. The detailed preparation conditions for the nanodrilling procedure and the structural features of the nanotunneled mesoporous carbon nanofibers have already been reported [17].

3.2.3
Preparation of Fuel Cell Catalysts

The 40 wt% Pt–Ru/CNF (1:1 mol/mol) catalysts were prepared by chemical reduction. $RuCl_3 \cdot nH_2O$ and $H_2PtCl_6 \cdot 6H_2O$ (both from Wako) were mixed in deionized water and the resulting solution was dropped into a CNF-dispersed mixture with gentle stirring. The dissolved metal salts were reduced using 0.5 M $NaBH_4$ and the resulting slurry was subsequently filtered off, washed and dried. The anode catalyst was prepared by mixing this slurry with 20 wt% Nafion solution (Wako, 5% Nafion dispersion solution). The slurry was hand-brushed on carbon paper so as to give a metal loading of 5 mg cm^{-2}, then dried at room temperature for 12 h.

3.2.4
Performance Characterization of Fuel Cell Catalysts

The oxidation of methanol was evaluated using a typical three-electrode electrochemical half cell test kit (K0235 Flat Cell, EG&G Instruments, Princeton Applied Research) using a solution containing 1 M MeOH and 1 M H_2SO_4. The catalyst, hand-brushed on 1 cm^2 carbon paper, was used as the working electrode, and platinum mesh and Ag/AgCl were used as counter and reference electrodes, respectively. The same electrode as in the half cell test was used as the anode in the single cell test. The methanol oxidation was evaluated by cyclic voltammetry (CV) at 25 °C with a scan speed of 20 mV s^{-1} (Hokudo Denko, HZ-3000).

The catalyst performances were compared with those of commercial Johnson Matthey and E-TEK catalysts. The cathode catalyst was made of Pt.

The cathode and anode were hot-pressed on both sides of Nafion 115 at 135 °C under a pressure of 100 kg cm^{-2} for 10 min.

Current–potential curves were obtained in a single cell test with a flow rate of 2 mL min^{-1} in 2 M methanol and 200 mL min^{-1} pure oxygen at 30, 60 and 90 °C. The single cell performances of the Pt–Ru/CNF 40 wt% catalysts (Pt 1.33, Ru 0.67 and carbon 3 mg cm^{-2}) were compared with those of E-TEK 60 wt% and Johnson Matthey 60 wt% (Hispec10 000) catalysts (Pt 2, Ru 1 and carbon: 2 mg cm^{-2}). Pt black (7 mg cm^{-2}, Johnson Matthey, Hispec1000) was used as a cathode catalyst.

3.3
Results

3.3.1
Structural Effects of CNFs

Figure 3.1 shows FE-SEM and FE-TEM photographs of (a) T-, (b) thick H-, (c) P-), (d) thin H- and (e) very thin H-CNFs. T-CNFs with hexagonal transverse shape showed a relatively higher aspect ratio (L/D). Such tubular nanofibers exhibited a high degree of graphitization, an interlayer distance d_{002} of 0.337 nm and a height of

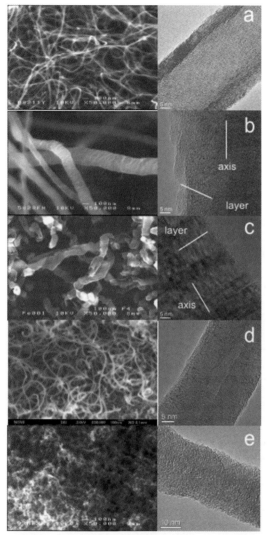

Figure 3.1 FE-SEM and FE-TEM images of (a) T-, (b) thick H-, (c) P-, (d) thin H- (40 nm) and (e) very thin (10 nm) H-CNFs.

graphene stacking Lc (002) of 11.7 nm [X-ray diffraction (XRD) evidence]. The diameter of the T-CNFs was fairly homogeneous at ∼40 nm. P-CNFs exhibited a ribbon-type transverse shape with the longest width of 150 nm. A relatively low aspect ratio was observed for this P-CNF, which was highly graphitic, having a d_{002} of 0.3363 nm, an Lc (002) of 28 nm and a graphene size [La (110)] of 22 nm. The diameter of thick H-CNFs was in the range 100–350 nm. This material showed a relatively higher aspect ratio than that of P-CNFs. Thin H-CNFs showed a rather winding shape with many nodes on the surface with a diameter from 30 to 50 nm. Very thin H-CNFs showed a very winding shape and a diameter from 7 to 15 nm.

TEM images of T-CNF showed the graphene layers to be parallel to the fiber axis. Thick H-CNF had the graphene alignment angled at about 60° to the fiber axis, as shown in Figure 3.1b. The graphene layer of P-CNF was perpendicular to the fiber axis. Thin and very thin H-CNFs showed the graphene alignment angled at 40–60° to the fiber axis. The surface areas of P-, thick H-, T-, thin H- and very thin H-CNFs were 90, 250, 90, 120 and 98 $m^2 g^{-1}$, respectively.

3.3.2
Catalytic Performance of CNFs in Half and Single Cells

The catalyst was supported on dispersed CNFs with a conventional procedure. Figure 3.2 shows cyclic voltammograms of methanol oxidation recorded at the 10th cycle. All the CNF-supported catalysts exhibited higher peak currents than the E-TEK Vulcan XC-72 catalyst, and the highest current peak appeared at slightly higher potentials. The highest peak potential decreased in the order thick H-CNF < P-CNF < T-CNF. The catalyst supported on thick H-CNF showed the best activity among the CNF-supported catalysts, being superior to the E-TEK catalyst. The catalyst supported on very thin H-CNF was the least active, which may be due to its poorer dispersion.

Figure 3.3 shows the single cell performance of catalysts supported on (a) T-CNF, (b) P-CNF), (c) thick H-CNF, (d) thin H-CNF and (e) E-TEK catalyst. The current density and the power density were normalized by unit electrode area (1 cm^2). The maximum power densities of T-, P-, thick H-, very thin H-CNF and E-TEK catalyst examined at 30 °C were 33, 52, 46, 28 and 41 mW cm^{-2}, respectively, at current densities of 156.3, 234.4, 234.5, 117.2 and 195.3 mA cm^{-2}. The maximum power densities examined at 60 °C were 82, 108, 113, 81 and 112 mW cm^{-2} at current densities of 312.5, 390.6, 429.7, 273.4 and 429.6 mA cm^{-2}, respectively. The maximum power observed at 90 °C were 112, 157, 165, 98 and 140 mW cm^{-2} at current

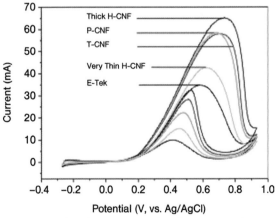

Figure 3.2 Cyclic voltammograms of catalysts supported on various CNFs: 1 M MeOH + 1 M H_2SO_4 at 25 °C.

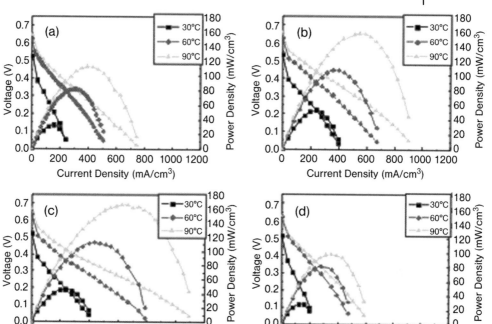
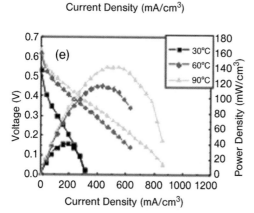

Figure 3.3 Single cell performances of catalysts supported on various CNFs: (a) T-CNF, (b) P-CNF, (c) thick H-CNF, (d) very thin H-CNF and (e) E-TEK catalyst. CNF-supported catalysts, Pt–Ru 40 wt% (Pt 1.33, Ru 0.67 and CNF 3 mg cm^{-2}); E-TEK catalyst, Pt–Ru 60 wt% (Pt 2, Ru 1 and CNF 2 mg cm^{-2}).

densities of 390.6, 546.9, 625.0, 351.6 and 507.8 mA cm^{-2}, respectively. The maximum power densities of the catalysts are summarized in Table 3.1.

The highest maximum power density examined at 30 °C was produced by the catalyst supported on P-CNF. However, the catalyst supported on thick H-CNF showed the highest maximum power density at 60 and 90 °C. The single cell

Table 3.1 Maximum power densities in single cell and average particle sizes of catalysts calculated by XRD.

CNF	Average diameter (nm)	BET surface area ($m^2 g^{-1}$)	Maximum power density ($mW\ cm^{-2}$)			Particle size (nm)
			30 °C	60 °C	90 °C	
T-CNF	40	90	33	82	112	3.46
P-CNF	150	90	52	108	157	3.35
Thick H-CNF	150	250	46	113	165	3.29
Thin H-CNF	40	120				
Very thin H-CNF	10	98	28	81	98	3.42
E-TEK			41	112	140	2.96

performances of the catalysts supported on the CNFs were higher than that obtained with the E-TEK catalyst, in spite of only a 36 wt% noble metal content. The very thin H-CNF-supported catalyst showed a lower single cell performance than that of the E-TEK catalyst. The particle size of E-TEK Pt–Ru, calculated by XRD, was smaller (2.96 nm) than those observed for the same alloy supported on CNFs (3.29–3.46 nm).

Since high current density at the maximum power density and the cost of the noble metals are important parameters for the commercialization of DMFCs, H-CNF-supported Pt–Ru alloys may be classified among the most efficient and cost-effective anode catalysts. It is also worth mentioning that the CNF-supported catalysts feature superior catalytic activity at the high temperatures where the mass transfer of methanol and oxygen is more favorable due to the fibrous network of CNFs.

3.3.3
Structure of Nanotunneled Mesoporous Thick H-CNF

Figure 3.4 shows high-resolution SEM and TEM photographs of nanotunneled mesoporous H-CNFs. No mesoporosity of as-prepared thick H-CNF was observed in the photographs. In contrast, well-aligned mesopores along the fiber axis of diameter 10–30 nm were observed in nanotunneled thick H-CNF obtained by 38 wt% burn-off. The N_2 BET isotherm of the nanotunneled thick H-CNF showed a clear hysteresis loop at P/P_0 0.48–0.53, which suggests the development of mesopores, as described in a previous paper [17]. Epitaxial nanodrilling of thick H-CNF with nanosized iron led to the predominant formation of mesopores of size 10–50 nm.

3.3.4
Catalytic Performance of Nanotunneled Mesoporous Thick H-CNF

Figure 3.5 compares the CV results for catalysts supported on thick H-CNF and on nanotunneled mesoporous thick H-CNF with those for the E-TEK catalyst at 25 °C. The current produced by the E-TEK catalyst (Pt–Ru 60 wt%) was 114.7 mA at a potential of 0.74 V. In comparison, the Pt–Ru 60 wt% catalysts supported on thick

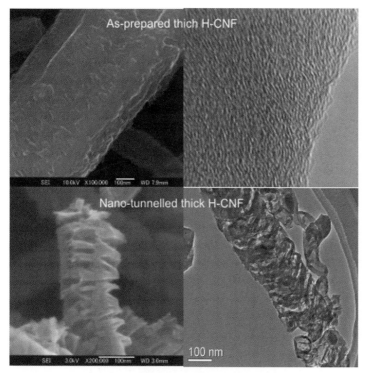

Figure 3.4 High-resolution FE-SEM and FE-TEM photographs of as-prepared and nanotunneled mesoporous thick H-CNFs (burn-off 32%).

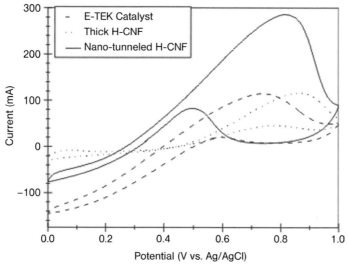

Figure 3.5 Cyclic voltammograms of Pt–Ru 60 wt% catalysts supported on H-CNF, nanotunneled H-CNF and E-TEK catalyst: 1 M MeOH + 1 M H_2SO_4 at 25 °C.

H-CNF and nanotunneled mesoporous H-CNF showed current intensities of 116.2 and 285.9 mA at a potential of 0.86 and 0.81 V, respectively. The catalyst supported on the nanotunneled mesoporous H-CNF showed about three times higher activity for methanol oxidation than those of the E-TEK and thick H-CNF-supported catalysts, although the peak potential of the CNF-supported catalysts was slightly higher. Three catalysts exhibited a similar value of peak current at a 20 wt% Pt–Ru content. However, the P-Ru 40 wt% catalysts supported on nanotunneled mesoporous H-CNF, thick H-CNF and 60 wt% E-TEK gave currents of 184, 101 and 80 mA, respectively. The oxidation of methanol on the 40 wt% catalyst supported on nanotunneled H-CNF was twice as efficient as those on the E-TEK and thick H-CNF-supported catalysts. The Pt–Ru 30 wt% catalyst supported on nanotunneled H-CNF showed comparable activity to the E-TEK 60 wt% catalyst.

The I–V curves from single cell tests with the Pt–Ru 40 wt% catalysts supported on thick H-CNF and nanotunneled H-CNF (Pt 1.33, Ru 0.67 and CNF 3 mg cm^{-2}) and with the Pt–Ru 60 wt% E-TEK catalyst (Pt 2, Ru 1 and C 2 mg cm^{-2}) (5 mg slurry hand-brushed on carbon paper) are shown in Figure 3.6, and their maximum power densities are listed in Table 3.2. The maximum power densities produced by the E-TEK catalyst at 30 and 90 °C were 41 and 140 mW cm^{-2}, respectively. The maximum

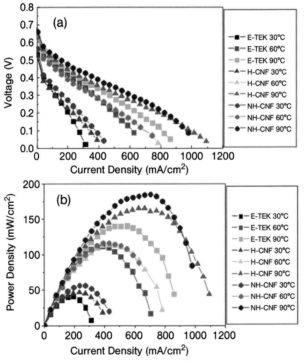

Figure 3.6 (a) I–V curves and (b) single cell performances of Pt–Ru 40 wt% catalysts supported on as prepared H-CNF, nanotunneled H-CNF and 60 wt% E-TEK catalyst examined at 30, 60 and 90 °C.

Table 3.2 Maximum power densities in single cell and average particle sizes, calculated by XRD, of catalysts supported on thick H-CNF, nanotunneled H-CNF and E-TEK.

CNF (Pt–Ru content)	Maximum power density (mW cm^{-2})			Average particle size (nm)
	30 °C	60 °C	90 °C	
Thick H-CNF (40 wt%)	46	113	165	3.39
Nanotunneled H-CNF (40 wt%)	56	116	184	3.13
E-TEK (60 wt%)	41	112	140	2.96

power densities given by the 40 wt% Pt–Ru catalyst supported on nanotunneled mesoporous H-CNF at 30 and 90 °C were 56 and 184 mW cm^{-2}, respectively, with a metal loading reduced by 36 wt%. The particle size of the Pt–Ru catalyst supported on nanotunneled mesoporous H-CNF (XRD analysis) was much smaller than that of thick H-CNF, but still larger than that of the E-TEK catalyst (Table 3.2).

The single cell performance of the catalyst supported on nanotunneled mesoporous H-CNF was found to be affected by the pH used in the impregnation procedure, which was controlled by adding 1 M NaOH solution. The highest maximum power density of 169 mW cm^{-2} was produced with the catalyst obtained at pH 3–4. Accurate control of the pH in the impregnation procedure is therefore an important factor for improving the catalytic activity.

Figure 3.7 shows the effect of the reduction temperature during the catalyst preparation on the performance of nanotunneled mesoporous H-CNF-supported catalysts. The maximum power densities of the Johnson Matthey Pt–Ru 60 wt% catalyst (HiSpec 10 000) were 55, 121 and 162 mW cm^{-2} at 30, 60 and 90 °C, respectively. The highest maximum power density of the 40 wt% Pt–Ru catalyst supported on nanotunneled mesoporous H-CNF was given by the catalyst chemically reduced at 0 °C, which gave maximum power densities of 62, 122 and 197 mW cm^{-2} at 30, 60 and 90 °C, respectively. A lower temperature during the chemical reduction, which is a highly exothermic process, apparently favors the formation of small metal particles, with enhanced catalytic activity. The maximum power densities obtained with the nanotunneled mesoporous H-CNF-supported catalysts prepared at different reduction temperatures are summarized in Table 3.3.

Table 3.4 reports the maximum power density in a single cell using nanotunneled thick mesoporous H-CNF obtained at a burn-off varying from 10 to 32%. The maximum power density was obtained with the highest burn-off.

3.3.5
Effect of the Dispersion of Thin and Very Thin H-CNFs on the Catalyst Activity

Very thin (average diameter 10 nm) and thin (average diameter 40 nm) H-CNFs were obtained in a highly dispersed form using nanodispersion equipment (T.K. FILMICS Model 56–50, Primix, Japan) to undo the entangled network with an impeller

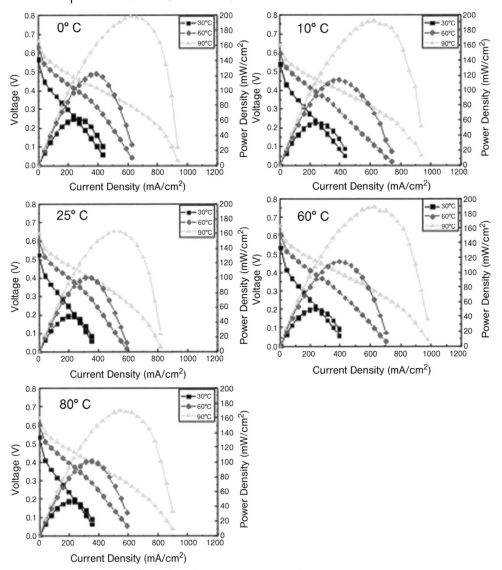

Figure 3.7 Effect of the reduction temperature on the performance of the catalyst supported on nanotunneled mesoporous H-CNF.

agitation of 16 500 rpm for 30 min in distilled water. The impeller was specifically designed to disperse nanosized materials with large shear force.

The single cell performance and the maximum power density of the Pt–Ru 40 wt% catalyst supported on thin H-CNF are shown in Figure 3.8. The maximum power densities were 76, 140 and 246 mW cm^{-2} at 30, 60 and 90 °C, respectively. The present nanodispersion treatment is therefore very effective in dispersing thin CNFs.

Table 3.3 Effects of the reduction temperature during catalyst preparation on the performance of the catalyst supported on nanotunneled mesoporous H-CNF.

Catalyst preparation temperature (°C)	Maximum power density (mW cm^{-2})		
	30°	60°C	90°C
0	62	122	197
10	56	114	191
25	48	101	163
60	50	115	188
80	46	101	169
Johnson Matthey (60 wt%)	55	121	162

Table 3.4 Maximum power density as a function of the gasification burn-off of nanotunneled mesoporous thick H-CNFs.

Gasification burn-off (%)	Maximum power density (mW cm^{-2})		
	30°C	60°C	90°C
32	56	116	184
27	51	108	178
19	50	103	169
10	48	102	165

Table 3.5 compares the maximum power densities of the catalysts supported on very thin and thin H-CNFs following the nanodispersion procedure described above. Nanodispersion seems to be effective in improving the power density for thin H-CNF-supported catalysts, probably due to an increased number of supporting

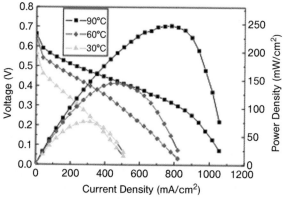

Figure 3.8 Single cell performance of the catalyst supported on highly dispersed thin H-CNF examined at 30, 60 and 90°C.

Table 3.5 Maximum power densities in single cell and particle size of catalysts supported on highly dispersed very thin and thin H-CNFs.

Reduction temperature (°C)	Maximum power density (mW cm^{-2})			Average particle size (nm)
	30 °C	60 °C	90 °C	
Very thin H-CNF	52	108	182	2.64
Thin H-CNF	76	144	246	2.90

sites for the noble metals. Better dispersion was also effective for very thin H-CNF, increasing the maximum power density from 28, 81 and 98 mW cm^{-2} (see Table 3.1) to 52, 108 and 182 mW cm^{-2} at 30, 60 and 90 °C, respectively. The catalyst particle size, calculated by XRD, was smaller after the nanodispersion procedure: from 3.42 nm for non-dispersed very thin H-CNF to 2.64 nm for highly dispersed very thin H-CNF (Tables 3.1 and 3.5). Very thin H-CNF is expected to give a better performance as a support material than thin H-CNF, provided that the dispersion of the former is fully achieved.

3.4
Discussion

The electrochemical oxidation of methanol on a catalyst supported on conductive carbon materials releases CO_2 from the anode catalyst surface in contact with the proton conductor. The catalyst must be highly dispersed to oxidize methanol effectively. On the other hand, a high activity must be balanced with appropriate contact with the Nafion membrane in the single cell system as no liquid ionomer is used. Hence the catalyst must be located on the free surface or on the surface composed of relatively larger pores of the carbon supports. For this purpose, small-sized carbons with large outer surface areas are appropriate materials. In addition, a good electron conductivity within the carbon network is necessary. The highly graphitic property and the fibrous forms, with small diameter, of CNFs are excellent properties of a catalyst support. The carbon has been recognized as having two surfaces from the different alignments of graphene: the basal and the edge surfaces. The metal catalyst is assumed to be more activated on the edge because of the increased electron density on the edge surface [45].

The present study examined a series of CNFs with well-defined diameters (10–200 nm) and graphene alignments. The alignment allows the occurrence of different exposures of basal and edge surfaces which are selectively found in tubular CNFs and platelet- and herringbone-type CNFs, respectively. In addition, artificial pores can be subsequently introduced epitaxtially along the graphene alignment from the surface in herringbone-type CNF. The diameter and depth of the artificial pores can be controlled by the extent of the catalytic gasification and the size of the gasification catalyst.

The catalyst impregnation on the dispersed CNFs must be carefully optimized to obtain a sufficient dispersion of noble metals. Dispersion of the CNF in the particular solvent is desirable for uniform impregnation of the catalyst. CNFs with very small diameter are very important for effective dispersion. Sophisticated procedures for dispersion must be applied. In the present study, 'nanodispersion' at an impeller agitation of 16 500 rpm was applied to disperse the thin CNFs better.

In the present study, the thick herringbone-type CNF showed the highest activity in the single cell among five types of CNFs with simple conventional dispersion in their catalyst preparation steps. The herringbone-type alignment may allow more effective dispersion of noble metals in a fairly large quantity of 40%. The thinner herringbone fiber failed to show the expected activity because of the poor dispersion by the conventional dispersion procedure. The activity of the thick herringbone fiber is comparable to that of the commercial catalyst of carbon black support with 60% metal catalyst. The superiority of the angled edge at the graphene is required to activate the metal catalyst. The activity obtained here is not remarkable but significant by decreasing the Pt–Ru amount by 36 wt%.

To increase the activity, the artificial introduction of mesoporosity for the thick herringbone CNF has been achieved. An optimum pore size, introduced by the controlled gasification procedure, increases substantially the activity of catalysts supported on mesoporous H-CNFs. Small-pore material may show higher activity in the half cell, but it is less effective in the single cell due to insufficient contact of the metal. We have shown here that the harsher the gasification procedure, the better is the performance in the half and single cells.

Nanodispersion can allow better dispersion to be obtained with thin CNFs as the best activity was obtained with the highly dispersed thin herringbone-type fiber of 40 nm diameter. The activity with this material is much higher than that of the commercial catalyst and can be further improved by a more delicate way of supporting the noble metals. Nevertheless, the present activity is double that of commercial catalysts taking into account the very low metal loading. A thinner fiber of 10 nm diameter might provide much higher activity, provided that sufficient dispersion is achieved. For this purpose, an improved dispersion procedure should be developed, also in terms of solvent and surface pretreatment.

The experimental conditions for the reduction of the metal salts were also investigated. Apparently, both the reduction temperature and the pH are critical for increasing the performance of the materials. A reduction temperature of 0 °C seems to be crucial to obtain highly dispersed metal particles.

Acknowledgments

The authors gratefully acknowledge the CREST Program 'Nano Catalyst' of JST and NEDO for financial support. They are also grateful to Sumitomo Trading Co. and Mr Uemura Masaaki for financial support and continuous cooperation during this study.

References

1 Acres, G.J.K. (2001) *Journal of Power Sources*, **100**, 60–66.
2 Heinzel, A., Hebling, C., Muller, M., Zedda, M. and Muller, C. (2002) *Journal of Power Sources*, **105**, 250–255.
3 Chu, D., Jiang, R., Gardner, K., Jacobs, R., Schmidt, J., Quakenbush, T. and Stephens, J. (2001) *Journal of Power Sources*, **96**, 174–178.
4 Wang, M. (2002) *Journal of Power Sources*, **112**, 307–321.
5 Ilic, D., Holl, K., Birke, P., Wohrle, T., Birke-Salam, F., Perner, A. and Haug, P. (2006) *Journal of Power Sources*, **155**, 72–76.
6 Munch, W., Frey, H., Edel, M. and Kessler, A. (2006) *Journal of Power Sources*, **155**, 77–82.
7 Cacciola, G., Antonucci, V. and Freni, S. (2001) *Journal of Power Sources*, **155**, 67–79.
8 Cropper, M.A.J., Geiger, S. and Jollie, D.M. (2004) *Journal of Power Sources*, **131**, 57–61.
9 Tanaka, A. Yoon, S.H. and (2004) *Carbon*, **42**, 591–597.
10 Lim, S.Y., Shimizu, A., Yoon, S.H., Korai, Y. and Mochida, I. (2004) *Carbon*, **42**, 1279–1283.
11 Tanaka, A., Yoon, S.H. and Mochida, I. (2004) *Carbon*, **42**, 1291–1298.
12 Lim, S.Y., Yoon, S.H., Shimizu, A., Jung, H. and Mochida, I. (2004) *Langmuir*, **20**, 5559–5563.
13 Lim, S.Y., Yoon, S.H., Korai, Y. and Mochida, I. (2004) *Carbon*, **42**, 1765–1781.
14 Lim, S.Y., Yoon, S.H. and Mochida, I. (2004) *Carbon*, **42**, 1773–1781.
15 Yoon, S.H., Lim, S.Y., Hong, Sh., Mochida, I., An, B. and Yokogawa, K. (2004) *Carbon*, **42**, 3087–3095.
16 Yoon, S.H., Lim, S.Y., Hong, Sh., Qiao, W., Whitehurst, D., Mochida, I., An, B. and Yokogawa, K. (2005) *Carbon*, **43**, 1828–1838.
17 Lim, S.Y., Hong, S.H., Qiao, W., Whitehurst, D., Yoon, S.H., Mochida, I., An, B. and Yokogawa, K. (2007) *Carbon*, **45**, 173–179.
18 Bessel, C.A., Laubernds, K., Rodriguez, N.M. and Baker, R.T.K. (2001) *Journal of Physical Chemistry B*, **105**, 1115–1118.
19 Ocampo, A.L., Miranda-Hernandez, M., Morgado, J., Montoya, J.A. and Sebastian, P.J. (2006) *Journal of Power Sources*, **160**, 915–924.
20 Wang, H.J., Yu, H., Peng, F. and Lv, P. (2006) *Electrochemistry Communications*, **8**, 499–504.
21 Park, I.S., Park, K.W., Choi, J.H., Park, C.R. and Sung, Y.E. (2007) *Carbon*, **45**, 28–33.
22 Tang, H., Chen, J., Nie, L., Liu, D., Deng, W., Kuang, Y. and Yao, S. (2004) *Journal of Colloid and Interface Science*, **269**, 26–31.
23 Li, W., Liang, C., Zhou, W., Qiu, J., Li, H., Sun, G. and Xin, Q. (2004) *Carbon*, **42**, 436–439.
24 Li, W., Liang, C., Qiu, J., Zhou, W., Han, H., Wei, Z., Sun, G. and Xin, Q. (2002) *Carbon*, **40**, 791–794.
25 Ismagilov, Z.R., Kerzhentsev, M.A., Shikina, N.V., Lisitsyn, A.S., Okhlopkova, L.B., Barnakov, Ch.N., Sakashita, M., Iijima, T. and Tadokoro, K. (2005) *Catalysis Today*, **102–103**, 58–66.
26 Danilov, M.O. and Melezhyk, A.V. (2006) *Journal of Power Sources*, **163**, 376–381.
27 He, Z., Chen, J., Liu, D., Tang, H., Deng, W. and Kuang, Y. (2004) *Materials Chemistry and Physics*, **85**, 396–401.
28 Guo, J., Sun, G., Wang, Q., Wang, G., Zhou, Z., Tang, S., Jiang, L., Zhou, B. and Xin, Q. (2006) *Carbon*, **44**, 152–157.
29 Wang, C.H., Shih, H.C., Tsai, Y.T., Du, H.Y., Chen, L.C. and Chen, K.H. (2006) *Electrochimica Acta*, **52**, 1612–1617.
30 Hacker, V., Wallnofer, E., Baumgartner, W., Schaffer, T., Besenhard, J.O., Schröttner, H. and Schmied, M. (2005) *Electrochemistry Communications*, **7**, 377–382.
31 Yuan, F., Yu, H.K. and Ryu, H. (2004) *Electrochimica Acta*, **50**, 685–691.
32 Coker, E.N., Steen, W.A. and Miller, J.E. (2007) *Microporous and Mesoporous Materials*, **104**, 236–247.

33 Calvillo, L., Lazaro, M.J., Garcia-Bordeje, E., Moliner, R., Cabot, P.L., Esparbe, I., Pastor, E. and Quintana, J.J. (2007) *Journal of Power Sources*, **169**, 59–64.

34 Coker, E.N., Steen, W.A., Miller, J.T., Kropf, A.J. and Miller, J.E. (2007) *Microporous and Mesoporous Materials*, **101**, 440–444.

35 Figueiredo, J.L., Pereira, M.F.R., Serp, P., Kalck, P., Samant, P.V. and Fernandes, J.B. (2006) *Carbon*, **44**, 2516–2522.

36 Joo, S.H., Pak, C.H., You, D.J., Lee, S.A., Lee, H.I., Kim, J.M., Chang, H. and Seung, D.Y. (2006) *Electrochimica Acta*, **52**, 1618–1626.

37 Choi, J.S., Chung, W.S., Ha, H.Y., Lim, T.H., Oh, I.H., Hong, S.A. and Lee, H.I. (2006) *Journal of Power Sources*, **156**, 466–471.

38 Joo, J.B., Kim, P., Kim, W.Y., Kim, J.S. and Yi, J.H. (2006) *Catalysis Today*, **111**, 171–175.

39 Samant, P.V., Rangel, C.M., Romero, M.H., Fernandes, J.B. and Figueiredo, J.L. (2005) *Journal of Power Sources*, **151**, 79–84.

40 Ding, J., Chan, K.Y., Ren, J. and Xiao, F.S. (2005) *Electrochimica Acta*, **50**, 3131–3141.

41 Nam, J.H., Jang, Y.Y., Kwon, Y.U. and Nam, J.D. (2004) *Electrochemistry Communications*, **6**, 737–741.

42 Han, S.G., Yun, Y.K., Park, K.W., Sung, Y.E. and Hyeon, T.H. (2003) *Advanced Materials*, **15**, 1922–1925.

43 Hyeon, T.H., Han, S.J., Sung, Y.E., Park, K.W. and Kim, Y.W. (2003) *Angewandte Chemie International Edition*, **42**, 4352–4356.

44 Park, K.W., Sung, Y.E., Han, S.J., Yun, Y.K. and Hyeon, T.H. (2004) *Journal of Physical Chemistry B*, **108**, 939–944.

45 Kim, T., Lim, S., Kwon, K., Hong, S.-H., Qiao, W., Rhee, C.K., Yoon, S.-Ho and Mochida, I. (2006) *Langmuir*, **22**, 9086–9088.

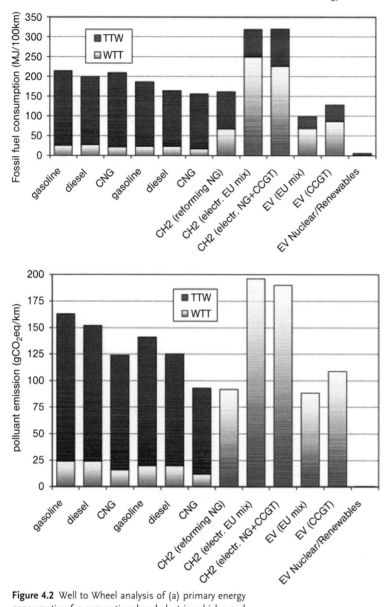

Figure 4.2 Well to Wheel analysis of (a) primary energy consumption for conventional and electric vehicles and (b) CO_2 emissions for conventional and electric vehicles.

Figure 4.3 shows the results for covering a total of 10 NDEC cycles (100 km). While a small-sized car of total weight 800 kg and $SC_x = 0.55\,\text{m}^2$ require $\sim$14 kW h, a medium–large-sized car of 1500 kg and $SC_x = \text{m}^2$ would require an energy of 27 kW h, almost twice as much.

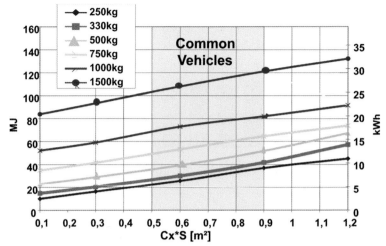

Figure 4.3 Influence of weight and shape on the energy needed to run an electric efficient car.

When designing an electric vehicle, the first question is the dimensions of the electric motor(s) in relation to the required performance. With that in mind, in Figure 4.4 the iso-power lines refer to the total nominal power required at the motors to reach a speed of 120 km h^{-1}. The typical peak power of an electric motor could be as much as 2–3 times the nominal power and the graph should then be used for

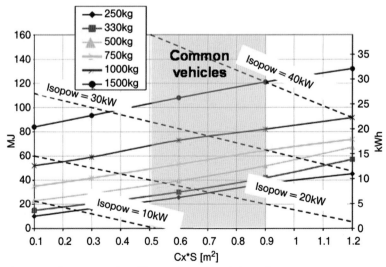

Figure 4.4 The isopower dotted lines indicate that a lightweight electric efficient car can be run with a total nominal power in the region of 10 kW; in contrast, a 1500 kg car would require as much as three to four times that power.

reference only, but in any case it is clear that a narrow superlight vehicle can be driven robustly with a total nominal power of less than 10 kW whereas a medium–large vehicle may require as much as 30–40 kW.

The control of both mass and shape is crucial for radical energy savings and also to reduce the overall complexity in motor design, cooling, electronic control and overall electric cabling.

4.3.4
A Roadmap of Feasibility with Batteries and Supercapacitors

To understand the evolution of the electric accumulators, let us consider a vehicle with a mass of 500 kg and aerodynamic factor $SC_x = 0.6\,\text{m}^2$. Figure 4.5 shows that for a range of 400 km in NDEC cycles the necessary energy to be delivered by the batteries is of about 34 kW h. In the hypothesis of full depth of charge and discharge by state of-the-art methods available in the 1997, the overall weight of the batteries needed was 675 kg (a trailer behind the car), whereas in 2007 it was 168 kg, more technically feasible, and conservative projections anticipate only 84 kg before 2020.

Referring to the Battery Association of Japan [8], in the period 1990–2005 the energy density of commercial batteries increased 5.2-fold (Figure 4.6). The typical commercial automotive Li-P batteries rank at 200 W h kg^{-1} and continuing with only half the current rate of advance the electric powertrain from 2017 on is very likely to be lighter and cheaper than the best solutions based on ICEs, fuel, tank, mechanical transmissions and gears.

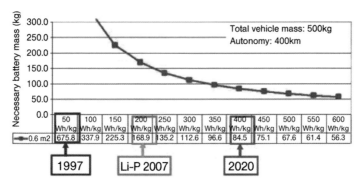

Figure 4.5 On the basis of the current predictions on battery development, the necessary battery mass for a 500 kg vehicle for a range of 400 km in NDEC cycles (120 km h^{-1}) is projected to be lower than the overall weight of a conventional powertrain based on an internal combustion engine.

List of Symbols

C_p	specific heat of convective medium
$\dot{m}C_p$	specific heat rate of the convective medium
η_{tot}	system total efficiency
η_{exc}	heat exchanger efficiency ε-NTU
η_{TE}	thermoelectric element efficiency
η_{Carnot}	Carnot contribution to the efficiency
η_{ZT}	contribution of thermoelectric material to the efficiency
H_{in}	inlet enthalpy
T_h	hot side temperature
T_{in}	convective medium inlet temperature
T_{out}	outgoing convective medium temperature
W	electric power generated
W^I	specific power of the fluid
ΔT	temperature gap between flows in heat exchanger

Combustor

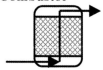

In this component the combustion of the air–fuel mixture takes place; also in this case the vertical line is proportional to the enthalpy feeding

Thermoelectric series

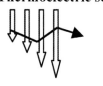

This symbol represents a heat flux (white arrows) from which an electric power (black arrow) is extracted by TE elements

Acknowledgments

Nicola Corino, Francesco Pitzalis and Daniel Zanello wish to acknowledge the Fondazione Cassa di Risparmio di Torino (FCRT) for providing fellowship support to complete their PhDs.

References

1 Energy Balances of OECD Countries, 2004–2005, 2007 edition, http://www.iea.org/Textbase/publications/free_new_Desc.asp?PUBS_ID=1931.
2 European Commission, Energy for the Future: Renewable Energy Sources, http://ec.europa.eu/energy/library/599fi_fr.pdf.
3 http://www.euresearch.ch/fileadmin/documents/PdfDocuments/Presentations/Pr_sentation_Transport January07.pdf.
4 http://www.ertrac.org/pdf/publications/ertrac_RF_brochure_june2006.pdf.
5 Shukla, A.K. (2001) *Resonance – Journal of Science Education*, **6** (11), 49–62;

Genta, G. Meccanica dell'Autoveicolo 2000, Levrotto e Bella edition.

6 Bossel, U. Well-to-Wheel Studies, Heating Values and Energy Conservation, www.efcf.com/reports/E10.pdf.
Bossel, U. Phenomena, Fact and Physics of a Sustainable Energy Future, www.efcf.com.

7 http://www.eucar.be/start.html/publications/well to wheel study.

8 Battery Association of Japan, http://www.baj.or.jp/e/index.html.

9 Bell, L.E.PhD BSST, LLC5462 Irwindale Avenue, Irwindale CA 91706lbell@amerigon.com, 626.815.7430. Alternate Thermoelectric Thermodynamic Cycles with Improved Power Generation Efficiencies.

10 Rowe, D.M. (2006) *Thermoelectrics Handbook*, Taylor & Francis, Boca Raton, FL.

11 Goldsmid and Ioffe, A.F. (1957) *Semiconductor Thermoelements and Thermoelectric Cooling*, Infosearch, London.

12 Goldsmid, H.J., Sheard, A.R. and Wright, D.A. (1958) *British Journal of Applied Physics*, **9**, 365.

13 Hsu, K.F., Loo, S., Guo, F., Chen, W., Dyck, J.S., Uher, C., Hogan, T., Polychroniadis, E.K. and Kanatzidis, M.G. (2004) *Science*, **303**, 818.

14 Snyder, G.J. (2005) Thermoelectric power generation: efficiency and compatibility, in *Thermoelectrics Handbook: Macro to Nano* (ed. D.M. Rowe), CRC Press, Taylor & Francis, Boca Raton, FL, Chapter. 9.

15 Ziggiotti, A. (2007) Hydrogen thermoelectric microcombustors, PhD thesis.

16 Cengel, Y.A. (1996) *Introduction to Thermodynamics and Heat Transfer*, McGraw-Hill, New York.

17 http://www.nrel.gov/pv/thin_film/.

18 http://www1.eere.energy.gov/solar/thin_films.html.

19 Perlo, P. (2008) Smart systems integration for the forthcoming electric mobility, Proceedings of the 2008 Annual Forum of the IEEE-ISSCC Society, Power Systems from the GigaWATT to the MicroWatt – Generation, Distribution, Storage and Efficient Use of Energy, 8 February 2008, San Francisco.

20 Burke, A.F. (2007) Batteries and ultracapacitors for electric, hybrid and fuel cell vehicles, *Proceedings of the IEEE*, **95** (4), 806–820.

21 Ehsani, M., Gao, Y. and Miller, J.M. (2007) Hybrid electric vehicles: architecture and motor drives, *Proceedings of the IEEE*, **95** (4), 719–729.

22 Burke, A.F. (2007) Supercapacitors in hybrid vehicle powertrains, in Proceedings of TransAlpine Workshop on Hybrid, Electric and fuel Cell Systems, October 2007, Pollein, Italy.

Part Two
Hydrogen Storage

Catalysis for Sustainable Energy Production. Edited by P. Barbaro and C. Bianchini
Copyright © 2009 WILEY-VCH Verlag GmbH & Co. KGaA, Weinheim
ISBN: 978-3-527-32095-0

5
Materials for Hydrogen Storage

Andreas Züttel

5.1
The Primitive Phase Diagram of Hydrogen

The triple point of hydrogen is at $T = 13.803$ K and 7.0 kPa. The density of solid and liquid hydrogen at the triple point are 86.5 and 77.2 kg m^{-3}, respectively. The boiling point at normal pressure ($p = 101.3$ kPa) is 20.3 K and the critical point is at $T_c = 33$ K and $p_c = 1293$ kPa (Figure 5.1).

At zero pressure, hydrogen (H_2, D_2) solidifies in the hexagonal close-packed (hcp) structure. Data for p-H_2 are $a = 375$ pm, $c/a = 1.633$, molar volume 22.56 cm^3 mol^{-1} [2]. The spherical $J = 0$ species (p-H_2, o-D_2) undergo a structural transition below 4 K where the rotational motion of the molecules is quenched and the molecules are located on the face-centered cubic (fcc) lattice (space group $Pa3$) with their axes oriented along the body diagonals. At very high pressures ($>2 \times 10^{11}$ Pa), solid hydrogen is expected to transform from a diatomic molecular phase to a monatomic metallic phase with a density >1000 kg m^{-3} [3]. This phase may become a high-temperature superconductor [4] (Table 5.1).

Due to the low critical temperature of hydrogen, liquefaction by compression at room temperature is not possible.

5.2
Hydrogen Storage Methods

Hydrogen storage basically implies a reduction in the enormous volume of hydrogen gas; 1 kg of hydrogen at ambient temperature and atmospheric pressure has a volume of 11 m^3. Three parameters allow the density of hydrogen to be increased: (i) increased pressure, (ii) lower temperature and (iii) decreased oscillation amplitude of the hydrogen atoms or molecules by interaction with other materials. The crucial

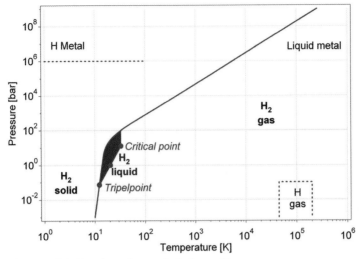

Figure 5.1 Primitive phase diagram for hydrogen [1].

parameter is the ratio of the energy (hydrogen) released to the energy stored in the system (Figure 5.2).

There are basically six methods in order to store hydrogen reversibly with a high volumetric and gravimetric density.

Table 5.1 Physical properties of *para*-hydrogen (*p*-H$_2$) and normal hydrogen (n-H$_2$) at the triple- and normal boiling point.

	p-H$_2$	n-H$_2$
Triple point ($T = 13.803$ K, $p = 7.04$ kPa)		
Temperature (K)	13.803	13.957
Pressure (kPa)	7.04	7.2
Density (solid) (kg m^{-3})	86.48	86.71
Density (liquid) (kg m^{-3})	77.03	77.21
Density (vapor) (kg m^{-3})	0.126	0.130
Heat of melting ΔH_m (J mol^{-1})	117.5	
Heat of sublimation ΔH_v (J mol^{-1})	1022.9	
Enthalpy $\Delta H°$ (J mol^{-1})	−740.2	
Entropy $\Delta S°$ (J mol^{-1} K^{-1})	1.49	
Thermal conductivity (W m^{-1} K^{-1})	0.9	
Dielectric constant	1.286	
Boiling point at $p = 101.3$ kPa		
Temperature T_b (K)	20.268	20.39
Density (liquid) (kg m^{-3})		70.811
Density (vapor) (kg m^{-3})		1.316
Heat of vaporization ΔH_V (J mol^{-1})	898.30	899.1

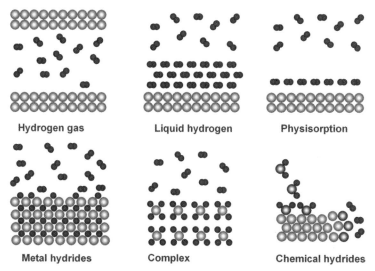

Figure 5.2 The six basic hydrogen storage methods and phenomena. From top left to bottom right: compressed gas (molecular H_2); liquid hydrogen (molecular H_2); physisorption (molecular H_2) on materials, for example, carbon with a very large specific surface area; hydrogen (atomic H) intercalation in host metals, metallic hydrides working at RT are fully reversible; complex compounds ($[AlH_4]^-$ or $[BH_4]^-$), desorption at elevated temperature, adsorption at high pressures; chemical oxidation of metals with water and liberation of hydrogen.

The volumetric hydrogen density describes the mass of hydrogen in a material or a system divided by the volume of the material or storage system:

$$\rho_V = \frac{m_H}{V} = [\text{kg m}^{-3}] \quad (5.1)$$

The gravimetric hydrogen density describes the ratio of the mass of hydrogen to the mass of the material or storage system:

$$\rho_m = \frac{m_H}{m_{\text{tot}}} = [\text{mass\%}] \quad (5.2)$$

5.3
Pressurized Hydrogen

The most established storage systems are high-pressure gas cylinders with a maximum pressure of 20 MPa. New lightweight composite cylinders have been developed that are able to withstand pressures up to 80 MPa and so the hydrogen can reach a volumetric density of 36 kg m^{-3}, approximately half that in its liquid form at the normal boiling point.

Table 5.2 Adiabatic coefficient γ and normal volume V_0 at $p_0 = 1.013 \times 10^5$ Pa and 273.15 K.

Gas	$\gamma = c_p/c_v$	V_0 (m^3 kg^{-1})
H_2	1.41	11.11
He	1.66	5.56
CH_4	1.31	1.39

5.3.1
Properties of Compressed Hydrogen

Compression of hydrogen consumes energy depending on the thermodynamic process. The ideal isothermal compression requires the least amount of energy (just compression work) and the adiabatic process requires the maximum amount of energy. The compression energy W depends on the initial pressure p_i and the final pressure p_f, the initial volume V_i and the adiabatic coefficient γ:

$$W_{iso} = p_i V_i \ln\left(\frac{p_f}{p_i}\right) \quad \text{isothermal compression work} \quad (5.3)$$

$$W_{adi} = \left(\frac{\gamma}{\gamma-1}\right) p_i V_i \left[\left(\frac{p_f}{p_i}\right)^{\frac{\gamma-1}{\gamma}} - 1\right] \quad \text{adiabatic compression work} \quad (5.4)$$

The compression work depends on the nature of the gas (Table 5.2).
Real compressors work close to the isothermal limit (Figure 5.3).

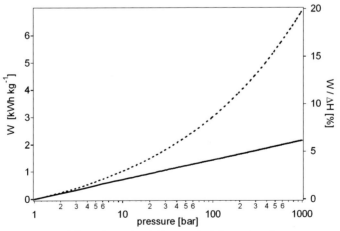

Figure 5.3 Work for isothermal (solid line) and adiabatic (dotted line) compression of hydrogen from an initial pressure of $p_i = 1$ bar on the left axis. Compression work as a percentage of the higher heating value (39.4 kW h kg^{-1}) of hydrogen on the right axis.

5.3 Pressurized Hydrogen

The density (ρ) of pressurized hydrogen is approximated by means of the van der Waals equation for a real gas, with the parameters $R = 8.314\,\mathrm{J\,K^{-1}\,mol^{-1}}$, $a(H_2) = 2.476 \times 10^{-2}\,\mathrm{m^6\,Pa\,mol^{-2}}$ and $b(H_2) = 2.661 \times 10^{-5}\,\mathrm{m^3\,mol^{-1}}$.

$$p(V) = \frac{nRT}{V - nb} - a\frac{n^2}{V^2} \qquad \text{van der Waals equation} \tag{5.5}$$

The density (ρ) is given by

$$\rho = \frac{nM}{V} \tag{5.6}$$

where $M(H_2) = 2\,\mathrm{g\,mol^{-1}}$.

5.3.2
Pressure Vessel

The ideal shape of a pressure vessel is spherical. However, for technical reasons cylindrical pressure vessels are often preferable (Figure 5.4).

The wall thickness of a cylinder capped with two hemispheres is given by the following equation [5]:

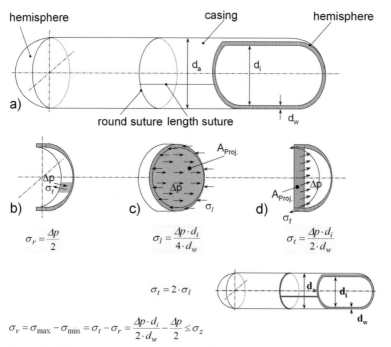

Figure 5.4 Schematic representation of a cylindrical pressure vessel and calculation of the radial, longitudinal and tangential stress.

Table 5.3 Density, melting temperature, Young's modulus and tensile strength for selected materials (L = low carbon, composition in %) [103].

Material	Density (g cm^{-3})	T_m (°C)	Young's modulus (GPa)	Tensile strength (MPa)
Stainless steel AISI 304 Fe/Cr18/Ni10	7.93	1400–1455	190–210	460–1100
Stainless steel AISI 316 Fe/Cr18/Ni10/Mo3	7.96	1370–1400	190–210	460–860
Copper	8.96	1083	129.8	224–314
Aluminum	2.70	660.4	70.6	50–195
Vanadium	6.1	1890	127.6	260–730
Spider silk (protein)	1.3			1300
Kevlar (polyaramid)	1.44			2760
Hexcel carbon fiber AS4D (12 000 filaments)	1.79			4280

$$d_w = \frac{\Delta p d_i}{2\sigma_v + \Delta p} \quad \text{wall thickness} \tag{5.7}$$

$$\frac{d_w}{d_o} = \frac{\Delta p}{2\sigma_v f_1 + \Delta p} + f_2 \quad \text{minimum wall thickness for a pressure cylinder} \tag{5.8}$$

where d_w is the wall thickness, d_o the outer diameter of the cylinder, Δp the overpressure, σ_v the tensile strength of the material and f_1 ($f_1 = 0.5$) and f_2 ($f_2 = 0.1$ mm) safety factors that depend strongly on the application.

The tensile strength of materials varies from 50 MPa for aluminum to more than 1100 MPa for high-quality steel. Future developments of new composite materials have a potential to increase the tensile strength above that of steel with a materials density that is less than half of the density of steel (Table 5.3).

Most pressure cylinders today use austenitic stainless steel (e.g. AISI 316 and 304 and AISI 316L and 304L above 300 °C to avoid carbon grain-boundary segregation) or copper or aluminum alloys, which are largely immune to hydrogen effects at ambient temperatures. Many other materials are subject to embrittlement and should not be used, for example, alloy or high-strength steels (ferritic, martensitic and bainitic), titanium and its alloys and some nickel-based alloys.

5.3.3
Volumetric and Gravimetric Hydrogen Density

The volumetric density of hydrogen in a pressure vessel increases with pressure and reaches a maximum above 1000 bar, depending on the tensile strength of the material. However, the gravimetric hydrogen density of the pressure cylinder

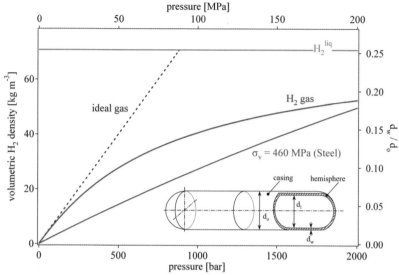

Figure 5.5 Volumetric density of compressed hydrogen gas as a function of gas pressure including the ideal gas and liquid hydrogen. The ratio of the wall thickness to the outer diameter of the pressure cylinder is shown on the right-hand side for steel with a tensile strength of 460 MPa. A schematic drawing of the pressure cylinder is shown as an inset.

decreases with increasing pressure due to the increasing thickness of the walls of the pressure cylinder, and the maximum gravimetric density is found for zero overpressure! Therefore, the increase in volumetric storage density is sacrificed by the reduction in the gravimetric density in pressurized gas systems (Figures 5.5–5.7).

5.3.4
Microspheres

The ideal pressure vessel is spherical and has a small diameter. Hydrogen storage in microspheres is based on the strong temperature dependence of the diffusion coefficient of hydrogen in silica. The closed microspheres are exposed at elevated temperature (300–400 °C) to pressurized hydrogen (400–500 bar) [7]. Hydrogen diffuses through the walls of the microspheres and builds up the same pressure inside. After cooling to room temperature, the diffusion coefficient of hydrogen is drastically reduced and an inner pressure of about 200 bar remains inside the microspheres even when the outer pressure is reduced to atmospheric pressure. In order to release the hydrogen, the microspheres are heated again to 300–400 °C.

Silica-based microspheres are typically between 5 and 200 μm in diameter, have wall thicknesses of 0.5–20 μm and can be filled with up to 100 MPa of H_2. The spheres are formed by melting spray-dried microparticles in free fall and the evolving gases

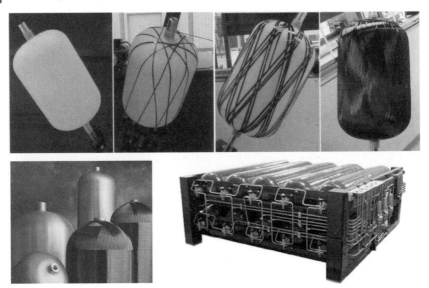

Figure 5.6 Dynetek composite cylinders consisting of an aluminum cylinder wrapped with carbon fibers (top and left) in an epoxy resin. Module with 10 cylinders (right) [6]. Companies that produce these tanks are Quantum Technologies, Lincoln Composites, Dynetek Industries and Advanced Lightweight Engineering (ALE).

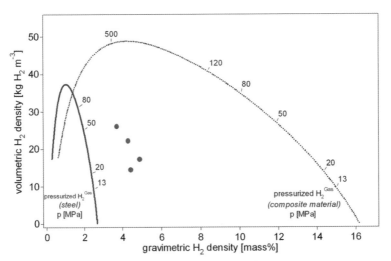

Figure 5.7 Volumetric and gravimetric hydrogen storage density for pressurized gas. Steel (tensile strength $\sigma_v = 460$ MPa, density 6500 kg m^{-3}) and a hypothetical composite material ($\sigma_v = 1500$ MPa, density 3000 kg m^{-3}). The circles represent pressure cylinders from Dynetek.

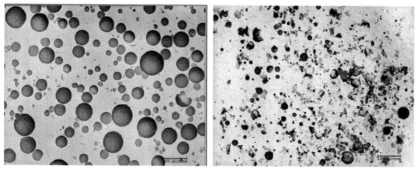

Figure 5.8 Scanning electron micrograph of glass spheres before pressurizing (left) and after pressurizing to 200 bar.

blow the particles into the hollow sphere. H$_2$ penetrates the walls of the hollow spheres rapidly at elevated temperatures and sufficient pressure differences. At ambient temperature, the penetration rate is so small that the spheres can safely be used as high-pressure containers.

The nature of the material defines the tensile strength (σ), while the wall thickness defines the maximum pressure (p_{max}) difference that the sphere can withstand:

$$p_{max} = 4\sigma \frac{d_w}{d_i} \tag{5.9}$$

It has been shown that the maximum strength depends on the material composition, the strongest material being associated with quartz glass materials (Figure 5.8).

It has been reported that the hydrogen and helium permeability of glasses increases with increasing concentration of glass formers (SiO$_2$, B$_2$O$_5$ and P$_2$O$_5$) and decreases with increasing concentration of network formers (Na$_2$O, CaO, MgO, SrO and NaO) [8]. The concentration of glass formers directly influences the packing density and chain length of the glass-forming units, leading to a randomly unorganized network of irregularly shaped pores and holes. The presence of network formers, such as Na$^+$ or Ca^{2+}, results in partial blockage of these openings, which leads to a decrease in the hydrogen permeability.

Glass microspheres offer only a limited volumetric hydrogen storage density of less than 20 kg m^3 [9]. Furthermore, a glass sphere with the pressurized system is not in equilibrium and only kinetically hindered in diffusion.

5.4
Liquid Hydrogen

Liquid hydrogen is stored in cryogenic tanks at 21.2 K at ambient pressure. Due to the low critical temperature of hydrogen (33 K), liquid hydrogen can only be stored in open systems, because there is no liquid phase existing above the critical temperature. The pressure in a closed storage system at room temperature could increase to about 10^4 bar. The volumetric density of liquid hydrogen is 70.8 kg m^{-3} and lower than that of solid hydrogen (86.7 kg m^{-3}). The challenges of liquid hydrogen storage are the energy-efficient liquefaction process and the thermal insulation of the cryogenic storage vessel in order to reduce the boil-off of hydrogen.

The hydrogen molecule is composed of two protons and two electrons. The combination of the two electron spins only leads to a binding state if the electron spins are antiparallel. The wavefunction of the molecule has to be antisymmetric in view of the exchange of the space coordinates of two fermions (spin = $\frac{1}{2}$). Therefore, two groups of hydrogen molecules exist according to the total nuclear spin ($I = 0$, antiparallel nuclear spin; $I = 1$, parallel nuclear spin). The first group with $I = 0$ is called *para*-hydrogen and the second group with $I = 1$ is called *ortho*-hydrogen. Normal hydrogen at room temperature contains 25% of the *para* form and 75% of the *ortho* form. The *ortho* form cannot be prepared in the pure state. Since the two forms differ in energy, the physical properties also differ. The melting and boiling points of *p*-hydrogen are about 0.1 K lower than those of normal hydrogen. At 0 K, all the molecules must be in a rotational ground state, that is, in the *para* form.

5.4.1
Liquefaction Process

When hydrogen is cooled from room temperature (RT) to the normal boiling point (nbp = 21.2 K), the *o*-hydrogen converts from an equilibrium concentration of 75% at RT to 50% at 77 K and 0.2% at nbp. The self-conversion rate is an activated process and very slow; the half-life of the conversion is more than 1 year at 77 K (Figure 5.9).

The conversion reaction from *o*- to *p*-hydrogen is exothermic and the heat of conversion is also temperature dependent. At 300 K, the heat of conversion is 270 kJ kg^{-1} and increases as the temperature decreases, reaching 519 kJ kg^{-1} at 77 K. At temperatures lower than 77 K, the enthalpy of conversion is 523 kJ kg^{-1} and almost constant. The enthalpy of conversion is greater than the latent heat of vaporization ($H_V = 451.9$ kJ kg^{-1}) of normal and *p*-hydrogen at the nbp. If the unconverted normal hydrogen is placed in a storage vessel, the enthalpy of conversion will be released in the vessel, which leads to evaporation of the liquid hydrogen. The transformation from *o*- to *p*-hydrogen can be catalyzed by a number of surface-active and paramagnetic species; for example, normal hydrogen can be adsorbed on charcoal cooled with liquid hydrogen and desorbed in the equilibrium mixture. The conversion may take only a few minutes if a highly active form of charcoal is used. Other suitable *ortho–para* conversion catalysts are metals such as tungsten or nickel or any paramagnetic oxides such as chromium or gadolinium oxide. The nuclear spin is reversed without breaking the H–H bond.

The simplest liquefaction cycle is the Joule–Thompson cycle (Linde cycle). The gas is first compressed and then cooled in a heat exchanger, before it passes through a throttle valve where it undergoes an isenthalpic Joule–Thomson expansion, producing some liquid. The cooled gas is separated from the liquid and returned to the compressor via the heat exchanger [10]. The Joule–Thompson cycle works for gases, such as nitrogen, with a inversion temperature above room temperature. Hydrogen, however, warms

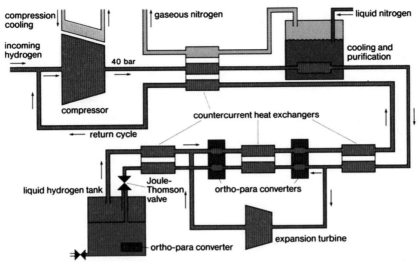

Figure 5.9 The Joule–Thompson cycle (Linde cycle). The gas is first compressed and then cooled in a heat exchanger, before it passes through a throttle valve where it undergoes an isenthalpic Joule–Thomson expansion, producing some liquid. The cooled gas is separated from the liquid and returned to the compressor via the heat exchanger.

upon expansion at room temperature. In order for hydrogen to cool upon expansion, its temperature must be below its inversion temperature of 202 K. Therefore, hydrogen is usually precooled using liquid nitrogen (78 K) before the first expansion step occurs. The free enthalpy change between gaseous hydrogen at 300 K and liquid hydrogen at 20 K is 11 640 kJ kg^{-1} [11]. The necessary theoretical energy (work) to liquefy hydrogen from RT is $W_{th} = 3.23$ kW h kg^{-1}; the technical work is about 15.2 kW h kg^{-1}, almost 40% of the higher heating value of the hydrogen combustion [12].

5.4.2
Storage Vessel

The boil-off rate of hydrogen from a liquid hydrogen storage vessel due to heat leaks is a function of the size, the shape and the thermal insulation of the vessel. Theoretically, the best shape is a sphere since it has the least surface-to-volume ratio and because stress and strain are distributed uniformly. However, large-sized spherical containers are expensive because of their manufacturing difficulty. Since boil-off losses due to heat leaks are proportional to the surface to volume ratio, the evaporation rate diminishes drastically as the size of the storage tank increases. For double-walled vacuum-insulated spherical dewars, boil-off losses are typically 0.4% per day for tanks which have a storage volume of 50 m^3, 0.2% for 100 m^3 tanks and 0.06% for 20 000 m^3 tanks (Figure 5.10).

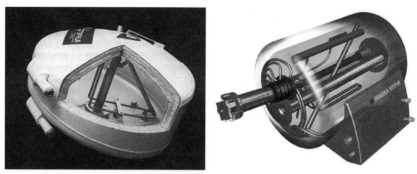

Figure 5.10 Two types of liquid hydrogen storage systems.

Low-temperature *p*-hydrogen requires the use of materials that retain good ductility at low temperatures. Austenitic stainless steel (e.g. AISI 316L and 304L) or aluminum and aluminum alloys (Series 5000) are recommended. Polytetrafluoroethylene (PTFE, Teflon) and 2-chloro-1,1,2-trifluoroethylene (Kel-F) can also be used.

5.4.3
Gravimetric and Volumetric Hydrogen Density

The gravimetric and volumetric hydrogen density depend strongly on the size of the storage vessel since the surface-to-volume ratio decreases with increasing size. Therefore, only the upper limit is defined (Figure 5.11).

The large amount of energy necessary for liquefaction, that is, 40% of the upper heating value, makes liquid hydrogen not an energy-efficient storage medium. Furthermore, the continuous boil-off of hydrogen limits the possible applications

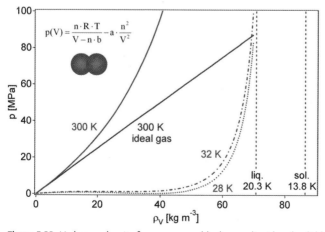

Figure 5.11 Hydrogen density for compressed hydrogen, liquid and solid hydrogen.

5.5
Physisorption

5.5.1
Van der Waals Interaction

The adsorption of a gas on a surface is a consequence of the field force at the surface of the solid, called the adsorbent, which attracts the molecules of the gas or vapor, called the adsorbate. The origin of the physisorption of gas molecules on the surface of a solid is resonant fluctuations of the charge distributions, which are therefore called dispersive interactions or van der Waals interactions (Figure 5.12).

In the physisorption process, a gas molecule interacts with several atoms at the surface of the solid. The interaction is composed of two terms: an attractive term which diminishes with the distance between the molecule and the surface to the power of -6 and a repulsive term which diminishes with the distance to the power of -12.

Therefore, the potential energy of the molecule shows a minimum at a distance of approximately one molecular radius of the adsorbate. The energy minimum is of the order of 0.01–0.1 eV (1–10 kJ mol^{-1}) [13]. Due to the weak interaction, a significant physisorption is observed only at low temperatures (<273 K).

If a surface with a finite number of sites is exposed to a gas, the number of molecules hitting the surface is given by the Hertz–Knudsen equation:

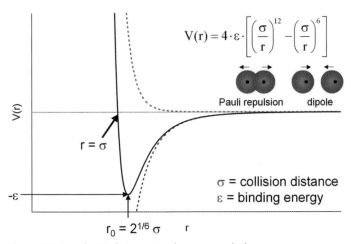

Figure 5.12 Van der Waals interaction between two hydrogen molecules and the resulting Lennard-Jones potential, $V(r)$.

$$F = \frac{p}{\sqrt{2\pi m k T}} \qquad (5.10)$$

The rate of adsorption depends on the flux F, the number of unoccupied sites and a parameter for the interaction between the gas molecules and the surface, $R_{abs} = k_a F (1 - \Theta)$. The rate of desorption depends on the number of occupied sites and a parameter for the activation energy of the desorption, $R_{des} = k_d \Theta$.

Equilibrium is reached when the absorption and the desorption rate are equal. The maximum occupation is then given by the Langmuir isotherm:

$$\Theta_{max}(T) = \frac{k_a F}{k_a F + k_d} \qquad (5.11)$$

with the equilibrium constant $K = F(k_a/k_d)$, we find $K = \Theta_{max}/(1 - \Theta_{max})$.

5.5.2
Adsorption Isotherm

Once a monolayer of adsorbate molecules is formed, the gaseous molecules interact with the surface of the liquid or solid adsorbate. Therefore, the binding energy of the second layer of adsorbate molecules is similar to the latent heat of sublimation or vaporization (ΔH_V) of the adsorbate. Consequently, the adsorption of the adsorbate at a temperature greater than the boiling point at a given pressure leads to the adsorption of a single monolayer [14] (Figure 5.13).

The model of Brunauer, Emmett and Teller (BET) assumes that the enthalpy of adsorption for the first monolayer of molecules is ΔH_{ads} and for all additional layers ΔH_V. Furthermore, it assumes that all layers are in equilibrium. With the following definitions:

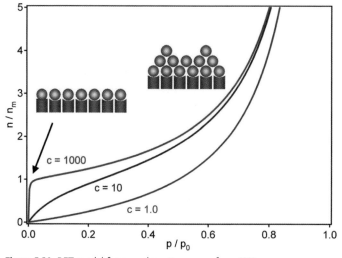

Figure 5.13 BET model for gas adsorption on surfaces [15].

$$c = e^{\frac{\Delta H_{ads} - \Delta H_V}{kT}} \tag{5.12}$$

and

$$\beta = F \frac{k_a^0}{k_d^0} e^{\frac{\Delta H_V}{kT}} = \frac{p}{p_0} \tag{5.13}$$

the number of adsorbed molecules is given by

$$\frac{n}{n_m} = \frac{c\beta}{(1-\beta)[1+(c-1)\beta]} \tag{5.14}$$

In order to estimate the quantity of adsorbate in the monolayer, the density of the liquid adsorbate and the volume of the molecule must be used. If the liquid is assumed to consist of a close-packed fcc structure, the minimum surface area S_{ml} for 1 mol of adsorbate in a monolayer on a substrate can be calculated from the density of the liquid ρ_{liq} and the molecular mass of the adsorbate M_{ads}:

$$S_{ml} = \frac{\sqrt{3}}{2} \left(\sqrt{2N_A} \frac{M_{ads}}{\rho_{liq}} \right)^{\frac{2}{3}} \tag{5.15}$$

where N_A is Avogadro's number ($N_A = 6.022 \times 10^{23}\,\text{mol}^{-1}$). The monolayer surface area for hydrogen is $S_{ml}(H_2) = 85\,917\,\text{m}^2\,\text{mol}^{-1}$. The amount of adsorbate m_{ads} on a substrate material with a specific surface area S_{spec} is then given by $m_{ads} = M_{ads} S_{spec}/S_{ml}$. In the case of carbon as the substrate and hydrogen as the adsorbate, the maximum specific surface area of carbon is $S_{spec} = 1315\,\text{m}^2\,\text{g}^{-1}$ (single-sided graphene sheet) and the maximum amount of adsorbed hydrogen is $m_{ads} = 3.0\,\text{mass\%}$. From this theoretical approximation, we may conclude that the amount of adsorbed hydrogen is proportional to the specific surface area of the adsorbent with $m_{ads}/S_{spec} = 2.27 \times 10^{-3}\,\text{mass\%}\,\text{m}^{-2}\,\text{g}$ and can only be observed at very low temperatures.

5.5.3
Hydrogen and Carbon Nanotubes

The main difference between carbon nanotubes and high surface area graphite is the curvature of the graphene sheets and the cavity inside the tube. In microporous solids with capillaries which have a width not exceeding a few molecular diameters, the potential fields from opposite walls will overlap so that the attractive force which acts upon adsorbate molecules will be increased in comparison with that on a flat carbon surface [16]. This phenomenon is the main motivation for the investigation of the interaction of hydrogen with carbon nanotubes (Figure 5.14).

Rzepka et al. used a grand canonical ensemble Monte Carlo program to calculate the amount of absorbed hydrogen for a slit pore and a tubular geometry [17]. The amount of absorbed hydrogen depends on the surface area of the sample; the maximum is at 0.6 mass% ($p = 6\,\text{MPa}$, $T = 300\,\text{K}$). The calculation was verified experimentally, with excellent agreement. At a temperature of 77 K the amount of absorbed hydrogen is about one order of magnitude higher than at 300 K (Figure 5.15).

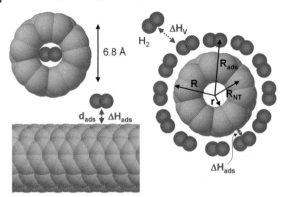

Figure 5.14 Hydrogen molecule on the surface of a SWNT (5,5). The van der Waals interaction between H_2 and the curved surface of the nanotube is weaker on the outer surface and stronger on the inner surface. Furthermore, the specific surface area of the adsorbate layer (R_{ads}) is larger than that of the nanotube (R_{NT}).

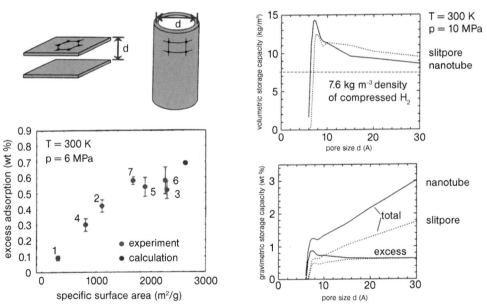

Figure 5.15 Comparison of the hydrogen adsorption in a slit and cylindrical pore [18]. The amount of absorbed hydrogen correlates with the specific surface area of the sample; the maximum is at 0.6 mass% ($p = 6$ MPa, $T = 300$ K). No significant difference was found in the calculated amount of hydrogen between the slit and cylindrical pores. The calculation was verified experimentally with excellent agreement.

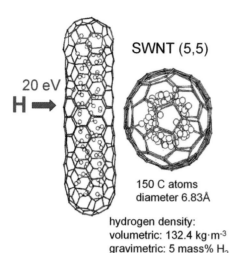
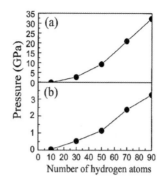

SWNT (5,5)

20 eV
H ➡

150 C atoms
diameter 6.83Å

hydrogen density:
volumetric: 132.4 kg·m⁻³
gravimetric: 5 mass% H_2

(a) Static pressure caused by the molecular repulsive force from the confined H_2 molecules
(b) dynamic pressure caused by the collisions between the H_2 molecules and the wall of the tube at 300 K.

Figure 5.16 Molecular dynamics simulation for H implantation (20 eV) in an SWNT (5,5).

Ma *et al.* performed a molecular dynamics simulation for H implantation [18]. The hydrogen atoms (20 eV) were implanted through the side walls of a single-walled carbon nanotube (SWNT) (5,5) consisting of 150 atoms and having a diameter of 0.683 nm. They found that the hydrogen atoms recombine to molecules inside the tube and arrange themselves into a concentric tube. The hydrogen pressure inside the SWNT increases as the number of injected atoms increases and reaches 35 GPa for 90 atoms (5 mass%). This simulation does not exhibit condensation of hydrogen inside the nanotube (Figure 5.16).

The measurement of the latent heat of condensation of nitrogen on carbon black showed that the heat for the adsorption of one monolayer is between 11 and 12 kJ mol⁻¹ (0.11–0.12 eV) and decreases for subsequent layers to the latent heat of condensation for nitrogen, which is 5.56 kJ mol⁻¹ (0.058 eV) [19]. If we assume that hydrogen behaves in a similar manner to nitrogen, then hydrogen would only form one monolayer of liquid at the surface of carbon at temperatures above the boiling point. Geometric considerations of the nanotubes lead to the specific surface area and, therefore, to the maximum amount of condensed hydrogen in a surface monolayer [20]. The maximum amount of adsorbed hydrogen is 2.0 mass% for an SWNT with a specific surface area of 1315 m² g⁻¹ at a temperature of 77 K (Figure 5.17).

In addition to the carbon nanostructures, other nonporous materials have been investigated for hydrogen absorption. The hydrogen absorption of zeolites of different pore architecture and composition, for example A, X, Y, was analyzed in the temperature range 293–573 K and pressure range 2.5–10 MPa [22]. Hydrogen was absorbed at the desired temperature and pressure. The sample was then cooled to RT and evacuated. Subsequently, the hydrogen release upon heating of the sample to the

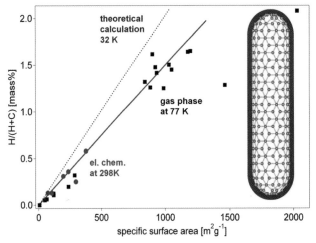

Figure 5.17 Reversible amount of hydrogen (electrochemical measurement at 298 K) versus the BET surface area (circles) of a few carbon nanotube samples including two measurements on high surface area graphite (HSAG) samples together with the fitted line [21]. Hydrogen gas adsorption measurements at 77 K from Nijkamp et al. (squares) are included [21]. The dotted line represents the calculated amount of hydrogen in a monolayer at the surface of the substrate.

absorption temperature was detected. The amount of hydrogen absorbed increased with increase in temperature and increase in absorption pressure. The maximum amount of desorbed hydrogen was found to be 0.08 mass% for a sample loaded at a temperature of 573 K and a pressure of 10 MPa. The adsorption behavior indicates that the absorption is due to a chemical reaction rather than physisorption. At liquid nitrogen temperature (77 K), the zeolites physisorb hydrogen proportionally to the specific surface area of the material. A maximum of 1.8 mass% of adsorbed hydrogen was found for a zeolite (NaY) with a specific surface area of 725 m^2 g^{-1} [23] (Figure 5.18).

The low-temperature physisorption (type I isotherm) of hydrogen in zeolites is in good agreement with the adsorption model mentioned above for nanostructured carbon. The desorption isotherm followed the same path as the adsorption, which indicates that no pore condensation occurred. The hydrogen adsorption in zeolites depends linearly on the specific surface areas of the materials and is in very good agreement with the results on carbon nanostructures [24].

A microporous meta–organic framework of the composition $Zn_4O(1,4$-benzene-dicarboxylate$)_3$ has been proposed as hydrogen storage material [25]. The material absorbs hydrogen at a temperature of 298 K proportional to the applied pressure. The slope of the linear relationship between the gravimetric hydrogen density and the hydrogen pressure was found to be 0.05 mass% bar^{-1}. No saturation of the hydrogen absorption was found, which is very unlikely for any kind of a hydrogen absorption process. At 77 K, the amount of adsorbed hydrogen was found to be 3.7 mass% already at very low hydrogen pressures, with a slight almost linear increase with increase in pressure. This behavior is not a type I isotherm as the authors claimed and the results should be treated with caution (Figure 5.19).

5.5 Physisorption

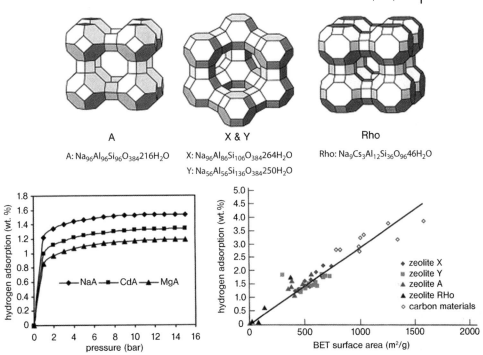

Figure 5.18 Hydrogen adsorption isotherms for zeolites (left) and amount of adsorbed hydrogen versus the specific surface area of the samples for zeolites and carbon nanostructures (right).

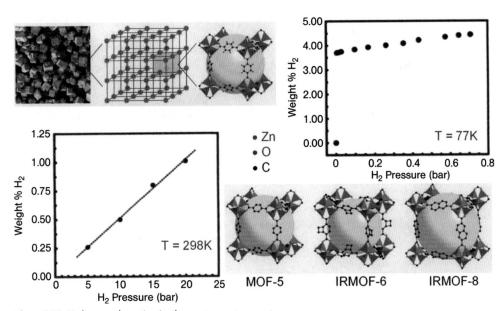

Figure 5.19 Hydrogen adsorption isotherm at room temperature (left) and at liquid nitrogen temperature (right) on metal organic frameworks (MOFs).

The great advantages of physisorption for hydrogen storage are the low operating pressure, the relatively low cost of the materials involved and the simple design of the storage system. The rather small amount of hydrogen adsorbed on carbon together with the low temperatures necessary are significant drawbacks of hydrogen storage based on physisorption.

5.6
Metal Hydrides

5.6.1
Interstitial Hydrides

Metals, intermetallic compounds and alloys generally react with hydrogen and form mainly solid metal–hydrogen compounds. Hydrides exist as ionic, polymeric covalent, volatile covalent and metallic hydrides.

Hydrogen reacts at elevated temperatures with many transition metals and their alloys to form hydrides. The electropositive elements are the most reactive, that is, scandium, yttrium, the lanthanides, the actinides and members of the titanium and vanadium groups (Figure 5.20).

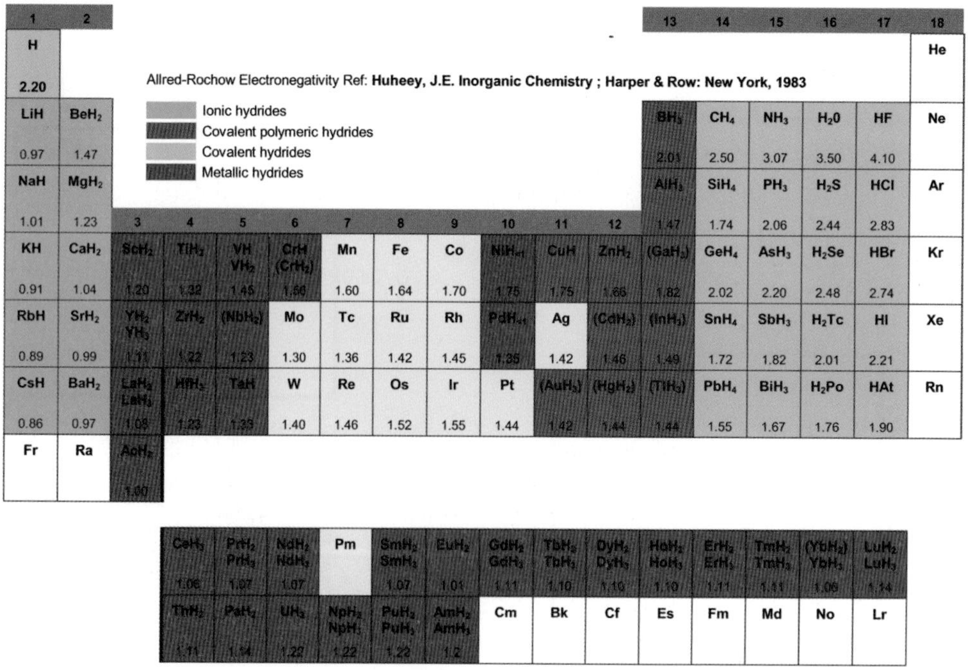

Figure 5.20 Table of the binary hydrides and the Allred–Rochow electronegativity [26]. Most elements react with hydrogen to form ionic, covalent or metallic binary hydrides.

Structure	fcc & hcp		bcc	
Site	O	T	O	T
Number	1	2	3	6
Size	0.414	0.255	0.155	0.291

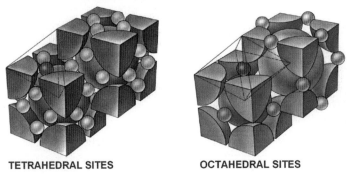

TETRAHEDRAL SITES   **OCTAHEDRAL SITES**

Figure 5.21 Tetrahedral (T) and octahedral (O) interstitial sites occupied by hydrogen atoms [27]. The number of sites per host metal atom and the size, that is, the diameter of the largest possible sphere on the interstitial site, are given in the table.

The binary hydrides of the transition metals are predominantly metallic in character and are usually referred to as metallic hydrides. They are good conductors of electricity and have a metallic or graphite-like appearance.

Many of these compounds (MH_n) show large deviations from ideal stoichiometry ($n = 1, 2, 3$) and can exist as multi-phase systems. The lattice structure is that of a typical metal with atoms of hydrogen on the interstitial sites; for this reason, they are also called interstitial hydrides. This type of structure has the limiting compositions MH, MH_2 and MH_3; the hydrogen atoms fit into octahedral or tetrahedral holes in the metal lattice or a combination of the two types (Figure 5.21).

The hydrogen carries a partial negative charge depending on the metal; an exception is, for example, $PdH_{0.7}$ [28]. Only a small number of the transition metals are without known stable hydrides. A considerable 'hydride gap' exists in the periodic table, beginning at group 6 (Cr) up to group 11 (Cu), in which the only hydrides are palladium hydride ($PdH_{0.7}$), the very unstable nickel hydride ($NiH_{<1}$) and the poorly defined hydrides of chromium (CrH, CrH_2) and copper (CuH). In palladium hydride, the hydrogen has high mobility and probably a very low charge density. In the finely divided state, platinum and ruthenium are able to adsorb considerable quantities of hydrogen, which thereby becomes activated. These two elements, together with palladium and nickel, are extremely good hydrogenation catalysts, although they do not form hydrides [29].

Especially interesting are the metallic hydrides of intermetallic compounds, in the simplest case the ternary system AB_xH_n, because the variation of the elements allows one to tailor the properties of the hydrides (Table 5.4).

The A element is usually a rare earth or an alkaline earth metal and tends to form a stable hydride. The B element is often a transition metal and forms only

Table 5.4 The most important families of hydride forming intermetallic compounds including the prototype and the structure.

Intermetallic compound[a]	Prototype	Hydrides	Structure
AB_5	$LaNi_5$	$LaNiH_6$	Haucke phases, hexagonal
AB_2	ZrV_2, $ZrMn_2$, $TiMn_2$	$ZrV_2H_{5.5}$	Laves phase, hexagonal or cubic
AB_3	$CeNi_3$, YFe_3	$CeNi_3H_4$	Hexagonal, $PuNi_3$ type
A_2B_7	Y_2Ni_7, Th_2Fe_7	$Y_2Ni_7H_3$	Hexagonal, Ce_2Ni_7 type
A_6B_{23}	Y_6Fe_{23}	$Ho_6Fe_{23}H_{12}$	Cubic, Th_6Mn_{23} type
AB	TiFe, ZrNi	$TiFeH_2$	Cubic, CsCl or CrB type
A_2B	Mg_2Ni, Ti_2Ni	Mg_2NiH_4	Cubic, $MoSi_2$ or Ti_2Ni type

[a]A is an element with a high affinity to hydrogen and B is an element with a low affinity to hydrogen.

unstable hydrides. Some well-defined ratios of B to A in the intermetallic compound $x = 0.5, 1, 2, 5$ have been found to form hydrides with a hydrogen to metal ratio of up to two.

5.6.2
Hydrogen Absorption

The reaction of hydrogen gas with a metal is called the absorption process and can be described in terms of a simplified one-dimensional potential energy curve (one-dimensional Lennard-Jones potential) [30] (Figure 5.22).

Far from the metal surface, the potentials of a hydrogen molecule and of two hydrogen atoms are separated by the dissociation energy ($\frac{1}{2}H_2 \rightarrow H$, $E_D = 218$ kJ mol^{-1} H). The first attractive interaction of the hydrogen molecule approaching the metal surface is the van der Waals force leading to the physisorbed state ($E_{Phys} \approx 5$ kJ mol^{-1} H) approximately one hydrogen molecule radius (~0.2 nm) from the metal surface. Closer to the surface, the hydrogen has to overcome an activation barrier for dissociation and formation of the hydrogen metal bond. The height of the activation barrier depends on the surface elements involved. Hydrogen atoms sharing their electron with the metal atoms at the surface are thus in the chemisorbed state ($E_{Chem} \approx -50$ kJ mol^{-1} H). The chemisorbed hydrogen atoms may have a high surface mobility, interact with each other and form surface phases at sufficiently high coverage. In the next step, the chemisorbed hydrogen atom can jump in the subsurface layer and finally diffuse on the interstitial sites through the host metal lattice. The hydrogen atoms contribute with their electron to the band structure of the metal (Figure 5.23).

The hydrogen is, at small hydrogen to metal ratios (H:M < 0.1), exothermically dissolved (solid solution, α-phase) in the metal. The metal lattice expands proportionally to the hydrogen concentration by approximately 2–3 Å^3 per hydrogen atom [31]. At greater hydrogen concentrations in the host metal (H:M > 0.1), a strong

5.6 Metal Hydrides

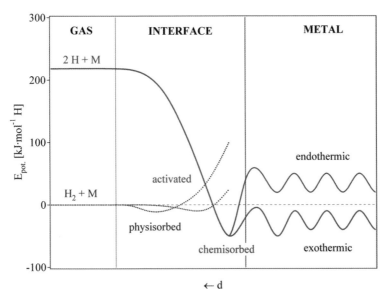

Figure 5.22 Lennard-Jones potential of hydrogen approaching a metallic surface. Far from the metal surface the potential of a hydrogen molecule and of two hydrogen atoms is separated by the dissociation energy. The first attractive interaction of the hydrogen molecule is the van der Waals force leading to the physisorbed state. Closer to the surface the hydrogen has to overcome an activation barrier for dissociation and formation of the hydrogen–metal bond. Hydrogen atoms sharing their electron with the metal atoms at the surface are then in the chemisorbed state. In the next step, the chemisorbed hydrogen atom can jump in the subsurface layer and finally diffuse on the interstitial sites through the host metal lattice.

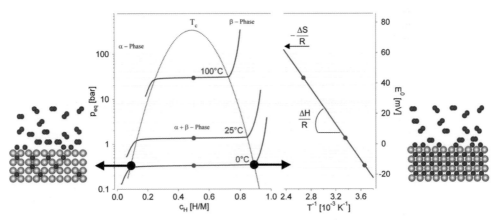

Figure 5.23 Pressure composition isotherms for hydrogen absorption in a typical metal (left). The solid solution (α-phase), the hydride phase (β-phase) and the region of the coexistence of the two phases. The coexistence region is characterized by the flat plateau and ends at the critical temperature T_c. The construction of the van't Hoff plot is shown on the right. The slope of the line is equal to the enthalpy of formation divided by the gas constant and the intercept with the axis is equal to the entropy of formation divided by the gas constant.

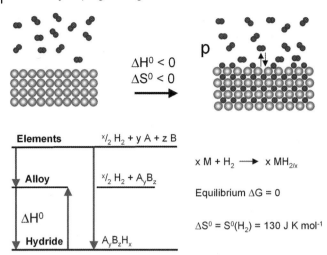

Figure 5.24 Formation of an interstitial hydride from an intermetallic compound. The enthalpy of the hydride formation is measured between the intermetallic compound and the metal hydride.

H–H interaction due to the lattice expansion becomes important and the hydride phase (β-phase) nucleates and grows. The hydrogen concentration in the hydride phase is often found to be H:M = 1. The volume expansion between the coexisting α- and β-phases corresponds in many cases to 10–20% of the metal lattice. Therefore, at the phase boundary high stress is built up and often leads to decrepitation of brittle host metals such as intermetallic compounds. The final hydride is a powder with a typical particle size of 10–100 μm (Figure 5.24).

The thermodynamic aspects of hydride formation from gaseous hydrogen are described by means of pressure–composition isotherms in equilibrium ($\Delta G = 0$). While the solid solution and hydride phase coexist, the isotherms show a flat plateau, the length of which determines the amount of H_2 stored. In the pure β-phase, the H_2 pressure rises steeply with increase in concentration. The two-phase region ends in a critical point T_c, above which the transition from the α- to the β-phase is continuous. The equilibrium pressure p_{eq} as a function of temperature is related to the changes $\Delta H°$ and $\Delta S°$ of enthalpy and entropy:

$$\Delta G = G°_{MH} - G°_{H_2} = 0 = \Delta G° - RT \ln\left(\frac{p_{eq}}{p°_{eq}}\right) = \Delta H° - T\Delta S° - RT \ln\left(\frac{p_{eq}}{p°_{eq}}\right)$$

(5.16)

and we find (van't Hoff equation)

$$\ln\left(\frac{p_{eq}}{p°_{eq}}\right) = \frac{\Delta H°}{R}\frac{1}{T} - \frac{\Delta S°}{R}$$

(5.17)

As the entropy change corresponds mostly to the change from molecular hydrogen gas to dissolved solid hydrogen, it amounts approximately to the standard entropy of hydrogen ($S° = 130$ J K^{-1} mol^{-1}) and is therefore $\Delta S_f \approx -130$ J K^{-1} mol^{-1} H$_2$ for all metal–hydrogen systems. The enthalpy term characterizes the stability of the metal hydrogen bond. The decomposition temperature for $P = p° = 1$ bar (usually) is

$$T_{dec} = \frac{\Delta H°}{\Delta S°} \tag{5.18}$$

To reach an equilibrium pressure of 1 bar at 300 K, ΔH should amount to 39.2 kJ mol^{-1} H$_2$. The entropy of formation term of metal hydrides (change of the hydrogen gas to a solid) leads to a significant heat evolution $\Delta Q = T \Delta S°$ (exothermic reaction) during the hydrogen absorption. The same heat has to be provided to the metal hydride to desorb the hydrogen (endothermic reaction). If the hydrogen desorbs below room temperature, this heat can be delivered by the environment. However, if the desorption is carried out above room temperature, the heat has to be delivered at the necessary temperature from an external source, which may be the combustion of the hydrogen. The ratio of the hydriding heat (ΔQ) to the upper heating value (ΔH_V) is approximately constant:

$$\frac{\Delta Q}{\Delta H_V} = \frac{\Delta S°}{\Delta H_V} T = 4.6 \times 10^{-4} T \tag{5.19}$$

5.6.3
Empirical Models

Several empirical models allow the estimation of the stability and the concentration of hydrogen in an intermetallic hydride. The maximum amount of hydrogen in the hydride phase is given by the number of interstitial sites in the intermetallic compound, for which the following two criteria apply. The distance between two hydrogen atoms on interstitial sites is at least 2.1 Å and the radius of the largest sphere on an interstitial site touching all the neighboring metallic atoms is at least 0.37 Å (Westlake criterion) [32, 33]. The theoretical maximum volumetric density of hydrogen in a metal hydride, assuming close packing of the hydrogen, is therefore 254 kg m^{-3}, which is 3.6 times the density of liquid hydrogen.

As a general rule, it can be stated that all elements with electronegativity in the range 1.35–1.82 do not form stable hydrides [34]. Exemptions are vanadium (1.45) and chromium (1.56), which form hydrides, and molybdenum (1.30) and technetium (1.36), where hydride formation would be expected. The adsorption enthalpy can be estimated from the local environment of the hydrogen atom on the interstitial site.

According to the rule of imaginary binary hydrides, the stability of hydrogen on an interstitial site is the weighted average of the stability of the corresponding binary hydrides of the neighboring metallic atoms [35].

$$\Delta H(AB_n H_{2m}) = \Delta H(AH_m) + \Delta H(B_n H_m) - (1-F)\Delta H(AB_n) \quad \text{Miedema model} \tag{5.20}$$

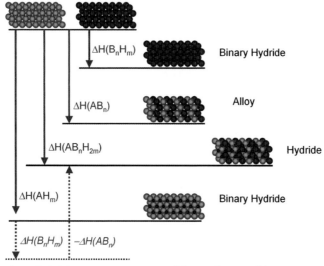

Figure 5.25 Schematic representation of the Miedema model [37].

More general is the rule of reversed stability (Miedema model): the more stable an intermetallic compound, the less stable is the corresponding hydride, and the other way around [36]. This model is based on the fact that hydrogen can only participate on a bond with a neighboring metal atom if the bonds between the metal atoms are at least partially broken (Figures 5.25 and 5.26).

The binding energy of a hydrogen atom as a function of the electron density of the environment can be calculated by means of the effective medium theory [38]. The binding energy of the hydrogen atoms shows a maximum at low electron densities and decreases with increasing electron density. When a hydrogen atom approaches a metal surface, an electron density with maximum binding energy is always found. On the way through the metal lattice, the electron density is usually too high except on vacancies.

Hydrogen absorption is electronically an incorporation of electrons and protons into the electronic structure of the host lattice. The electrons have to fill empty states at the Fermi energy E_F while the protons lead to the hydrogen-induced s-band approximately 4 eV below E_F. The heat of formation of binary hydrides MH_x is related linearly to the characteristic band energy parameter $\Delta E = E_F - E_s$, where E_F is the Fermi energy and E_s the center of the host metal electronic band with a strong s character at the interstitial sites occupied by hydrogen. For most metals E_s can be taken as the energy that corresponds to one electron per atom on the integrated density-of-states curve [39].

The semiempirical models mentioned above allow an estimation of the stability of binary hydrides provided that the rigid band theory can be applied. However, the interaction of hydrogen with the electronic structure of the host metal in some binary hydrides and especially in the ternary hydrides is often more complicated. In many cases, the crystal structure of the host metal and therefore also the electronic structure

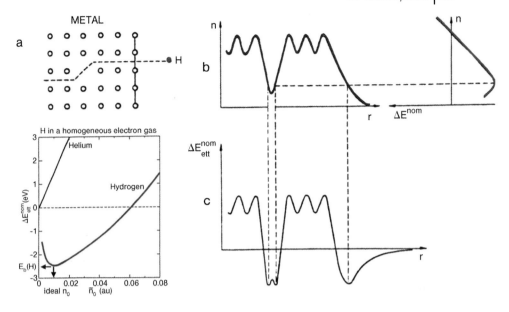

Figure 5.26 Hydrogen absorption in the effective medium theory. (a) Hydrogen atom on a path to the surface of a metal and through the lattice; (b) electron density along the path of the hydrogen atom; (c) potential energy of the hydrogen atom along the path.

changes upon the phase transition and the theoretical calculation of the stability of the hydride becomes very complicated, if not impossible (Figures 5.27 and 5.28).

The stability of metal hydrides is presented in the form of van't Hoff plots. The most stable binary hydrides have enthalpies of formation of $\Delta H_f = -226\,\text{kJ}\,\text{mol}^{-1}$ H_2, for example, HoH$_2$. The least stable hydrides are FeH$_{0.5}$, NiH$_{0.5}$ and MoH$_{0.5}$, with enthalpies of formation of $\Delta H_f = +20, +20$ and $+92\,\text{kJ}\,\text{mol}^{-1}\,H_2$, respectively [42].

Due to the phase transition upon hydrogen absorption, metal hydrides have the very useful property of absorbing large amounts of hydrogen at a constant pressure, that is, the pressure does not increase with the amount of hydrogen absorbed as long as the phase transition takes place. The characteristics of the hydrogen absorption and desorption can be tailored by partial substitution of the constituent elements in the host lattice. Some metal hydrides absorb and desorb hydrogen at ambient temperature and close to atmospheric pressure.

One of the most interesting features of the metallic hydrides is the extremely high volumetric density of the hydrogen atoms present in the host lattice. The highest volumetric hydrogen density know today is $150\,\text{kg}\,\text{m}^{-3}$, found in Mg$_2FeH_6$ and Al(BH$_4$)$_3$. Both hydrides belong to the complex hydrides and will be discussed in section 5.7. Metallic hydrides reach a volumetric hydrogen density of $115\,\text{kg}\,\text{m}^{-3}$, for example LaNi$_5$. Most metallic hydrides absorb hydrogen up to a hydrogen to metal ratio of H:M = 2.

$$\Delta \overline{H}_\infty = a \cdot \Delta E \cdot \sqrt{W} \cdot \sum_j R_j^{-4} + b$$

$a = 18.6 \text{ kJ} \cdot \text{mol}^{-1}\text{H}\text{Å}^4\text{eV}^{-3/2}$

$b = -90 \text{ kJ} \cdot \text{mol}^{-1}\text{H}$

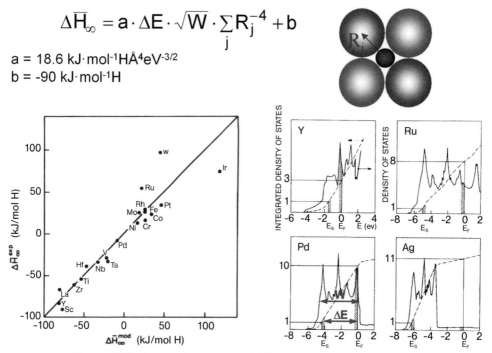

Figure 5.27 The local band-structure model [40, 41].

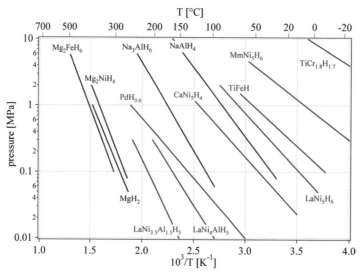

Figure 5.28 Van't Hoff plots for some selected hydrides. The stabilization of the hydride of LaNi$_5$ by the partial substitution of nickel with aluminum in LaNi$_5$ is shown, and also the substitution of lanthanum with mischmetal (e.g. 51% La, 33% Ce, 12% Nd, 4% Pr).

Greater ratios up to H:M = 4.5, for example in BaReH$_9$, have been found [43]; however, all hydrides with a hydrogen to metal ratio of more than 2 are ionic or covalent compounds and belong to the complex hydrides.

5.6.4
Lattice Gas Model

One of the simplest models for the description of an intercalation compound is the solid metal lattice with mobile hydrogen atoms on interstitial sites (mean field theory of lattice gas) [44]. The hydrogen in a metal hydride is not a free gas, and therefore the lattice gas model is applied [45]. It is assumed that an interstitial site is occupied by only one hydrogen atom at a given time with a binding energy ε_0. Furthermore, the hydrogen atoms in the lattice interact with an interaction energy ε. The energy of the hydrogen in the lattice is

$$E = N_H \varepsilon_0 + N_{HH} \varepsilon \tag{5.21}$$

where N_H is the number of hydrogen atoms, ε_0 is the binding energy of a hydrogen atom to the lattice, N_{HH} is the number of nearest-neighbor H–H pairs and ε is the H–H pair interaction energy [46] (Figure 5.29).

In an open system, the most probable state $[P = c\exp(E/kT)]$ is the one with the smallest value for the free energy $F = U - TS$.

If N is the number of interstitial sites then $N_H < N$ in all cases. The free energy will be calculated using the following relation ($U = 0$ and $S = k \ln P$):

$$F = -kT \ln \sum \exp\left(-\frac{E}{kT}\right) = -kT \ln \sum \exp\left(-\frac{N_H \varepsilon_0 + N_{HH} \varepsilon}{kT}\right) \tag{5.22}$$

Applying the Bragg–William approximation (assuming that the H–H interaction is given by the 'mean field' of the H-atoms) and with the concentration $c_H = N_H/N$, the free energy is

$$F_H = kTN[c_H \ln c_H + (1-c_H)\ln(1-c_H)] + N\left(\varepsilon_0 c_H + \varepsilon \frac{n}{2} c_H^2\right) \tag{5.23}$$

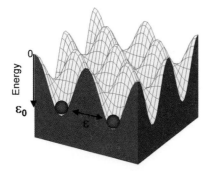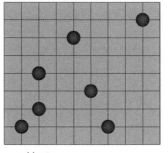

Figure 5.29 Potential for the hydrogen atom in the metal lattice (left) and occupation of the interstitial sites of the metal lattice by hydrogen atoms (right).

The concentration c_H is not necessarily equal to x for a metal hydride MH_x, because in some structures the number of interstitial sites N of a certain type is not always equal to the number of metal atoms.

The chemical potential μ_H of a hydrogen atom in a metal is (5.24)

$$\mu_H(c_H, T) = \frac{\partial F}{\partial N_H}\bigg|_{T,V} = \frac{1}{N}\frac{\partial F}{\partial c_H}\bigg|_{T,V} \tag{5.24}$$

and therefore

$$\mu_H = kT \ln\left(\frac{c_H}{1-c_H}\right) + \varepsilon_0 + \varepsilon n c_H \tag{5.25}$$

In equilibrium of the lattice with the gas phase

$$\frac{1}{2}kT \ln\left[\frac{p}{p_0(T)}\right] = \varepsilon_0 + \varepsilon n c_H - \frac{1}{2}\varepsilon_b + kT \ln\left(\frac{c_H}{1-c_H}\right) \tag{5.26}$$

This is the hydrogen solubility isotherm. For low concentrations c_H this equation approaches Sievert's equation:

$$\frac{1}{2}kT \ln\left[\frac{p}{p_0(T)}\right]_{p \to 0} \cong kT \ln c_H \bigg|_{c_H \to 0} \tag{5.27}$$

For low temperatures, that is, when the temperature is below the critical temperature ($T < T_c$) of the MH_x system, the equation loses its physical meaning (Figure 5.30).

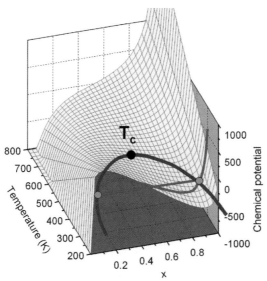

Figure 5.30 Chemical potential as a function of the hydrogen concentration x and the temperature. The coexistence curve (T_c) and the Maxwell construction between the two points where f (c_H) = 0.

$$\frac{1}{2}\left\{\varepsilon_b - \varepsilon n + kT \ln\left[\frac{p}{p_0(T)}\right] - 2\varepsilon_0\right\} = f(c_H) \tag{5.28}$$

with

$$f(c_H) = kT \ln\left(\frac{c_H}{1-c_H}\right) + \varepsilon n\left(c_H - \frac{1}{2}\right) \tag{5.29}$$

The function $y = f(c_H)$ is asymmetric in view of $c_H = 0.5$ and $y = 0$.

$$kT \ln\left(\frac{c_i}{1-c_i}\right) + \varepsilon n\left(c_i - \frac{1}{2}\right) = 0 \quad \text{coexistence curve} \tag{5.30}$$

and the dissociation pressure p_{dis} is given by

$$kT \ln\left[\frac{p_{dis}}{p_0(T)}\right] = 2\varepsilon_0 + \varepsilon n - \varepsilon_b \tag{5.31}$$

For PdH_x, $2\varepsilon_0 + \varepsilon n \approx -4.9$ eV. For most metal hydrides $2\varepsilon_0 + \varepsilon n < -5$ eV and the value can reach approximately -6.4 eV for LaH_x.

The values of ε_0 and εn can be determined from the isotherm for single-phase systems:

$$\ln\left[\frac{p}{p_0(T)}\left(\frac{1-c_H}{c_H}\right)^2\right] = \frac{2\varepsilon_0 - \varepsilon_b + 2\varepsilon n c_H}{kT} \tag{5.32}$$

The coexistence curve exhibits, for $T = T_c$, a maximum ($c_H = \frac{1}{2}$). Therefore, T_c is calculated from the derivative of the coexistence curve:

$$k\frac{dT}{dc}\bigg|_{T=T_c} \ln\left(\frac{c_c}{1-c_c}\right) + kT_c \frac{1}{c_c(1-c_c)} + \varepsilon n = 0 \tag{5.33}$$

Because of the symmetry of the $T(c)$ curve, the maximum $T = T_c$ corresponds to the concentration $c_c = \frac{1}{2}$. Therefore,

$$T_c = -\frac{\varepsilon n}{4k} \tag{5.34}$$

The physical interpretation is that the phase separation only exists up to the temperature T_c, which corresponds to the H–H interaction.

PdH_x : $\varepsilon n = -0.20$ eV; $T_c = 566$ K
NbH_x : $\varepsilon n = -0.16$ eV; $T_c = 443$ K.

The H–H pair energy is, therefore,

$\underset{\substack{\text{fcc}\\\text{octahedral sites}}}{n = 12} \to \varepsilon \approx -0.02$ eV/pair $\quad (PdH_x)$

$\underset{\substack{\text{bcc}\\\text{tetrahedral sites}}}{n = 4} \to \varepsilon \approx -0.09$ eV/pair $\quad (NbH_x)$.

The approximation of the coexistence curve for low concentrations $c_H \to 0$ is given by

$$kT \ln c_\alpha - \frac{\varepsilon n}{2} = 0 \quad \to \quad c_\alpha = e^{\frac{\varepsilon n}{2kT}} \tag{5.35}$$

At low temperatures and for a system in equilibrium, the solubility of hydrogen is very low.

The set of all points of concentrations which corresponds to the points B and C is the spinodal line and is given by the following conditions:

$$\frac{\partial^2 G_H}{\partial c_H^2} = 0 \tag{5.36}$$

which corresponds to

$$\frac{\partial \mu_H}{\partial c_H} = 0 \tag{5.37}$$

From the equation for the chemical potential of the lattice gas,

$$\frac{\partial \mu_H}{\partial c_H} = kT_{sp} \frac{1}{c_H(1-c_H)} + \varepsilon n = 0 \tag{5.38}$$

where T_{sp} is the spinodal temperature,

$$T_{sp} = -\frac{\varepsilon n}{k} c_H(1-c_H) \quad \text{spinodal curve} \tag{5.39}$$

which is a parabola with a maximum at the critical temperature T_c (Figure 5.31).

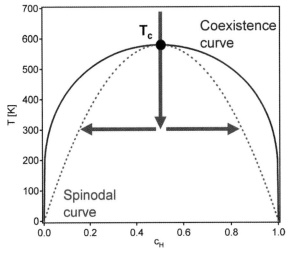

Figure 5.31 Coexistence curve and spinodal line for the MH system.

5.7
Complex Hydrides

All the elements of group 13 (boron group) form polymeric hydrides $(MH_3)_x$. The monomers MH_3 are strong Lewis acids and are unstable. Borane (BH_3) achieves electronic saturation by dimerization to form diborane (B_2H_6). All other hydrides in this group attain closed electron shells by polymerization. Aluminum hydride, alane $(AlH_3)_x$, has been extensively investigated [47], the hydrides of gallium, indium and thallium much less so [48, 49] (Figure 5.32).

The hydrogen in the p-element complex hydrides is often located in the corners of a tetrahedron with boron or aluminum in the center. The bonding character and the properties of the complexes $M^+[BH_4]^-$ and $M^+[AlH_4]^-$ are largely determined by the difference in electronegativity between the cation and the boron or aluminum atom, respectively. IUPAC has recommended the names tetrahydroborate for $[BH_4]^-$ and tetrahydroaluminate for $[AlH_4]^-$ (Figure 5.33).

The alkali metal tetrahydroborates are ionic, white, crystalline, high-melting solids that are sensitive to moisture but not to oxygen. Group 3 and transition metal

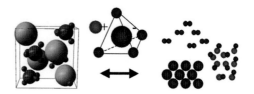

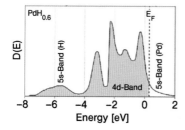

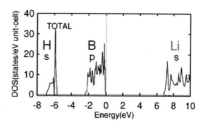

Figure 5.32 Comparison of metal hydrides with complex hydrides [50]. Metal hydrides consist of an almost unchanged metal lattice with the interstitial sites filled with hydrogen. The additional electrons from hydrogen are added at the Fermi level of the metallic electron density of states and an additional hydrogen-induced band several eV below the Fermi level accommodates some electrons. The complex hydrides have structures where the hydrogen builds a negatively charged complex ion with one metal and a second metal ion compensates the charge. Complex hydrides exhibit a significant cap of several eV between the valence band and the conduction band.

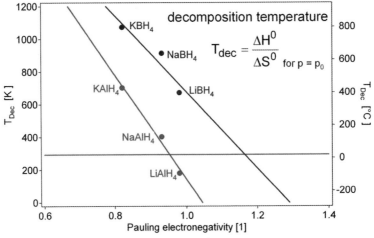

Figure 5.33 Decomposition temperature ($p = 1$ bar H_2) as a function of the electronegativity of the cation [51].

tetrahydroborates are covalently bonded and are either liquids or sublimable solids. The alkaline earth tetrahydroborates are intermediate between ionic and covalent. The tetrahydroaluminates are very much less stable than the tetrahydroborates and therefore considerably more reactive. The difference between the stabilities of the tetrahydroaluminate and the tetrahydroborates is due to the different Pauling electronegativities of B and Al, which are 2.04 and 1.61, respectively. The properties of the complex hydrides can be varied by partial substitution of the boron or aluminum atom (Table 5.5).

Table 5.5 The group 1, 2 and 3 light elements, for example Li, Mg, B and Al, form a large variety of metal–hydrogen complexes.

Complex	CAS No.	M (g mol^{-1})	Density (g cm^{-3})	H (mass%)	H (kg m^{-3})	T_m (°C)	ΔH_f° (kJ mol^{-1})
LiBH$_4$	16949-15-8	21.78376	0.66	18.36	18.5	268	−194
NaBH$_4$	16940-66-2	37.83253	1.07	10.57	10.6	505	−191
KBH$_4$	13762-51-1	53.94106	1.17	7.42	7.4	585	−229
Be[BH$_4$]$_2$	17440-85-6	38.697702	0.702	20.67		123	
Mg[BH$_4$]$_2$	16903-37-0	53.99052		14.82		320	
Ca[BH$_4$]$_2$	17068-95-0	69.76352		11.47		260	
Al[BH$_4$]$_3$	16962-07-5	71.509818	0.549	16.78		−64.5 / 44.5	
LiAlH$_4$	16853-85-3	37.954298	0.917	10.54		190^d	−119
NaAlH$_4$	13770-96-2	54.003068	1.28	7.41		178	−113
KAlH$_4$		70.111598		5.71			
Be[AlH$_4$]$_2$		71.038778		11.26			
Mg[AlH$_4$]$_2$	17300-62-8	86.331596		9.27			
Ca[AlH$_4$]$_2$	16941-10-9	102.10460		7.84		>230	

In contrast to the interstitial hydrides, where the metal lattice hosts the hydrogen atoms on interstitial sites, the desorption of the hydrogen from the complex hydride leads to a complete decomposition of the complex hydride and a mixture of at least two phases is formed. For alkali metal tetrahydroborates and tetrahydroaluminates, the decomposition reaction is described according to the following equation:

$$A(BH_4) \rightarrow \text{``}ABH_2 + H_2\text{''?} \rightarrow AH + B + \frac{3}{2}H_2$$

and

$$A(AlH_4) \rightarrow \frac{1}{3}A_3AlH_6 + H_2 \rightarrow AH + Al + \frac{3}{2}H_2 \tag{5.40}$$

For alkaline earth metal tetrahydroborates and tetrahydroaluminates, the decomposition reaction is described according to the following equations:

$$E(BH_4)_2 \rightarrow EH_2 + B \tag{5.41}$$

and

$$E(AlH_4)_2 \rightarrow ? \rightarrow EH_2 + 2Al + 3H_2 \tag{5.42}$$

The physical properties and the hydrogen sorption mechanism of the tetrahydroborates and the tetrahydroalanates are to a large part still not known (Figure 5.34).

The most industrially important complex hydrides are sodium tetrahydridoborate, $NaBH_4$, and lithium tetrahydridoaluminate, $LiAlH_4$. These are produced in tonnage quantities and used mainly as reducing agents in organic chemistry. Many other hydridoborates and hydridoaluminates are produced using these as starting materials [52, 53].

5.7.1
Tetrahydroalanates

The $NaAlH_4$ and Na_3AlH_6 mixed ionic–covalent complex hydrides have been known for many years. $NaAlH_4$ consists of an Na^+ cation and a covalently bonded

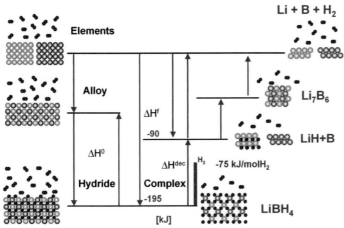

Figure 5.34 Schematic enthalpy diagram for a classical metal hydride (left) and a complex hydride (right).

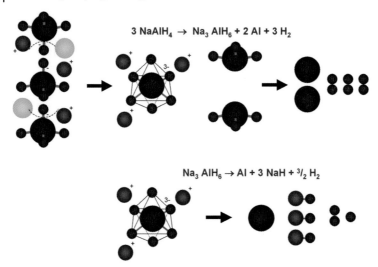

Figure 5.35 Schematic mechanism for the hydrogen desorption of NaAlH$_4$. The catalyst permits the transfer of the two H$^-$ ions to form the hexahydride.

[AlH$_4$]$^-$ complex. In the case of Na$_3$AlH$_6$, there is a related [AlH$_6$]$^{3-}$ complex (Figure 5.35).

These alanates have been synthesized by both indirect and direct methods and used as chemical reagents (see [54] for a historical review of alanate synthesis). However, the practical key to using the Na alanates for hydrogen storage is to be able to accomplish easily the following reversible two-step dry-gas reaction:

$$3\text{NaAlH}_4 \rightarrow \text{Na}_3\text{AlH}_6 + 2\text{Al} + 3\text{H}_2 \qquad (3.7 \text{ wt\% H}) \qquad (5.43)$$

$$\text{Na}_3\text{AlH}_6 \rightarrow 3\text{NaH} + \text{Al} + \frac{3}{2}\text{H}_2 \qquad (3.0 \text{ wt\% H}) \qquad (5.44)$$

Stoichiometrically, the first step consists of 3.7 wt% H$_2$ release and the second step 1.9 wt% H$_2$ release, for a theoretical net reaction of 5.6 wt% reversible gravimetric H storage.

Although Dymova *et al.* showed that the reversibility of Reaction 5.43 was possible, the conditions required were impractical in their severity [55, 56]. For example, the formation of NaAlH$_4$ from the elements required temperatures of 200–400 °C (i.e. above the 183 °C melting temperature of the tetrahydride) and H$_2$ pressures of 100–400 bar (Figure 5.36).

The thermodynamics of Reaction 5.43 are comparable to those of other metallic, ionic and covalent hydrides. NaAlH$_4$ exhibits typical low hysteresis and a two-plateau absorption and desorption isotherm.

The enthalpy changes ΔH for the NaAlH$_4$ and Na$_3$AlH$_6$ decompositions are about 37 and 47 kJ mol^{-1} H$_2$, respectively. Therefore, NaAlH$_4$ is thermodynamically comparable to those of classic low-temperature hydrides, in the range useful for a

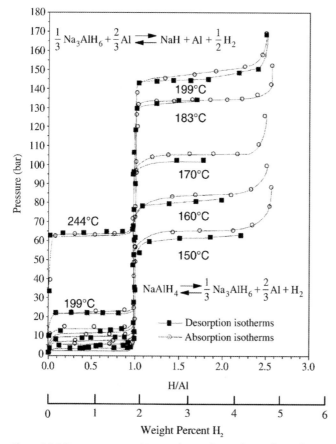

Figure 5.36 Pressure–composition isotherms for $NaAlH_4$ and Na_3AlH_{6}.2.

near-ambient temperature hydrogen store. Na_3AlH_6 requires about 110 °C for H_2 liberation at atmospheric pressure (Figure 5.37).

The practical use of the desorption reaction requires a catalyst for the improvement of the kinetics. The first work on catalyzed alanates at MPI – Mülheim was derived from studies that used transition-metal catalysts for the preparation of MgH_2. The $NaAlH_4$ was doped with Ti by solution chemistry techniques whereby nonaqueous liquid solutions or suspensions of $NaAlH_4$ and either $TiCl_3$ or the alkoxide $Ti(OBu^n)_4$ [titanium(IV) n-butoxide] catalyst precursors were decomposed to precipitate solid Ti-doped $NaAlH_4$ [57, 58].

An alternative approach was taken by Jensen and co-workers at the University of Hawaii, whereby the liquid $Ti(OBu^n)_4$ precursor was simply ball-milled with the solid $NaAlH_4$ [59, 60]. They also added $Zr(OPr^i)_4$ (Pr^i = isopropyl) to help stimulate the kinetics of the second step (Na_3AlH_6 decomposition) of Reaction 5.43. Zaluska et al. also obtained positive results by ball-milling with carbon (i.e. not using a transition metal catalyst) [61]. Ball-milling with alkoxide catalyst precursors results in in situ decomposition during at least the first several hydriding/dehydriding cycles,

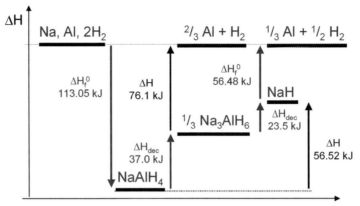

Figure 5.37 Energy diagram for NaAlH$_4$.

resulting in significant contamination of the H$_2$ with hydrocarbons; using the inorganic TiCl$_3$ catalyst precursor is clearly better if one wishes to use the ball-milling catalyst-doping approach [62]. More detailed reviews of the recent history of catalyzed alanate work are available [63, 64] (Figure 5.38).

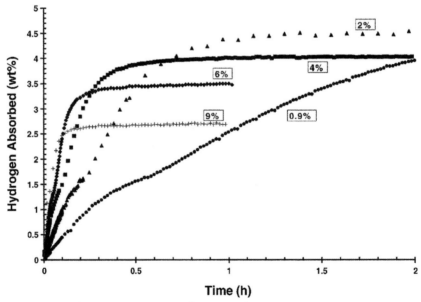

Figure 5.38 H$_2$ absorption curves starting from NaH+Al (dehydrided NaAlH$_4$), as a function of added TiCl$_3$ (using the dry TiCl$_3$–NaAlH$_4$ ball-milling/doping technique [65]) catalyst precursor (expressed in mol%). Initial temperature $T_i = 125\,°C$. Applied hydrogen pressure $p_{H_2} = 81–90$ bar [66].

The kinetics are improved and the capacity decreased with increasing catalyst level. Nearly identical catalyst effects are seen with desorption experiments, although the two steps of Reaction 5.43 are more clearly sequential in desorption [67]. The catalysis affects both of the reaction directions of the reaction and also both steps. The capacity is decreased with increasing catalyst level because the catalyst precursor reacts with some of the Na as part of its process to produce the catalyst itself:

$$(1-x)\text{NaAlH}_4 + x\text{TiCl}_3 \rightarrow (1-4x)\text{NaAlH}_4 + 3x\text{NaCl} + x\text{Ti} + 3x\text{Al} + 6x\text{H}_2$$
(5.45)

Reaction 5.45 is at least partly hypothetical. Evidence that the Cl does react with the Na component of the alanate to form NaCl was found by means of X-ray diffraction (XRD), but the final form of the Ti catalyst is not clear [68]. Ti is probably metallic in the form of an alloy or intermetallic compound (e.g. with Al) rather than elemental. Another possibility is that the transition metal dopant (e.g. Ti) actually does not act as a classic surface 'catalyst' on NaAlH_4, but rather enters the entire Na sublattice as a variable valence species to produce vacancies and lattice distortions, thus aiding the necessary short-range diffusion of Na and Al atoms [69]. Ti, derived from the decomposition of TiCl_4 during ball-milling, seems to also promote the decomposition of LiAlH_4 and the release of H_2 [70]. In order to understand the role of the catalyst, Sandrock *et al.* performed detailed desorption kinetics studies (forward reactions, both steps, of the reaction) as a function of temperature and catalyst level [71] (Figure 5.39).

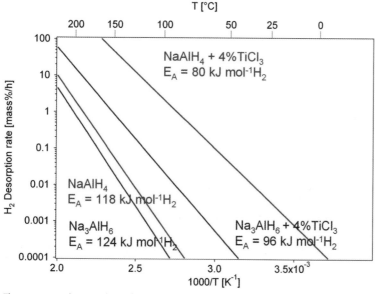

Figure 5.39 Arrhenius plots of the catalyzed (4% TiCl_3) and pure NaAlH_4 and Na_3AlH_6.

Both the $NaAlH_4$ and Na_3AlH_6 decomposition reactions obey thermally activated behavior consistent with the Arrhenius equation. Undoped and 4 mol% Ti-doped Na alanate samples were investigated. It was found that catalysis results in multiple order-of-magnitude increases in kinetics for both reactions. The rate increase is a combined result of changes in both the activation energy and the pre-exponential factor. The smallest Ti addition (0.9 mol%) has marked effects on lowering the activation energy for both steps of the reaction. However, further increases in Ti addition have no further effects on the activation energy. Further increases in Ti level lead to increases in desorption kinetics, but only through increases in the pre-exponential factor of the rate constant.

The identity of the effective catalyst is not known yet. XRD and microscopic analyses have not led to its identification, probably because it is amorphous, too fine or is substantially located in the alanate lattice itself. Can we produce effective catalysts only by the *in situ* decomposition of a precursor?

The history of intermetallic hydrides is deeply laced with thermodynamic tailoring by means of partial substitution of secondary and higher order components. Li can be partially substituted for Na in $NaAlH_4$. Numerous $M[AlH_4]_x$ species are known, but their thermodynamic stabilities are largely unknown.

The alanates have numerous isostructural counterparts in the borohydrides (e.g. $NaBH_4$). The borohydrides tend to be more thermodynamically stable than the alanates, yet less water reactive [72].

5.7.2
Tetrahydroborates

The first report of a pure alkali metal tetrahydroborate appeared in 1940 by Schlesinger and Brown, who synthesized lithium tetrahydroborate (lithium borohydride) ($LiBH_4$) by the reaction of ethyllithium with diborane (B_2H_6) [73]. The direct reaction of the corresponding metal with diborane in ethereal solvents under suitable conditions produces high yields of the tetrahydroborates, $2MH + B_2H_6 \rightarrow 2MBH_4$, where M = Li, Na, K, and so on [74]. Direct synthesis from the metal, boron and hydrogen at 550–700 °C and 30–150 bar H_2 has been reported to yield the lithium salt and it has been claimed that such a method is generally applicable to group 1A and 2A metals [75]. The reaction involving either the metal or the metal hydride or the metal together with triethylborane in an inert hydrocarbon has formed the basis of a patent, $M + B + 2H_2 \rightarrow MBH_4$, where M = Li, Na, K, and so on.

The stability of metal tetrahydroborates has been discussed in relation to their percentage ionic character and those compounds with less ionic character than diborane are expected to be highly unstable [76]. Steric effects have also been suggested to be important in some compounds [77, 78]. The special feature exhibited by the covalent metal hydroborides is that the hydroboride group is bonded to the metal atom by bridging hydrogen atoms similar to the bonding in diborane, which may be regarded as the simplest of the so-called 'electron-deficient' molecules. Such molecules possess fewer electrons than those apparently required to fill all the bonding orbitals, based on the criterion that a normal bonding orbital involving two

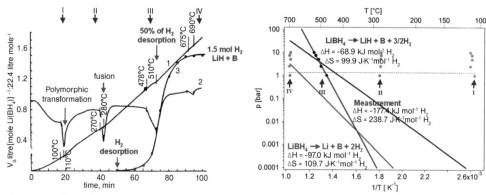

Figure 5.40 Thermal desorption spectroscopy and corresponding van't Hoff plot [81].

atoms contains two electrons. The molecular orbital bonding scheme for diborane has been discussed extensively (Figure 5.40).

The compound with the highest gravimetric hydrogen density at room temperature known today is LiBH$_4$ (18 mass%). Therefore, this complex hydride could be the ideal hydrogen storage material for mobile applications. The first report by Fedneva of the thermal hydrogen desorption from LiBH$_4$ appeared in 1964 [79]. Stasinevich and Egorenko investigated thermal hydrogen desorption under various hydrogen pressures [80]. LiBH$_4$ desorbs three of the four hydrogens in the compound upon melting at 280 °C and decomposes into LiH and boron.

Thermal desorption spectroscopy exhibits four endothermic peaks. The peaks are attributed to a polymorphic transformation around 110 °C, melting at 280 °C, the hydrogen desorption [50% of the hydrogen was desorbed (LiBH$_2$) around 490 °C] and when three of the four hydrogens are desorbed at 680 °C. Only the third peak (hydrogen desorption) is pressure dependent; all other peak positions (temperature) do not vary with pressure. The calculated enthalpy $\Delta H = -177.4$ kJ mol^{-1} H$_2$ and entropy $\Delta S = 238.7$ J K^{-1} mol^{-1} H$_2$ of decomposition are not in agreement with the values deduced from indirect measurements of the stability [81]. Especially the value of the entropy is far too high and cannot be explained, because the standard entropy for hydrogen is 130 J K^{-1} mol^{-1} H$_2$. The thermal properties of LiBH$_4$ between room temperature and 300 °C were investigated by means of differential scanning calorimetry (DSC) (Figures 5.41 and 5.42).

An endothermic (4.18 kJ mol^{-1}) structural transition is observed at 118 °C from the orthorhombic low-temperature structure to the hexagonal high-temperature structure. The high-temperature structure melts at 287 °C with a latent heat of 7.56 kJ mol^{-1}. Both transitions are reversible, although a small hysteresis in temperature was observed.

Harris and Meibohm suggested the space group *Pcmn* for LiBH$_4$ at room temperature, which turned out to be wrong [82]. The low- and high-temperature structures were investigated by means of synchrotron X-ray powder diffraction [83, 84].

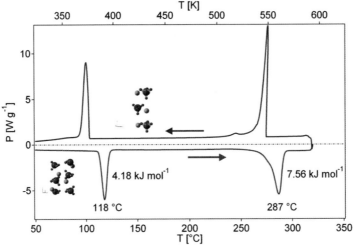

Figure 5.41 Thermal power as a function of temperature for a small LiBH$_4$ sample measured at constant volume.

The thermal hydrogen desorption of a pure LiBH$_4$ sample into vacuum starts upon the melting of the sample. The desorption process can be catalyzed by adding SiO$_2$ and significant thermal desorption was observed starting at 200 °C [85, 86] (Figure 5.43).

The thermal desorption spectrum at very low heating rates shows three distinct desorption peaks. This is an indication that the desorption mechanism involves more than one intermediate step. This can be explained by the formation of polymeric borohydrides (Figure 5.44).

Recently it has been shown that the hydrogen desorption reaction is reversible and the end products lithium hydride and boron absorb hydrogen at 690 °C and 200 bar to

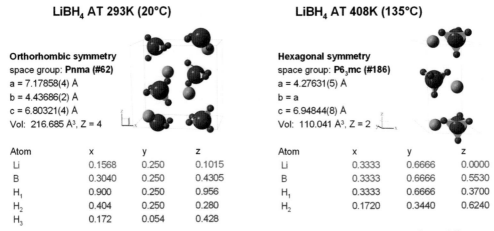

Figure 5.42 Low- and high-temperature structure of LiBH$_4$ determined by means of X-ray diffraction.

5.7 Complex Hydrides | 151

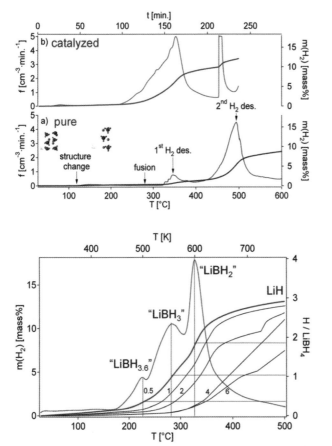

Figure 5.43 Thermal desorption spectroscopy of pure and catalyzed LiBH$_4$.

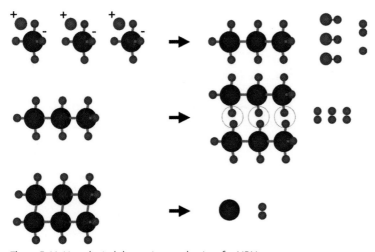

Figure 5.44 Hypothetical desorption mechanism for LiBH$_4$.

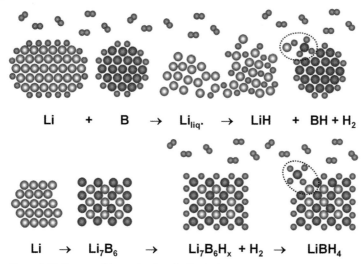

Figure 5.45 Two possible mechanism for the hydrogen absorption by lithium and boron to form $LiBH_4$.

form $LiBH_4$ [87]. A temperature >680 °C is necessary in order to melt LiH and therefore improve the intermixing of LiH and boron. The understanding of the mechanism of the thermal hydrogen desorption from $LiBH_4$ and the absorption remains a challenge and more research work is necessary (Figure 5.45).

The enthalpy diagram of $LiBH_4$ shows the hydrogen desorption steps and also a possible path for the reabsorption of hydrogen. It is essential that three hydrogen atoms bind to boron and subsequently an H^- is transferred from the LiH to the 'BH_3' in order to form $[BH_4]^-$ and the charge is compensated by the remaining Li^+. The last step of the reaction is energy driven (Figures 5.46 and 5.47).

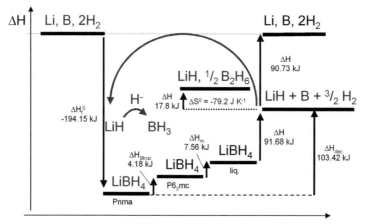

Figure 5.46 Schematic enthalpy diagram of $LiBH_4$.

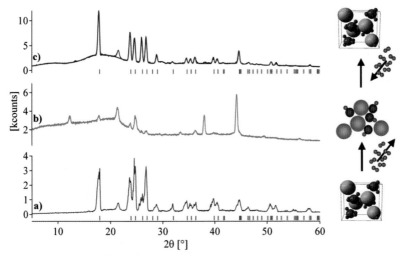

Figure 5.47 (a) X-ray diffraction pattern (Cu Kα) of a commercial LiBH$_4$ sample; (b) X-ray diffraction patter after the thermal desorption of hydrogen from the sample (a); (c) X-ray diffraction pattern of the sample upon reabsorption of the hydrogen.

The stability of the tetrahydroborates correlates with the desorption temperature according to $T_d = \Delta H/\Delta S$ for $p = p^0$. Furthermore, the desorption temperature correlates with the Pauling electronegativity of the atom forming the cation (Figures 5.48 and 5.49).

Very little is known today about Al(BH$_4$)$_3$, a complex hydride with a very high gravimetric hydrogen density of 17 mass% and the highest known volumetric hydrogen density of 150 kg m^{-3}. Al(BH$_4$)$_3$ has a melting point of −65 °C and is liquid at room temperature. Apart from the covalent hydrocarbons, this is the only liquid hydride at room temperature.

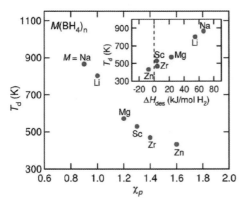

Figure 5.48 Decomposition temperature ($p = 1$ bar H$_2$) as a function of the electronegativity of the cation [88].

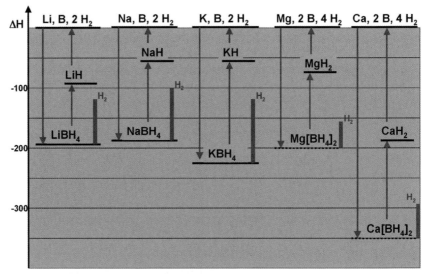

Figure 5.49 Enthalpy diagram of the alkali and alkaline earth metal borates. The enthalpy for the desorption of H_2 is indicated by the red bar.

5.8
Chemical Hydrides (Hydrolysis)

Hydrogen can be generated from metals and chemical compounds reacting with water. The common experiment – shown in many chemistry classes – where a piece of sodium floating on water produces hydrogen, demonstrates such a process. The sodium is converted to sodium hydroxide in this reaction. The reaction is not directly reversible but the sodium hydroxide could later be removed and reduced in a solar furnace back to metallic sodium. Two sodium atoms react with two water molecules and produce one hydrogen molecule. The hydrogen molecule produces a water molecule again in the combustion, which can be recycled to generate more hydrogen gas. However, the second water molecule necessary for the oxidation of the two sodium atoms has to be added. Therefore, sodium has a gravimetric hydrogen density of 4.4 mass%. The same process carried out with lithium leads to a gravimetric hydrogen density of 6.3 mass%. The major challenge with this storage method is the reversibility and the control of the thermal reduction process in order to produce the metal in a solar furnace (Table 5.6).

5.8.1
Zinc Cycle

The process has been successfully demonstrated with zinc [89] (Figure 5.50).

The first, endothermic step is the thermal dissociation of $ZnO(s)$ into $Zn(g)$ and O_2 at 2300 K using concentrated solar energy as the source of process heat. The

5.8 Chemical Hydrides (Hydrolysis)

Table 5.6 Some reactions with metals and water[a].

Reaction	H (mass% including H_2O)	H (mass%)
$Li + H_2O \rightarrow LiOH + \frac{1}{2}H_2$	4.0	6.3
$Na + H_2O \rightarrow NaOH + \frac{1}{2}H_2$	2.4	4.4
$Mg + 2H_2O \rightarrow Mg(OH)_2 + H_2$	3.3	4.7
$Al + 3H_2O \rightarrow Al(OH)_3 + 1\frac{1}{2}H_2$	3.7	5.6
$Zn + 2H_2O \rightarrow Zn(OH)_2 + H_2$	1.97	3.0

[a]The hydrogen density is calculated based on the metal and water and based on the metal only (reuse of the water from the combustion).

second, non-solar, exothermic step is the hydrolysis of Zn(l) at 700 K to form H_2 and ZnO(s); the ZnO(s) separates naturally and is recycled to the first step. Hydrogen and oxygen are derived in different steps, thereby eliminating the need for high-temperature gas separation. A second-law analysis performed on the closed cyclic process indicates a maximum energy conversion efficiency of 29% [ratio of ΔG ($H_2 + 0.5O_2 \rightarrow H_2O$) for the H_2 produced to the solar power input], when using a solar cavity receiver operated at 2300 K and subjected to a solar flux concentration ratio of 5000. The major sources of irreversibility are associated with the re-radiation losses from the solar reactor and the quenching of Zn(g) and O_2 to avoid their recombination. An economic assessment for a large-scale chemical plant, having a solar thermal power input into a solar reactor of 90 MW and a hydrogen production output from the hydrolyzer of 61 GW h per year, indicates that the cost of solar hydrogen ranges between 0.13 and $0.15 per kW h (based on its low heating value and a heliostat feld cost of $100–150 m^{-2}) and, thus, might be competitive with respect to other renewables-based routes such as electrolysis of water using solar-generated electricity. The economic feasibility of the proposed solar process is strongly dependent on the development of an effective Zn–O_2 separation technique (either by quenching or by *in situ* electrolytic separation) that eliminates the need for an inert gas (Figure 5.51).

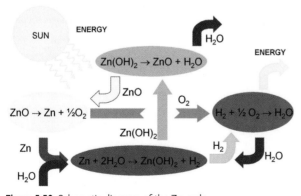

Figure 5.50 Schematic diagram of the Zn cycle.

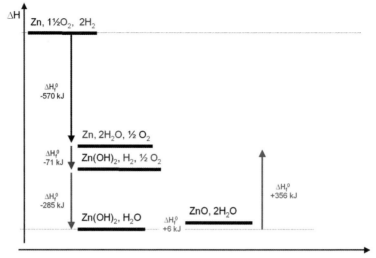

Figure 5.51 Energy diagram of the Zn cycle. The oxidation of Zn with water releases only a small amount (71 kJ mol^{-1} Zn) of energy. Each Zn atom requires two water molecules and produces one H$_2$ molecule.

The small energy difference between Zn + 2H$_2$O and Zn(OH)$_2$ + H$_2$ is responsible for the energy efficiency of the process. However, the fact that two H$_2$Os are necessary to produce one H$_2$ lowers the storage density significantly.

5.8.2
Borohydride

Complex hydrides react with water to give hydrogen, a metal hydroxide and borax [90]. Very high hydrogen densities are reached if the water from the combustion of the hydrogen is reused (Table 5.7).

Hydrolysis reactions involve the oxidation reaction of chemical hydrides with water to produce hydrogen. The reaction of sodium borohydride has been the most studied to date:

$$NaBH_4 + 2H_2O \rightarrow NaBO_2 + 4H_2 \tag{5.46}$$

Table 5.7 Some reactions with chemical hydrides.

Reaction	H (mass%)	H (mass%)^a
NaAlH$_4$ + 4H$_2$O → NaOH + Al(OH)$_3$ + 4H$_2$	6.3	14.8
NaBH$_4$ + 3H$_2$O → NaOH + HBO$_2$ + 4H$_2$	8.7	21.3
NaBH$_4$ + 4H$_2$O → NaOH + HBO$_3$ + 5H$_2$	9.1	26.6
LiBH$_4$ + 3H$_2$O → LiOH + HBO$_2$ + 4H$_2$	10.6	37.0
LiBH$_4$ + 4H$_2$O → LiOH + HBO$_3$ + 5H$_2$	10.7	46.2

^{a}The hydrogen density is calculated based on the reuse of the water from the combustion.

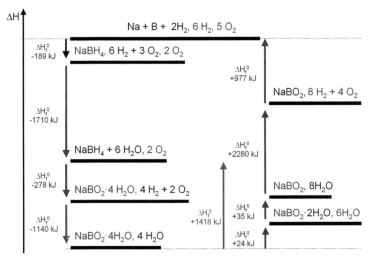

Figure 5.52 Energy diagram for the reaction of NaBH$_4$ with water.

In the first embodiment, a slurry of an inert stabilizing liquid protects the hydride from contact with moisture and makes the hydride pumpable. At the point of use, the slurry is mixed with water and the consequent reaction produces high-purity hydrogen.

The reaction can be controlled in an aqueous medium via pH and the use of a catalyst [91]. While the material hydrogen capacity can be high and the hydrogen release kinetics fast, the borohydride regeneration reaction must take place off-board. Regeneration energy requirements, cost and life-cycle impacts are key issues currently being investigated.

Millennium Cell has reported that their NaBH$_4$-based 'hydrogen on demand' system possesses a system gravimetric capacity of about 4 wt%. Similarly to other material approaches, issues include system volume, weight and complexity and water availability.

Another hydrolysis reaction that is currently being investigated by Safe Hydrogen is the reaction of MgH$_2$ with water to form Mg(OH)$_2$ and H$_2$. In this case, particles of MgH$_2$ are contained in a non-aqueous slurry to inhibit premature water reactions when hydrogen generation is not required. Material-based capacities for the MgH$_2$ slurry reaction with water can be as high as 11 mass%. However, similarly to the sodium borohydride approach, the slurry and the Mg(OH)$_2$ must be carried on-board the vehicle regenerated off-board (Figure 5.52).

5.9
New Hydrogen Storage Materials

Hydrogen will be stored in various ways depending on the application, for example, mobile or stationary. Today we know of several efficient and safe ways to store

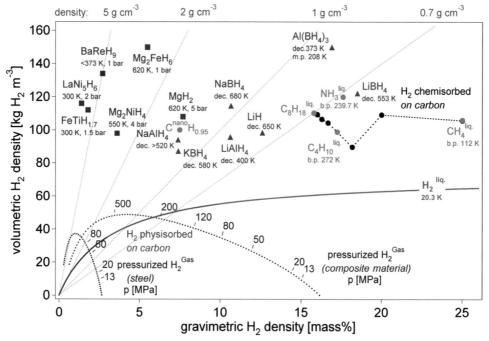

Figure 5.53 Volumetric and gravimetric hydrogen density of some selected hydrides. Mg_2FeH_6 shows the highest known volumetric hydrogen density of 150 kg m^{-3}, which is more than double that of liquid hydrogen. $BaReH_9$ has the largest H:M ratio of 4.5, that is, 4.5 hydrogen atoms per metal atom. $LiBH_4$ exhibits the highest gravimetric hydrogen density of 18 mass%. Pressurized gas storage is shown for steel (tensile strength $\sigma_v = 460$ MPa, density 6500 kg m^{-3}) and a hypothetical composite material ($\sigma_v = 1500$ MPa, density 3000 kg m^{-3}).

hydrogen; however, there are many other new potential materials and methods possible to increase the hydrogen density significantly (Figure 5.53).

The material science challenge is to understand better the electronic behavior of the interaction of hydrogen with other elements, especially metals. Complex compounds such as $Al(BH_4)_3$ have to be investigated and new compounds formed from lightweight metals and hydrogen will be discovered.

New material systems have been investigated recently which are based on new effects and interactions.

5.9.1
Amides and Imides (–NH$_2$, =NH)

The thermal desorption of hydrogen from lithium nitride ($LiNH_2$) was investigated by Chen et al. [92, 93]. The thermal desorption of pure lithium amide mainly evolves NH_3 at elevated temperatures following the reaction (Figure 5.54)

$$2LiNH_2 \rightarrow Li_2NH + NH_3 \tag{5.47}$$

5.9 New Hydrogen Storage Materials

$LiNH_2 + 2LiH \rightarrow Li_3N + 2H_2$ (10.3 mass%, $\Delta H = 40$ kJ mol^{-1} H)

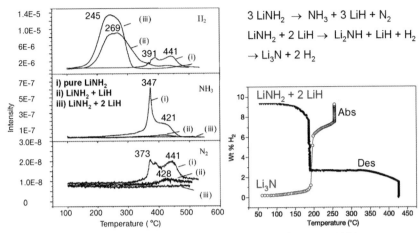

Figure 5.54 Thermal desorption of pure $LiNH_2$, $LiNH_2 + LiH$ and $LiNH_2 + 2LiH$.

The reaction enthalpy ΔH is around 83.68 kJ mol^{-1}. The calculated weight loss from the above reaction is about 37%. However, when $LiNH_2$ is in contact with LiH, the reaction path changes completely. When $LiNH_2 + LiH$ is used, H_2 is released at a much lower temperature and imide is formed:

$$LiNH_2 + LiH \rightarrow Li_2NH + H_2 \tag{5.48}$$

The reaction enthalpy ΔH is around 45 kJ mol^{-1}. In this case, the weight loss was calculated to be 6.5%. When $LiNH_2 + 2LiH$ was used, H_2 and an imide-like structure formed at temperatures below 320 °C and H_2 and Li_3N formed at higher temperatures:

$$LiNH_2 + 2LiH \rightarrow Li_xNH_{3-x} + (x-1)H_2 + (3-x)LiH \rightarrow Li_3N + 2H_2 \tag{5.49}$$

The reaction heat of the first stage is estimated to be much less than 83 kJ mol^{-1}; the overall reaction heat is around 161 kJ mol^{-1}. At temperatures below 320 °C, 6.5 mass% H_2 was released, which equals the theoretical value of 7.0 mass%. Thus, the intermediate structure Li_xNH_{3-x} has the formula $Li_{2.2}NH_{0.8}$. It might be treated as an Li-rich imide.

The H in $LiNH_2$ is partially positively charged, but in LiH it is negatively charged. The redox of $H^{\delta+}$ and $H^{\delta-}$ to H_2 and simultaneously combination of $N^{\delta-}$ and $Li^{\delta+}$ (in LiH) are energetically favorable.

The combination of hydrides where the hydrogen is partially positive charged with hydrides where the hydrogen is partially negative charged leads to destabilization of the hydrides.

5.9.2
bcc Alloys

Hydrogen-absorbing alloys with body-centered cubic (bcc) structures, such as Ti–V–Mn, Ti–V–Cr, Ti–V–Cr–Mn and Ti–Cr–(Mo, Ru), have been developed since 1993 [94]. These alloys offer more interstitial sites than fcc and hcp structures and have a higher hydrogen capacity (about 3.0 mass%) than conventional intermetallic hydrogen-absorbing alloys. Generally, bcc metals and alloys exhibit two plateaus in pressure–composition isotherms, but the lower plateau is far below atmospheric pressure at room temperature. Many efforts have been made to increase the hydrogen capacity and raise the equilibrium pressure of this lower plateau (Figure 5.55).

5.9.3
AlH$_3$

The potential for using aluminum hydride, AlH$_3$, for vehicular hydrogen storage was explored by Sandrock *et al.* [95]. It was shown that particle size control and doping of AlH$_3$ with small levels of alkali metal hydrides (e.g. LiH) result in accelerated desorption rates. For AlH$_3$–20 mol% LiH, 100 °C desorption kinetics are nearly high enough to supply vehicles. It is highly likely that 2010 gravimetric and

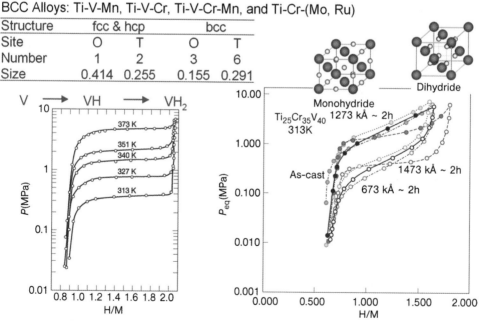

Figure 5.55 Examples of bcc alloys and corresponding pressure–composition isotherms. The bcc structure offers three times more interstitial sites than the fcc and hcp structures.

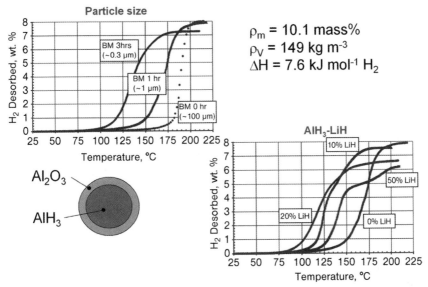

Figure 5.56 Thermal hydrogen desorption from AlH$_3$ stabilized by an oxide layer.

volumetric vehicular system targets (6 wt%H$_2$ and 0.045 kg L^{-1}) can be met with on-board AlH$_3$. However, a new, low-cost method of off-board regeneration of spent Al back to AlH$_3$ is needed (Figure 5.56).

5.9.4
Metal Hydrides with Short H-H-Distance

First-principle studies on the total energy, electronic structure and bonding nature of RNiIn (R$_5$La, Ce and Nd) and their saturated hydrides (R$_3$Ni$_3$In$_3$H$_4$ = RNiInH$_{1.333}$) are performed using a full-potential linear muffin-tin orbital approach. This series of phases crystallizes in a ZrNiAl-type structural framework. When hydrogen is introduced in the RNiIn matrix, anisotropic lattice expansion is observed along [001] and lattice contraction along [100]. In order to establish the equilibrium structural parameters for these compounds, we have performed force minimization and also volume and c/a optimization. The optimized atomic positions, cell volume and c/a ratio are in very good agreement with recent experimental findings. From the electronic structure and charge density, charge difference and electron localization function analyses, the microscopic origin of the anisotropic change in lattice parameters on hydrogenation of RNiIn has been identified. The hydrides concerned, with their theoretically calculated interatomic H–H distances of 1.57 Å, violate the '2 Å rule' for H–H separation in metal hydrides. The shortest internuclear Ni–H separation is almost equal to the sum of the covalent radii. H is bonded to Ni in an H–Ni–H dumb-bell-shaped linear array, with a character of NiH$_2$ subunits. Density of states, valence charge density, charge transfer plot and electron localization

RTInH$_{1.333}$ (R = La, Ce, Pr, or Nd; T = Ni, Pd, or Pt)

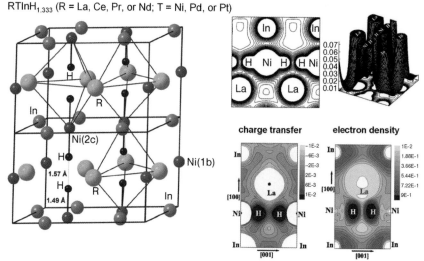

Figure 5.57 RNiInH$_{1.333}$ structure and charge density distribution.

function analyses clearly indicate significant ionic bonding between Ni and H and weak metallic bonding between H–H. The paired and localized electron distribution at the H site is polarized towards La and In, which reduces the repulsive interaction between negatively charged H atoms. This could explain the unusually short H–H separation in these materials. The calculations show that all of these materials have a metallic character [96] (Figure 5.57).

5.9.5
MgH$_2$ with a New Structure

The comparison of the structure of TiO$_2$ with that of MgH$_2$ shows that the anatase-type structure has not yet been observed for MgH$_2$ [97]. However, theoretical calculations of the assumed anatase structure exhibit a destabilized hydride compared with α-MgH$_2$ (Figure 5.58).

5.9.6
Destabilization of MgH$_2$ by Alloy Formation

Alloying with Si destabilizes the strongly bound hydrides LiH and MgH$_2$ [98]. For the LiH–Si system, an Li$_{2.35}$Si alloy forms upon dehydrogenation, causing the equilibrium hydrogen pressure at 490 °C to increase from approximately 5×10^{-5} to 1 bar. For the MgH$_2$–Si system, Mg$_2$Si forms upon dehydrogenation, causing the equilibrium pressure at 300 °C to increase from 1.8 to >7.5 bar.

Thermodynamic calculations indicate equilibrium pressures of 1 bar at approximately 20 °C and 100 bar at approximately 150 °C. These conditions indicate that the MgH$_2$–Si system, which has a hydrogen capacity of 5.0 mass%, could be practical for

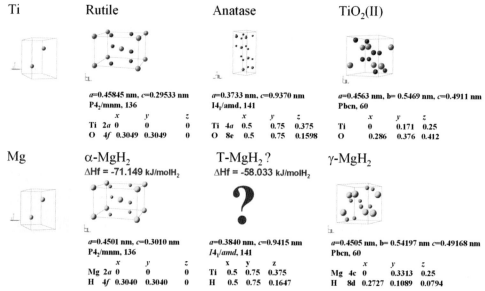

Figure 5.58 Comparison of the structures of TiO_2 and MgH_2 [98].

hydrogen storage at reduced temperatures. The LiH–Si system is reversible and can be cycled without degradation. Absorption–desorption isotherms, obtained at 400–500 °C, exhibited two distinct flat plateaus with little hysteresis. The plateaus correspond to the formation and decomposition of various Li silicides. The MgH_2–Si system was not readily reversible. Hydrogenation of Mg_2Si appears to be kinetically limited because of the relatively low temperature, <150 °C, required for hydrogenation at 100 bar. These two alloy systems show how hydride destabilization through alloy formation upon dehydrogenation can be used to design and control equilibrium pressures of strongly bound hydrides (Figure 5.59).

5.9.7
Ammonia Storage

Metal–ammine complexes represent a promising new solid form of storage. Using $Mg(NH_3)_6Cl_2$ as an example, it was shown by Christensen et al. that it can store up to 9.1 wt% hydrogen in the form of ammonia [99]. The storage is completely reversible and by combining it with ammonia decomposition catalysts, hydrogen can be delivered at temperatures below 650 K. The complexes can be handled easily and safely in air (Figure 5.60).

5.9.8
Borazane

Chemical hydrides with empirical formula BH_2NH_2 and $B_xN_xH_y$ have been probed as high-capacity hydrogen storage materials for a long time [100]. Earlier, H_2 gas-

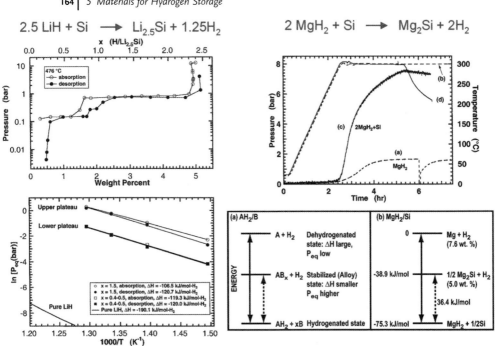

Figure 5.59 Hydrogen desorption from LiH + Si and MgH$_2$ + Si.

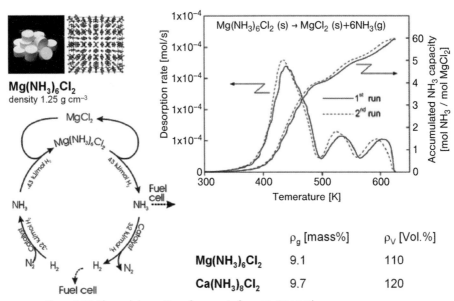

	ρ_g [mass%]	ρ_V [Vol.%]
Mg(NH$_3$)$_6$Cl$_2$	9.1	110
Ca(NH$_3$)$_8$Cl$_2$	9.7	120

Figure 5.60 Thermal desorption of ammonia from Mg(NH$_3$)$_6$Cl$_2$.

$H_3BNH_3(s) \rightarrow H_3BNH_3(l)$ 114°C
$H_3BNH_3(l) \rightarrow H_2BNH_2(s) + H_2(g)$ 137°C
7.4 mass% H_2, $\Delta H = -21.7$ kJ mol^{-1}
$x\, H_2BNH_2(s) \rightarrow [H_2BNH_2]_x(s)$ 125°C
$[H_2BNH_2]_x(s) \rightarrow [HBNH]_x(s) + H_2(g)$ 155°C
$[HBNH]_x(s) \rightarrow B_3N_3H_6 +$ others
$[HBNH]_3(s) \rightarrow BN + 3H_2(g)$ 500°C

Reversibility?

$LiBH_4 + NH_4Cl \rightarrow H_3BNH_3 + LiCl + H_2$

$\Delta H_f\,(H_3BNH_3) = -178$ kJ mol^{-1}

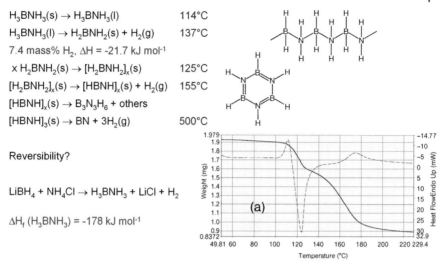

Figure 5.61 Thermal desorption of hydrogen from borazane.

generating formulations were prepared using amine boranes (or their derivatives), mixed and ball-milled together with a reactive heat-generating compound, such as LiAlH$_4$, or a mixture, such as NaBH$_4$ and Fe$_2$O$_3$, until a uniform mixture was obtained. Upon ignition, the heat-generating compound in the mixture reacted and the energy released pyrolyzed the amine borane(s) forming boron nitride (BN) and H$_2$ gas. A Nichrome heating wire was used to initiate a self-sustaining reaction within these gas-generating compounds. Ammoniaborane (H$_3$BNH$_3$), the simplest stable AB, has been used in these gas generators and, more recently, in fuel cell applications. Another stable AB used in the gas generators was diborane diammoniate, H$_2$B(NH$_3$)$_2$BH$_4$ [101].

Ammoniaborane has a very high hydrogen content (about 19.6 wt%) and a volumetric energy density of about 4.94 kW h L^{-1}, far surpassing that of liquid hydrogen (2.36 kW h L^{-1}). At room temperature and pressure, ammoniaborane is a white crystalline solid, stable in water and ambient air. Although, over the years, several synthetic procedures have been developed to prepare this compound, the preparation costs are still high ($\sim$\$10 g^{-1}) and, unlike other AB complexes, ammoniaborane is not made commercially in large quantities [102]. This is due to difficulties involved in its preparation and especially in its purification (Figure 5.61).

References

1 Leung, W.B., March, N.H. and Motz, H. (1976) *Physics Letters*, **56**, 425–426.
2 Silvera, I.F. (1980) *Reviews of Modern Physics*, **52**, 393.
3 Wigner, E. and Huntington, H.B. (1935) *Journal of Chemical Physics*, **3**, 764.
4 Asheroft, N.W. (1968) *Physical Review Letters*, **21**, 1748.

5 Matek, W., Muhs, D., Wittel, H. and Becker, M. (1994) Roloff/Matek Maschinenelemente, Viewegs Fachbücher der Technik.
6 Dynetek Europe, Ratingen, http://www.dynetek.de.
7 Teitel, R.J. (1981) *Microcavity Hydrogen Storage*, Final Progress Report, Brookhaven National Laboratory, New York, BNL 51439 UC-94 d, pp. 1–40.
8 Barton, J.L. and Morain, M. (1970) *Journal of Non-Crystalline Solids*, **3**, 115–126.
9 Rambach, G. and Hendricks, C. (1996) Proceedings of the US DOE Hydrogen Program Review, Miami, Vol. 2, pp. 765–772
10 Flynn, T.M. (1992) Liquification of gases, in *McGraw-Hill Encyclopedia of Science & Technology*, Vol. 10, 7th edn, McGraw-Hill, New York, pp. 106–109.
11 NASA/NIST databases, http://www.inspi.ufl.edu/data/h_prop_package.html.
12 von Ardenne, M., Musiol, G. and Reball, S. (1990) *Effekte der Physik*, Verlag Harry Deutsch, pp. 712–715.
13 London, F. (1930) *Zeitschrift fur Physik*, **63**, 245; *Zeitschrift fur Physikalische Chemie*, **11**, 222.
14 Brunauer, S., Emmett, P.H. and Teller, E. (1938) *Journal of the American Chemical Society*, **60**, 309.
15 Critchfield, C. and Teller, E. (1938) *Physical Review*, **53**, 812–818.
16 Gregg, S.J. and Sing, K.S.W. (1967) *Adsorption, Surface Area and Porosity*, Academic Press, New York.
17 Rzepka, M., Lamp, P. and de la Casa-Lillo, M.A. (1998) *Journal of Physical Chemistry B*, **102**, 10849.
18 Ma, Y., Xia, Y., Zhao, M., Wang, R. and Mei, L. (2001) *Physical Review B: Condensed Matter*, **63**, 115422.
19 Beebe, R.A., Biscoe, J., Smith, W.R. and Wendell, C.B. (1947) *Journal of the American Chemical Society*, **69**, 95.
20 Züttel, A., Sudan, P., Mauron, Ph., Kyiobaiashi, T., Emmenegger, Ch. and Schlapbach, L. (2002) *International Journal of Hydrogen Energy*, **27**, 203.
21 Nijkamp, M.G., Raaymakers, J.E.M.J., van Dillen, A.J. and de Jong, K.P. (2001) *Applied Physics A: Materials Science and Processing*, **72**, 619.
22 Weitkamp, J., Fritz, M. and Ernst, S. (1995) *International Journal of Hydrogen Energy*, **20**, 967–970.
23 Langmi, H.W., Walton, A., Al-Mamouri, M.M., Johnson, S.R., Book, D., Speight, J.D., Edwards, P.P., Gameson, I., Anderson, P.A. and Harris, I.R. (2003) *Journal of Alloys and Compounds*, **356–357**, 710–715.
24 Harris, R., Book, D., Anderson, P. and Edwards P. (2004) *The Fuel Cell Review*, **1**, 17–23.
25 Rosi, N.L., Eckert, J., Eddaoudi, M., Vodak, D.T., Kim, J., O'Keeffe, M. and Yaghi, O.M. (2003) *Science*, **300**, 1127–1129.
26 Huheey, J.E. (1983) *Inorganic Chemistry*, Harper & Row, New York.
27 Reilly, J.J. and Sandrock, G.D. (1980) Metallhydride als Wasserstoff-Speicher, Spektrum der Wissenschaften, pp. 53–59.
28 Pearson, G.R. (1985) *Chemical Reviews*, **85**, 41–49.
29 Mueller, W.M., Blackledge, I.R. and Libowitz, G.G.(eds) (1968) *Metal Hydrides*, Academic Press, New York.
30 Lennard-Jones, J.E. (1932) *Transactions of the Faraday Society*, **28**, 333.
31 Fukai, Y. (1989) *Zeitschrift fur Physikalische Chemie*, **164**, 165.
32 Switendick, A.C. (1979) *Zeitschrift fur Physikalische Chemie*, **117**, 89.
33 Westlake, D.J. (1983) *Journal of the Less-Common Metals*, **91**, 275–292.
34 Rittmeyer, P. and Wietelmann, U.(2002) Hydrides, in *Ullmann's Encyclopedia of Industrial Chemistry*, 5th edn, Vol. A13: High-performance Fibers to Imidazole and Derivatives, VCH Verlag GmbH, Weinheim, pp. 199–226.
35 Miedema, A.R. (1973) *Journal of the Less-Common Metals*, **32**, 117.
36 Van Mal, H.H., Buschow, K.H.J. and Miedema, A.R. (1974) *Journal of the Less-Common Metals*, **35**, 65.

37 Buschow, K.H.J., Kuijpers, F.A., Miedema, A.R. and Van Mal, H.H. (1974) Proceedings of the 11th Rare Earth Resesrch Conference, Vol. 1, pp. 417–429.

38 Nordlander, P., Norskov, J.K. and Besenbacher, F. (1986) *Journal of Physics F: Metal Physics*, **16**, 1161–1171; Norskov, J.K. and Besenbacher, F. (1987) *Journal of the Less-Common Metals*, **130**, 475–490.

39 Griessen, R. and Driessen, A. (1984) *Physical Review B: Condensed Matter*, **30**, 4372–4381.

40 Griessen, R. (1988) *Physical Review B: Condensed Matter*, **38**, 3690–3698.

41 Moruzzi, V.L., Janak, J.F. and Williams, A.R. (1978) *Calculated Electronic Properties of Metals*, Pergamon Press, New York.

42 Griessen, R. and Riesterer, T. (1988) Heat of formation models. in *Hydrogen in Intermetallic Compounds I: Electronic, Thermodynamic and Crystallographic Properties, Preparation*, Vol. 63 (ed. L. Schlapbach), Springer Series Topics in Applied Physics, Springer, Berlin, pp. 219–284.

43 Yvon, K. (1998) *Chimia*, **52**, 613–619.

44 Griessen, R. (2003) Lecture Script Science and Technology of Hydrogen, Vrije Universiteit Amsterdam.

45 Lee, T.D. and Yang, C.N. (1952) *Physical Review*, **87**, 404 and 410.

46 Lacher, J.R. (1937) *Proceedings of the Royal Society of London Series A*, **161**, 525; *Proceedings of the Cambridge Philosophical Society*, **34**, 518.

47 Fauroux, J.C. and Teichner, S.J. (1966) *Bulletin de la Société Chimique de France*, **9**, 3014–3016.

48 Wiberg, E. and Amberger, E. (1971) *Hydrides of the Elements of Main Groups I–IV*, Elsevier, Amsterdam.

49 James, B.D. and Wallbridge, M.G.H. (1970) *Progress in Inorganic Chemistry*, **11**, 99–231.

50 Miwa, K., Ohba, N., Towata, S., Nakamori, Y. and Orimo, S. (2004) *Physical Review B: Condensed Matter*, **69**, 245120.

51 Orimo, S., Nakamori, Y. and Züttel, A. (2004) *Materials Science and Engineering B*, **108**, 51–53.

52 Ashby, E.C., Schwartz, R.D. and James, B.D. (1970) *Inorganic Chemistry*, **9**, 325.

53 Ashby, E.C. (1966) *Advances in Inorganic Chemistry and Radiochemistry*, **8**, 283–338.

54 Bogdanović B., Brand, R.A., Marjanović A., Schwickardi, M. and Tölle, J. (2000) *Journal of Alloys and Compounds*, **302**, 36.

55 Dymova, T.N., Eliseeva, N.G., Bakum, S.I. and Dergachev, Yu.M. (1974) *Doklady Akademii Nauk SSSR*, **215**, 1369.

56 Dymova, T.N., Dergachev, Yu.M., Sokolov, V.A. and Grechanaya, N.A. (1975) *Doklady Akademii Nauk SSSR*, **224**, 591.

57 Bogdanović B. and Schwickardi, M.J. (1997) *Journal of Alloys and Compounds*, **253**, 1.

58 Bogdanović B., Brand, R.A., Marjanović A., Schwickardi, M. and Tölle, J. (2000) *Journal of Alloys and Compounds*, **302**, 36.

59 Jensen, C.M., Zidan, R.A., Mariels, N., Hee, A.G. and Hagen, C. (1999) *International Journal of Hydrogen Energy*, **24**, 461.

60 Zidan, R.A., Takara, S., Hee, A.G. and Jensen, C.M. (1999) *Journal of Alloys and Compounds*, **285**, 119.

61 Zaluska, A., Zaluski, L. and Ström-Olsen, J.O. (2000) *Journal of Alloys and Compounds*, **298**, 125.

62 Sandrock, G., Gross, K., Thomas, G., Jensen, C., Meeker, D. and Takara, S. (2002) *Journal of Alloys and Compounds*, **330–332**, 696.

63 Jensen, C.M. and Gross, K.J. (2001) *Applied Physics A: Materials Science and Processing*, **72**, 213.

64 Gross, K.J., Thomas, G.J. and Jensen, C.M. (2002) *Journal of Alloys and Compounds*, **330–332**, 683.

65 Sandrock, G., Gross, K. and Thomas, G. (2002) *Journal of Alloys and Compounds*, **339**, 299.

66 Sandrock, G., Gross, K., Thomas, G., Jensen, C., Meeker, D. and Takara, S. (2002) *Journal of Alloys and Compounds*, **330–332**, 696–701.

67. Gross, K.J., Sandrock, G. and Thomas, G. (2002) *Journal of Alloys and Compounds*, **330–332**, 691.
68. Thomas, G.J., Gross, K.J. and Yang, N. (2002) *Journal of Alloys and Compounds*, **330–332**, 702.
69. Sun, D., Kiyobayashi, T., Takeshita, H.T., Kuriyama, N. and Jensen, C.M. (2002) *Journal of Alloys and Compounds*, **337**, L8.
70. Balema, V.P., Dennis, K.W. and Pecharsky, V.K. (2000) *Chemical Communications*, 1665.
71. Sandrock, G., Gross, K. and Thomas, G. (2002) *Journal of Alloys and Compounds*, **339**, 299.
72. Sullivan, E.A. (1995) *Kirk-Othmer Encyclopedia of Chemical Technology*, Vol. 13, John Wiley & Sons, Inc., New York, p. 606.
73. Schlesinger, H.J. and Brown, H.C. (1940) *Journal of the American Chemical Society*, **62**, 3429–3435.
74. Schlesinger, H.J., Brown, H.C., Hoekstra, H.R. and Ra, L.R. (1953) *Journal of the American Chemical Society*, **75**, 199–204.
75. Goerrig, D. (1958) German Patent 1,077,644.
76. Schrauzer, G.N. (1955) *Naturwissenschaften*, **42**, 438.
77. Liard, S.J. and Ucko, D.A. (1968) *Inorganic Chemistry*, **7**, 1051.
78. Lipscomb, W.N. (1963) *Boron Hydrides*, Benjamin, New York.
79. Fedneva, E.M., Alpatova, V.L. and Mikheeva, V.I. (1964) *Russian Journal of Inorganic Chemistry*, **9**, 826–827.
80. Stasinevich, D.S. and Egorenko, G.A. (1968) *Russian Journal of Inorganic Chemistry*, **13**, 341–343.
81. Davis, W.D., Mason, L.S. and Stegeman, G. (1949) *Journal of the American Chemical Society*, **71**, 2775–2781.
82. Harris, P.M. and Meibohm, E.P. (1947) *Journal of the American Chemical Society*, **69**, 1231–1232.
83. Soulié, J.-Ph., Renaudin, G., Cerny, R. and Yvon, K. (2002) *Journal of Alloys and Compounds*, **346**, 200–205.
84. Züttel, A. et al. (2003) *Journal of Alloys and Compounds*, **356–357**, 515–520.
85. Brown, W.G., Kaplan, L. and Wilzbach, K.E. (1952) *Journal of the American Chemical Society*, **74**, 1343–1344.
86. Züttel, A., Wenger, P., Rensch, S., Sudan, P., Mauron, P. and Emmenegger, C. (2003) *Journal of Power Sources*, **5194**, 1–7.
87. Sudan, P., Zuttel, A., Mauron, P., Emmenegger, C., Wenger, P. and Schlapbach, L. (2003) *Carbon*, **41**, 2377–2383.
88. Nakamori, Y., Miwa, K., Ninomiya, A., Li, H., Ohba, N., Towata, S.-I., Züttel, A. and Orimo, S.-I. (2006) *Physical Review B*, **74**, 045126.
89. Steinfeld, A. (2002) *International Journal of Hydrogen Energy*, **27**, 611–619.
90. Aiellot, R., Matthews, M.A., Reger, D.L. and Collins, J.E. (1998) *International Journal of Hydrogen Energy*, **23**, 1103–1108.
91. Liu, B.H., Li, Z.P. and Suda, S. (2006) *Journal of Alloys and Compounds*, **415**, 288–293.
92. Chen, P., Xiong, Z.R., Luo, J., Lin, J. and Tan, K.L. (2003) *Journal of Physical Chemistry B*, **107**, 10967–10970.
93. Chen, P., Xiong, Z.R., Luo, J., Lin, J. and Tan, K.L. (2002) *Nature*, **420**, 302–304.
94. Akiba, E. and Okada, M. (2002) Metallic hydrides III: body-centered-cubic solid-solution alloys, *MRS Bulletin*, September, 699–703.
95. Sandrock, G., Reilly, J., Graetz, J., Zhou, W.-M., Johnson, J. and Wegrzyn, J. (2005) *Applied Physics A: Materials Science and Processing*, **80**, 687–690.
96. Vajeeston, P. et al. (2003) *Physical Review B: Condensed Matter*, **67**, 014101.
97. Song, Y., Guo, Z.X. and Yang, R. (2004) *Materials Science and Engineering A – Structural Materials Properties Microstructure and Processing*, **365**, 73–79.
98. Vajo, J.J., Mertens, F., Ahn, C.C., Bowman, R.C. Jr, and Fultz, B. (2004) *Journal of Physical Chemistry B*, **108**, 13977–13983.

99 Christensen, C.H., Sorensen, R.Z., Johannessen, T., Quaade, U.J., Honkala, K.E., Tobias, D., Kohler, R. and Norskov, J.K. (2005) *Journal of Materials Chemistry*, **15**, 4106–4108.

100 Mohajeri, N., Robertson, T. and T-Raissi, A. (2003) FSEC Final Report for Task III-B, Hydrogen Storage in Amine Borane Complexes.

101 Grant, L.R. and Flanagan, J.E. (1983) US Patent 4,381,206.

102 Shore, S.G. and Parry, R.W. (1955) *Journal of the American Chemical Society*, **77**, 6084; Shore, S.G. and Parry, R.W. (1958) *Journal of the American Chemical Society*, **80**, 8; Shore, S.G. and Böddeker, K.W. (1964) *Inorganic Chemistry*, **3**, 914;Mayer, E. (1973) *Inorganic Chemistry*, **12**, 1954–1955.

103 a) http://www.goodfellow.com.
b) Vollrath, F. and Knight, D.P. (2001) *Nature*, **410**, 541–549.

**Part Three
H$_2$ and Hydrogen Vectors Production**

6
Catalyst Design for Reforming of Oxygenates*

Loredana De Rogatis and Paolo Fornasiero

6.1
Introduction

Global energy consumption in the last half century has increased very rapidly and is expected to continue to grow dramatically in the coming years due to population increase, urbanization and rising living standards. The development of new energy strategies that could be economical and environmentally sustainable and be able to meet the demands for a broad range of services (household, commerce, industry and transportation needs) is a major challenge. There is no unique solution which is able to solve all energy-related problems. Indeed, there must be a global strategy which is based on local solutions: each option shows its own advantages, handicaps and social–economic impact. Nowdays, it is generally accepted that diversification of energy sources is essential [1]. Energy security, economic growth and protection of the environment are the national energy policy drivers of any country in the world [2].

The world's current energy systems have been built around the many advantages of fossil fuels. Every aspect of modern existence is made from, powered with or affected by them. Fossil fuels such as oil, coal and natural gas are extremely attractive as energy sources because they are highly concentrated, enabling large amounts of energy to be stored in relatively small volumes, and they are relatively easy to distribute. However, fossil energy sources are non-renewable, being an irreplaceable endowment produced from millennia of biological and geological processes. The inevitability of exhaustion of fossil fuel reserves is becoming increasingly appreciated. Predictions based on extrapolations of energy consumption show that demand will soon exceed supply. No matter how long the fossil fuels may last, their amount is finite. The slowdown in oil production may already be beginning. In 1956, a geophysicist named Marion King Hubbert developed a theory to predict future oil production. He assumed that for any given oil field, production follows a bell-shaped curve. After the discovery of the first well, production quickly ramps up as

---

*A List of Abbreviations can be found at the end of this chapter.

new wells are added, but eventually, as the oil is drained from the underground reservoirs, the production rate hits a peak, after which it begins to decline, eventually returning to zero. What is true of an individual oil field should, Hubbert reasoned, also be true for the entire planet. This does not mean that oil will suddenly run out, but the supply of cheap, conventional oil will drop and prices will rise, perhaps dramatically. Some observers believe that because of the high dependence of most modern industrial transport, agricultural and industrial systems on inexpensive oil, the post-peak production decline and possible severe increases in the price of oil will have negative implications for the global economy. Predictions as to what exactly these negative effects will be vary greatly. More optimistic outlooks, delaying the peak of production to the 2020s or 2030s, assuming that major investments in alternative fuels occur before the crisis, show the price first escalating and then retreating as other types of fuel sources are used as transport fuels and fuel substitution in general occurs. More pessimistic predictions suggest that the peak will occur soon or has already occurred, and predict a global depression and even the collapse of industrial global civilization as the various feedback mechanisms of the global market cause a disastrous chain reaction. However, there is another aspect which has to be taken into account. Extraction energy costs would become higher than the actual energy yield due to increased energy costs for research, deep drilling and lower quality and accessibility of the still available oil storages. Consequently, the era of oil would decline before it ends. At this point the oil would not be used for energetic purposes, but for example as a source of building blocks for the chemical industry, if alternative energy sources exist.

Added to this there is clearly a problem of worldwide energy dependence. Since the fossil fuels were created in specific circumstances where the geological conditions were favorable, the largest deposits of oil, gas and coal tend to be concentrated in particular regions of the globe (e.g. two-thirds of the world's proven oil reserves are located in the Middle East and North Africa), often characterized by political instability in their international relationships. Therefore, the price of oil is subjected to important fluctuations due to economic and geopolitical reasons [3].

The potentially damaging environmental effect of continuous fossil fuel usage is another factor which has to be considered. Although there is considerable disagreement as to whether increased fossil fuel consumption is the primary cause of global climate change (e.g. the Earth's temperature increase and rises in sea levels), there is general agreement that a strong correlation exists between localized and regional air pollution and fossil fuel consumption. The exploitation of fossil fuel resources entails significant health hazards in the course of their extraction, for example coal mining accidents and fires on oil or gas drilling rigs. They can also occur during distribution, for example oil spillages from tankers that pollute beaches and kill wildlife, or on evaporation processs or on combustion, which generate atmospheric pollutants such as sulfur dioxide, carbon monoxide, fine particulate matter, nitrogen oxides, hydrocarbons and very large quantities of carbon dioxide, which contributes to the well-known greenhouse effect.

Despite great efforts in the realization of sophisticated end-of-pipe strategies to reduce air pollution such as three-way catalysts (TWCs) [4, 5], surely the solution to these environmental and also social–economic aspects is to wean ourselves quickly off of fossil fuels.

An important means of mitigating the environmental impacts of current fuel use is to improve the energy conversion efficiency of fuel-based energy supply systems, so that less fuel is required to achieve a given level of energy output [6]. This entails such technologies as the high-efficiency combined cycle gas turbine (CCGT) for electricity generation, the use of combined heat and power (CHP) to enable the waste heat from electricity generation to be usefully employed, and the use of high-efficiency condensing boilers in buildings or high-efficiency engines in vehicles to allow less fuel to be burned for a given level of useful output [1]. Furthermore, significant and promising efficiency improvements can be achieved using an energy conversion device, namely the fuel cell, for the generation of electric power for both electric vehicles and distributed electric power plants [7]. The key advantage is that electricity is produced directly from fuel (e.g. natural gas, ethanol, hydrogen). This means that its efficiency can be higher than in conventional engines. Depending on the type of cell, from 40 to, in some cases, 80% of the energy content of the input fuel is converted to electricity. It also means that there are no emissions of the gaseous pollutants that are associated with combustion processes. If pure hydrogen is used, there are no CO_2 emissions and the only emission, apart from heat, is water vapor. Although still costly, in recent years there have been substantial falls in fuel cell prices and major improvements in performance, trends that lead many to believe that they could soon become competitive with more traditional energy conversion systems.

Although the improvements in the energy efficiency in buildings, appliances and industrial processes offer impressive savings, no plan to reduce greenhouse gas emissions substantially can succeed through increases in energy efficiency alone. Because economic growth continues to boost the demand for energy (e.g. more coal for powering new factories, more oil for fuelling new cars, more natural gas for heating new homes), carbon emissions will continue to climb despite the introduction of more energy-efficient vehicles, buildings and appliances.

We can increase our use of nuclear power, which now supplies some 7% of world primary energy. A major advantage of nuclear power plants, in contrast to fossil fuelled plants, is that they do not emit greenhouse gases. Also, supplies of uranium, the principal nuclear fuel, are sufficient for many decades of supply at current use rates. However, the use of nuclear energy leads to problems arising from the routine emissions of radioactive substances, difficulties with radioactive waste disposal and danger from the proliferation of nuclear weapons material. To these must be added the possibility of major nuclear accidents, which, slthough highly unlikely, could be catastrophic in their effects. Although some of these problems may be amenable to solution in the longer term, such solutions have not yet been fully developed [1].

Continuing concerns about the sustainability of both fossil and nuclear fuel use have been a major catalyst of the renewed interest in renewable energy sources such as biomass in recent decades [8].

Sustainable development within a society demands a sustainable supply of energy resources (that, in the long term, are readily and sustainably available at reasonable cost and can be utilized for all required tasks without causing negative societal impacts) and effective and efficient utilization of energy resources. In this regard, the intimate connection between renewable energy sources and sustainable development comes out. Renewable sources can be considered highly responsive to overall energy policy guidelines and environmental, social and economic goals: they can help to reduce dependence on energy imports or do not create a dependence on energy imports, thereby ensuring a sustainable security of supply. Furthermore, renewable energy sources can help to improve the competitiveness of industries, at least in the long run, and can have a positive impact on regional development and employment. Renewable energy technologies are suitable for off-grid services, serving those in remote areas of the world without having to build or extend expensive and complicated grid infrastructures. They can enhance diversity in energy supply markets, secure long-term sustainable energy supplies and reduce local and global atmospheric emissions. They can also provide commercially attractive options to meet specific needs for energy services (particularly in developing countries and rural areas), create new employment opportunities and offer possibilities for local manufacturing of equipment [9].

A wide variety of technologies are available, or under development, to provide inexpensive, reliable and sustainable energy services from renewables. The stage of development and the competitiveness of those technologies differ greatly. Although often commercially available, most are still at an early stage of development and not technically mature, so they demand continuing research, development and demonstration efforts. Moreover, performance and competitiveness are determined by local physical and socioeconomic conditions.

The interest in the use of biomass, the most versatile non-petroleum renewable resource, has grown rapidly over the last few years [10–12]. Among other aspects, such as the replacement of scarce fossil resources and the generation of alternative markets for agricultural products, environmental considerations turned out to be the key drivers for this development. On the other hand, options for using biomass to produce energy, fuels and materials as substitutes for those currently manufactured from petrochemicals, may be limited because cultivation, extraction and processing of biomass are by no means environmentally neutral. It is often assumed that the carbon dioxide originating from biomass is equivalent to the amount that was previously withdrawn from the atmosphere during the growth period of crops and, therefore, does not contribute to global warming. However, the consumption of fossil fuels and the application of fertilizers and hazardous chemicals for transport and processing of biomass must also be taken into account for a correct life cycle assessment.

The list of plants, by-products and waste materials that can potentially be used as feedstock is almost endless. Major resources in biomass include agricultural crops and their waste by-products, lignocellulosic products such as wood and wood waste, waste from food processing and aquatic plants and algae and effluents produced in the human habitat. Moderately dried wastes such as wood residue, wood scrap and urban garbage can be directly burned as fuel. Energy from water-containing biomass such as sewage sludge, agricultural and livestock effluents and animal excreta is recovered mainly by microbial fermentation [13].

Large-scale biomass production requires agricultural land and might cause adverse environmental effects such as degradation of soils, eutrophication of ground and surface waters and fragmentation of ecosystems. Biomass cultivation for non-food applications might be limited due to the demand for land for other purposes such as infrastructure, housing and recreation.

A frequently expressed concern about shifting agriculture towards the energy sector is the potential impact on food production. Securing a safe and inexpensive food supply is the keystone of agricultural policy in several countries. Many people oppose the use of agricultural land for the production of bioenergy and biobased products on the grounds that, at best, domestic food prices will rise and, at worst, starvation will increase in developing countries that are dependent on agricultural imports. Agricultural production in excess of domestic use and export demands exists in a growing number of countries. To minimize conflicts with global food supply, however, agricultural and municipal waste should be the preferred biomass for energy production. This is possible but more difficult compared with the conversion of crops such as corn and sugarcane that are rich in starch and sugar [14]. Using waste as an energy source both reduces the CO_2 emission and exploits an otherwise unused product.

In this respect, hydrogen is emerging as the energy carrier of the future [15] that can bridge the transition from fossil fuels to renewable energy like biomass. It has been proposed as a convenient way of storing the energy originating from other sources. However, before a hydrogen-fuelled future can become a reality, many complex challenges must be overcome. Before it can be used in, for instance, fuel cell systems, hydrogen needs to be extracted in a clean and efficient way from the other compounds within which it is normally bonded in Nature and this separation requires energy [16, 17]. Moreover, there are problems related to storage technology [18] (see also Chapter 5) and the creation of a distribution and transport network for this new energy carrier.

The production of hydrogen can be achieved by selective reforming of oxygenates such as carbohydrates (e.g. sugars, starch, hemicelluloses and cellulose) and polyols (e.g. methanol, ethanol, ethylene glycol, glycerol and sorbitol). They may be derived from renewable biomass which consists principally of C_6- and C_5-sugars from cellulose (40–50 wt%) and hemicellulose (15–30 wt%). There is a third component, lignin (16–33 wt%), which is a highly cross-linked polymer built of substituted phenols and, together with cellulose and hemicellulose, gives strength to plants. In addition to those components, plants are also able to elaborate energy storage products such as lipids, sugars and starches and other products relatively rich in hydrogen and carbon (terpenes) that are found in essential oils, which are components of resins, steroids and rubber. They offer advantages such as low toxicity, low reactivity and compatibility with the current infrastructure for transportation and storage. Water-soluble biomass-derived oxygenates (including carbohydrates, polysaccharides, furfural and lignin-derived compounds) can be produced from cellulosic biomass by, for instance, enzymatic decomposition, acid hydrolysis and liquefaction processes [19–23]. Bio-oils, produced by fast pyrolysis or liquefaction from biomass, are a mixture of more than 300 highly oxygenated compounds. Glycerol also can be produced from biomass through fermentation of sugars and transesterification of vegetable oils during biodiesel production.

Although technologies have been developed over past 50 years to process petroleum-based feedstock efficiently to generate hydrogen, its production from renewable biomass-derived resources remains a major challenge, because conversion processes often suffer from low hydrogen production rates and/or complex processing requirements [24, 25].

Biomass can be used to produce hydrogen or hydrogen rich gas via different technical pathways such as anaerobic digestion, fermentation, metabolic processing, high-pressure supercritical conversion, gasification and pyrolysis [26]. Compared with other pathways, gasification and pyrolysis appear techno-economically viable at the current stage.

The use of lignocellulosic biomass as a feedstock for hydrogen production via gasification followed by shift conversion has received considerable attention in the past [27, 28]. However, a more interesting and promising approach to the production of hydrogen from biomass involves fast pyrolysis of biomass to generate a liquid product, bio-oil, and reforming of pyrolytic oils so as to produce a gaseous stream rich in hydrogen [24, 29–34]. Bio-oil has a higher energy density than biomass, it can be readily stored and transported and can be used either as a renewable liquid fuel or for chemical production. Bio-oils are thermally unstable and must be egraded if they have to be used as fuels.

Moreover, an integrated process, in which biomass is partially used to produce more valuable materials or chemicals, while the residual fractions are utilized for the generation of hydrogen, may be economically viable in today's energy market.

Although the molecular composition of pyrolysis oil varies significantly with the type of biomass and pyrolysis conditions used (pyrolysis severity and media), its major components are oxygenates belonging to the following groups: acids, aldehydes, alcohols, ketones and substituted furans derived from cellulose and hemicellulose and phenolic and cyclic oxygenates derived from lignin [32, 35–38]. These molecules are not thermally stable at typical reforming temperatures, and some of them may not be stable even at much lower temperatures [36]. Hence there is significant competition between catalytic reforming reactions and thermal decomposition for most oxygenates [32].

Sugars and polyols can be efficiently converted over appropriate catalysts by different reforming processes (e.g. catalytic steam reforming, catalytic partial oxidation, autothermal reforming, aqueous phase reforming) which involve the rupture of C–C, C–O, C–H and O–H bonds and the formation of H–H and new C–O and O–H bonds. Each reaction offers unique opportunites, benefits and disadvantages. Local availability of feedstock, maturity of technology, market applications and demand, policy issues and costs will influence the choice and timing of the various options for hydrogen production.

6.2
Catalyst Design

Heterogeneous catalysts are highly complex materials with respect to both their composition and structure. Most catalysts used in industry consist of one or several catalytically active component(s) in the form of very small particles (typically in the size range of a few nanometers) deposited on the surface of a support (e.g. oxide), a highly porous and thermostable material with a high surface area and suitable mechanical strength. The main purpose of using a support is to achieve optimal dispersion of the catalytically active component(s) and to stabilize it (them) against sintering and hence to increase the catalyst life. Furthermore, in several reactions, the support is not inert and the overall process is actually a combination of two catalytic functions: that of the active component(s) and that of the support.

The choice of a catalyst is based on different criteria such as activity, selectivity and stability. In addition to these fundamental properties, industrial applications require that the catalyst is mechanically and thermally stable, regenerable and inexpensive and possesses suitable morphological characteristics [39].

A high activity will be reflected either in high productivity from relatively small reactors and catalyst volumes or in mild operating conditions, particularly temperature. High selectivity produces high yields of a desired product while suppressing undesirable reactions. The catalyst must be able to sustain the desired reaction at an acceptable rate under conditions of temperature and pressure that are practicable and over prolonged periods. A catalyst with good stability will modify only very slowly under operating conditions. Indeed, it is only in theory that a catalyst remains unaltered during reaction. All catalysts age. The chief causes of deterioration in use are reversible poisoning due to impurities in the reactants (e.g. sulfur) or to the formation of coke deposits, which reduces the number of active sites and hence the reaction rates, and the irreversible physical changes including loss of surface area (sintering) or mechanical failure. The external morphological characteristics of a catalyst (e.g. its form, grain size) must be suited to the corresponding process. For certain catalysts, thermal conductivity and specific heat require consideration. High thermal conductivity of the catalytic mass leads to reduced temperature gradients within the grain, and also in the catalytic bed, by improving heat transfer. When the activity and/or selectivity of catalytic systems have become insufficient, they must be regenerated through a treatment that will restore part or all of their catalytic properties. The mechanical strength must be preserved during successive regenerations. Reproducibility has been indicated as another criterion to be considered. It characterizes the catalyst preparation, which generally takes place in several rather complex stages depending on a large number of variables that are difficult to control simultaneously. The result is that it is essential to verify rapidly that the reproducibility of the preparation is feasible, and also to keep in mind that the formula developed in the laboratory should be extrapolated to the pilot scale and to the industrial scale under acceptable economic conditions [40, 41].

Even when a catalyst possesses all the properties and characteristics discussed above, there still remains another requirement: it must withstand comparison with competitive catalysts or processes with equivalent functions from the point of view of cost; or at least its cost should not place too heavy a burden on the economics of the process in which it will be used.

None of the above properties and characteristics act independently. When one among them is changed with a view to improvement, the others are also modified and not necessarily in the direction of an overall improvement. As a result, industrial catalysts are never ideal. Fortunately, however, the ideal is not altogether indispensable. Certain properties, such as activity and reproducibility, are always necessary, but selectivity, for example, has hardly any meaning in reactions such as ammonia synthesis, and the same holds true for thermal conductivity in an isothermal reaction. Stability is always of interest but becomes less important in processes that include continuous catalyst regeneration. Regenerability must be optimized in this case.

Catalyst selection for a commercial process is not based on one reaction or one parameter, but rather on many, which are different for each catalyst and which need to be assessed before a catalyst can be proposed as an attractive option for a new or an existing commercial process [42].

The compositional and structural complexity of these systems is their principal advantage. It is this feature which allows surface properties to be tuned in order to optimize selectivity and activity with respect to a specific reaction. At the same time, complexity is the reason of the fact that at a molecular level, an understanding of reaction kinetics at heterogeneous and porous interfaces is difficult to achieve. Consequently, the reaction kinetics on their surfaces depend sensitively on a number of structural and chemical factors including the particle size and structure, the support and the presence of poisons and promoters.

It is clear that the need to formulate new catalysts, which exhibit enhanced performance with respect to those currently employed for specific reactions, represents a difficult undertaking. The goal, therefore, is not an ideal catalyst but the optimum, which may be defined by economic feasibility studies concerning not only the catalyst but also the rest of the process [41]. Depending on the use and the economic competition, optimization studies establish a hierarchy among the properties and characteristics of a catalyst.

Considering this complexity, the approach to an optimum catalyst is mostly an experimental procedure advancing step-by-step through trial and error. The discovery and optimization of catalytic processes can be effectively approached using chemical intuition and experience with related catalytic processes [39]. However, recent and innovative steps forward in the design and working principles result from a detailed atomic-scale understanding. Indeed, nowadays, researchers start to simulate *ab initio* nanoscale materials to prediction the structural, morphological, compositional, electronic and chemical aspects of a catalyst, with the final goal of identifying specific guidelines for improved reactivity, selectivity and stability.

Once the potential catalytic materials have been selected and synthesized, they must be tested in catalytic reactors and fully characterized by various physical, chemical and spectroscopic techniques [43]. For the case of developing catalysts with improved selectivity, an understanding of the factors controlling the selectivity is achieved by elucidating the reaction pathways that operate over the catalyst at various reaction conditions. With the advent of surface science techniques in past decades, the promise was perceived of turning an increased molecular level understanding of reaction mechanisms and surface sites into principles of catalyst design. Surface science alone has not proven to be sufficient for this purpose. Over the past decade, the rise of powerful, computationally efficient theoretical methods has shown promise, not just for identifying catalytic intermediates and reaction pathways accessible to experiments, but of providing quantitative predictions of energetics for elementary reaction processes not easily accessible experimentally.

6.2.1
Impregnated Catalysts: the Role of Metal, Support and Promoters

The severe working conditions often encountered in an H_2 production process, such as high temperature and high space velocity, combined with the necessity for a long catalyst lifetime, impose the development of an appropriate synthetic procedure to stabilize the catalyst. The reforming activity and product distribution over supported metal catalysts depend on the choice of metal and its content, the presence of promoters, the type of support and method of catalyst preparation.

Supported metal catalysts are usually prepared through impregnation of a porous support with a solution of the metal (or metal oxide) precursor followed by suitable chemical and thermal treatment. The most attractive feature of this route is its simplicity in practical execution both in the laboratory and on an industrial scale. Although the practical execution is seemingly simple, the fundamental phenomena involved are extremely complex. The interaction between the metal precursor and the support is often rather weak, thereby allowing redistribution of the active phase over the support body during drying. Since evaporation of the solvent takes place at the exterior of the support particles, capillary flow of solution to its exterior surface may take place, thereby causing the production of so-called egg-shell catalysts, often with poor dispersion of the active phase.

Reforming reactions should preferably be carried out at as low as possible temperatures and at atmospheric pressure to reduce the operating costs. The catalyst should provide high selectivity to H_2 formation and inhibit the formation of by-products which are different according to the type of reaction and the nature of the reactants (e.g. ethanol, ethylene glycol, glycerol, sorbitol, glucose). Furthermore, it should not be poisoned by impurities produced during biomass fermentation.

Transition metals are used extensively as reforming catalysts and the variation in the catalytic activity can be determined by the differences in the strength of the adsorbate–surface interaction with various metals. One of the fundamental properties of a metal surface is in fact its ability to bond or to interact with surrounding atoms and molecules. The bonding ability determines the state of the metal surface when exposed to a gas or liquid and it determines the ability of the surface to act as a catalyst. During catalysis, the surface forms chemical bonds to the reactants and it helps in this way the breaking of intramolecular bonds and the formation of new bonds.

Among the many catalysts available, Ni-based systems are universal and widely used in the laboratory and industry in several reforming processes, mainly due to the low cost of Ni. Although it shows high reforming activity, the main problem with Ni is that the rate of C–C bond formation is high, promoting the rapid growth of a carbon deposit, which leads to catalyst deactivation and reactor plugging. Catalyst coking depends strongly on the reaction conditions (e.g. steam quantities in the steam reforming reaction). Moreover, Ni-based catalysts easily deactivate to sintering the metallic and support phase. To overcome this problem, high metal loadings are commonly used: commercial Ni-based systems may contain up to 25 wt% of metal. It is clear that the same approach cannot be used for precious metals.

Among noble metals such as Pd, Pt, Ru and Rh, Rh is the most suitable choice to break the C–C bonds involved in reforming processes [44]. Rh is known also to activate the C–H bonds [45] and it shows good resistance towards deactivation due to coke deposition [46]. The high catalytic activity on Rh permits a low metal loading (0.1–1 wt%), which is a significant economic advantage for the commercialization of such catalysts and low operating temperatures. However, it is rare and prohibitively expensive. It has limited water gas shift (WGS) activity. In comparison, Pt has a relatively higher WGS activity and also has good thermal stability. Among the noble metals, Ru is the cheapest and hence a catalyst based on Ru is expected to be far less expensive.

The physical and chemical properties required by reforming processes dictate also the choice, design and manufacture of the support. Careful choice of materials, synthesis and appropriate calcinations are necessary to achieve high catalytic performance. The general criteria for a good support are as follows: (i) it needs to be mechanically strong with a high surface area; (ii) it should exhibit high impact and abrasion/attrition resistance; and (iii) it should not deactivate. Moreover, in many applications the surface properties, for example acidity and chemical composition, are important. The nature of the support plays a key role in determining the selectivity to H_2 formation. A bifunctional mechanism for some reforming reactions such as steam reforming has been proposed: oxygenates to be reformed would be activated on the metal particle while the water would be activated on the support as hydroxyl groups (see Section 6.4).

Alumina is one the best known catalyst support materials and is frequently used in both research and industrial applications not only for its relatively high surface area on which active metal atoms/crystallites can spread out as reaction sites, but also for its enhancement of productivity and/or selectivity through metal–support interaction and spillover phenomena.

The use of promoters in formulating a catalyst is often critical to the performance. Promoters can provide an extra edge to the performance of a catalyst by improving its operation. Promoters can be many and varied. They can be additives to stabilize a particular oxidation state of the catalyst, to optimize a particular phase or structure of the active ingredient(s), to provide additional pathways for facilitating reactions, to alter the concentration of a particular oxidation state in the active phase of the catalyst, to increase the activity or selectivity (chemical), to preserve mechanical strength and limit sintering (structural), to increase the surface area of the catalyst or to alter the texture of a catalyst. Structural promoters may be added to control the porosity of the final catalyst, to maximize the exposure of the active sites within a catalyst and/or to control crystallite size. For example, promoters such as La, Ca, K and their combinations were reported to show a positive effect on the activity and stability of nickel catalysts either by decreasing the metal sintering and/or coke deposition or by increasing the metal dispersion, resulting in intimate contact between the reacting gases and the catalyst [47–51]. Acidic oxides might be added to neutralize basic oxide components within a catalyst. Some oxides can exert electronic control on the active sites, modifying impacting adsorption of reactants. Low levels of poisons (such as sulfur) can be added deliberately to lower the effectiveness of very active sites, which might lead to undesired secondary reactions.

A highly interesting class of catalysts is represented by bimetallic systems, which in many important catalytic processes show improved activity or selectivity compared with catalysts involving only one metal. Understanding their better performance is still a challenge. One metal can tune and/or modify the catalytic properties of the other metal as the result of both electronic or/and structural effects. Several mechanisms for synergism can be proposed, but it is difficult to assess their relative importance. It is clear that each metal can play a very important role in proper circumstances [41].

6.2.2
Emerging Strategies: Embedded Catalysts

Recently, great attention has been dedicated to the development of novel synthetic procedures for the preparation of nanostructured catalysts with superior activity and thermal stability to those currently available.

The solid-phase crystallization (SPC) technique is one of these proposed approaches. The SPC strategy is based on the preparation of a crystalline oxide precursor (generally perovskite and hydrotalcite compounds) by a sol–gel or coprecipitation method in the presence of ions of the active metal. After calcination, the material will homogeneously contain species of the active metal dispersed inside the bulk. Reduction at high temperature leads to the migration of the metal atoms to the surface to form small metallic particles which are homogeneously dispersed. It is expected that the metal–support interaction is stronger than that obtainable by the usual impregnation or deposition methods. Using SPC, active and thermally stable catalysts have been produced for reforming reactions involving methane [52–63] and methanol [64].

Another synthetic route is represented by the microemulsion technique, which shows interesting advantages related to the possibility of controlling properties such as particle size, morphology and size distribution. Nanosized particles with a narrow size distribution can often be achieved with consequent benefits for catalytic reactions and also for support materials where a high surface area and thermal stability are required. Although this synthetic strategy is fairly successful in producing active and stable catalysts, it usually requires expensive reagents and large quantities of them, which have subsequently to be removed during post-synthesis treatments.

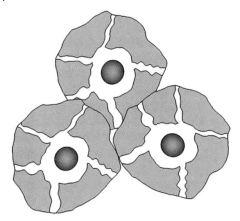

Figure 6.1 Ideal structure of an embedded catalyst.

An innovative and elegant approach, recently adopted for instance by Budroni and Corma [65], is based on the incorporation of the metal nanoparticles into an open shell of support (oxide) in order to limit the sintering of the particles at high temperatures, as depicted in Figure 6.1.

The porous nature of the inorganic matrix prevents the total occlusion of the particles, favoring accessibility of the catalytic sites to the reactants. Adopting a modified sol–gel procedure, Au nanoparticles embedded in silica have been prepared [65]. In particular, the synthesis involves the formation of a three-component metal–organic structure composed of Au nanoparticles that are capped with alkanethiols and partially functionalized groups and polymerized with tetraethyl orthosilicate. Alkanethiols reduce the tendency of Au particles to aggregate [66]. Moreover, the particles reveal a narrow size distribution centered at approximately 2 nm. The reduced dimensions of Au particles together with interaction strength with the support is a important factor to convert the inert gold into highly active catalyst [67–70]. The material obtained in this way shows high activity in the oxidation of CO and in the WGSR [65].

Using the same catalyst design, our group has developed a simple and low-cost strategy for the synthesis of efficient and stable embedded Rh-based catalysts for the catalytic partial oxidation of methane and steam reforming of ethanol [71–74]. The method offers the possibility of modulating the nature of the support and its texture, and the inclusion of extra components (e.g. ceria-based mixed oxides as promoters) in the catalyst formulation, thereby leading a strong flexibility to the approach.

The catalysts can be obtained by a coprecipitation method consisting of two steps (Figure 6.2). In the first step, a stable suspension of protected metal nanoparticles is obtained according to the method reported by Schulz and co-workers [75–77]. The metal particles are prepared in the presence of a highly water-soluble ionic surfactant which is able, due to its nature, to modulate the particle size and to prevent their aggregation. Modifying parameters such as pH, temperature and surfactant concentration, it is possible to tune the metal particle size [71]. Moreover, the role of the

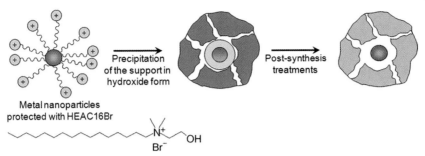

Figure 6.2 Embedded catalyst design developed by our group.

surfactant is also to control the encapsulation of preformed metal particles in the second part of the synthesis. During this step, growth of the porous oxide layers around metal nanoparticles also takes place. In order to remove organic materials, post-synthesis treatments are required, leading to the final catalyst.

The embedded catalyst synthesized in this way shows higher thermal stability than the traditional catalyst obtained by incipient wetness impregnation.

6.3
Reforming Reactions: Process Principles

Catalytic conversion of biomass-derived feedstocks, such as ethanol, sugars, sugar alcohols, polyols and less refined hemicellulose or cellulose, is becoming more and more important with a view to sustainable hydrogen production through the use of renewable sources. Because of the relatively high cost of current technologies for utilizing biomass feedstocks compared with fossil fuels, however, biomass to hydrogen conversion processes must be highly efficient in terms of high hydrogen productivity and selectivity.

Concerning the reforming technology, there are four main processes to consider: catalytic steam reforming (CSR), catalytic partial oxidation (CPO), autothermal reforming (ATR) and aqueous phase reforming (APR). Despite their advantages, each of these processes has barriers such as design, fuel and operating temperature. The choice of reforming processes depends on several factors such as the operating characteristics of the application, the nature of the fuel and thermal management. A brief survey of these technologies is presented in the following.

6.3.1
Catalytic Steam Reforming

CSR has been applied mainly to hydrocarbons, especially methane. It is the most mature and the best established technology for hydrogen production. The process is endothermic and therefore a significant fuel penalty must be paid to provide the needed heat.

Steam reforming for an oxygenated compound with a generic chemical formula $C_nH_mO_k$ proceeds according to the following equation:

$$C_nH_mO_k + (n-k)H_2O \rightleftharpoons nCO + \left(n + \frac{m}{2} - k\right)H_2 \qquad (6.1)$$

The water gas shift reaction (WGSR) (Equation 6.2), which occurs simultaneously, constitutes an integral part of the reforming process:

$$nCO + nH_2O \rightleftharpoons nCO_2 + nH_2 \qquad (6.2)$$

If both reactions (Equations 6.1 and 6.2) go to completion, the overall process can be expressed by Equation 6.3:

$$C_nH_mO_k + (2n-k)H_2O \rightarrow nCO_2 + \left(2n + \frac{m}{2} - k\right)H_2 \qquad (6.3)$$

According to the stoichiometry of Equation 6.3, the maximum yield of hydrogen that can be obtained by reforming/WGS (corresponding to the complete conversion of organic carbon to CO_2) equals $2 + \frac{m}{2n} - \frac{k}{n}$ moles per mole of carbon in the feed material. In reality, this yield is always lower because both the steam reforming and WGS reactions are reversible. Furthermore, in some cases, undesirable by-products can be formed in appreciable quantities due to the low selectivity of the catalyst. This is the case, for instance, with methane. High steam-to carbon ratios shift these two equilibrium reactions towards hydrogen production. In addition, at the typical operating temperatures of a steam reformer, oxygenate molecules undergo homogeneous (gas-phase) thermal decomposition, and also cracking reactions (Equation 6.4) on the acidic sites of the support of the catalyst:

$$C_nH_mO_k \rightarrow C_xH_yO_z + (H_2,\ CO,\ CO_2,\ CH_4,\ldots) + coke \qquad (6.4)$$

These reactions, which occur in parallel with reforming, produce carbonaceous deposits resulting in catalyst deactivation and in lower yields of hydrogen.

Few of the primary products of biomass pyrolysis are thermally stable at the typical temperatures of the hydrocarbon reformer. Hence there is significant competition between catalytic reforming reactions (Equation 6.3) and thermally induced cracking decomposition (Equation 6.4).

Steam reforming is catalyzed by metal-based catalysts. The mechanism of steam reforming of oxygenates has not been sufficiently elucidated to date [78, 79], hampering the development of efficient and stable catalysts. Several studies on simple model compounds (see Section 6.4) suggest that, during the process, organic molecules adsorb dissociatively on metal sites, whereas water molecules are adsorbed on the support surface. Hydrogen can be produced through two chemical routes: the dehydrogenation of adsorbed organic molecules and the reaction of adsorbed organic fragments with hydroxyl groups which are able to move from the support to the metal/support interfaces, leading to carbon oxides production. Carbon deposit formation on the catalyst surface also takes place. With respect to hydrocarbons, the latter unwanted effect is enhanced in the case of oxygenated organic molecules due to their higher unsaturation, molecular weight and aromaticity. Biomass-derived liquids show generally greater reactivity than hydrocarbons due to the presence of C–O bonds in the molecules. At elevated temperatures, however, they also show a greater tendency to form carbonaceous residues due to the large size and thermal instability of the constitutive molecules. For this reason, the reforming of biomass-derived compounds requires suitable conditions that allow for good contact of the organic molecules with the catalyst and that minimize the formation of, or alternatively facilitate the removal, by steam gasification, of coke deposits from the catalyst surface.

Employing process conditions similar to those used for steam reforming of natural gas (e.g. fixed-bed reactors, temperatures in the 800–900 °C range) has been demonstrated to be inadequate for processing thermally unstable biomass liquids [29]. The most important problem is represented by coke formation, especially in the upper layer of the catalyst bed and in the reactor freeboard, that limits the operation time (e.g. 3–4 h on commercial Ni-based catalysts) and requires a long regeneration process for the catalyst (e.g. 6–8 h on commercial Ni-based catalysts).

In this respect, some improvements have been achieved in replacing a fixed-bed reactor with a fluidized-bed reformer. An advantage of a fluidized-bed system is that, in a well-mixed regime, the feedstock is in contact with all of the catalyst particles, not just with its upper layer, as is the case in the fixed-bed mode. Furthermore, carbon deposits on the particles are better exposed to steam and, therefore, can be gasified more quickly. Consequently, a fluidized-bed reactor should extend the duration of the reforming activity of the catalyst and shorten the regeneration cycle. In addition, if needed, the catalyst regeneration can be carried out continuously in a two-reactor system. However, it should be noted that commercial catalysts commonly used in the reforming of bio-oil are often designed for use in a fixed-bed configuration. An attrition problem has been observed when they have been employed in fluidized-bed reactors [29]. Consequently, significant improvements in catalyst activity and mechanical strength are needed.

Other important parameters in the steam reforming process are temperature, which depends on the type of oxygenate, the steam-to-carbon ratio and the catalyst-to-feed ratio. For instance, methanol and acetic acid, which are simple oxygenated organic compounds, can be reformed at temperatures lower than 800 °C. On the other hand, more complex biomass-derived liquids may need higher temperatures and a large amount of steam to gasify efficiently the carbonaceous deposits formed by thermal decomposition.

Table 6.1 lists the stoichiometric yields of hydrogen and percentage yields by weight from steam reforming of some representative model compounds present in biomass pyrolysis oils, and also several biomass and related materials. The table also shows the equilibrium yield of H_2, as a percentage of the stoichiometric yield, predicted by thermodynamic calculations at 750 °C and with a steam-to-carbon (S/C) ratio of 5 [32].

In general, compounds derived from the lignin portion of the biomass yield more hydrogen than those from the carbohydrates (cellulose and hemicellulose) in terms of both weight and mole(s). Oxygenated aromatics such as furans and phenolics produce higher yields of hydrogen than do anhydrosugars and other carbohydrate-derived

Table 6.1 Stoichiometric yields of hydrogen from complete steam reforming reactions (adapted from Ref. [34]).

Compound	Stoichiometric H_2 yield		Equilibrium H_2 yield (predicted at 750 °C, S/C = 5)	
	Moles[a]	% (by wt)[b]	% (of st.)[c]	% (+WGSR)[d]
Methanol	3.00	18.9	87.1	96.8
Ethanol	3.00	26.3	85.9	96.2
Acetone	2.67	27.8	85.2	96.2
Furan	2.25	23.5	84.5	96.5
Methylfuran	2.40	29.5	84.5	96.3
Dimethylfuran	2.50	31.5	84.5	96.2
Anisole, cresol	2.43	31.7	84.3	96.2
Phenol	2.33	30.0	84.3	96.3
Lignin	2.21	22.1	84.9	96.7
Furfural	2.00	21.0	84.6	96.9
Xylan	2.00	15.3	85.8	97.3
Cellulose	2.00	14.9	85.8	97.3
Cellobiose	2.00	13.4	86.3	97.5
Glucose	2.00	13.4	86.3	97.5
Acetic acid	2.00	13.4	86.3	97.5
Formic acid	1.00	4.4	87.7	98.9

[a] Moles of H_2 produced per mole of carbon.
[b] Amount of H_2 divided by the sample molecular weight.
[c] The equilibrium moles of H_2 predicted at 750 °C and S/C = 5 divided by the stoichiometric (st.) yield.
[d] With the additional H_2 arising from the contribution of the WGSR at 750 °C and S/C = 5.

products, such as acetic acid and hydroxyacetaldehyde ($C_4H_4O_2$). In reality, as mentioned above, the yield of hydrogen is lower than the stoichiometric maximum because undesired products are also formed during the process. For instance, 4 mol of hydrogen are lost for each mole of methane formed. High S/C ratios are usually employed to promote hydrogen production. Moreover, the formation of carbonaceous deposits can contribute to other missing hydrogen.

For some oxygenates, such as glucose, the application of steam reforming is complicated by the formation of large quantities of char at the high operating temperatures, causing the reactor to plug.

6.3.2
Catalytic Partial Oxidation

CPO has been suggested as an attractive alternative to steam reforming. In CPO, the fuel reacts with a quantity of oxidizer (O_2) which is less than the stoichiometric amount required for complete combustion. Compared with the conventional CSR, this technology offers many advantages which rest on the small size of the reactor, the rapid response to changes and in the possible mitigation of coking problem. The reforming reaction is less exothermic than CSR and it avoids the need for large amounts of expensive superheated steam, so it requires lower energy costs. The disadvantages are that fuel and oxygen must be premixed. The proportions are such that the mixture may be flammable or even explosive, particularly if small variations (e.g. as a result of pumping and vaporizing liquid fuels) are possible.

Hydrogen production from light alcohols (e.g. methanol and ethanol) via CPO has been the subject of several recent studies. Many catalysts were proposed which showed sufficient activity and stability to be considered further for practical applications (see Sections 6.4.1 and 6.4.2). Information on the CPO of more complex oxygenated molecules is scarce in the literature.

6.3.3
Autothermal Reforming

An alternative approach to CPO and CSR is ATR, which results from a combination of these two techniques.

Typically, ATR reactions are considered to be thermally self-sustaining and therefore do not produce or consume external thermal energy. In fact, since ATR consists of the combination of an exothermic reaction (CPO) which produces heat, with an endothermic reaction (CSR) where heat must be externally generated to the reformer, the balance of the specific heat for each reaction becomes a very distinctive characteristic of this process. This makes the whole process relatively more energy efficient since the heat produced from CPO can transfer directly to be used by CSR. However, other exothermic reactions may simultaneously occur, such as WGS and methanation reactions.

The composition of the gas produced is determined by the thermodynamic equilibrium of these reactions at the exit temperature, which is given by the adiabatic heat balance based on the composition and flow of the feed, steam and oxygen added to the reactor.

A typical ATR reactor consists of a burner, a combustion chamber and a refractory-lined pressure vessel where the catalyst or catalysts are placed. The key elements in the reactor are the burner and the catalyst. Different geometries for ATR reactors have been proposed, considering, for instance, fixed beds or fluidized beds.

The severe operating conditions in ATR necessitate catalysts with good mechanical properties and which are stable at the high temperatures of the reaction (650–900 °C) and at the high steam partial pressure applied.

Under ideal operating conditions with the precise amount of air, fuel and steam, the reaction's theoretical efficiency can even be higher than in the conventional CSR process. Potentially, ATR installations show superior performance to conventional CSR plants in terms of reduced size and weight, lower costs, faster starting time and improved transient time.

Optimization of an ATR reactor for maximum H_2 yield is not a trivial matter and requires complex and iterative calculations.

The design of novel autothermal reactors and catalysts holds promises for efficient reforming of oxygenates such as methanol and ethanol. However, up to now, only limited studies have appeared in the literature.

6.3.4
Aqueous Phase Reforming

APR is a remarkably flexible process which allows the production of hydrogen from biomass-derived oxygenated compounds such as sugars, glycerol and alcohols using a variety of heterogeneous catalysts based on supported metals or metal alloys. The APR process is a unique method that generates hydrogen from aqueous solutions of these oxygenated compounds in a single-step reactor process compared with more reaction steps required for hydrogen generation via conventional processes that utilize non-renewable fossil fuels. The key breakthrough of the APR process is that the reforming is done in the liquid phase. Moreover, hydrogen is produced without volatilizing water, which represents major energy savings. The APR process occurs at temperatures (typically from 150 to 270 °C) and pressures (typically from 15 to 50 bar) where the WGSR is favorable, making it possible to generate hydrogen with low amounts of CO in a single chemical reactor. The latter aspect makes it particularly suited for fuel cell applications and also for chemical process applications. The hydrogen-rich effluent can be effectively purified using pressure-swing adsorption or membrane technologies and the carbon dioxide can also be effectively separated for either sequestration or use as a chemical. By taking place at low temperatures, APR also minimizes undesirable decomposition reactions typically encountered when carbohydrates are heated to elevated temperatures. Various competing reaction pathways are also involved in the reforming process.

Good reviews dealing with different aspects of this process can be found in the literature [80, 81].

Various competing pathways are involved in the reforming process. Figure 6.3 depicts the possible reaction pathways involved in the formation of H_2 and alkanes from oxygenated molecules over a metal-based catalyst.

The oxygenated molecules (e.g. methanol, ethylene glycol, glycerol, xylose, glucose and sorbitol) can undergo reforming via C–C cleavage on metal active sites leading to the production of H_2 and CO. In the presence of water, the adsorbed CO can be removed further by the WGSR to give CO_2 and additional H_2. In fact, at temperatures near 230 °C and total pressures near 30 atm, the WGS equilibrium is favorable for the production of H_2 and CO_2, hence the effluent gas typically contains low levels of CO. Undesired parallel reactions catalyzed by metals can also occur. Indeed, CO_2 can undergo a series of methanation/Fischer–Tropsch reactions to form alkanes. Alternatively, C–O bond scission followed by hydrogenation leads to alcohols or even acids, which can then further react on the metal surface to form alkanes. In addition, undesired reaction pathways such as dehydrogenation–hydrogenation leading to alcohols and dehydrogenation–rearrangement to form acids (mainly during aqueous phase processing of highly oxygenated molecules) can take place in solution or on the support aided by metal catalysis. These intermediates can then react in solution or on the catalyst to make more alkanes. Hence good catalysts for the production of H_2 by APR reactions must be highly active for C–C bond cleavage and also capable of removing adsorbed CO by the WGSR, but they must not facilitate C–O bond cleavage and hydrogenation of CO_x. Moreover, in order to achieve high selectivity towards

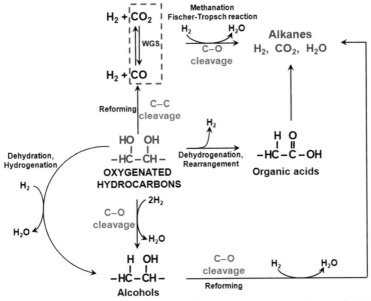

Figure 6.3 Reaction pathways involved in APR of oxygenates (adapted from Ref. [83]).

hydrogen, the choice of the source of oxygenated feedstock is important. As a general trend, H_2 selectivity decreases with increase in the size of the feed molecules.

Notably, once the oxygenated hydrocarbons have been converted into synthesis gas, it is then possible to carry out the subsequent conversion of synthesis gas into a variety of liquid products by well-established catalytic processes, such as the production of long-chain alkanes by Fischer–Tropsch synthesis and/or the production of methanol.

For some oxygenates extracted from biomass such as carbohydrates, the production of H_2 and CO_2 requires APR conditions because of their low volatility. The advantage of the APR process is that it can be used to produce H_2 and CO_2 from oxygenated compounds, such as glucose, that have low vapor pressures at the temperatures that can be achieved without leading to excessive decomposition of the feed. However, the need to maintain water in the liquid state requires that the APR process be operated at pressures that are higher than the vapor pressure of water. Hence the practical range of temperatures that can be employed is limited by the pressures that can be tolerated safely in the reactor system. Although the need to operate at elevated pressures can be a disadvantage for system design, it is an advantage for the subsequent separation of H_2 from CO_2. In particular, because the H_2–CO_2 gas mixture is produced at elevated pressure, the H_2 and CO_2 can be readily separated using a membrane or by pressure-swing adsorption.

The leaching of catalyst components into the aqueous phase during the reaction represents a possible disadvantage of the process. Therefore, the choice of catalyst support materials has to be limited to those that exhibit long-term hydrothermal stability (e.g. carbon, titania, zirconia).

Feed concentration is another important parameter which has to be taken into account. It has been observed that using solutions more concentrated than 1 wt% leads to a decrease in both conversions and selectivities for H_2 production [82]. However, this dependency changes with the nature of the oxygenated molecules. Processing such dilute solutions is not economically practical, even though reasonably high hydrogen yields could be obtained. In this respect, more improvements are needed in terms of catalysts and reactor configurations.

It is also interesting that if the goal is to produce H_2–CO gas mixtures, for instance in the Fischer–Tropsch synthesis, one needs to suppress the WGSR operating in the vapor phase or to operate at higher concentrations of the oxygenated reactant in water or to use a catalyst on which the rate of the WGSR is slow.

The selectivity of the reforming process depends on various factors such as the nature of the catalytically active metal, support, solution pH, feed and reaction conditions, as shown in Figure 6.4. By manipulating the process conditions, it is possible to control the product distribution.

Metals such as Pt, Pd and Ni–Sn alloys show high selectivity for hydrogen production and a very low tendency for alkane formation. On the other hand, metals such as Ru and Rh are more active towards alkane formation. Acidic supports have high selectivity for this reaction, whereas the more basic/neutral supports favor hydrogen generation. Moreover, oxide supports play a key role in the activation of water molecules, resulting in inhibition or promotion of the WGSR. The acidity of the solution also affects the performance of the aqueous phase reformer. Depending on the nature of the by-product/intermediate compounds formed during the process, the aqueous solution in contact with the catalyst can be acidic, neutral or basic. Acidic solutions (pH = 2–4) promote alkane formation, due to acid-catalyzed dehydration reactions that occur in solution (followed by hydrogenation on the metal). In contrast, neutral and basic solutions lead to high hydrogen selectivity and low alkane selectivity.

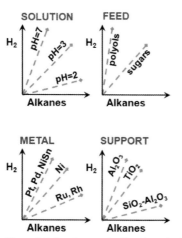

Figure 6.4 Reactors controlling the selectivity of the APR process (adapted from Ref. [85]).

Some energy companies are currently commercializing the APR process. Virent Energy Systems has recently developed APR systems for two applications: on-demand hydrogen generation for PEM and SOFC fuel cells and on-demand generation of hydrogen-rich fuel gas from biomass-derived glycerol and sorbitol to fuel a stationary internal combustion engine-driven generator [84].

6.4
Key Examples of Oxygenate Reforming Reactions

Bio-oils obtained by biomass pyrolysis are a complex mixture of a large number of organic compounds. Although the molecular composition can vary significantly with the type of biomass and the pyrolysis conditions, its major components are oxygenated, including aldehydes, alcohols, ketones and acids, in addition to more-complex carbohydrate- and lignin-derived oligomeric materials emulsified with water. In addition to the extremely heterogeneous composition of the bio-oils, the thermal instability of the oxygenated compounds at typical temperatures of the reformer has to be carefully taken into account.

In order to design efficient and stable reforming catalysts and to select the best operating conditions for hydrogen production, several experimental and theoretical studies have been performed using model compounds which are representative of the major classes of components present in bio-oils.

6.4.1
Methanol

Methanol as a hydrogen carrier has received great attention in the literature due to its high hydrogen-to-carbon ratio and its molecular simplicity (absence of C–C bond), resulting in a relatively low reforming temperature (250–350 °C). It has been demonstrated that hydrogen production from methanol is possible through several alternative processes.

Catalytic production of hydrogen from methanol has been studied for the last 30–40 years and decomposition of methanol to CO and H_2 (Equation 6.5) has been utilized commercially in the steel industry for decades as a method for providing carbon monoxide for the carbonization of steel.

$$CH_3OH \rightarrow 2H_2 + CO \tag{6.5}$$

Catalytic hydrogen generation by decomposition of methanol received much attention during the 1980s, when a significant number of investigations were performed. The goal was the use of the methanol decomposition reaction for improving the efficiency and decreasing emissions of internal combustion engines, in addition to improving the cold start of alcohol engines [85]. The decomposition process is unsuitable, however, for fuel cell applications as CO is one of the main products, which has detrimental effects on the fuel cells.

The processes that are most feasible for the on-board production of hydrogen from methanol for fuel cell applications are: catalytic steam reforming of methanol (CSRM), catalytic partial oxidation of methanol (CPOM) and combined reforming. Although not described in this chapter, photocatalytic production of hydrogen from water–methanol solution is becoming an interesting and promising topic. Since direct splitting of water into H_2 and O_2 has a very low efficiency due to the rapid reverse reaction, a much higher hydrogen production rate can be obtained by addition of so-called sacrificial reagents [86] such as alcohols, which are oxidized to products that are less reactive towards hydrogen [87–90].

Recently, a laser based method for photocatalytic reforming/transformation of methanol at ambient temperature using WO_3–TiO_2 semiconductor-based catalysts was investigated. A mixture containing hydrogen, CO and CH_4 with a high concentration of hydrogen was produced. The process appears to be highly promising [91, 92], although further improvements are necessary for industrial-scale applications.

CSRM (Equation 6.6) is a highly developed and thoroughly studied process [93–99]:

$$CH_3OH + H_2O \rightarrow 3H_2 + CO_2 \tag{6.6}$$

The reaction can yield a gas containing up to 75% hydrogen while maintaining high selectivity towards carbon dioxide. The main drawback of the steam reforming process is that it is slow and endothermic. The high energy requirement for the reaction is a major obstacle for the implementation of a reformer based on this process in automotive applications. There are, however, several commercial solutions available based on steam reforming [100].

The CSRM is often operated using a 30% excess steam in the feed stream in order to lower the CO concentrations [93, 96].

A wide variety of catalysts have been reported to be active for CSRM. The kinetics and reaction paths depend on the catalytic materials used. The majority of these systems have been copper based.

Two major pathways for CSRM have been suggested using copper-based catalysts: (i) a decomposition–WGSR sequence and (ii) dehydrogenation of methanol to methyl formate (Equation 6.7).

In the decomposition–WGSR pathway, methanol decomposes initially to CO and H_2 and then the CO reacts further with water to form CO_2 and H_2. This mechanism was proposed by several groups [96, 101–104] and was thoroughly studied over both commercial and novel catalysts.

The methyl formate reaction route was shown to be dependent on the nature of the support. CO does not form and methyl formate and formic acid are the only intermediates [93]. The suggested reaction path over γ-alumina is as follows [101, 105, 106]:

$$2CH_3OH \rightarrow CH_3OCHO + 2H_2 \tag{6.7}$$

$$CH_3OCHO + H_2O \rightarrow HCOOH + CH_3OH \tag{6.8}$$

$$HCOOH \rightarrow CO_2 + H_2 \tag{6.9}$$

The formation of by-products in the steam reforming reaction over copper-based catalysts is generally low. The formation of products such as CO, formic acid and methyl formate, which was reported by some researchers [103, 105–107], is significant as it poses a threat to the performance of the fuel cells. It is possible to minimize the formation of CO by operating the CSRM in an excess of steam, thereby integrating the WGSR into the reformer.

Commercial Cu/ZnO catalysts for the WGSR and methanol synthesis [106–108] were found to be active for the steam reforming reaction. Shimokawabe et al. [109] reported that a highly active Cu/ZrO$_2$ can be prepared by impregnation of a ZrO$_2$ substrate with aqueous solutions of the complex [Cu(NH$_3$)$_4$][(NO$_3$)$_2$], which proved to be more active than the corresponding Cu/SiO$_2$ catalysts. The good performance of ZrO$_2$ as a substrate for the copper-based phases led to the preparation of highly active Cu/ZrO$_2$ catalysts based on the screening of the best preparation methods, including impregnation of copper salts on the ZrO$_2$ support [110, 111], precipitation of copper [110–114], formation of amorphous aerogels [115, 116], a microemulsion technique [117] and CuZr alloys [111, 118]. The central idea in all of these approaches is to maintain the zirconia support in the amorphous state under the calcination and reaction conditions in order to retain a high level of activity. The major drawback of zirconia crystallization is the drop in both the copper surface area and the support specific surface area. Additionally, a high copper/zirconia interfacial area must be maintained to prevent catalyst deactivation. Tetragonal zirconia can be stabilized by incorporation of aluminum, yttrium and lanthanum oxides [119], thus preventing, or at least minimizing, its crystallization. Breen and Ross [93] found that Cu/ZnO/ZrO$_2$ catalysts are active at temperatures as low as 170 °C but that they become severely deactivated above 320 °C. However, deactivation is inhibited upon incorporation of Al$_2$O$_3$. The improvement of catalyst stability brought about by Al$_2$O$_3$ incorporation comes from the increase in the temperature of crystallization of ZrO$_2$, which remains amorphous at the reaction temperature. Furthermore, the incorporation of alumina increases both the copper and support surface areas, also increasing the catalyst's activity.

The catalytic properties of copper catalysts for CSRM are significantly different from those of other transition metals. Several investigations have been performed on the behavior of group 9–10 transition metals in the conversion of alcohols [120–122]. The major difference between copper and other transition metals is the CO$_2$ selectivity. Several investigations [120, 123, 124] showed that CO concentrations up to 25% could be achieved during CSRM, a result comparable to the decomposition route. The influence of the support was demonstrated to be significant. The high CO concentrations obtained for these transition metals make them highly unsuitable for fuel cell applications.

Catalytic partial oxidation of methanol can be expressed by the following equation:

$$CH_3OH + \frac{1}{2}O_2 \rightarrow 2H_2 + CO_2 \tag{6.10}$$

Although there are advantages related to the possibility of highly dynamic and fast reforming systems [97, 125–127], the formation of hot-spots is one of the main drawbacks from using the partial oxidation process, as the formation of these hot-zones in the catalyst can result in its sintering, thus lowering the catalyst activity.

The partial oxidation process can theoretically, at complete conversion, generate a gaseous product containing up to 67% hydrogen. However, for automobile applications, the oxygen will, most likely, be supplied using compressed air, which results in dilution of the product with nitrogen and lowering of the maximum hydrogen concentration to 41%. The performance of the fuel cells is dependent on the hydrogen concentration and therefore partial oxidation itself may not be suitable.

Two types of catalysts have been proposed for the CPOM reaction: copper and palladium. The catalytic properties of these materials show significant discrepancies with respect to by-product formation and the effect of oxygen partial pressure. The Cu-based catalysts display high selectivity for the CPOM reaction whereas for the Pd-based catalysts CO formation is significant.

The reaction path for CPOM over Cu-based catalysts is complex. Several reactions were observed which are catalyzed by copper, for example, steam reforming, partial oxidation, decomposition of methanol, WGSR and total oxidation [125–131]. The selectivity for H_2 formation over Cu-based catalysts was shown to have a strong dependence on the methanol conversion, suggesting that the oxidation and reforming take place in two consecutive reactions. Formaldehyde was proved to be an intermediate in the CPOM reaction over Cu-based catalysts by studying its formation and subsequent decomposition into CO and H_2. Investigations also indicated that on operating CPOM under fuel-rich conditions, a product mixture containing formaldehyde, CO and H_2 is produced. There are several studies that suggest that the CPOM reaction involves SRM and decomposition of methanol.

Based on these results, one can conclude that steam reforming is part of the partial oxidation scheme.

Reaction temperature also strongly influences the activity and selectivity of the CPOM. It has been often observed that the conversion of methanol increases with temperature, whereas the CO_2 selectivity decreases.

Copper–zinc-based catalysts were found to be very active for the CPOM [129]. The partial oxidation reaction starts at temperatures as low as 215 °C and the rate of methanol and oxygen conversion increases strongly with temperature to produce selectively H_2 and CO_2. As a general rule, methanol conversion to H_2 and CO_2 increases with copper content, reaching a maximum with $Cu_{40}Zn_{60}$ catalysts (40:60 atomic percent) and decreasing with higher copper loadings. The $Cu_{40}Zn_{60}$ catalyst with the highest copper metal surface area was found to be the most active and selective for the CPOM. Unreduced copper–zinc oxide catalysts display very low activity, mainly producing CO_2 and H_2O and only traces of H_2, although the catalysts could be eventually reduced under reaction conditions at high temperatures. For Cu–Zn catalysts, with Cu concentrations in the 40–60 wt% range, the copper metal surface area seems to be the main factor in determining the reaction rate [132]. However, these systems tend to deactivate quickly during operation. The use of alumina as support will stabilize the material at the cost of lower activity, indicating that aluminum has an inhibiting effect on the reaction. It was also observed that the $O_2:CH_3OH$ molar ratio in the feed has a strong influence on catalyst performance.

This effect is related to the active species (Cu^+ and Cu^0) present on the catalyst, which change according to the reaction environment. Cu^+ sites appear to be responsible for the partial oxidation reaction of methanol, whereas Cu^0 has low reactivity towards methanol and so the activity is optimized at intermediate surface coverages by oxygen.

Morphological characteristics of the Cu-based catalyst surface play a central role in the evolution of the oxidation state and structural morphology during the reaction, because the dynamic behavior of the catalyst surface is determined by the conditions of the gaseous atmosphere during the reaction.

The group 10 metals, such as palladium and platinum, are active for the conversion of methanol. However, they are much less selective than the copper-based catalysts, yielding primarily the decomposition products [123, 124, 133]. This catalytic property makes them less feasible for fuel cell applications. The only exception found is for Pd/ZnO, which showed selectivity close to that of a copper catalyst [105, 121].

The idea of combining steam reforming with partial oxidation was first initiated in 1986 by Huang and Wang [134], who proposed a new reaction route for producing H_2 from methanol. The combined reforming reaction (Equation 6.11) provides a method for producing hydrogen with relatively high selectivity and low CO concentrations while maintaining a dynamic system.

$$CH_3OH + xH_2O + \frac{1}{2}yO_2 \rightarrow (3x + 2y)H_2 + CO_2 \tag{6.11}$$

When operating the combined process under stoichiometric conditions, the sum of the coefficients of water (x) and oxygen (y) equals the molar coefficient of methanol ($x + y = 1$).

The combined reforming reaction can be operated under endothermic, exothermic or thermally neutral conditions, depending on the chosen oxygen-to-methanol ratio. The number of research groups focusing on combined reforming has increased rapidly during recent years, and also the number of publications [97, 131, 134–138]. The main by-product for the combined reforming reaction has been shown to be CO; however, the formation of methyl formate and formic acid was also reported [131, 134]. The oxidative steam reforming of methanol has been studied using various catalysts such as $CuO/ZnO/Al_2O_3$ [139–141], Cu/Zn/Zr/Al oxide [136], CuO/CeO_2 [142–144], Pd/ZnO [145], Pd–Zn/Cu–Zn–Al [146] and $ZnO-Cr_2O_3/CeO_2-ZrO_2$ [147]. Turco et al. [139, 140] observed 100% methanol conversion over $CuO/ZnO/Al_2O_3$ at temperatures higher than 350 °C, with a low yield of hydrogen. The product also contained methane and oxygenates in addition to H_2, CO_2 and CO. For the preferential oxidation (PROX) of carbon monoxide, copper–ceria-based catalysts were widely used [148, 149]. Ceria is also known to improve the stability of catalysts [150].

6.4.2
Ethanol

Ethanol has been indicated as one of the most important first-generation biofuels [151, 152]. Since it can be easily produced in fermentation processes, and is safe to handle, transport and store, it is extensively used in Brazil in direct combustion engines [153]. Ethanol is also used as a gasoline additive to increase the octane number [154]. Furthermore, the absence of sulfur and metals and the high oxygen content lead to exhaust emissions from the combustion engine that are cleaner than those from the gasoline engine.

In addition to the direct use of ethanol as a fuel, its use as a source of H_2 to be used with high efficiency in fuel cells has been thoroughly investigated. H_2 production from ethanol has advantages compared with other H_2 production techniques, including steam reforming of hydrocarbons and methanol. Unlike hydrocarbons, ethanol is easier to reform and is also free of sulfur, which is a well-known catalyst poison. Furthermore, unlike methanol, ethanol is completely renewable and has lower toxicity.

Despite the advantages mentioned above, there are important limitations. In fact, agricultural cultivation devoted to ethanol production requires large areas and consumes significant water resources, as in the case of sugar cane. It has been indicated that overcropping of the soil can be a problem in some cases. For other cultivations, such as maize, a possible competition between food and energy was indicated.

Ethanol can be converted directly to hydrogen through two main processes, steam reforming of ethanol (SRE) and partial oxidation of ethanol (POE). These two reforming techniques are described by the following equations:

$$C_2H_5OH + 3H_2O \rightarrow 6H_2 + 2CO_2 \tag{6.12}$$

$$C_2H_5OH + \frac{3}{2}O_2 \rightarrow 3H_2 + 2CO_2 \tag{6.13}$$

Whereas POE offers exothermicity and a rapid response (Equation 6.13), SRE is an endothermic process and produces greater amounts of hydrogen, resulting in higher efficiency. There is also a third option, which combines the advantages of the first two approaches by co-feeding oxygen, steam and ethanol simultaneously through an oxidative reforming process (Equation 6.14):

$$C_2H_5OH + xO_2 + (3-2x)H_2O \rightarrow (6-2x)H_2 + 2CO_2$$
$$0 < x < 0.5 \tag{6.14}$$

Compared with partial oxidation, steam reforming has received more attention due to its relatively higher conversion efficiency [155].

The reaction network involved in SRE is very complex. Many pathways are possible, as shown in Figure 6.5 [156–165].

The process includes several steps that require catalytic sites able to dehydrogenate ethanol, to break the C–C bonds of surface intermediates producing CO and CH_4 and to promote the steam reforming of CH_4. Moreover, the WGSR is also involved, contributing to reducing the CO concentration and increasing the H_2 production. Some of these steps can be favored depending on the catalyst used [157, 166–171]. However, other secondary reactions can be involved. A dehydration reaction leads to the formation of ethylene, especially when acid supports are used (such as Al_2O_3 [171]). Ethylene is easily transformed into carbon that is deposited on the

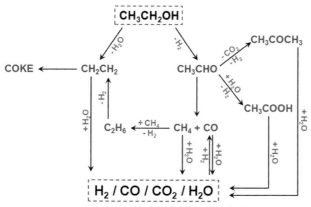

Figure 6.5 Reaction pathways for steam reforming of ethanol.

active phase, leading to deactivation of the catalyst. Also ethane, formed by methane coupling, acts as a very strong promoter for carbon formation. Acetone could be produced from acetaldehyde through a series of reactions involving aldol condensation, oxidation and decarboxylation. Significant formation of this product is observed when the support may provide structural oxygen for the oxidation step (such as CeO_2 [162, 164] or $Ce_xZr_{1-x}O_2$ [172]). The formation of by-products such as methane, ethane, acetaldehyde, acetic acid and acetone is undesirable because they decrease the hydrogen production efficiency and they can reduce the operational lifetime of the catalyst.

Catalysts play a crucial role in the reactivity towards complete conversion of ethanol. Each catalyst, however, induces different pathways and, therefore, the selection of a suitable catalyst is difficult. Most of the catalysts used are Al_2O_3-supported metals [71, 161, 163, 167, 173–182], noble metals [72, 156, 161, 176, 182–187] or alloys [188, 189].

Extensive experimental and theoretical studies on hydrogen production from SRE have been reported. In the thermodynamic studies carried out by Vasudeva et al. [190], it was reported that in all ranges of conditions considered, there is nearly complete conversion of ethanol and only traces of acetaldehyde and ethylene are present in the reaction equilibrium mixture. Methane formation is inhibited at high water-to-ethanol ratios or at high temperatures [191].

Garcia and Laborde [192] reported that it is possible to obtain hydrogen by SRE at temperatures greater than 280 °C with methane being an unwanted product. Hydrogen production is favored, however, by high temperature, low pressure and high water-to-ethanol feed ratio.

The thermodynamic analysis of Ioannides [193] on SRE in a solid polymer fuel cell indicated that the ethanol steam reforming reaction needs to be carried out in two steps: a high-temperature endothermic step (steam reforming), in which ethanol is converted to a gaseous mixtures of H_2, CO, CO_2, CH_4 and unreacted H_2O, and a subsequent, low-temperature step (WGSR) in which CO reacts with water to form H_2 and CO_2.

Cavallaro and Freni [181] investigated SRE over many catalytic systems such as CuO/ZnO/Al$_2$O$_3$, NiO/CuO/SiO$_2$, Cu/Zn/Cr/Al$_2$O$_3$, Pt/Al$_2$O$_3$, Pt/La$_2$O$_3$/Al$_2$O$_3$, Pt/TiO$_2$, Pt/MgO/Al$_2$O$_3$, Rh/SiO$_2$, Rh/Al$_2$O$_3$ and Rh/MgO/Al$_2$O$_3$. Intermediate products such as acetic acid, acetaldehyde and ethyl acetate were observed only at low temperature. The selectivity with respect to hydrogen, carbon dioxide and carbon monoxide increases with increasing temperature.

Haga *et al.* [194] investigated the effects of crystallite size on different alumina-supported cobalt catalysts in SRE. It was reported that SRE over cobalt catalysts proceeds via the formation of acetaldehyde at temperatures below 400 °C. The production of acetaldehyde gradually increases with temperature, reaching a maximum at around 330 °C. Above this temperature, acetaldehyde is converted into carbon dioxide and hydrogen. The ethanol conversion reaches 100% at 400 °C. The activity was found to be independent of the nature of the Co starting material (cobalt acetate, cobalt carbonyl and cobalt chloride). Carbon monoxide evolution increases with temperature, leading to a maximum at about 380 °C, after which it decreases sharply due to the occurrence of the WGSR. Methane selectivity reaches a maximum of 20% at 400 °C, after which it decreases gradually to 10% at 450 °C. The activity of γ-alumina-supported copper–nickel catalysts for hydrogen production from SRE was studied by Marino *et al.* [174]. The catalysts exhibit acceptable activity, stability and hydrogen selectivity when the reaction occurs at 300 °C. It has been reported that copper is the active agent; nickel promotes C–C bond rupture and increases hydrogen selectivity and potassium neutralizes the acidic sites of the γ-alumina and improves the general performance of the catalyst. The results of catalyst activity and selectivity measurements together with studies on catalyst structure indicate that the catalyst must have a high dispersion of the active phase in order to maximize ethanol conversion per copper unit mass. Fatsikostas *et al.* [195] investigated the ESR reaction over Ni-based catalyst supported on yttria-stabilized-zirconia (YSZ), La$_2$O$_3$, MgO and Al$_2$O$_3$. It was reported that Ni/La$_2$O$_3$ catalyst exhibits high activity and selectivity towards hydrogen production and also has long-term stability. The long-term stability of Ni/La$_2$O$_3$ was attributed to the scavenging ability of coke deposition on the Ni surface by lanthanum oxycarbonate species. Results obtained from time-on-stream over an Ni/Al$_2$O$_3$ catalyst are comparable to those for Ni/La$_2$O$_3$, but the H$_2$ selectivity of the latter is higher. In the case of Ni/YSZ catalyst, the selectivity towards hydrogen production is constant during operation; however, the selectivity towards CO$_2$ and CO decreases with time, reaching a steady-state value after 20 h on-stream. Ni/MgO catalysts are very stable under the prevailing conditions, but have poor selectivity compared with the above-mentioned catalysts. Velu *et al.* [196] studied the SRE over a Cu–Ni–Zn–Al mixed oxide catalyst in the presence or absence of oxygen. The ethanol conversion increases with increase in oxygen-to-ethanol ratio and reaches 100% at a ratio of 0.6. Also, the selectivity of both CO and CO$_2$ increases until an oxygen-to-ethanol ratio of 0.4 is reached; CO selectivity, however, drops at a ratio of 0.6. The hydrogen yield decreases from 3 to 2 mols mol^{-1} of reacted ethanol in the absence of oxygen. They concluded that addition of oxygen improves the ethanol conversion and also the oxidation of CH$_3$CHO to CH$_4$ and CO$_2$. It has also been observed that a Cu-rich catalyst favors the dehydrogenation of ethanol to acetaldehyde, whereas the addition of nickel to the Cu/Al_2O_3 system favors the rupture of C–C bonds, enhances the ethanol gasification and reduces the selectivity in acetaldehyde and acetic acid.

Cavallaro et al. [156] reported that rhodium impregnated on γ-alumina is highly suitable for SRE. The catalyst stability was investigated with and without oxygen. It was observed that the catalyst deactivates very rapidly in the absence of oxygen. The presence of oxygen enhances the catalyst stability. Breen et al. [187] investigated SRE over a range of oxide-supported metal catalysts. They concluded that the support plays an important role in the reaction. In fact, they observed that alumina-supported catalysts are very active at low temperatures for dehydration of ethanol to ethylene, which at higher temperatures (550 °C) is converted into H_2, CO and CO_2 as major products and CH_4 as a minor product. The activities of the metal decrease in the order of Rh > Pd > Ni ≈ Pt. Ceria/zirconia-supported catalysts are more active and exhibit 100% conversion of ethanol at high space velocity and high temperature (650 °C). The order of activity at higher temperatures is Pt ≈ Rh > Pd. By using a combination of a ceria/zirconia-supported metal catalyst with the alumina support, it was observed that the formation of ethene does not inhibit the steam reforming reaction at higher temperatures.

Freni [159] examined ESR reactions over Rh/Al_2O_3 catalyst. The results indicated that the catalytic activity of alumina (Al_2O_3) is not negligible; ethylene and water are produced at 347 °C and their production increases and reaches equilibrium at 600 °C. It was observed that the water content does not influence ethylene formation. When 5% Rh is added to alumina, the main steam reforming reaction occurs above 460 °C and the products include hydrogen, carbon dioxide, carbon monoxide and methane. Freni et al. [197] also examined the SRE for hydrogen production on Ni/MgO. They reported that the catalyst exhibits very high selectivity to hydrogen and carbon dioxide. The CO methanation and ethanol decomposition are considerably reduced. In addition, coke formation is strongly depressed because of the benefits induced by the use of the basic support, which modify positively the electronic properties of Ni.

Llorca et al. [198] examined the hydrogen production process by SRE over several cobalt-supported catalysts. It was observed that negligible SRE occurred over Co/Al_2O_3 catalyst. The dehydration of ethanol to ethylene takes place to a large extent, which was attributed to the acidic behavior of Al_2O_3 under similar conditions. Co/MgO catalyst shows a low conversion of ethanol (30%) and the main reaction was dehydrogenation of ethanol to acetaldehyde. Co/SiO_2 shows dehydrogenation of ethanol to acetaldehyde as the main reaction. At low temperature, 100% ethanol conversion has been obtained on Co/V_2O_5. Co/ZnO exhibits promising catalytic performance. It was reported that 100% ethanol conversion was achieved and the highest selectivity of hydrogen and carbon dioxide per mole of ethanol reacted was obtained without catalyst deactivation [198].

Aupretre et al. [182] also studied the effects of different metals (Rh, Pt, Ni, Cu, Zn and Fe) and role of the support (γ-Al_2O_3, 12%CeO_2-Al_2O_3, CeO_2 and $Ce_{0.63}Z_{0.37}O_2$) on SRE. At 700 °C, γ-Al_2O_3-supported Rh and Ni catalysts appear to be the most active and selective catalysts. Ni/Al_2O_3 gives a higher yield but lower selectivity to CO_2

compared with Rh/Al$_2$O$_3$. There is a significant effect of the support on the reactivity. The results at 600 °C show that the catalyst activity decreases in the order Rh/Ce$_{0.63}$Z$_{0.37}$O$_2$ > Rh/CeO$_2$–Al$_2$O$_3$ > Rh/CeO$_2$ > Rh/γ-Al$_2$O$_3$. A similar trend was obtained for Ni: Ni/Ce$_{0.63}$Z$_{0.37}$O$_2$ > Ni/CeO$_2$ > Ni/CeO$_2$–Al$_2$O$_3$ > Ni/γ-Al$_2$O$_3$. Comas et al. [165] studied the ESR over Ni/Al$_2$O$_3$ catalyst. They concluded that temperatures above 500 °C and a high water-to-ethanol molar ratio (6:1) are necessary to obtain a high hydrogen yield and selectivity. The excess of water in the feed enhances methane steam reforming and depresses carbon deposition. In the comprehensive study carried out by Llorca et al. [198], various metallic oxides such as MgO, γ-Al$_2$O$_3$, TiO$_2$, V$_2$O$_5$, CeO$_2$, ZnO, Sm$_2$O$_3$, La$_2$O$_3$ and SiO$_2$ were used as catalysts for SRE at temperatures between 300 and 450 °C. The ethanol conversion increases with increase in temperature in all cases. However, significant differences were observed in terms of activity, stability and selectivity of the catalysts. It was observed that γ-Al$_2$O$_3$ and V$_2$O$_5$, although showing high conversion of ethanol at lower temperatures, are not suitable for H$_2$ production as both are highly selective for ethylene production by dehydration of ethanol (being acidic in nature). It was also shown that MgO and SiO$_2$ give conversions less than 10% and are also selective for dehydrogenation of ethanol to form acetaldehyde, and La$_2$O$_3$ and CeO$_2$ give a conversion of approximately 20%. Other oxides such as TiO$_2$ and Sm$_2$O$_3$ show strong deactivation processes. This was attributed to carbon deposition during the reaction, which could be responsible for the drop in activity of the catalysts. ZnO enhances the SRE and shows high selectivity for H$_2$ and CO$_2$.

Galvita et al. [199] investigated the SRE for syngas production in a two-layer fixed-bed catalytic reactor. In the first bed, the ethanol was converted to a mixture of methane, carbon oxides and hydrogen on a Pd/C catalyst and then this mixture was converted to syngas over an Ni-based catalyst for methane steam reforming. It was observed that ethanol conversion increased with increase in temperature. They concluded that the use of a two-layer fixed-bed reactor prevents coke formation and gives a yield close to equilibrium. Good results were also obtained on combining the catalytic activity of Cu-based catalysts with the activity of Ni-based catalysts [200]. In particular, ethanol passes through the first layer containing Cu catalyst at low temperatures (300–400 °C) to perform dehydrogenation to acetaldehyde and hydrogen, followed by acetaldehyde steam reforming or decomposition. The intermediate species are then passed through the second layer containing an Ni-based catalyst, to enhance hydrogen production [201]. This novel concept of a double-bed reactor offers an economical method to enhance hydrogen production and catalyst stability.

In this respect, an improvement in the catalyst lifetime was obtained by our group, using the novel synthetic approach described in Section 6.2.2. In this way, active and stable catalysts were obtained, combining the high reactivity of nanosized noble metal particles with the excellent high-temperature stability of Al$_2$O$_3$-based nanocomposites [71, 72]. In particular, stable Rh–ceria–zirconia–alumina nanocomposites for ethanol steam reforming were prepared by embedding preformed Rh nanoparticles in a ceria–zirconia–alumina matrix. In the absence of rare earth doping, ethanol is converted at low temperature mainly to ethylene via a dehydration process.

At higher temperatures, ESR to H_2 and CO/CO_2 is operative. The introduction of ceria–zirconia favors the dehydrogenation of ethanol to acetaldehyde, oxidation of which leads to some acetone formation. This is consistent with a significant coverage of the alumina by $Ce_xZr_{1-x}O_2$ mixed oxides and the subsequent reduction of the alumina acid sites responsible for the dehydration reaction in the low-temperature region. All catalysts show stable catalytic activity at high temperatures for at least 160 h, indicating excellent thermal stability. The ceria-based oxides play a key role both in preventing coke deactivation and in favoring its removal.

The partial oxidation of ethanol was investigated, but with less intensity than in the case of steam reforming. The reason is that the use of the pure partial oxidation process is not advised for bioethanol reforming because bioethanol is an ethanol–water mixture in which removal of all the water entails a significant cost. Therefore, for bioethanol partial oxidation, the process is combined with steam reforming in autothermal schemes with the stoichiometry shown in Equation 6.18.

The generation of hydrogen from ethanol via catalytic autothermal partial oxidation (Equation 6.18) has been performed at temperatures of 430–730 °C using catalytic systems based on noble metals [202, 203]. Ethanol oxidation follows a very complex pathway, including several reaction intermediates formed and decomposed on both the supports and active metals that integrate the catalytic systems [204, 205]. In the light of the above studies, it has been claimed that the ethoxy species generated on the metal and on the support can be decomposed on the metal sites, forming CH_4, H_2 and CO, whereas part of the ethoxy species generated on the supports is further oxidized to acetate species, which decomposes to CH_4 and/or oxidizes to CO_2 via carbonate species [202]. Hence supports with redox properties that help the oxidation of ethoxy species and metals with a high capacity to break C–C bonds and to activate C–H bonds are suitable for use in catalysts applied to the partial oxidation of ethanol. Salge *et al.* [203] studied the effect of the nature of the metal (Rh, Pd, Pt) on the performance of catalysts supported on Al_2O_3 and CeO_2. The order of effectiveness in hydrogen production for catalysts supported on Al_2O_3 was Rh–Ru > Rh > Pd > Pt. Rh supported on CeO_2 was the most stable and gave greater hydrogen selectivity than noble metals supported on Al_2O_3. The better activity and stability associated with the presence of CeO_2 can be related to the capacity of CeO_2 to store oxygen and make it available for reaction via a redox reaction.

6.4.3
Dimethyl Ether

As a general statement, direct-fed liquid fuel cells are ideal for portable applications due to high energy density fuel storage and the absence of humidification, reforming and cooling subsystems. Although direct liquid methanol fuel cells (DMFCs) are becoming more developed and are indeed promising for portable applications, major drawbacks include the use of parasitic fluid pumps, methanol crossover oxidation reaction at the cathode and mild toxicity of methanol vapor [7]. Several alternative fuels have been proposed to alleviate various drawbacks associated with either methanol or hydrogen. In this context, dimethyl ether (DME) has been suggested as a suitable and attractive alternative fuel to methanol for both portable power and transportation applications. Independently of fuel cell development, however, the introduction of DME in the energy sector can address energy security and environmental concerns immediately in a cost-effective manner with current commercialized technology without relying on future technologies where the timeframe of market penetration is uncertain [206]. The use of DME as a residential fuel for household heating and cooking is also being considered.

The flexibility of DME, considering both source and application, makes it a strong case in a long-term scenario.

DME, $H_3C–O–CH_3$, is a colorless, non-corrosive gaseous ether at room temperature and atmospheric pressure. It has a high hydrogen-to-carbon ratio. It is biologically degradable and, in low amount, non-hazardous from a health point of view (non-carcinogenic, non-teratogenic, non-mutagenic, non-toxic). It burns with a very low tendency to produce soot, which makes it very interesting as a diesel substitute. In addition, the high cetane number (around 68) makes DME an interesting option for use in compression ignition engines. DME has physical properties similar to those of liquid petroleum gas (LPG) fuels (e.g. propane and butane), resulting in similar handling, storage and transport considerations [207].

One major advantage of DME use is that it can be stored as a high-density liquid phase at modest pressures (around 5 atm) and delivered as a gas-phase fuel in a pumpless operation. Therefore, the use of DME can potentially combine the advantages of easy fuel delivery of pressurized hydrogen and the high energy density storage of liquid fuel. In addition, DME is less toxic than methanol.

Currently, the major application of DME is as an aerosol propellant, for example in hairsprays and paintsprays, where it has replaced the formerly used ozone-destroying chlorofluorocarbons (CFCs).

DME can be produced from any carbonaceous material, such as natural gas, coal, crude oil and regenerable resources such as biomass [208–212].

Before the 1990s, most DME was synthesized by an expensive process in which syngas (typically generated from the steam reforming of methane) was first converted to methanol, followed by methanol dehydration to DME over solid acid catalysts (e.g. ZrO_2 [213], γ-Al_2O_3 [214–217] and zeolites [215, 218–220]) and it was therefore not considered as a potential fuel. The use of DME as a fuel gained renewed interest only recently, due to the development of a low-cost production process as demonstrated in several papers [221–230] and patents [231–234]. DME can now be produced directly in a single step from synthesis gas via autothermal reactors and slurry-phase reactors [206]. In the direct DME synthesis process, CO and H_2 react according to Equation 6.15 or 6.16 to form methanol that subsequently forms DME in the same reactor [206, 235].

$$3CO + 3H_2 \rightarrow CH_3OCH_3 + CO_2 \tag{6.15}$$

$$2CO + 4H_2 \rightarrow CH_3OCH_3 + H_2O \tag{6.16}$$

Fundamental research on DME synthesis from renewable sources is ongoing. In Sweden, for example, Chemrec developed a process for producing DME through the gasification of black liquor, a by-product from the pulp and paper industry [236]. The Växjö Värnamo Biomass Gasification Centre is part of the EU-funded project CHRISGAS (Clean Hydrogen-Rich Synthesis GAS) with the aim of demonstrating the possibility of the production of clean synthesis gas from biomass gasification which can be processed into 'renewable' fuels such as DME [237].

However, until efficient production methods based on renewables can be established, fossil-based DME could bridge the gap.

There are two major technology options for producing hydrogen-rich fuel cell feeds from DME: steam reforming and partial oxidation. Another option could be ATR, but only a few studies are available [238].

DME can be reformed by the processes mentioned above at lower temperatures (200–450 °C) than hydrocarbons due to the absence of C–C bonds.

The steam reforming of DME has been demonstrated to occur through a pair of reactions in series, where the first reaction is DME hydration followed by methanol steam reforming to produce a hydrogen-rich stream, as expressed in Equations 6.17 and 6.18, respectively:

$$CH_3OCH_3 + H_2O \rightarrow 2CH_3OH \tag{6.17}$$

$$CH_3OH + H_2O \rightarrow CO_2 + 3H_2 \tag{6.18}$$

The overall reaction is

$$CH_3OCH_3 + 3H_2O \rightarrow 2CO_2 + 6H_2 \tag{6.19}$$

DME hydration occurs over acid catalysts, whereas the methanol steam reforming reaction proceeds over metal catalysts. Consequently, DME steam reforming requires a multi-component catalyst. Two approaches have been proposed in the literature: (a) physical mixtures of a DME hydrolysis catalyst and a methanol steam reforming catalyst; (b) supported catalysts that combine the DME hydrolysis and methanol steam reforming components into a single catalyst.

DME hydrolysis is an equilibrium-limited reaction and is considered as the rate-limiting step of overall DME steam reforming. The equilibrium conversion of hydration of DME is low at low temperatures (e.g. about 20% at 275 °C). However, when methanol formed in the first step is rapidly converted into H_2 and CO_2 by methanol steam reforming catalysts, high DME conversion is expected. Therefore, enhancement of DME hydrolysis is an important factor to obtain high reforming conversion.

In addition to DME steam reforming, the reverse water gas shift reaction (r-WGSR), Equation 6.20, generally takes place over such metal catalysts during the reforming process:

$$CO_2 + H_2 \rightleftharpoons H_2O + CO \tag{6.20}$$

Notably, since high-temperature steam reforming enhances the r-WGSR, which would produce CO, the undesirable poison of fuel cells, low-temperature reforming is preferable. Low temperatures can be achieved over strong acid catalysts, although the strong acid at the same time tends to cause deactivation by coke formation.

Notably, a strongly acidic catalyst or high reforming temperatures favor methane formation via DME decomposition (Equation 6.21). DME decomposition is generally suppressed in the presence of water.

$$CH_3OCH_3 \rightarrow CH_4 + CO + H_2 \qquad (6.21)$$

Several catalytic systems are described in the literature.

The most commonly used catalyst for hydrolysis of DME is γ-Al_2O_3. Acid site density, strength of the acid sites and hydrophobicity have been suggested to influence the activity of the hydrolysis catalyst [225, 239]. Different zeolites have also been employed but they show a tendency to generate long-chain hydrocarbons at temperatures above 300 °C through the methanol-to-gasoline reaction [239, 240]. ZrO_2 has also been proposed but it does not reach the predicted equilibrium conversions of DME during experiments on DME hydrolysis. Many studies concern the solid super acid WO_3/ZrO_2 system [241–249]. Tungsten oxide on a ZrO_2 surface is present as dispersed clusters, the size of which changes according to the surface amount of W, with influence on the catalytic performance [250]. The valence state of surface W has not yet been clarified. WO_3/ZrO_2 combined with CuO/CeO_2 was presented as a valid alternative in terms of durability and amounts of CO produced [251] to the combination of a zeolite such as H-mordenite and CuO/CeO_2 [223], which shows higher efficiency but a shorter lifetime. Durability of the DME reforming catalyst is one of the key factors for practical use. Carbon formation and metal sintering are the main reasons for deactivation. Operating the reforming reaction at appropriate conditions can suppress such deactivation. Furthermore, catalyst durability is also related to the amount of WO_3 loading on ZrO_2. The observed deactivation is due principally to the formation of carbon deposits, which can easily be removed by treatment with oxygen at 300 °C, leading to catalyst regeneration. The deactivation occurs on the CuO/CeO_2 side. Notably, carbonaceous precursors are formed on WO_3/ZrO_2, then transferred to CuO/CeO_2 and deposited as coke [251].

Concerning the methanol steam reforming component, copper is the metal predominately used. Copper catalysts are widely known to have high activity for steam reforming of methanol. This reaction occurs at temperatures of 200–300 °C but the hydrolysis of DME to methanol occurs in the temperature range 300–400 °C [225] and therefore the durability of the copper catalyst will be a concern. Copper catalysts are prone to sintering at temperatures above approximately 300–350 °C [252], leading to loss of activity. Therefore, they must be made thermally stable for use in reforming of DME. Cu–Mn, Cu–Fe or Cu–Cr interactions in spinel oxides have been reported to suppress sintering of copper at temperatures up to 400 °C during steam reforming of DME. Moreover, the selectivity towards hydrogen and CO_2 can be easily controlled by optimizing the ratio of the two metals [228, 252]. Palladium catalysts have also been shown to exhibit high activity for DME steam reforming, but associated with the formation of large amounts of CO [223, 227]. This is not a surprise since Pd is known to be selective for methanol decomposition [253, 254].

A high DME steam reforming activity in terms of conversion and H_2 yield is also reported for the Ga_2O_3/TiO_2 system at 400 °C. Its good activity is attributed to an electronic interaction between Ga and Ti which favors, with a synergetic effect, simultaneous performance of DME hydrolysis and methanol steam reforming. Moreover, this interaction is also helpful for suppressing the direct decomposition of DME, avoiding methane formation [255].

Partial oxidation of DME (Equation 6.22) has been studied over various metal catalysts supported on Al_2O_3 and over nickel on different supports [229, 256].

$$CH_3OCH_3 + \frac{1}{2}O_2 \rightarrow 2CO + 3H_2 \qquad (6.22)$$

The most active and selective catalysts are Ni supported on $LaGaO_3$ and Rh supported on Al_2O_3. The reaction was suggested to proceed through the oxidation of methyl ($-CH_3$) or decomposition of methoxy species ($-OCH_3$) formed by dissociative adsorption of DME [229].

Although most research groups have investigated the reforming of DME at low temperatures, it should be noted that it can match the wide range of operating temperatures of fuel cells. An indirect internal reforming operation has recently been developed that can reform DME efficiently at SOFC working temperatures, ~900 °C [257]. The successful development of this operation is to eliminate the requirement for an external reformer installation, making SOFC fueled by DME more efficient and attractive. DME can be decomposed homogeneously without the requirement for a catalyst at high temperature, producing methane and methanol with small amounts of carbon monoxide, carbon dioxide and hydrogen. The use of CeO_2–ZrO_2 as the pre-reforming catalyst along with Ni/Al_2O_3 is an efficient way to catalyze the steam reforming of DME, producing high contents of hydrogen and carbon monoxide with low selectivity in by-products (i.e. methane). The role of CeO_2–ZrO_2 in the steam reforming of DME is to decompose methanol and some methane (generated by the homogeneous decomposition of DME), while the role of Ni/Al_2O_3 is to decompose methane left from the reforming over CeO_2–ZrO_2 to hydrogen and carbon monoxide.

6.4.4
Acetic Acid

Acetic acid (CH_3COOH) is one of the major components of bio-oils (up to 12 wt%). Furthermore, being non-flammable, it is a safe hydrogen carrier. Nevertheless, only a few studies have dealt with the production of hydrogen from acetic acid.

The steam reforming of acetic acid is highly endothermic ($\Delta H \approx 135$ kJ mol^{-1} at 25 °C) and it can be expressed by the following equation:

$$CH_3COOH + 2H_2O \rightarrow 2CO_2 + 4H_2 \qquad (6.23)$$

The network of reactions which may be take place during the process is fairly complex. Several reactions can occur simultaneously, such as thermal decomposition (Equations 6.24–6.26) and ketonization (Equation 6.27) of acetic acid:

$$CH_3COOH \rightarrow 2CO + 2H_2 \qquad (6.24)$$

$$CH_3COOH \rightarrow CH_4 + CO_2 \tag{6.25}$$

$$CH_3COOH \rightarrow C_2H_4, C_2H_6, C_3H_4, \text{coke}, \ldots \tag{6.26}$$

$$2CH_3COOH \rightarrow (CH_3)_2CO + H_2O + CO_2 \tag{6.27}$$

Steam reforming of CH_4 (Equation 6.28), methanation of CO (Equation 6.29) and CO_2 (Equation 6.30), the Boudourd reaction (Equation 6.31) and the WGSR are also involved:

$$CH_4 + H_2O \rightarrow CO + 3H_2 \tag{6.28}$$

$$CO + 3H_2 \rightleftharpoons CH_4 + H_2O \tag{6.29}$$

$$CO_2 + 4H_2 \rightleftharpoons CH_4 + 2H_2O \tag{6.30}$$

$$2CO \rightleftharpoons C + CO_2 \tag{6.31}$$

The complete steam reforming of acetic acid can be achieved over commercial Ni-based catalysts [79]. The operating temperature of these systems is always higher than 650 °C. The robustness of the catalysts based on Ni guarantees operation over thousands of hours, but this metal leads to extensive coke formation. In order to improve the stability, La_2O_3 was introduced in the catalyst formulation [258].

Although noble metal-based catalysts are typically more expensive than traditional Ni-based formulations, their employment has been proposed due to their higher catalytic activity per unit volume. In this respect, the steam reforming of acetic acid has been investigated over Pt and Rh supported on Al_2O_3 or ceria–zirconia mixed oxides in the temperature range 650–950 °C [30]. Alumina-supported catalysts are less active than those supported on ceria–zirconia, in terms both of CO_x yield and hydrogen production, suggesting the key role of the support type in the catalytic performance. Moreover, the use of ceria–zirconia mixed oxides leads to higher H_2 yields than with alumina-supported catalysts. The increased activity observed over ceria–zirconia can be associated with the redox properties of this material, which would open up an additional reaction pathway as compared with the set of reactions occurring on the alumina-based materials. Takanabe et al. proposed the involvement of the support in the activation of steam [31]. On noble metals supported on CeO_2–ZrO_2, it has been supposed that the metal can activate the oxygenated molecules, the fragments of which can then react with lattice oxygen at the interface between the metal and the support. This bifunctional mechanism was recently demonstrated over Pt/ZrO_2 catalyst by the same group [31, 259, 260]. This system shows high activity at 600 °C, completely converting acetic acid, and it gives a hydrogen yield close to the thermodynamic equilibrium [261]. Pt and ZrO_2 also play an important role in the activation of reactants [259]. The active sites for the steam reforming process are suggested to be located at the Pt–ZrO_2 boundary. Acetic acid activation occurs on Pt particles, leading to the formation of acetate or acyl species which then may decompose to give CO_2, CO and CH_x. The CH_x species

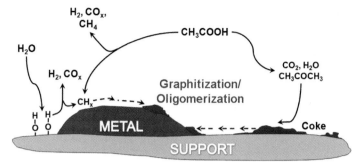

Figure 6.6 Proposed pathway for the steam reforming of acetic acid over Pt/ZrO$_2$ (adapted from Ref. [263]).

can be hydrogenated with adsorbed hydrogen to form CH$_4$ or it can also undergo graphitization and/or oligomerization, causing deactivation due to blockage of the Pt-related active sites by the deposits. On the support, the activation of H$_2$O leads to the formation of surface hydroxyl groups, which are able to react with CH$_x$ species on the Pt through the Pt–ZrO$_2$ boundary (Figure 6.6). Subsequently, the WGS reaction occurs.

In the deactivation mechanism, a key role is also played by acetone formed on the ZrO$_2$ through dehydration reactions (e.g. aldol-type condensation reactions) (Equation 6.32):

$$2CH_3COOH \rightarrow CH_3COCH_3 + CO_2 + H_2O \qquad (6.32)$$

It is well known that the acid–base properties of the support are very important in catalyzing such transformations [262]. In order to improve the catalyst lifetime, two aspects have to be considered in modifying the support: enhancing the ability to steam activation and minimizing the activity towards oligomer formation.

Novel and promising Ni–Co unsupported catalysts have been reported recently [263]. These systems exhibit high activity in a relatively low temperature range (350–550 °C), high selectivity to hydrogen and, most importantly, satisfactory long-term stability. For the steam reforming of acetic acid, single-metal Ni and Co catalysts are also active, but they are inferior to bimetallic Ni–Co catalyst, in terms of both conversion of acetic acid and selectivity to the products. It is important to note that an effective catalyst for reforming of oxygenate compounds should be active not only for cleavage of C–C bonds, but also for the WGSR to remove CO formed on the metal surface. Sinfelt and Yates [264] reported that C–C bond cleavage is faster over Ni than other group 8 metals. However, Ni had limited activity for the WGSR [165, 182], whereas Co had relatively higher activity for the reactions [265]. Hence the combination of Ni and Co in the catalysts can effectively enhance the activity for acetic acid reforming. The best performance is obtained with a Co:Ni molar ratio of 4 [263]. Acetic acid, converted with high selectivity at low temperature over this system, could be used as a fuel in the proton exchange membrane fuel cell (PEMFC).

Other bimetallic systems have been investigated. An Fe–Co-based catalyst [266] exhibits high activity (achieving complete acetic acid conversion at 400 °C), high H_2 selectivity and good stability.

6.4.5
Sugars

Biomass carbohydrates, which represent the most abundant renewable sources available, are currently viewed as a feedstock for the 'green chemistry' of the future [267–269].

Concerning the problems related to global energy supply, carbohydrates, such as sugars, starch, hemicelluloses and cellulose, can contribute to creating new and promising options, as depicted in Figure 6.7. Carbohydrates can be converted by fermentation to alcohols, which can then be used as an energy carrier for the future hydrogen society, as discussed in Sections 6.4.1 and 6.4.2. Although steam reforming of renewable alcohols has been studied widely in recent times (see Sections 6.4.1 and 6.4.2), there are several processing steps before biomass is converted to them which make the process energy intensive and costly. An alternative route is the direct conversion to hydrogen by reaction with water in the liquid phase (APR process). Furthermore, it is possible to convert them into hydrocarbons that are of interest for use as fuels.

Dumesic and co-workers, using Pt-based catalysts, have demonstrated that the catalytic aqueous phase reforming process is a suitable way to produce hydrogen from sugars at low temperature [270, 271]. Hydrogen, CO_2, CO and light alkanes can be obtained by APR of aqueous sugar feeds. Despite some limitations related to the feed concentration used, this technology, as discussed in Section 6.3.4, has been commercialized by Virent Energy Systems.

Two types of sugars are present in biomass: hexoses (C_6-sugars), of which glucose is the most common, and pentoses (C_5-sugars), of which xylose is the most common. Glucose is of particular interest since it constitutes the major energy reserves in plants and animals. Furthermore, it is attractive as a fuel since it can be produced in

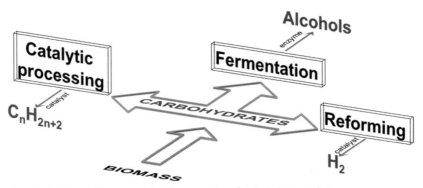

Figure 6.7 Scheme of process routes for conversion of carbohydrates to fuels.

large quantities from agricultural wastes. Unlike methanol, glucose is safe and can be easily transported, making it a suitable fuel for portable micro fuel cells. In this respect, it is interesting to mention the development of a catalytic system for direct conversion of glucose to methanol for use in a direct methanol fuel cell [272]. Pt- and Ni-based catalysts on microporous, mesoporous and macroporous supports have been studied. Platinum supported on macroporous silica shows the highest conversion with good selectivity for methanol. Addition of nickel catalyst suppresses coke formation during the reaction. Further selectivity improvement is necessary to limit the production of higher alcohols (i.e. mainly ethanol and diols) [272].

Reforming of glucose and sorbitol to H_2 and CO_2 can be described according to the following stoichiometric reactions:

$$C_6H_{12}O_6 + 6H_2O \rightarrow 12H_2 + 6CO_2 \qquad (6.33)$$

$$C_6H_{14}O_6 + 6H_2O \rightarrow 13H_2 + 6CO_2 \qquad (6.34)$$

Although the conversion is highly thermodynamically favorable, the selectivity towards hydrogen through this route can be low due to consecutive reactions between the products (CO_2 and H_2) which lead to the formation of alkanes (C_nH_{2n+2}) and water.

The possible reaction pathways which take place during APR of glucose and sorbitol are depicted in Figure 6.8. H_2 and CO_2 are formed through cleavage of C–C bonds catalyzed by metal, followed by the WGSR. C–O bond scission on metal catalysts and dehydration reactions which can be promoted by the support also take place, leading to the formation of undesired products such as alkanes. Furthermore, undesirable homogeneous side-reactions can be involved, resulting in a decrease in hydrogen selectivity [270].

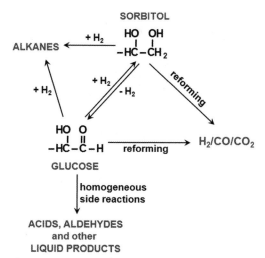

Figure 6.8 Reaction pathways involved in glucose and sorbitol reforming (adapted from Ref. [274]).

Several catalysts have been reported in the literature that are based on noble metals, especially Pt and Pd, and transition metals such as Ni.

Dumesic and co-workers observed that glucose is significantly less reactive than sorbitol on Pt/Al_2O_3. Aqueous phase reforming of glucose leads to hydrogen selectivity (~10–13%) that is significantly lower than that of sorbitol (~60%) under similar conditions. Higher alkane selectivity has also been detected [270]. Furthermore, the rate of H_2 formation by glucose reforming depends considerably on the metal supported on the Al_2O_3. In particular, the order of activity is Pt > Pd > Ni [273].

It is interesting that whereas the selectivity for hydrogen production is insensitive to the liquid-phase concentration of sorbitol, the selectivity of hydrogen formation from reforming of glucose decreases as the liquid concentration increases from 1 to 10 wt% because of undesired hydrogen-consuming side-reactions. In this case, it has been observed that the formation of organic acids, aldehydes and carbonaceous deposits [274] (Figure 6.8) is first order in glucose concentration, whereas the desirable reforming reactions on the catalyst surface are of fractional order [275]. This is the main reason for the poor hydrogen selectivity at high glucose concentrations. An important reaction which can occur during APR is the hydrogenation of glucose to sorbitol (Figure 6.9), which occurs on metal catalysts with high selectivity at low temperatures and high pressures [276]. This decrease in selectivity is a serious limitation, because the processing of dilute aqueous solutions requires handling of large quantities of water. Higher selectivity for hydrogen by APR of glucose can be achieved by improvements in reactor design. It was reported by Davda and Dumesic

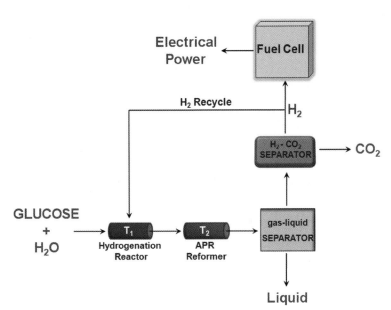

Figure 6.9 Schematic representation of the dual reactor system proposed for hydrogen generation by glucose APR (adapted from Ref. [281]).

that high hydrogen selectivity can be achieved from high liquid-phase concentrations of glucose employing a system which combines a hydrogenation reactor with an APR reformer [277]. This dual reactor system is illustrated in Figure 6.9. According to this novel approach, glucose is first completely hydrogenated to sorbitol before being sent to the reformer. The first reactor operates at relatively low temperature ($\sim$120 °C), in order to minimize the glucose decomposition reaction in the liquid phase, and at relatively high H_2 pressure ($\sim$50 bar) to favor the conversion of glucose to sorbitol. The aqueous solution of sorbitol and gaseous H_2 are then fed to the reforming reactor, which is operated at higher temperature ($\sim$265 °C), necessary to convert sorbitol with high selectivity to H_2 and CO_2. Finally, the liquid and gaseous effluents from the reformer are cooled and sent to a separator situated downstream of the reformer, which is maintained at a low temperature (e.g. 25 °C) [277]. A fraction of the purified H_2 at high pressure can then be recycled to the hydrogenation reactor and the remaining hydrogen may be directed to a fuel cell for conversion to electric power.

An alternative approach for the utilization of biomass resources for energy applications is the production of clean-burning liquid fuels. In this respect, current technologies to produce liquid fuels from biomass are typically multi-step and energy-intensive processes. Aqueous phase reforming of sorbitol can be tailored to produce selectively a clean stream of heavier alkanes consisting primarily of butane, pentane and hexane. The conversion of sorbitol to alkanes plus CO_2 and water is an exothermic process that retains approximately 95% of the heating value and only 30% of the mass of the biomass-derived reactant [278].

Production of alkanes by aqueous phase reforming of sorbitol takes place by a bifunctional reaction pathway involving first the formation of hydrogen and CO_2 on the appropriate metal catalyst (e.g. Pt or Pd) and the dehydration of sorbitol on a solid acid catalyst (e.g. silica–alumina). These initial steps are followed by hydrogenation of the dehydrated reaction intermediates on the metal catalyst. When these steps are balanced properly, the hydrogen produced in the first step is fully consumed by hydrogenation of dehydrated reaction intermediates, leading to the overall conversion of sorbitol to alkanes plus CO_2 and water. The selectivities for the production of alkanes can be varied by changing the catalyst composition and the reaction conditions and by modifying the reactor design. In addition, these selectivities can be modified by co-feeding hydrogen with the aqueous sorbitol feed, leading to a process in which sorbitol can be converted to alkanes and water without CO_2. As another variation, the production of alkanes can be accomplished by modifying the support with a mineral acid (such as HCl) that is co-fed with the aqueous sorbitol reactant. In general, the selectivities to heavier alkanes increase as more acid sites are added to a non-acidic Pt/alumina catalyst by making physical mixtures of Pt/alumina and silica–alumina. The alkane selectivities are similar for an acidic Pt/silica–alumina catalyst and a physical mixture of Pt/alumina and silica–alumina components, both having the same ratio of Pt to acid sites, indicating that the acid and metal sites need not be mixed at the atomic level. The alkane distribution also shifts to heavier alkanes for the non-acidic Pt/alumina catalyst when the pH of the aqueous sorbitol feed is lowered by addition of HCl. The advantages of using a solid acid are that the liquid products do not need to be neutralized and the reactor does not need to be constructed from special corrosion-resistant materials. On the other hand, acids may already be present in liquid streams following hydrolysis of biomass feeds, thereby promoting the formation of heavier alkanes during aqueous phase reforming [81, 151, 278].

Alkane production from sugars by aqueous phase dehydration/hydrogenation reactions has the advantage that most of the alkane fraction is spontaneously separated from the aqueous phase. Unfortunately, the major compound produced by this process is hexane, which has a low value as a gasoline additive due to its relatively high volatility. This limitation has been partially overcome by promoting a base-catalyzed aldol condensation step which links carbohydrate-derived units via formation C–C bonds to form heavier alkanes ranging from C_7 to C_{15} [151].

Since the APR process often requires long reaction times and high pressures, several studies have also been focused on catalytic steam reforming of sugar to produce hydrogen. In this respect, platinum group catalysts such as Rh/Al_2O_3, Pt/Al_2O_3, Pd/Al_2O_3 and Ir/Al_2O_3 were investigated. The development of a steam or autothermal reforming catalyst optimized for hydrogen production from fermentation products represents an opportunity for on-site hydrogen production that can be incorporated into a local fuel cell system for energy generation in small-scale operations. Autothermal reforming of glucose, using a palladium/nickel/copper-based catalyst, at different steam-to-carbon ratios and oxygen-to-carbon ratios gives hydrogen yields as high as 60% at 750 °C [279]. High rates of char formation have been observed, causing reactor plugging after 6–8 h of operation. Indeed, one of the major challenges for reforming of glucose is the formation of large amounts of char. Although it has been widely reported that the main reaction leading to char formation is C–H bond cleavage, it has been observed that the addition of H_2 to the feed does not significantly reduce the formation of char [279]. On the other hand, increasing the oxygen or steam concentration in the feed decreases the rate of char formation due to the promoted combustion and WGSR. As already discussed, this problem is minimized in aqueous phase reforming. However, the amount of hydrogen produced in APR is very small relative to that in autothermal reforming [279].

6.4.6
Ethylene Glycol

Ethylene glycol (EG), $HOCH_2–CH_2OH$, which is readily available from renewable biomass [280, 281], is an interesting feed molecule for the generation of hydrogen due to its low volatility, which makes it convenient for transport and storage.

Aqueous phase reforming of EG has been the subject of several reported studies [275, 282–288]. This great interest is due to the fact that this molecule can be considered as a model of larger and more complex polyols, because it contains the same functionalities, including C–C, C–O, C–H and O–H bonds, and also OH groups on adjacent carbon atoms. Reforming of EG therefore serves as model reactions that may occur during the direct production of hydrogen from biomass-derived sugars (e.g. glucose) and sugar-alcohols (e.g. sorbitol).

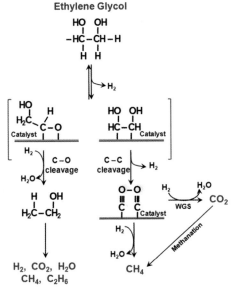

Figure 6.10 Reaction pathways for APR of EG (adapted from Ref. [288]).

Aqueous phase reforming of EG takes place according to the following reaction:

$$HOCH_2CH_2OH + 2H_2O \rightarrow 5H_2 + 2CO_2 \tag{6.35}$$

It has been reported [288] that the reaction starts, as illustrated in Figure 6.10, with the reversible dehydrogenation of EG to form adsorbed $C_2H_{6-x}O_2$ species (Equation 6.36) on the catalyst surface. These adsorbed intermediates can be formed on the metal surface by the formation of either metal–carbon bonds and/or metal–oxygen bonds. The dehydrogenation step must be followed by rapid and irreversible cleavage of the C–C bond (Equation 6.37) to avoid parallel undesired reactions which lower the selectivity towards H_2.

$$C_2H_6O_2 \rightleftharpoons C_2H_{6-x}O_2 + \frac{x}{2}H_2 \tag{6.36}$$

$$C_2H_{6-x}O_2 \rightleftharpoons 2CH_yO \tag{6.37}$$

The adsorbed CH_yO species, such as formyl or methoxy, can undergo further dehydrogenation to form adsorbed CO, which can subsequently be converted by the WGSR into CO_2 and H_2. CO_2 can also be formed via reaction of water or hydroxyl groups with more hydrogenated surface intermediates formed during the process.

Undesirable alkanes, such as methane and ethane, can form on the catalyst surface by C–O bond scission, followed by hydrogenation of the resulting adsorbed species (Equations 6.38 and 6.39) [81]:

$$C_2H_6O_2 + 3H_2 \rightarrow 2CH_4 + 2H_2O \tag{6.38}$$

$$C_2H_6O_2 + 2H_2 \rightarrow 2C_2H_6 + 2H_2O \tag{6.39}$$

Moreover, alkanes can also be produced from the reaction of reforming products, H_2 and CO/CO_2, via methanation and Fischer–Tropsch processes.

Another undesired pathway is the cleavage of C–O bonds through dehydration reactions, which is typically favored on acidic catalyst supports, to form aldehydes, followed in some cases by hydrogenation over metals to produce alcohols.

The reforming of EG can be accompanied by significant production of acetic acid through bifunctional dehydrogenation/isomerization and dehydration/hydrogenation routes over the metal and the support.

These dehydrogenation/isomerization reactions are typically undesirable because the product acids are generally low in value, highly stable in aqueous solution and corrosive towards catalysts and equipment.

By altering the nature of catalytically active metal and by choice of a suitable catalyst support, it is possible to control hydrogen selectivity in the process.

Several experimental and theoretical studies have identified Pt as one of the most promising metals for the reforming of EG, showing good selectivity for H_2 production [81, 275, 282, 287]. Among the noble metals, Pt shows the highest catalytic activity. Pd is also selective towards H_2, whereas metals such as Ni and Ru exhibit good catalytic performance but favor the formation of alkanes. Rh, Ir, Co, Cu, Ag, Au and Fe have low catalytic activity for aqueous phase reforming reactions. In this respect, over SiO_2-supported group 8 metal catalysts, it has been found that, around 230 °C, the selectivity for H_2 production decreases in the following order [289]: Pd > Pt > Ni > Ru > Rh (Figure 6.11) The opposite trend has been observed for the

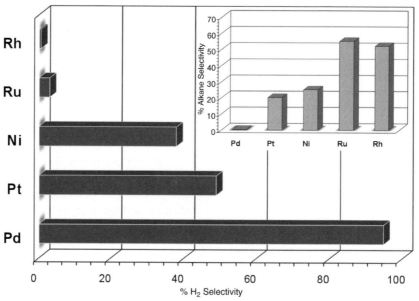

Figure 6.11 Hydrogen and alkane selectivity for reforming of 10 wt% aqueous EG at 225 °C and 22 bar (adapted from Ref. [289]).

alkane selectivity. Whereas Pt, Ni and Ru exhibit relatively high activities for the reforming reaction, only Pt and Pd also show relatively high selectivity for hydrogen generation. Moreover, Pt and Pd catalysts show low activity for C–O cleavage reactions and the series methanation and Fischer–Tropsch reactions between the reforming products, CO/CO_2 and H_2. Catalysts based on Pt/Pd can be used effectively for hydrogen production over a wider range of operating conditions, since they suffer less from selectivity issues.

The support can also influence the selectivity for H_2 production by catalyzing parallel dehydration pathways leading to the formation of alkanes. For example, the H_2 selectivity observed over silica-supported Pt is significantly lower than that with alumina-supported Pt [270]. This can be explained considering the acid–base properties of the support: the higher acidity of silica compared with alumina can facilitate acid-catalyzed dehydration reactions of EG, followed by hydrogenation on the metal surface to form an alcohol. The alcohol can subsequently undergo surface reactions to form alkanes (CH_4, C_2H_6), CO_2, H_2 and H_2O. This bifunctional dehydration/hydrogenation pathway consumes H_2, leading to decreased hydrogen selectivity and increased alkane selectivity.

In order to stabilize the effect of the support on the activity and selectivity for hydrogen production, several supported platinum catalysts have been prepared and tested [282]. At 225 °C, Pt supported on TiO_2 is the most active catalyst. In addition, Pt on alumina and activated carbon also show high activity, whereas Pt supported on SiO_2–Al_2O_3 and ZrO_2 show moderate catalytic activity for H_2 production. Lower activity has been observed for Pt supported on CeO_2 and ZnO, mainly due to the deactivation caused by hydrothermal degradation of the supports.

In addition to H_2 selectivity, there is an important aspect to be considered in the design of the support, namely its physical stability in the reaction environment. For example, supports such as ZrO_2, CeO_2 and TiO_2 are known to degrade or to dissolve in the aqueous reforming solution over a long period through sintering and phase transformation from metastable high to low surface area phases [290]. Therefore, they are unsuitable for long-term use. Addition of dopants such as Y [291] (for TiO_2 and ZrO_2) was shown to inhibit phase transition and sintering processes, but the effects on the chemistry of reforming reactions are still unknown. Among the supports mentioned above, Al_2O_3 shows better hydrothermal stability.

Although Pt-based catalysts showed the best catalytic performance for EG reforming, Ni-based catalysts have also been investigated owing to their low cost. One of the main problems with nickel-based catalysts is their tendency to produce alkanes. However, it was reported by Dumesic and co-workers that it is possible to control the selectivity of Ni-based systems through the introduction of modifiers in the catalyst formulation. In this way, it is possible to select the reaction conditions under which catalysts perform as well as or better than Pt catalysts for aqueous phase reforming reactions. The addition of Sn, Au or Zn to Ni, for example, improves the hydrogen selectivity of the catalyst, with Ni–Sn/Al_2O_3 being the most active and selective [292]. However, these materials are not stable with respect to time on-stream during aqueous phase reforming, exhibiting severe deactivation phenomena. Sn-modified Raney Ni catalysts have been developed due to the relatively stable performance of

Raney Ni-based systems under reaction conditions [293]. In this case, the introduction of Sn significantly decreases the rate of methane formation from C–O bond cleavage, while maintaining sufficiently high rates of C–C bond scission required for hydrogen formation. The alkane selectivity is also reduced [294]. For example, for aqueous reforming of 2 wt% ethylene glycol solutions, the addition of only 0.5 wt% Sn (Ni:Sn atomic ratio 400:1) reduces the rate of methane production by about half, whereas the rate of hydrogen production increases slightly. Methane production is nearly eliminated by the addition of more than 10 wt% Sn (Ni:Sn atomic ratio 18:1), while the rate of hydrogen production is decreased only slightly [292]. The beneficial effect of Sn on the selectivity for H_2 production may be caused by the presence of Sn at Ni-defect sites and by the formation of an Ni–Sn alloy surface through Sn migration into Ni particles after reduction in hydrogen and exposure to reaction conditions. The decoration of defect sites by Sn may thereby suppress methanation reactions. It is also possible that methanation reactions are suppressed by geometric effects caused by the presence of Sn on Ni–Sn alloy surfaces, for example by decreasing the number of surface ensembles composed of multiple nickel atoms necessary for CO or H_2 dissociation. Cleavage of C–O bonds may also be slowed on Sn-promoted Ni surfaces by electronic effects, because Sn has been reported to weaken the adsorption of carbide fragments (coke) that are known to be intermediates in the production of methane [294].

The addition of Sn to Raney Ni catalysts also improves the stability and corrosion resistance of the catalyst under aqueous phase reforming conditions.

Although addition of Sn to Raney Ni catalysts greatly reduces the rate of methane formation, to achieve the highest selectivities for production of H_2 it is also essential to minimize the partial pressures of the gases produced and their residence time in the reactor. In contrast, over unpromoted Raney Ni catalysts it is impossible to achieve this high selectivity under any conditions [292].

Several new bimetallic catalysts, such as PtNi, PtCo, PtFe and PdFe, have recently been identified for aqueous phase reforming of EG by testing over 130 metal combinations [285]. These new bimetallic catalysts are significantly more active than monometallic Pt and Pd catalysts. Density functional theory (DFT) calculations show an electronic modification of Pt upon alloying with Ni, Co or Fe and these calculations predict that Pt should segregate to the surface in these systems [295]. Experimental studies of PtNi, PtCo and PtFe supported on carbon have shown that the surfaces of these alloys are covered with Pt, leading to surface properties that are different than those of pure Pt [296, 297]. Alloying Pt with Ni, Co or Fe improves the activity for H_2 production by lowering the d-band center [283, 287], which causes a decrease in the adsorption energy of CO and hydrogen, thereby increasing the fraction of the surface available for reaction with EG. On Pd-based catalysts, the rate-limiting step for APR of EG is the WGSR and addition of a WGS promoter such as Fe_2O_3 to Pd can improve the catalytic activity for APR.

EG can be reformed to synthesis gas under autothermal conditions over noble metal catalysts:

$$C_2H_6O_2 + H_2O + \frac{1}{2}O_2 \rightarrow 2CO_2 + \frac{7}{4}H_2 \quad (6.40)$$

Rhodium catalysts with the addition of ceria on a γ-Al_2O_3 washcoat layer exhibit the best combination of high fuel conversion and high selectivity to H_2 near equilibrium. High selectivity to H_2 is achieved by adjusting the fuel-to-air and fuel-to-steam feed ratios, and also the catalyst. The addition of steam significantly suppresses CO selectivity while increasing selectivity to H_2 to as high as 92% near equilibrium. Minor products observed in the reactor effluent include methane, acetaldehyde and trace amounts of ethane and ethylene [298].

6.4.7
Glycerol

Glycerol, $HOCH_2$–$CH(OH)$–CH_2OH, also known as glycerin, is an important biomass-derived product. It can be produced by fermentation of sugars such as glucose. Furthermore, it is the major by-product of biodiesel production, which, over the last decade, has emerged as an important fuel for blending petroleum diesel [299]. Biodiesel and glycerol are produced from the catalytic transesterification of vegetable oils and fats with alcohol [300]. About 10 wt% of vegetable oil is converted into glycerol during the transesterification process. For about 4 L of biodiesel formed, approximately 0.3 kg of crude glycerol is produced. Such crude glycerol is of very low value due its high content of impurities [301]. The purification of crude glycerol from the biodiesel plants is a major issue. As the demand for and production of biodiesel grow, the quantity of crude glycerol generated is becoming considerable [302, 303]. Currently, glycerol is used in many applications but mainly in personal care, food, oral care, tobacco and polyurethane production and so on (Figure 6.12). Alternative applications need to be found [304].

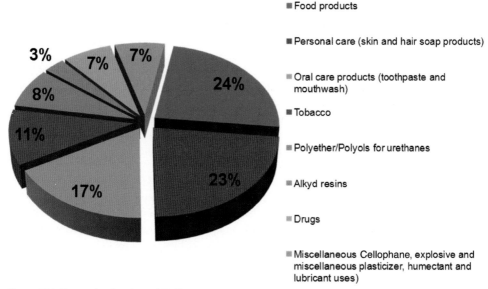

Figure 6.12 The market for glycerol [320].

An increase in biodiesel production not only will reduce the world market price of glycerol but also will generate environmental concerns associated with contaminated glycerol disposal. Although blending glycerol with gasoline is a possibility, the immiscibility of glycerol in gasoline hinders this option. Several attempts have been made to convert glycerol into a product that could be blended with gasoline. Glycerol can be catalytically converted into a mixture of lower alcohols with catalysts such as sulfated ZrO_2 or with microorganism such as *Klebsiella pneumoniae, Clostridium pasteurianum, Citrobacter freundii* and *Enterobacter agglomerans* [305–308]. Converting glycerol to a mixture of alcohols with specific concentrations would enable glycerol to be blended with gasoline [309]. Glycerol can also be processed into more valuable compounds which at the moment do not have a developed market due to their still too high production costs.

An attractive option is the use of glycerol as a hydrogen source. Theoretically, from 1 mol of glycerol it is possible to produce up to 4 mol of hydrogen. When glycerol is cracked at high temperature, syngas can be obtained and used as a feedstock in the Fischer–Tropsch synthesis to produce long-chain hydrocarbons ($-CH_2-$; green diesel) [310, 311]. Gases which are produced from thermal cracking of glycerol would have medium heating value and can be used as a fuel gas. Therefore, it has been proposed to produce value-added products, such as hydrogen or syngas, and medium heating value gases from glycerol using a fixed-bed reactor without a catalyst. Non-catalytic processes such as pyrolysis and steam gasification are technologies that can be used for this purpose. Pyrolysis is the high-temperature thermal cracking process of organic liquids or solids in the absence of oxygen [312]. Steam gasification produces a gaseous fuel with a higher hydrogen content than the pyrolytic process in the presence of oxygen and it reduces the diluting effect of nitrogen, used as a carrier gas in the pyrolysis, in the gas produced [313].

To date, only a few studies have been conducted on catalytic reforming of glycerol for hydrogen production. Dumesic and co-workers [81, 270, 314] produced hydrogen from glycerol by an aqueous phase reforming process. Czernik *et al.* [29] produced hydrogen via steam reforming of crude glycerol using a commercial nickel-based naphtha-reforming catalyst. Recently, other studies [298, 315, 316] reported the performance of noble metal-based catalysts for glycerol reforming.

The overall reaction of hydrogen production via steam reforming of glycerol is

$$C_3H_8O_3 + 3H_2O \rightarrow 3CO_2 + 7H_2 \tag{6.41}$$

which takes into account the contribution of the WGSR.

Steam reforming of glycerol for hydrogen production involves complex reactions. As a result, several intermediate by-products are formed and end up in the product stream, affecting the final purity of the hydrogen produced. Furthermore, the yield of hydrogen depends on several process variables, such as the system pressure, temperature and water-to-glycerol feed ratio.

Detailed thermodynamic analyses of the process were carried out recently [317, 318]. These studies reveal that the best conditions for producing hydrogen are a working temperature greater than 650 °C, atmospheric pressure and a molar ratio of water to glycerol of 9:1. Under these conditions, methane production is minimized and carbon formation is thermodynamically inhibited.

Hydrogen generation via steam reforming of glycerol has been studied at 500 and 600 °C over nickel catalysts supported on alumina modified with MgO, ZrO_2, CeO_2 or La_2O_3. The feed composition was varied from 1 to 10 wt% of glycerol in water, which is close to that obtained in the first phase of glycerol separation from biodiesel. The results indicate that all catalysts are able to convert the glycerol completely. The addition of promoters allows the hydrogen selectivity to be improved significantly, avoiding the formation of undesirable by-products. This improvement achieved its maximum for a 5 wt% La_2O_3-promoted catalyst [315]. Group 8–10 metals were loaded on the optimized La_2O_3-doped system and showed the following order of catalytic activity: Ru ≈ Rh > Ni > Ir > Co > Pt > Pd > Fe. Among the investigated catalysts there is little difference in the selectivity of the gaseous products. In fact, the support promotes the WGSR leading to high CO_2 selectivity. Among all metals, however, Ru exhibits the highest H_2 yield. For this metal, the influence of the support on the glycerol conversion was further examined. Al_2O_3, Y_2O_3, ZrO_2 and MgO were selected as support. The results indicate that Ru supported on Y_2O_3 gives the best performance in terms of high glycerol conversion, high H_2 yield and high resistance against carbon deposition [315].

A catalyst containing Cu, Ni and Pd has also been found to be effective under atmospheric pressure within the temperature range 500–600 °C towards both steam and autothermal reforming of glycerol [279].

As shown in Figure 6.13, the hydrogen yield for both reactions increases with temperature (in the range 550–850 °C), and for temperatures of 600 °C or higher the H_2 yield in autothermal reforming is greater than that in steam reforming.

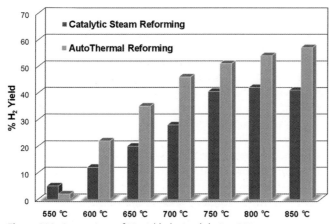

Figure 6.13 Comparison of H_2 yield obtained during CSR (S/C = 3) and ATR (S/C = 3, O/C = 0.3) of glycerol over Pd/Ni/Cu/K catalyst, as a function of reaction temperature (adapted from Ref. [279]).

The addition of oxygen as air leads to an increase in the H_2 yield through an increase in the conversion of glycerol to gaseous products. Moreover, it inhibits the formation of coke precursor.

The ATR of glycerol has also been examined over Pt- and Rh-based catalysts supported on alumina foams [298]. High selectivity to H_2 can be achieved by adjusting the fuel-to-air and fuel-to-steam feed ratios, and also the catalyst formulation. The addition of ceria to Al_2O_3 provides a catalyst which shows high fuel conversion and high selectivity to H_2.

Aqueous phase reforming of glycerol in several studies by Dumesic and co-workers has been reported [270, 275, 277, 282, 289, 292, 294, 319]. The first catalysts that they reported were platinum-based materials which operate at relatively moderate temperatures (220–280 °C) and pressures that prevent steam formation. Catalyst performances are stable for a long period. The gas stream contains low levels of CO, while the major reaction intermediates detected in the liquid phase include ethanol, 1,2-propanediol, methanol, 1-propanol, propionic acid, acetone, propionaldehyde and lactic acid. Novel tin-promoted Raney nickel catalysts were subsequently developed. The catalytic performance of these non-precious metal catalysts is comparable to that of more costly platinum-based systems for the production of hydrogen from glycerol.

6.5
Conclusions

The Kyoto protocol, together with the desire and need to reduce society's dependence on fossil fuels, has directed researchers' attention to the use of renewable resources. An alternative fuel should have superior environmental benefits over a fossil fuel, should be economically competitive, producible in such quantities as to make a meaningful impact on energy demands and to provide a net energy gain over the energy sources used to produce it. Biomass has been suggested as a potential candidate that can meet the challenges of sustainable and green energy systems.

The idea of creating 'biorefineries', plants similar in principle to petroleum refineries, has been proposed to produce both useful chemicals and fuels from biomass. The biorefinery can provide a means of transition to a more energy-efficient and environmentally sustainable chemical and energy economy through a highly integrated system of new chemical, biological and mechanical technologies that are optimized for energy efficiency and resource utilization. A biorefinery can produce fuels (low-value products) in high volumes, and high-value chemicals in low volumes, while generating electricity and process heat for its own use, and also it can produce surplus electricity for sale to the power grid.

The major components of biomass are organic oxygenates belonging to the following groups: acids, aldehydes, alcohols, ketones and phenols. Notably, the extent of matter did not allow a comprehensive discussion and we focused our attention here on some key oxygenated compounds.

A survey of the main routes to convert biomass-derived oxygenates selectively to hydrogen or alkanes over a variety of heterogeneous catalysts has been presented.

It was shown how strongly the efficiency of each process depends on the nature of the oxygenated compound. Furthermore, we stressed the concept that the successful development of one process lies largely in the design of a suitable catalytic material in conjunction with the optimization of the working conditions. In our opinion, the development of an efficient catalyst able to cope with the complex chemical nature of the oxygenated reactant is the real challenge. Moreover, improvements are needed to control and direct the reaction pathway to the desired products.

6.6
List of Abbreviations

APR	aqueous phase reforming
ATR	autothermal reforming
CCGT	combined cycle gas turbine
CFC	chlorofluorocarbon
CHP	combined heat and power
CPO	catalytic partial oxidation
CPOM	catalytic partial oxidation of methanol
CSR	catalytic steam reforming
CSRM	catalytic steam reforming of methanol
DFT	density functional theory
DME	dimethyl ether
DMFC	direct liquid methanol fuel cell
EG	ethylene glycol
LPG	liquid petroleum gas
SPC	solid-phase crystallization
PEM(FC)	proton exchange membrane (fuel cell)
PROX	preferential oxidation
r-WGSR	reverse water gas shift reaction
SOFC	solid oxide fuel cell
TWC	three-way catalyst
WGS(R)	water gas shift (reaction)
YSZ	yttria-stabilized-zirconia

Acknowledgments

Professor Mauro Graziani (University of Trieste) is warmly thanked for critical reading of the manuscript and for suggestions. Dr F. Colombo, Mr R. Crevatin, Mr E. Merlach and F. Pianigiani (University of Trieste) are acknowledged for qualified technical assistance. The University of Trieste, INSTM, PRIN2007 'Sustainable processes of 2nd generation for the production of H_2 from renewable resources' and FISR2002, are gratefully acknowledged for financial support.

References

1 Boyle, G., Everett, B. and Ramage, J. (eds) (2003) *Energy Systems and Sustainability. Power for a Sustainable Future*, Oxford University Press.
2 Ayres, R.U., Turton, H. and Casten, T. (2007) *Energy*, **32**, 634–648.
3 Campbell, C.J. and Laherrere, J.H. (1998) *Scientific American, March*, 60–65.
4 Kaspar, J., Fornasiero, P. and Hickey, N. (2003) *Catalysis Today*, **77**, 419–449.
5 Fornasiero, P. (2008) *Catalysis for the Protection of the Environment and Quality of Life* (ed. G. Centi), Chemical Sciences Engineering and Technology Resources. Eolss Publishers, Oxford,
6 Jaccard, M. (2005) *Sustainable Fossil Fuels*, Cambridge University Press.
7 Carrette, L., Friedrich, K.A. and Stimming, U. (2000) *ChemPhysChem*, **1**, 163–193.
8 Graziani, M. and Fornasiero, P. (eds) (2007) *Renewable Resources and Renewable Energy: a Global Challenge*, Taylor and Francis, New York.
9 Boyle, G. (ed.) (2004) *Renewable Energy. Power for a Sustainable Future*, Oxford University Press.
10 Corma Canos, A., Iborra, S. and Velty, A. (2007) *Chemical Reviews*, **107**, 2411–2502.
11 Lucia, L.A., Argyropoulos, D.S., Adamopoulos, L. and Gaspar, A.R. (2006) *Canadian Journal of Chemistry*, **84**, 960–970.
12 Metzger, J.O. (2006) *Angewandte Chemie International Edition*, **45**, 696–698.
13 Nath, K. and Das, D. (2003) *Current Science*, **85**, 265–271.
14 Kapdan, I.K. and Kargi, F. (2006) *Enzyme and Microbial Technology*, **38**, 569–582.
15 Momirlan, M. and Veziroglu, T.N. (2005) *International Journal of Hydrogen Energy*, **30**, 795–802.
16 Momirlan, M. and Veziroglu, T. (1999) *Renewable and Sustainable Energy Reviews*, **3**, 219–231.
17 Kothari, R., Buddhi, D. and Sawhney, R.L. (2008) *Renewable and Sustainable Energy Reviews*, **12**, 553–563.
18 Zhou, L. (2005) *Renewable and Sustainable Energy Reviews*, **9**, 395–408.
19 Elliott, D.C., Beckman, D., Bridgwater, A.V., Diebold, J.P., Gevert, S.B. and Solantausta, Y. (1991) *Energy and Fuels*, **5**, 399–410.
20 Huber, G.W., Iborra, S. and Corma, A. (2006) *Chemical Reviews*, **106**, 4044–4098.
21 Mohan, D., Pittman, J. and Steele, P.H. (2006) *Energy and Fuels*, **20**, 848–889.
22 Wyman, C.E., Dale, B.E., Elander, R.T., Holtzapple, M., Ladisch, M.R. and Lee, Y.Y. (2005) *BioreSource Technology*, **96**, 2026–2032.
23 Wyman, C.E., Dale, B.E., Elander, R.T., Holtzapple, M., Ladisch, M.R. and Lee, Y.Y. (2005) *BioreSource Technology*, **96**, 1959–1966.
24 Garcia, L., French, R., Czernik, S. and Chornet, E. (2000) *Applied Catalysis A–General*, **201**, 225–239.
25 Woodward, J., Orr, M., Cordray, K. and Greenbaum, E. (2000) *Nature*, **405**, 1014–1015.
26 Ni, M., Leung, D.Y.C., Leung, M.K.H. and Sumathy, K. (2006) *Fuel Processing Technology*, **87**, 461–472.
27 Aznar, M.P., Corella, J., Delgado, J. and Lahoz, J. (1993) *Industrial and Engineering Chemistry Research*, **32**, 1–10.
28 Hauserman, W.B. (1994) *International Journal of Hydrogen Energy*, **19**, 413–419.
29 Czernik, S., French, R., Feik, C. and Chornet, E. (2002) *Industrial and Engineering Chemistry Research*, **41**, 4209–4215.
30 Rioche, C., Kulkarni, S., Meunier, F.C., Breen, J.P. and Burch, R. (2005) *Applied Catalysis B–Environmental*, **61**, 130–139.
31 Takanabe, K., Aika, K.I., Seshan, K. and Lefferts, L. (2004) *Journal of Catalysis*, **227**, 101–108.

32 Wang, D., Montané, D. and Chornet, E. (1996) *Applied Catalysis A–General*, **143**, 245–270.
33 Wang, D., Czernik, S. and Chornet, E. (1998) *Energy and Fuels*, **12**, 19–24.
34 Wang, D., Czernik, S., Montané, D., Mann, M. and Chornet, E. (1997) *Industrial and Engineering Chemistry Research*, **36**, 1507–1518.
35 Demirbas, A. (2002) *Energy Conversion Management*, **43**, 1801–1809.
36 Oasmaa, A. and Kuoppala, E. (2003) *Energy and Fuels*, **17**, 1075–1084.
37 Oasmaa, A., Kuoppala, E. and Solantausta, Y. (2003) *Energy and Fuels*, **17**, 433–443.
38 Oasmaa, A., Kuoppala, E., Gust, S. and Solantausta, Y. (2003) *Energy and Fuels*, **17**, 1–12.
39 Bartholomew, C.H. and Ferrauto, R.J. (2006) *Fundamentals of Industrial Catalytic Processes*, John Wiley and Sons, Inc., New York.
40 Bond, G.C. (1974) *Heterogeneous Catalysis: Principles and Applications*, Clarendon Press, Oxford.
41 Ribeiro, F.H. and Somorjai, G.A. (1997) *Handbook of Heterogeneous Catalysis* (eds G. Ertl, H. Knozinger and J. Weitkamp), VCH Verlag GmbH.
42 Richardson, J.T. (1989) *Principles of Catalyst Development*, Springer.
43 Dumesic, J.A. (2005) *Industrial Bioprocessing*, **27**, 1–2.
44 Au, C.T., Ng, C.F. and Liao, M.S. (1999) *Journal of Catalysis*, **185**, 12–22.
45 Wittborn, A.M.C., Costas, M., Blomberg, M.R.A. and Siegbahn, P.E.M. (1997) *Journal of Chemical Physics*, **107**, 4318–4328.
46 Hickman, D.A. and Schmidt, L.D. (1993) *Science*, **259**, 343–346.
47 Hou, Z.Y., Yokota, O., Tanaka, T. and Yashima, T. (2003) *Applied Catalysis A–General*, **253**, 381–387.
48 Martinez, R., Romero, E., Guimon, C. and Bilbao, R. (2004) *Applied Catalysis A–General*, **274**, 139–149.
49 Roh, H.S., Potdar, H.S., Jun, K.W., Kim, J.W. and Oh, Y.S. (2004) *Applied Catalysis A–General*, **276**, 231–239.
50 Juan-Juan, J., Rohán-Martínez, M.C. and Illán-Gómez, M.J. (2004) *Applied Catalysis A–General*, **264**, 169–174.
51 Dias, J.A.C. and Assaf, J.M. (2003) *Catalysis Today*, **85**, 59–68.
52 Basile, F., Basini, L., Fornasari, G., Gazzano, M., Trifiro, F. and Vaccari, A. (1996) *Chemical Communications*, 2435–2436.
53 Basile, F., Fornasari, G., Gazzano, M., Kiennemann, A. and Vaccari, A. (2003) *Journal of Catalysis*, **217**, 245–252.
54 Takehira, K., Shishido, T., Wang, P., Kosaka, T. and Takaki, K. (2003) *Physical Chemistry Chemical Physics*, **5**, 3801–3810.
55 Takehira, K., Shishido, T., Wang, P., Kosaka, T. and Takaki, K. (2004) *Journal of Catalysis*, **221**, 43–54.
56 Takehira, K., Shishido, T. and Kondo, M. (2002) *Journal of Catalysis*, **207**, 307–316.
57 Basile, F., Fornasari, G., Trifiro, F. and Vaccari, A. (2001) *Catalysis Today*, **64**, 21–30.
58 Basile, F., Fornasari, G., Rosetti, V., Trifiro, E. and Vaccari, A. (2004) *Catalysis Today*, **91–92**, 293–297.
59 Shishido, T., Wang, P., Kosaka, T. and Takehira, K. (2002) *Chemistry Letters*, 752–753.
60 Arpentinier, P., Basile, F., Del Gallo, P., Fornasari, G., Gary, D., Rosetti, V. and Vaccari, A. (2005) *Catalysis Today*, **99**, 99–104.
61 Arpentinier, P., Basile, F., Del Gallo, P., Fomasari, G., Gary, D., Rosetti, V. and Vaccari, A. (2006) *Catalysis Today*, **117**, 462–467.
62 Morioka, H., Shimizu, Y., Sukenobu, M., Ito, K., Tanabe, E., Shishido, T. and Takehira, K. (2001) *Applied Catalysis A–General*, **215**, 11–19.
63 Basile, F., Fornasari, G., Trifiro, F. and Vaccari, A. (2002) *Catalysis Today*, **77**, 215–223.
64 Shishido, T., Yamamoto, Y., Morioka, H., Takaki, K. and Takehira, K. (2004) *Applied Catalysis A–General*, **263**, 249–253.

65 Budroni, G. and Corma, A. (2006) *Angewandte Chemie–International Edition*, **45**, 3328–3331.

66 Martin, J.E., Wilcoxon, J.P., Odinek, J. and Provencio, P. (2000) *Journal of Physical Chemistry B*, **104**, 9475–9486.

67 Haruta, M., Tsubota, S., Kobayashi, T., Kageyama, H., Genet, M.J. and Delmon, B. (1993) *Journal of Catalysis*, **144**, 175–192.

68 Haruta, M. (1997) *Catalysis Today*, **36**, 153–166.

69 Hickey, N., Arneodo Larochette, P., Gentilini, C., Sordelli, L., Olivi, L., Polizzi, S., Montini, T., Fornasiero, P., Pasquato, L. and Graziani, M. (2007) *Chemistry of Materials*, **19**, 650–651.

70 Sakurai, H., Ueda, A., Kobayashi, T. and Haruta, M. (1997) *Chemical Communications*, 271–272.

71 De Rogatis, L., Montini, T., Casula, M.F. and Fornasiero, P. (2008) *Journal of Alloys and Compounds*, **451**, 516–520.

72 Montini, T., De Rogatis, L., Gombac, V., Fornasiero, P. and Graziani, M. (2007) *Applied Catalysis B–Environmental*, **71**, 125–134.

73 Montini, T., Condò, A.M., Hickey, N., Lovey, F.C., De Rogatis, L., Fornasiero, P. and Graziani, M. (2007) *Applied Catalysis B–Environmental*, **73**, 84–97.

74 Fornasiero, P., Montini, T. and De Rogatis, L. (2007) *Diffusion and Reactivity of Solids* (ed. J.Y. Murdoch), Nova Science Publishers, pp. 65–109.

75 Roucoux, A., Schulz, J. and Patin, H. (2002) *Chemical Reviews*, **102**, 3757–3778.

76 Roucoux, A., Schulz, J. and Patin, H. (2003) *Advanced Synthesis and Catalysis*, **345**, 222–229.

77 Schulz, J., Roucoux, A. and Patin, H. (2000) *Chemistry – A European Journal*, **6**, 618–624.

78 Bridgwater, A.V. (1996) *Catalysis Today*, **29**, 285–295.

79 Galdamez, J.R., Garcia, L. and Bilbao, R. (2005) *Energy and Fuels*, **19**, 1133–1142.

80 Chheda, J.N., Huber, W. and Dumesic, J.A. (2007) *Angewandte Chemie–International Edition*, **46**, 2–22.

81 Simonetti, D.A. and Dumesic, J.A. (2008) *ChemSusChem*, **1**, 725–733.

82 Cortright, R.D., Davda, R.R. and Dumesic, J.A. (2002) *Nature*, **418**, 964–967.

83 Davda, R.R., Shabaker, J.W., Huber, G.W., Cortright, R.D. and Dumesic, J.A. (2005) *Applied Catalysis B–Environmental*, **56**, 171–186.

84 www.virent.com.

85 Pettersson, L.J. and Sjostrom, K. (1991) *Combustion Science and Technology*, **80**, 265.

86 Chen, X. and Mao, S.S. (2007) *Chemical Reviews*, **107**, 2891–2959.

87 Dickinson, A., James, D., Perkins, N., Cassidy, T. and Bowker, M. (1999) *Journal of Molecular Catalysis A–Chemical*, **146**, 211–221.

88 Zalas, M. and Laniecki, M. (2005) *Solar Energy Materials and Solar Cells*, **89**, 287–296.

89 Wu, N.L. and Lee, M.S. (2004) *International Journal of Hydrogen Energy*, **29**, 1601–1605.

90 Wu, N.L., Lee, M.S., Pon, Z.J. and Hsu, J.Z. (2004) *Journal of PhotoChemistry and PhotoBiology A–Chemistry*, **163**, 277–280.

91 Gondal, M.A., Hameed, A. and Yamani, Z.H. (2004) *Journal of Molecular Catalysis A–Chemical*, **222**, 259–264.

92 Hameed, A. and Gondal, M.A. (2005) *Journal of Molecular Catalysis A–Chemical*, **233**, 35–41.

93 Breen, J.P. and Ross, J.R.H. (1999) *Catalysis Today*, **51**, 521–533.

94 Choi, Y. and Stenger, H.G. (2002) *Applied Catalysis B–Environmental*, **38**, 259–269.

95 Lwin, Y., Daud, W.R.W., Mohamad, A.B. and Yaakob, Z. (2000) *International Journal of Hydrogen Energy*, **25**, 47–53.

96 Peppley, B.A., Amphlett, J.C., Kearns, L.M. and Mann, R.F. (1999) *Applied Catalysis A–General*, **179**, 21–29.

97 Raimondi, F., Geissler, K., Wambach, J. and Wokaun, A. (2002) *Applied Surface Science*, **189**, 59–71.

98 Segal, S.R., Anderson, K.B., Carrado, K.A. and Marshall, C.L. (2002) *Applied Catalysis A–General*, **231**, 215–226.

99 Palo, D.R., Dagle, R.A. and Holladay, J.D. (2007) *Chemical Reviews*, **107**, 3992–4021.

100 Renzi, S. and Crawford, R. (2000) *Corporate Environmental Strategy*, **7**, 38–50.

101 Amphlett, J.C., Creber, K.A.M., Davis, J.M., Mann, R.F., Peppley, B.A. and Stokes, D.M. (1994) *International Journal of Hydrogen Energy*, **19**, 131–137.

102 Asprey, S.P., Wojciechowski, B.W. and Peppley, B.A. (1999) *Applied Catalysis A–General*, **179**, 51–70.

103 Jiang, C.J., Trimm, D.L., Wainwright, M.S. and Cant, N.W. (1993) *Applied Catalysis A–General*, **97**, 145–158.

104 Santacesaria, E. and Carra, S. (1983) *Applied Catalysis*, **5**, 345–358.

105 Iwasa, N., Masuda, S., Ogawa, N. and Takezawa, N. (1995) *Applied Catalysis*, **125**, 145–157.

106 Takahashi, K., Takezawa, N. and Kobayashi, H. (1982) *Applied Catalysis*, **2**, 363–366.

107 Jiang, C.J., Trimm, D.L., Wainwright, M.S. and Cant, N.W. (1993) *Applied Catalysis A–General*, **93**, 245–255.

108 Amphlett, J.C., Evans, M.J., Mann, R.F. and Weir, R.D. (1985) *Canadian Journal of Chemical Engineering*, **63**, 605–611.

109 Shimokawabe, M., Asakawa, H. and Takezawa, N. (1990) *Applied Catalysis*, **59**, 45–58.

110 Koppel, R.A., Baiker, A., Schild, Ch. and Wokaun, A. (1991) *Catalysis and Automotive Pollution Control IV*, **63**, 59.

111 Yao, C.Z., Wang, L.C., Liu, Y.M., Wu, G.S., Cao, Y., Dai, W.L., He, H.Y. and Fan, K.N. (2006) *Applied Catalysis A–General*, **297**, 151–158.

112 Matter, P.H., Braden, D.J. and Ozkan, U.S. (2004) *Journal of Catalysis*, **223**, 340–351.

113 Nitta, Y., Fujimatsu, T., Okamoto, Y. and Imanaka, T. (1993) *Catalysis Letters*, **17**, 157–165.

114 Nitta, Y., Suwata, O., Ikeda, Y., Okamoto, Y. and Imanaka, T. (1994) *Catalysis Letters*, **26**, 345–354.

115 Koppel, R.A., Stocker, C. and Baiker, A. (1998) *Journal of Catalysis*, **179**, 515–527.

116 Matter, P.H. and Ozkan, U.S. (2005) *Journal of Catalysis*, **234**, 463–475.

117 Ritzkopf, I., Vukojevic, S., Weidenthaler, C., Grunwaldt, J.D. and Schüth, F. (2006) *Applied Catalysis A–General*, **302**, 215–223.

118 Gasser, D. and Baiker, A. (1989) *Applied Catalysis*, **48**, 279–294.

119 Mercera, P.D.L., van Ommen, J.G., Doesburg, E.B.M., Burggraaf, A.J. and Ross, J.R.H. (1991) *Applied Catalysis*, **71**, 363–391.

120 Iwasa, N., Masuda, S., Ogawa, N. and Takezawa, N. (1995) *Applied Catalysis A–General*, **125**, 145–157.

121 Takezawa, N., Kobayashi, H., Hirose, A., Shimokawabe, M. and Takahashi, K. (1982) *Applied Catalysis*, **4**, 127–134.

122 Iwasa, N., Mayanagi, T., Nomura, W., Arai, M. and Takezawa, N. (2003) *Applied Catalysis A–General*, **248**, 153–160.

123 Takezawa, N. and Iwasa, N. (1997) *Catalysis Today*, **36**, 45–56.

124 Iwasa, N., Kudo, S., Takahashi, H., Masuda, S. and Takezawa, N. (1993) *Catalysis Letters*, **19**, 211–216.

125 Agrell, J., Hasselbo, K., Jansson, K., Järås, S.G. and Boutonnet, M. (2001) *Applied Catalysis A–General*, **211**, 239–250.

126 Velu, S., Suzuki, K. and Osaki, T. (1999) *Catalysis Letters*, **62**, 159–167.

127 Cubeiro, M.L. and Fierro, J.L.G. (1998) *Applied Catalysis A–General*, **168**, 307–322.

128 Alejo, L., Lago, R., Pena, M.A. and Fierro, J.L.G. (1997) *Catalysis and Automotive Pollution Control IV*, **110**, 623–632.

129 Alejo, L., Lago, R., Pena, M.A. and Fierro, J.L.G. (1997) *Applied Catalysis A–General*, **162**, 281–297.

130 Cubeiro, M.L. and Fierro, J.L.G. (1998) *Journal of Catalysis*, **179**, 150–162.

131 Reitz, T.L., Ahmed, S., Krumpelt, M., Kumar, R. and Kung, H.H. (2000) *Journal of Molecular Catalysis A–Chemical*, **162**, 275–285.

132 Fierro, J.L.G. (2000) *Catalysis and Automotive Pollution Control IV*, **130**, A, 177–186.

133 Karim, A.M., Conant, T. and Datye, A. (2008) *Physical Chemistry and Chemical Physics*, **10**, 5584–5590.

134 Huang, T.J. and Wang, S.W. (1986) *Applied Catalysis*, **24**, 287–297.

135 Velu, S., Suzuki, K., Okazaki, M., Kapoor, M.P., Osaki, T. and Ohashi, F. (2000) *Journal of Catalysis*, **194**, 373–384.

136 Velu, S., Suzuki, K., Kapoor, M.P., Ohashi, F. and Osaki, T. (2001) *Applied Catalysis A–General*, **213**, 47–63.

137 Schrum, E.D., Reitz, T.L. and Kung, H.H. (2001) *Catalysis and Automotive Pollution Control IV*, **139**, 229–235.

138 Reitz, T.L., Lee, P.L., Czaplewski, K.F., Lang, J.C., Popp, K.E. and Kung, H.H. (2001) *Journal of Catalysis*, **199**, 193–201.

139 Turco, M., Bagnasco, G., Costantino, U., Marmottini, F., Montanari, T., Ramis, G. and Busca, G. (2004) *Journal of Catalysis*, **228**, 43–55.

140 Turco, M., Bagnasco, G., Costantino, U., Marmottini, F., Montanari, T., Ramis, G. and Busca, G. (2004) *Journal of Catalysis*, **228**, 56–65.

141 Murcia, M., Navarro, R.M., mez-Sainero, L., Costantino, U., Nocchetti, M. and Fierro, J.L.G. (2001) *Journal of Catalysis*, **198**, 338–347.

142 Shan, W.J., Feng, Z.C., Li, Z.L., Jing, Z., Shen, W.J. and Can, L. (2004) *Journal of Catalysis*, **228**, 206–217.

143 Papavasiliou, J., Avgouropoulos, G. and Ioannides, T. (2007) *Applied Catalysis B–Environmental*, **69**, 226–234.

144 Papavasiliou, J., Avgouropoulos, G. and Ioannides, T. (2004) *Catalysis Communications*, **5**, 231–235.

145 Liu, S.T., Takahashi, K. and Ayabe, M. (2003) *Catalysis Today*, **87**, 247–253.

146 Chen, G., Li, S., Li, H., Jiao, F. and Yuan, Q. (2007) *Catalysis Today*, **125**, 97–102.

147 Wang, C., Liu, N., Pan, L., Wang, S., Yuan, Z. and Wang, S. (2007) *Fuel Processing Technology*, **88**, 65–71.

148 Papavasiliou, J., Avgouropoulos, G. and Ioannides, T. (2006) *Applied Catalysis B–Environmental*, **66**, 168–174.

149 Wang, J.B., Tsai, D.H. and Huang, T.J. (2002) *Journal of Catalysis*, **208**, 370–380.

150 Mattos, L.V., de Oliveira, E.R., Resende, P.D., Noronha, F.B. and Passos, F.B. (2002) *Catalysis Today*, **77**, 245–256.

151 Navarro, R.M., Pena, M.A. and Fierro, J.L.G. (2007) *Chemical Reviews*, **107**, 3952–3991.

152 Demirbas, A. (2007) *Progress in Energy and Combustion Science*, **33**, 1–18.

153 de Oliveira, J.A. (2002) *Renewable and Sustainable Energy Reviews*, **6**, 129–140.

154 Niven, R.K. (2005) *Renewable and Sustainable Energy Reviews*, **9**, 535–555.

155 Haryanto, A., Fernando, S., Murali, N. and Adhikari, S. (2005) *Energy and Fuels*, **19**, 2098–2106.

156 Cavallaro, S., Chiodo, V., Freni, S., Mondello, N. and Frusteri, F. (2003) *Applied Catalysis A–General*, **249**, 119–128.

157 Marino, F., Boveri, M., Baronetti, G. and Laborde, M. (2004) *International Journal of Hydrogen Energy*, **29**, 67–71.

158 Goula, M.A., Kontou, S.K. and Tsiakaras, P.E. (2004) *Applied Catalysis B–Environmental*, **49**, 135–144.

159 Freni, S. (2001) *Journal of Power Sources*, **94**, 14–19.

160 Sun, J., Qiu, X.P., Wu, F., Zhu, W.T., Wang, W.D. and Hao, S.J. (2004) *International Journal of Hydrogen Energy*, **29**, 1075–1081.

161 Fatsikostas, A.N. and Verykios, X.E. (2004) *Journal of Catalysis*, **225**, 439–452.

162 Nishiguchi, T., Matsumoto, T., Kanai, H., Utani, K., Matsumura, Y., Shen, W.J. and Imamura, S. (2005) *Applied Catalysis A–General*, **279**, 273–277.

163 Sun, J., Qiu, X.P., Wu, F. and Zhu, W.T. (2005) *International Journal of Hydrogen Energy*, **30**, 437–445.

164 Elliott, D.J. and Pennella, F. (1989) *Journal of Catalysis*, **119**, 359.

165 Comas, J., Marino, F., Laborde, M. and Amadeo, N. (2004) *Chemical Engineering Journal*, **98**, 61–68.

166 Marino, F., Baronetti, G., Jobbagy, M. and Laborde, M. (2003) *Applied Catalysis A–General*, **238**, 41–54.

167 Batista, M.S., Santos, R.K.S., Assaf, E.M., Assaf, J.M. and Ticianelli, E.A. (2003) *Journal of Power Sources*, **124**, 99–103.

168 Kaddouri, A. and Mazzocchia, C. (2004) *Catalysis Communications*, **5**, 339–345.

169 Batista, M.S., Santos, R.K.S., Assaf, E.M., Assaf, J.M. and Ticianelli, E.A. (2004) *Journal of Power Sources*, **134**, 27–32.

170 Frusteri, F., Freni, S., Chiodo, V., Spadaro, L., Di Blasi, O., Bonura, G. and Cavallaro, S. (2004) *Applied Catalysis A–General*, **270**, 1–7.

171 Llorca, J., de la Piscina, P.R., Sales, J. and Homs, N. (2001) *Chemical Communications*, 641–642.

172 Montini, T. (2005) PhD Thesis, University of Trieste, Italy.

173 Song, S.M., Akande, A.J., Idem, R.O. and Mahinpey, N. (2007) *Engineering Applications of Artificial Intelligence*, **20**, 261–271.

174 Marino, F.J., Cerrella, E.G., Duhalde, S., Jobbagy, M. and Laborde, M.A. (1998) *International Journal of Hydrogen Energy*, **23**, 1095–1101.

175 Osaki, T., Horiuchi, T., Sugiyama, T., Suzuki, K. and Mori, T. (1998) *Journal of Non-Crystalline Solids*, **225**, 111–114.

176 Cavallaro, S., Mondello, N. and Freni, S. (2001) *Journal of Power Sources*, **102**, 198–204.

177 Marino, F., Boveri, M., Baronetti, G. and Laborde, M. (2001) *International Journal of Hydrogen Energy*, **26**, 665–668.

178 Haga, F., Nakajima, T., Miya, H. and Mishima, S. (1997) *Catalysis Letters*, **48**, 223–227.

179 Fajardo, H.V. and Probst, L.F.D. (2006) *Applied Catalysis A–General*, **306**, 134–141.

180 Sahoo, D.R., Vajpai, S., Patel, S. and Pant, K.K. (2007) *Chemical Engineering Journal*, **125**, 139–147.

181 Cavallaro, S. and Freni, S. (1996) *International Journal of Hydrogen Energy*, **21**, 465–469.

182 Aupretre, F., Descorme, C. and Duprez, D. (2002) *Catalysis Communications*, **3**, 263–267.

183 Cavallaro, S., Chiodo, V., Vita, A. and Freni, S. (2003) *Journal of Power Sources*, **123**, 10–16.

184 Navarro, R.M., Alvarez-Galvan, M.C., Sanchez-Sanchez, M.C., Rosa, F. and Fierro, J.L.G. (2005) *Applied Catalysis B–Environmental*, **55**, 229–241.

185 Erdohelyi, A., Rasko, J., Kecskes, T., Toth, M., Domok, M. and Baan, K. (2006) *Catalysis Today*, **116**, 367–376.

186 Vaidya, P.D. and Rodrigues, A.E. (2006) *Industrial and Engineering Chemistry Research*, **45**, 6614–6618.

187 Breen, J.P., Burch, R. and Coleman, H.M. (2002) *Applied Catalysis B–Environmental*, **39**, 65–74.

188 Casanovas, A., Llorca, J., Homs, N., Fierro, J.L.G. and de la Piscina, P.R. (2006) *Journal of Molecular Catalysis A–Chemical*, **250**, 44–49.

189 Homs, N., Llorca, J. and de la Piscina, P.R. (2006) *Catalysis Today*, **116**, 361–366.

190 Vasudeva, K., Mitra, N., Umasankar, P. and Dhingra, S.C. (1996) *International Journal of Hydrogen Energy*, **21**, 13–18.

191 Freni, S., Maggio, G. and Cavallaro, S. (1996) *Journal of Power Sources*, **62**, 67–73.

192 Garcia, E.Y. and Laborde, M.A. (1991) *International Journal of Hydrogen Energy*, **16**, 307–312.

193 Ioannides, T. (2001) *Journal of Power Sources*, **92**, 17–25.

194 Haga, F., Nakajima, T., Yamashita, K. and Mishima, S. (1998) *Reaction Kinetics and Catalysis Letters*, **63**, 253–259.

195 Fatsikostas, A.N., Kondarides, D.I. and Verykios, X.E. (2002) *Catalysis Today*, **75**, 145–155.

196 Velu, S., Satoh, N., Gopinath, C.S. and Suzuki, K. (2002) *Catalysis Letters*, **82**, 145–152.

197 Freni, S., Cavallaro, S., Mondello, N., Spadaro, L. and Frusteri, F. (2002) *Journal of Power Sources*, **108**, 53–57.

198 Llorca, J., Homs, N., Sales, J. and de la Piscina, P.R. (2002) *Journal of Catalysis*, **209**, 306–317.

199 Galvita, V.V., Semin, G.L., Belyaev, V.D., Semikolenov, V.A., Tsiakaras, P. and Sobyanin, V.A. (2001) *Applied Catalysis A–General*, **220**, 123–127.

200 Vaidya, P.D. and Rodrigues, A.E. (2006) *Chemical Engineering Journal*, **117**, 39–49.

201 Batista, M.S., Assaf, E.M., Assaf, J.M. and Ticianelli, E.A. (2006) *International Journal of Hydrogen Energy*, **31**, 1204–1209.

202 Mattos, L.V. and Noronha, F.B. (2005) *Journal of Catalysis*, **233**, 453–463.

203 Salge, J.R., Deluga, G.A. and Schmidt, L.D. (2005) *Journal of Catalysis*, **235**, 69–78.

204 Mattos, L.V. and Noronha, F.B. (2005) *Journal of Power Sources*, **152**, 50–59.

205 Yee, A., Morrison, S.J. and Idriss, H. (2000) *Catalysis Today*, **63**, 327–335.

206 Semelsberger, T.A., Borup, R.L. and Greene, H.L. (2006) *Journal of Power Sources*, **156**, 497–511.

207 Semelsberger, T.A. and Borup, R.L. (2005) *International Journal of Hydrogen Energy*, **30**, 425–435.

208 Lee, S.H., Cho, W., Ju, W.S., Cho, B.H., Lee, Y.C. and Baek, Y.S. (2003) *Catalysis Today*, **87**, 133–137.

209 Sardesai, A., Gunda, A., Tartamella, T. and Lee, S. (2000) *Energy Sources*, **22**, 77–82.

210 Sun, K., Lu, W., Qiu, F., Liu, S. and Xu, X. (2003) *Applied Catalysis A–General*, **252**, 243–249.

211 Takashi, O., Norio, I., Tutomu, S. and Yotaro, O. (2003) *Journal of Natural Gas Chemistry*, **12**, 219.

212 Jia, M., Li, W., Xu, H., Hou, S., Yu, C. and Ge, Q. (2002) *Catalysis Letters*, **84**, 31–35.

213 Vishwanathan, V., Roh, H.S., Kim, J.W. and Jun, K.W. (2004) *Catalysis Letters*, **96**, 23–28.

214 Kim, S.M., Lee, Y.S., BAE, J.W., Potdar, H.S. and Jun, K.W. (2008) *Applied Catalysis A - General*, **348**, 113–120.

215 Schiffino, R.S. and Merrill, R.P. (1993) *Journal of Physical Chemistry*, **97**, 6425–6435.

216 Xu, M., Lunsford, J.H., Goodman, D.W. and Bhattacharyya, A. (1997) *Applied Catalysis A–General*, **149**, 289–301.

217 Yaripour, F., Baghaei, F., Schmidt, I. and Perregaard, J. (2005) *Catalysis Communications*, **6**, 147–152.

218 Jiang, S., Hwang, Y.K., Jhung, S.H., Chang, J.S., Cai, T.X. and Park, S.E. (2004) *Chemistry Letters*, **33**, 1048–1049.

219 Hytha, M., Tich, I., Gale, J.D., Terakura, K. and Payne, M.C. (2001) *Chemistry – A European Journal*, **7**, 2521–2527.

220 Vishwanathan, V., Jun, K.W., Kim, J.W. and Roh, H.S. (2004) *Applied Catalysis A–General*, **276**, 251–255.

221 Galvita, V.V., Semin, G.L., Belyaev, V.D., Yurieva, T.M. and Sobyanin, V.A. (2001) *Applied Catalysis A–General*, **216**, 85–90.

222 Mathew, T., Yamada, Y., Ueda, A., Shioyama, H. and Kobayashi, T. (2005) *Catalysis Letters*, **100**, 247–253.

223 Matsumoto, T., Nishiguchi, T., Kanai, H., Utani, K., Matsumura, Y. and Imamura, S. (2004) *Applied Catalysis A–General*, **276**, 267–273.

224 Semelsberger, T.A., Brown, L.F., Borup, R.L. and Inbody, M.A. (2004) *International Journal of Hydrogen Energy*, **29**, 1047–1064.

225 Semelsberger, T.A., Ott, K.C., Borup, R.L. and Greene, H.L. (2005) *Applied Catalysis B–Environmental*, **61**, 281–287.

226 Sobyanin, V.A., Cavallaro, S. and Freni, S. (2000) *Energy and Fuels*, **14**, 1139–1142.

227 Takeishi, K. and Suzuki, H. (2004) *Applied Catalysis A–General*, **260**, 111–117.

228 Tanaka, Y., Kikuchi, R., Takeguchi, T. and Eguchi, K. (2005) *Applied Catalysis B–Environmental*, **57**, 211–222.

229 Wang, S., Ishihara, T. and Takita, Y. (2002) *Applied Catalysis A–General*, **228**, 167–176.

230 Hansen, J.B., Voss, B., Joensen, F. and Siguroardottir, I.D. (2007) SAE Paper 950063.

231 Bhattacharyya, A. and Basu, A.K. (1997) U.S. Patent 5,626,794.

232 Carpenter, I. and Hayes, J. (1999) World Patent 9948804.

233 Topsoe, H., Dybkjaer, I., Nielsen, P. and Voss, B. (1999) European Patent 0931762.

234 Takeishi, K. and Yamamoto, K. (2004) European Patent 1452230.

235 Ogawa, T., Inoue, N., Shikada, T. and Ohno, Y. (2003) *Journal of Natural Gas Chemistry*, **12**, 219–227.

236 Landalv, I. and Lindblom, M. (2007) World Patent 0240768.

237 CHRISGAS, http://www.chrisgas.com/.

238 Nilsson, M., Pettersson, L.J. and Lindstrom, B. (2006) *Energy and Fuels*, **20**, 2164–2169.

239 Faungnawakij, K., Tanaka, Y., Shimoda, N., Fukunaga, T., Kawashima, S., Kikuchi, R. and Eguchi, K. (2006) *Applied Catalysis A–General*, **304**, 40–48.

240 Kawabata, T., Matsuoka, H., Shishido, T., Li, D., Tian, Y., Sano, T. and Takehira, K. (2006) *Applied Catalysis A–General*, **308**, 82–90.

241 Baertsch, C.D., Komala, K.T., Chua, Y.H. and Iglesia, E. (2002) *Journal of Catalysis*, **205**, 44–57.

242 Barton, D.G., Soled, S.L. and Iglesia, E. (1998) *Topics in Catalysis*, **6**, 87–99.

243 Barton, D.G., Soled, S.L., Meitzner, G.D., Fuentes, G.A. and Iglesia, E. (1999) *Journal of Catalysis*, **181**, 57–72.

244 Kuba, S., Heydorn, P.C., Grasselli, R.K., Gates, B.C., Che, M. and zinger, H. (2001) *Physical Chemistry Chemical Physics*, **3**, 146–154.

245 Occhiuzzi, M., Cordischi, D., Gazzoli, D., Valigi, M. and Heydorn, P.C. (2004) *Applied Catalysis A–General*, **269**, 169–177.

246 Ramu, S., Lingaiah, N., Prabhavathi Devi, B.L.A., Prasad, R.B.N., Suryanarayana, I. and Sai Prasad, P.S. (2004) *Applied Catalysis A–General*, **276**, 163–168.

247 Scheithauer, M., Cheung, T.K., Jentoft, R.E., Grasselli, R.K., Gates, B.C. and Zinger, H. (1998) *Journal of Catalysis*, **180**, 1–13.

248 Triwahyono, S., Yamada, T. and Hattori, H. (2003) *Applied Catalysis A–General*, **242**, 101–109.

249 De Rossi, S., Ferraris, G., Valigi, M. and Gazzoli, D. (2002) *Applied Catalysis A–General*, **231**, 173–184.

250 Barton, D.G., Shtein, M., Wilson, R.D., Soled, S.L. and Iglesia, E. (1999) *The Journal of Physical Chemistry. B*, **103**, 630–640.

251 Nishiguchi, T., Oka, K., Matsumoto, T., Kanai, H., Utani, K. and Imamura, S. (2006) *Applied Catalysis A–General*, **301**, 66–74.

252 Lindstrom, B. and Pettersson, L.J. (2001) *Catalysis Letters*, **74**, 27–30.

253 Matsumura, Y., Okumura, M., Usami, Y., Kagawa, K., Yamashita, H., Anpo, M. and Haruta, M. (1997) *Catalysis Letters*, **44**, 189–191.

254 Usami, Y., Kagawa, K., Kawazoe, M., Matsumura, Y., Sakurai, H. and Haruta, M. (1998) *Applied Catalysis A–General*, **171**, 123–130.

255 Mathew, T., Yamada, Y., Ueda, A., Shioyama, H., Kobayashi, T. and Gopinath, C.S. (2006) *Applied Catalysis A–General*, **300**, 58–66.

256 Zhang, Q., Li, X., Fujimoto, K. and Asami, K. (2005) *Applied Catalysis A–General*, **288**, 169–174.

257 Laosiripojana, N. and Assabumrungrat, S. (2007) *Applied Catalysis A–General*, **320**, 105–113.

258 Basagiannis, A.C. and Verykios, X.E. (2006) *Applied Catalysis A–General*, **308**, 182–193.

259 Takanabe, K., Aika, K., Inazu, K., Baba, T., Seshan, K. and Lefferts, L. (2006) *Journal of Catalysis*, **243**, 263–269.

260 Takanabe, K., Aika, K., Seshan, K. and Lefferts, L. (2006) *Chemical Engineering Journal*, **120**, 133–137.

261 Vagia, E. Ch. and Lemonidou, A.A. (2007) *International Journal of Hydrogen Energy*, **32**, 212–223.

262 Iglesia, E., Barton, D.G., Biscardi, J.A., Gines, M.J.L. and Soled, S.L. (1997) *Catalysis Today*, **38**, 339–360.

263 Hu, X. and Lu, G. (2007) *Journal of Molecular Catalysis A–Chemical*, **261**, 43–48.

264 Sinfelt, J.H. and Yates, D.J.C. (1967) *Journal of Catalysis*, **8**, 82–90.

265 Grenoble, D.C., Estadt, M.M. and Ollis, D.F. (1981) *Journal of Catalysis*, **67**, 90–102.

266 Hu, X. and Lu, G. (2006) *Chemistry Letters*, **35**, 452–453.

267 Lichtenthaler, F.W. (2002) *Accounts of Chemical Research*, **35**, 728–737.

268 Lichtenthaler, F.W. and Peters, S. (2004) *Comptes Rendus Chimie*, **7**, 65–90.

269 Lichtenthaler, F.W. (1998) *Carbohydrate Research*, **313**, 69–89.

270 Blommel, P.G., Keenan, G.R., Rozmiarek, R.T. and Cortright, R.D. (2008) *International Sugar Journal*, **110**, 676–679.

271 Huber, G.W., Chheda, J.N., Barrett, C.J. and Dumesic, J.A. (2005) *Science*, **308**, 1446–1450.

272 Cheong, C.H.T. and Lun, Y.K. (2005) AIChE Annual Meeting Conference Proceedings, p. 13145.

273 Tanksale, A., Wong, Y., Beltramini, J.N. and Lu, G.Q. (2007) *International Journal of Hydrogen Energy*, **32**, 717–724.

274 Kabyemala, B.M., Adschiri, T., Malaluan, R.M. and Arai, K. (1999) *Industrial and Engineering Chemistry Research*, **38**, 2888.

275 Shabaker, J.W., Davda, R.R., Huber, G.W., Cortright, R.D. and Dumesic, J.A. (2003) *Journal of Catalysis*, **215**, 344–352.

276 Gallezot, P., Cerino, P.J., Blanc, B., Fleche, G. and Fuertes, P. (1994) *Journal of Catalysis*, **146**, 93.

277 Davda, R.R. and Dumesic, J.A. (2004) *Chemical Communications*, 36–37.

278 Huber, G.W., Cortright, R.D. and Dumesic, J.A. (2004) *Angewandte Chemie–International Edition*, **43**, 1549–1551.

279 Swami, S.M. and Abraham, M.A. (2006) *Energy and Fuels*, **20**, 2616–2622.

280 Blanc, B., Bourrel, A., Gallezot, P., Haas, T. and Taylor, P. (2000) *Green Chemistry*, **2**, 89–91.

281 Tronconi, E., Ferlazzo, N., Forzatti, P., Pasquon, I., Casale, B. and Marini, L. (1992) *Chemical Engineering Science*, **47**, 2451–2456.

282 Shabaker, J.W., Huber, G.W., Davda, R.R., Cortright, R.D. and Dumesic, J.A. (2003) *Catalysis Letters*, **88**, 1–8.

283 Chen, J.G., Skoplyak, O., Barteau, M.A. and Menning, C. (2006) ACS National Meeting Book of Abstracts, p. 232.

284 Liu, X., Shen, K., Wang, Y., Guo, Y., Yong, Z. and Lu, G. (2008) *Catalysis Communications*, **9**, 2316–2318.

285 Huber, G.W., Shabaker, J.W., Evans, S.T. and Dumesic, J.A. (2006) *Applied Catalysis B–Environmental*, **62**, 226–235.

286 Sengwa, R.J. and Kaur, K. (1999) *Journal of Molecular Liquids*, **82**, 231–243.

287 Skoplyak, O., Barteau, M.A. and Chen, J.G.G. (2006) *Journal of Physical Chemistry. B*, **110**, 1686–1694.

288 Matsuoka, K., Iriyama, Y., Abe, T., Matsuoka, M. and Ogumi, Z. (2005) *Electrochimica Acta*, **51**, 1085–1090.

289 Davda, R.R., Shabaker, J.W., Huber, G.W., Cortright, R.D. and Dumesic, J.A. (2003) *Applied Catalysis B–Environmental*, **43**, 13–26.

290 Lin, Y.S., Chang, C.H. and Gopalan, R. (1994) *Industrial and Engineering Chemistry Research*, **33**, 860–870.

291 Johnson, M.F. (1990) *Journal of Catalysis*, **123**, 245–259.

292 Shabaker, J.W., Huber, G.W. and Dumesic, J.A. (2004) *Journal of Catalysis*, **222**, 180–191.

293 Xie, F.Z., Chu, X.W., Hu, H.R., Qiao, M.H., Yan, S.R., Zhu, Y.L., He, H.Y., Fan, K.N., Li, H.X., Zong, B.N. and Zhang, X.X. (2006) *Journal of Catalysis*, **241**, 211–220.

294 Huber, G.W., Shabaker, J.W. and Dumesic, J.A. (2003) *Science*, **300**, 2075–2077.

295 Ruban, A.V., Skriver, H.L. and Norskov, J.K. (1999) *Physical Review B: Condensed Matter and Materials Physics*, **59**, 15990–16000.

296 Igarashi, H., Fujino, T., Zhu, Y., Uchida, H. and Watanabe, M. (2001) *Physical Chemistry Chemical Physics*, **3**, 306–314.

297 Wan, L.J., Moriyama, T., Ito, M., Uchida, H. and Watanabe, M. (2002) *Chemical Communications*, 58–59.

298 Dauenhauer, P.J., Salge, J.R. and Schmidt, L.D. (2006) *Journal of Catalysis*, **244**, 238–247.

299 Pramanik, T. and Tripathi, S. (2005) *Hydrocarbon Processing*, **84**, 49–54.

300 Srivastava, A. and Prasad, R. (2000) *Renewable and Sustainable Energy Reviews*, **4**, 111–133.

301 Thompson, J.C. and He, B.B. (2006) *Applied Engineering in Agriculture*, **22**, 261–265.

302 Bournay, L., Casanave, D., Delfort, B., Hillion, G. and Chodorge, J.A. (2005) *Catalysis Today*, **106**, 190–192.

303 Ramadhas, A.S., Jayaraj, S. and Muraleedharan, C. (2005) *Fuel*, **84**, 335–340.

304 Pagliaro, M., Ciriminna, R., Kimura, H., Rossi, M. and la Pina, C. (2007) *Angewandte Chemie–International Edition*, **46**, 4434–4440.

305 Menzel, K., Zeng, A.-P. and Deckwer, W.-D. (1997) *Enzyme and Microbial Technology*, **20**, 82–86.

306 Himmi, E.H., Bories, A. and Barbirato, F. (1999) *BioreSource Technology*, **67**, 123–128.

307 Nakamura, C.E. and Whited, G.M. (2003) *Current Opinion in Biotechnology*, **14**, 454–459.

308 Malinowski, J.J. (1999) *Biotechnology Techniques*, **13**, 127–130.

309 Fernando, S., Adhikari, S., Kota, K. and Bandi, R. (2007) *Fuel*, **86**, 2806–2809.

310 Chaudhari, S.T., Bej, S.K., Bakhshi, N.N. and Dalai, A.K. (2001) *Energy Fuels*, **15**, 736–742.

311 Steynberg, A.P. and Nel, H.G. (2004) *Fuel*, **83**, 765–770.

312 Cutler, A.H. and Antal, J. (1987) *Journal of Analytical and Applied Pyrolysis*, **12**, 223–242.

313 Franco, C., Pinto, F., Gulyurtlu, I. and Cabrita, I. (2003) *Fuel*, **82**, 835–842.

314 Shabaker, J.W. and Dumesic, J.A. (2004) *Industrial and Engineering Chemistry Research*, **43**, 3105–3112.

315 Hirai, T., Ikenaga, N.O., Miyake, T. and Suzuki, T. (2005) *Energy and Fuels*, **19**, 1761–1762.

316 Luo, N., Fu, X., Cao, F., Xiao, T. and Edwards, P.P. (2008) *Fuel*, **87**, 3483–3489.

317 Adhikari, S., Fernando, S. and Haryanto, A. (2007) *Energy and Fuels*, **21**, 2306–2310.

318 Adhikari, S., Fernando, S., Gwaltney, S.R., To, S.D.F., Bricka, R.M., Steele, P.H. and Haryanto, A. (2007) *International Journal of Hydrogen Energy*, **32**, 2875–2880.

319 Davda, R.R. and Dumesic, J.A. (2003) *Angewandte Chemie–International Edition*, **42**, 4068–4071.

320 Pachauri, N. and He, B. (2006) Valued-added utilisation of crude glycerol from biodiesel production: a survey of current research, *American Society of Agricultural and Biological Engineers Annual Meeting*, Portland, USA.

7
Electrocatalysis in Water Electrolysis
Edoardo Guerrini and Sergio Trasatti

7.1
Introduction

Electrolytic water splitting was the first electrochemical process to be performed. Historically, the first experiment on water electrolysis was attributed to Nicholson and Carlisle, who in 1800, using the newly invented Volta's pile, observed the formation of gaseous products in the laboratory [1]. In reality, there are documents proving that Volta himself noted the phenomenon a few years earlier, although he never reported the observation in a publication [2].

Despite such an early observation, the production of hydrogen by water electrolysis on an industrial scale dates back only to the beginning of the 1900s, with the increasing need for large amounts of hydrogen for ammonia synthesis. At that time, water electrolysis was the most economic among the chemical processes able to produce hydrogen with purity close to that necessary for the catalytic synthesis of ammonia. Reasons were the low cost of hydroelectric power and the flexibility of the process itself that allowed the use of seasonal peaks of electric power.

It is clear from the above that the cost of electricity is an additional factor that contributes to determining the economic convenience of electrochemical processes. The successive development of water electrolysis depended essentially on the oscillatory behavior of such a variable, which is in turn linked to political and environmental factors. Actually, the electrolytic method was largely applied until the mid-1900s, when the rising costs of electric power and the abundant availability of hydrogen from natural hydrocarbons confined the use of electrolytic hydrogen to cases where the purity of hydrogen is a necessary requirement, such as in the pharmaceutical industry, hydrogenation in the food industry and so on.

In the 1970s, a first serious energy crisis related to a sharp increase in the cost of fossil fuels led to a revival of interest in water electrolysis. This time the interest was focused on hydrogen as an energy vector that could possibly replace fossil fuels in the world economy. The aim was twofold: to become relatively independent of the influence of the oil market and to reduce the environmental impact of the

Catalysis for Sustainable Energy Production. Edited by P. Barbaro and C. Bianchini
Copyright © 2009 WILEY-VCH Verlag GmbH & Co. KGaA, Weinheim
ISBN: 978-3-527-32095-0

use of fossil fuels whose combustion is responsible for the production of 'greenhouse' gases, such as CO_2 and NO_x, and of polluting compounds such as SO_2 (acid rain).

In that scenario, the idea of a 'hydrogen economy' was put forward, where hydrogen is the engine of the world producing electric power in fuel cells and replacing oil as a fuel in internal combustion engines [3]. In this context, it should not be forgotten that a technology cannot be cleaner than the energy source used to drive it. In other words, the use of electric cars and of electrolytic hydrogen entails the consumption of large amounts of electricity to recharge batteries and to feed electrolyzers. The solution of this circle necessarily lies in the use of clean, renewable energy sources that also avoid the risk of exhaustion implicit in the use of fossil fuels.

In 1974, the International Association for Hydrogen Energy was founded, whose official journal is the *International Journal of Hydrogen Energy*. At the same time, funding agencies promoted research projects along the same lines. In the electrochemical field, investigations were pushed to improve the technology of hydrogen electrolysis, reducing at the same time the investment and operational costs of the electrolytic process. In particular, R&D programs were financed by the European Community in two sections in the period 1975–85. The advances achieved during these programs have been reviewed in two articles and a book [4–6].

In the meantime, the sharp decrease in the cost of fossil fuels to the levels before the 1970s led to a weakening (again!) of interest in a hydrogen economy and to a slowdown of research on improving water electrolysis.

Nowadays, the concerns about pollution and the recent dramatic increase in the price of oil have definitely boosted interest in hydrogen as an energy vector and a clean fuel, although the problem of a clean energy source is still far from being solved.

The splitting of water into gaseous H_2 and O_2 by the action of electricity:

$$H_2O + 2F \rightarrow H_2 + \frac{1}{2}O_2 \quad (7.1)$$

where F is the Faraday constant measuring 1 mol of electricity (96 485 C), is definitely an extremely clean process since no polluting by-products are formed.. Equation (7.1) explicitly shows that the cost of electricity is an additional economic burden, even though the product yield is 100%, that is, no electrical energy is wasted. Moreover, Equation (7.1) implicitly shows that environmental drawbacks involved in the production of electricity bear negatively on the general economy of the process.

A definite advantage of electrochemical technology is its reversibility. The reverse of reaction (7.1):

$$H_2 + \frac{1}{2}O_2 \rightarrow H_2O + 2F \quad (7.2)$$

occurs in an H_2–O_2 fuel cell. In fact, the same technology can be used for both processes, that is, the same cell can work as an electrolyzer or as a fuel cell, depending on the operating conditions. This is not the case with other competing reactions for the production of H_2.

The term 'water electrolysis' implicitly means that the electrochemical reactor does not contain pure water only. Conventional electrolysis requires that the solution should be electrically conducting for the process to proceed. This implies that an electrolyte should be dissolved in water. Whereas in other cases, for example electrochemical organic or inorganic processes, the presence of an inert electrolyte may constitute a problem for the separation of products, this is not the case for water electrolysis since gaseous products are obtained. Nevertheless, the electrolyte can give other kinds of problems, such as corrosion phenomena, poisoning of electrodes and so on.

The conventional technology of water electrolysis makes use of alkaline solutions [7]. In particular, about 30% KOH is used at about 80 °C. The use of KOH, although more expensive than NaOH, is dictated by two reasons: (1) KOH is more conductive (about 1.3 times) than NaOH and (2) KOH is chemically less aggressive than NaOH. A 30% concentration is used because the conductivity exhibits a maximum there.

Traditionally, the electrode materials for the conventional technology have been iron or mild steel for cathodes and nickel or nickel-covered iron for anodes. Among metallic elements, these materials show excellent chemical resistance to KOH and satisfactory electrochemical activity. During the past three decades several advances have been achieved in the technology, which have lead to an appreciable decrease in the cost of the electrical energy consumed in the electrolytic processes. In synthesis, advances have involved (1) activation of electrodes, (2) cell design, (3) solid polymer electrolyte (SPE) technology and (4) steam electrolysis. These items will be discussed separately in the following.

7.2
Thermodynamic Considerations

An electrochemical system differs from a chemical system in the presence of an additional variable. In a chemical system, the variables are temperature (T), pressure (p) and composition (μ_i, the chemical potential of component i). In an electrochemical system, in addition to T, p and μ_i, its electrical state is an additional variable. Since an electrochemical system consists of two electron conductors in contact with an ionic conductor (electrolyte), the electrical state is measured by the 'electrode potential'.

A chemical system in equilibrium is expressed by the condition

$$\Delta G = 0 \tag{7.3}$$

A chemical system is therefore unable to perform work. The equilibrium condition for an electrochemical system is expressed by

$$\Delta G = -nF\Delta E \tag{7.4}$$

where n is the number of moles of electricity involved in the transformation of 1 mol of product and ΔE is the potential difference measured between the two electrodes. It is clear from Equation (7.4) that an electrochemical system has energy stored in it and can perform or absorb work each time the equilibrium conditions are perturbed.

Let us consider the reaction

$$H_2O \rightleftharpoons H_2 + \frac{1}{2}O_2 \qquad (7.5)$$

which proceeds to the right in electrolyzers and to the left in fuel cells. The minimum potential difference that can be applied between the electrodes of a water electrolyzer without producing electrolysis is

$$\Delta E = \frac{\Delta \vec{G}}{nF} \quad (n=2) \qquad (7.6)$$

where $\Delta \vec{G} = -\Delta G$ in Equation (7.4), that is, it is a positive quantity. Since

$$\Delta \vec{G} = \mu_{H_2} + \frac{1}{2}\mu_{O_2} - \mu_{H_2O} \qquad (7.7)$$

from Equations (7.6) and (7.7) it follows that

$$\Delta E = \frac{\Delta \vec{G}^\circ}{2F} + \frac{RT}{2F} \ln\left(\frac{p_{H_2} p_{O_2}^{\frac{1}{2}}}{a_{H_2O}}\right) \qquad (7.8)$$

where p_i is the partial pressure of component i and a_{H_2O} is the activity of water. ΔE measures the difference between the reversible potentials of the anode (the O_2 electrode) and the cathode (the H_2 electrode). Formally, it is possible to write

$$\Delta E = E_a - E_c \qquad (7.9)$$

where E_a and E_c are measured with respect to the same reference electrode, formally a standard hydrogen electrode (SHE), practically a saturated calomel electrode (SCE) whose observed potential is 0.241 V (SHE).

Under standard conditions at 25 °C with $\Delta \vec{G} = 237.178$ kJ mol^{-1} [8]:

$$\Delta E^\circ = \frac{237.178}{2 \cdot 96.485} = 1.229 V \qquad (7.10)$$

Considering Equation (7.8), it is evident that ΔE is affected by p, T and composition. In particular, ΔE increases as p is increased whereas it decreases as T is increased mainly through the temperature dependence of $\Delta \vec{G}^\circ$. Thermodynamic considerations therefore suggest that the best conditions for energy saving would be met at high temperatures and low pressures. However, due to solvent evaporation, these conditions would be impractical, so that temperature can be substantially increased only by working at enhanced pressure.

The limitation of solvent evaporation could be avoided by working at high temperature with water vapor in the presence of a solid electrolyte (steam electrolysis). Nominally, at about 4000 °C, at which $\Delta \vec{G} = 0$, electroless thermal water splitting would be realized. In practice, problems of material stability, necessary heat sources, separation of products and so on make this thermodynamic possibility very difficult to realize.

Whereas $\Delta \vec{G}$ represents the reference value for kinetic considerations and for calculating the minimum electrical energy required, the energy balance is referred to $\Delta \vec{H}$:

$$\Delta \vec{G} = \Delta \vec{H} - T\Delta \vec{S} \tag{7.11}$$

The energy required to break and to form molecular bonds and to bring reactants and products to their reference states is in fact measured by the enthalpy $\Delta \vec{H}$, which thus defines the so-called thermoneutral potential (ΔE is used only for equilibrium quantities, while ΔV are potential differences away from equilibrium conditions):

$$\Delta V_{tn} = \frac{\Delta \vec{H}}{nF} \tag{7.12}$$

Under standard conditions at 25 °C, $\Delta V_{tn} = 1.481$ V. From thermodynamic considerations, electrolysis at $\Delta V < \Delta E_{tn}$ proceeds with heat absorption from the environment, whereas the opposite is the case at $\Delta V > \Delta E_{tn}$. At $\Delta V = \Delta E_{tn}$, no net exchange takes place between the cell and the environment and the term 'thermoneutral' has been coined to emphasize such a situation.

The efficiency with respect to the electrical energy used is thermodynamically defined by

$$\varepsilon = \frac{\Delta \vec{H}}{\Delta \vec{G}} = \frac{\Delta V_{tn}}{\Delta E} \tag{7.13}$$

which can therefore potentially be >1.

7.3
Kinetic Considerations

ΔE, as defined by Equation (7.6), has no practical meaning. In order to drive water electrolysis at a practical rate, a $\Delta V > \Delta E$ must be applied. This implies that part of the electrical energy is spent to overcome reaction resistances:

$$\Delta V = \frac{\Delta \vec{G}}{nF} + \text{dissipation} \tag{7.14}$$

The consumption of electrical energy in a cell where electrolysis is carried out at a current I with a potential difference ΔV is given by:

$$\Delta VIt = \Delta VQ/\text{kW h} \tag{7.15}$$

The economic target of an electrochemical process is to keep the electrical consumption to a minimum. On the basis of Equation (7.15), this implies that a process should be performed at the maximum I with the minimum ΔV. In industrial processes, it is the current to be fixed rather than the electrode potential difference,

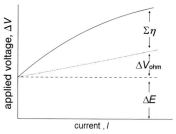

Figure 7.1 Variation of the voltage (ΔV) applied to an electrolysis cell with current flowing through the cell.

for two reasons: (i) it is technically easier and (ii) the current fixes the desired production rate.

The applied potential difference, according to Equation (7.14), includes a thermodynamic contribution and a kinetic contribution (dissipation). The latter is determined by the factors that govern the reaction resistances. To a first approximation, dissipation can be ascribed to three main factors (Figure 7.1):

$$\Delta V = \Delta E + \sum \eta + IR + \Delta V_t \tag{7.16}$$

7.3.1
Equilibrium Term (ΔE)

ΔE is the thermodynamic value that depends on the nature of the electrode reactions. It cannot be modified as long as the electrode reactions remain the same. Therefore, in principle, to reduce ΔE it would be necessary to build up a cell with a couple of electrode reactions with lower ΔE. Since the electrode reaction producing H_2 cannot be changed in that it is the target process of water electrolysis, it would in principle be possible to replace the anodic reaction of O_2 evolution with another oxidation reaction taking place at a more favorable potential (i.e. $< E_{O_2}$). On the other hand, the reaction of water electrolysis is not energetically symmetric, in the sense that O_2 evolution is much more demanding than H_2 evolution, so that energy dissipation is much more important at the anode than at the cathode. Therefore, getting rid of the reaction of O_2 evolution would open up interesting possibilities for energy saving.

In electrochemical terms, the replacement of an electrode reaction with another taking place at a more favorable electrode potential is called *depolarization*.

7.3.2
Ohmic Dissipation Term (IR)

In Equation (7.16), IR represents the energy dissipation related to ohmic drops in the electrolytic cell. These include a number of contributions: electrolyte, electrodes and electrical connections. The total resistance between the anode and

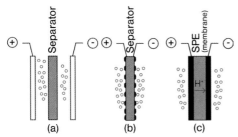

Figure 7.2 Cell designs. (a) Conventional alkaline electrolysis; (b) advanced alkaline electrolysis (zero gap); (c) SPE configuration (acid electrolysis).

the cathode of an electrolytic cell depends first on the conductivity of the electrolyte and is governed by the distance between the two electrodes. For this reason, conventional water electrolysis has traditionally been carried out in alkaline solutions under conditions of maximum conductivity (30% KOH, 80 °C).

The resistance between the two electrodes also includes other contributions. Conventional water electrolysis calls for a divided cell to keep the gaseous products separate. This entails the presence of a diaphragm, which is just a physical separation allowing electrical contact through pores (Figure 7.2a). The resistance of a diaphragm is certainly higher than that of a same volume of electrolyte. Therefore, diaphragms have to be kept as thin as possible, compatible with mechanical stability. Earlier diaphragms made of asbestos were later replaced by less noxious and mechanically more stable materials, such as ceramics, cermets and organic materials [9–11].

The problem of gas bubbles is to be added to the resistive effect of mechanical separators [12–14]. H_2 and O_2 are formed at the surface of the electrodes facing the separator. Hence the solution between electrode and diaphragm becomes saturated with gas bubbles that reduce the volume occupied by the electrolyte, thus incrementing the electrical resistance of the solution. In the conventional cell configuration, IR can be minimized, once the electrolyte and the separator are fixed, only by minimizing the distance between the anode and cathode. However, a certain distance between the electrode and separator must be necessarily maintained.

Electrode materials in principle should not bear on ohmic drop problems. In practice, they can do, if the conductivity is poor and the thickness of the active film is sizable. Thus, although in principle IR should not depend on electrode materials but only on cell design, in practice catalytically active materials with poor electrical performance cannot be used industrially since they would unacceptably increase the energy consumption for the product unit.

7.3.2.1 Cell Design

The problem of ohmic drops has recently been tackled by modifying the cell design. This is a matter of engineering, not of chemistry. The conventional design of divided cell has been replaced with a so-called *zero-gap* configuration [15], where the electrodes are directly pressed against the separator so that the interelectrode distance is minimized to the thickness of the separator (Figure 7.2b). Clearly, the interelectrode gap is zero only nominally, whereas the gap between the electrode and separator is effectively zero.

The new configuration calls for a modification also of the design of electrodes. These cannot be solid plates otherwise there would be no surface for electrical contact between anodic and cathodic electrolytes. Electrodes consist of perforated or stretched plates or meshes leaving free areas of separator available for electrical contacts.

Separators (diaphragms) never ensure complete mechanical separation between the two compartments of an electrolyzer. In particular, the possibility of free transport of a gas through the voids to the other compartment can create safety problems in the case of H_2 and O_2. For this reason, the zero-gap configuration has evolved to a completely new technology: the solid polymer electrolyte (SPE) technology (Figure 7.2c). Here the physical separator is replaced by a membrane permeable in particular only to the ions of water. Since an alkaline environment poses problems of chemical stability and correct functioning of an alkaline membrane, the SPE technology has been developed with a proton conductor as a membrane [16–19].

With the SPE configuration, electrodes are pressed against the membrane so that the thickness of the latter fixes the interelectrode gap. A membrane can be thinner than a physical separator, thus reducing such a kind of contribution to *IR*. The ensemble of electrode and membrane is usually referred to as a membrane–electrode assembly (MEA). This term is used irrespective of the operation of the cell, that is, for both electrolyzers and fuel cells.

More importantly, in the SPE technology gaseous H_2 and O_2 are liberated on the electrode surface on the side of the solution, thus solving the problem of the solution resistance due to the presence of bubbles. The membrane acts as an electrolyte. At the anode H_2O is oxidized to O_2 with liberation of H^+, which migrates through the membrane to the cathode, where it is reduced to H_2. In practice, a flow of solution is needed only at the anode to replace water molecules oxidized to O_2. However, the solution no longer needs to be conductive since no current passes through it. Actually, SPE electrolyzers are fed with plain water [20].

The SPE technology solves some problems but it poses others. In particular, the strong acid environment developed on the membrane calls for a complete change of electrode materials from those used in the conventional alkaline electrolysis. More specifically, especially the requirements for electrode materials for O_2 evolution are stringent since the anodic conditions are especially aggressive for corrosion problems.

7.3.3
Stability Term (ΔV_t)

The term ΔV_t in Equation (7.16) expresses a phenomenological observation: the potential difference applied to an electrolyzer tends to increase with time.

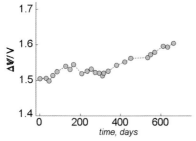

Figure 7.3 Performance of a conventional electrolyzer as a function of time. Data from [21].

In other words, ΔV_t measures the drift of ΔV with time as a consequence of performance degradation. The latter can be due to various reasons: loss of activity of electrode materials, increase in ohmic drops, that is, the various contributions to ΔV with the exception of ΔE, the magnitude of which is fixed by thermodynamics. As a concise term, ΔV_t can be called instability.

The most general case is that of instability of electrode materials that can deteriorate for various reasons (see later). A practical example of instability of a water electrolyzer is given in Figure 7.3 [21]. The applied voltage ΔV is seen to increase with time, which entails a higher consumption of electrical energy. Since energy consumption is the critical factor discriminating between electrochemical and non-electrochemical technology for a given process, the importance of ΔV_t can hardly be overemphasized.

Unfortunately, activity and stability for a given material as a rule do not proceed in parallel. Rather, they vary in opposite directions. Hence the search for more and more active materials must find a compromise with an acceptable stability. Academic research as a rule overemphasizes activity, whereas industry does so for stability. For this reason, it can happen that academic research reports very interesting activity properties but without any industrial interest. On the other hand, industrial breakthroughs may be emphasized that have in fact no scientific relevance. This is just to stress that a single point of view never offers an exhaustive picture of a problem.

7.3.4
Overpotential Dissipation Term ($\Sigma\eta$)

The term $\Sigma\eta$ in Equation (7.16) is the sum of anodic and cathodic overpotentials and expresses the way in which the fourth variable in electrochemistry, the electrical variable, operates to govern the rate of electrode reactions. Overpotentials are essentially determined by the activation energy of the electrode reaction. Overpotentials are the applied potential difference, in excess of the thermodynamic value, that is spent to overcome the activation energy.

The relationship between η and I, the reaction rate of an electrode reaction, is expressed by the Butler–Volmer equation, whose model describes a linear variation of the activation energy with the applied overpotential [22]. Hence, in electrochemistry, reaction rates can be controlled and varied by simply turning the knob of an appropriate instrument (galvanostat) without any need to operate on the other variables of chemical systems. That is the reason why in principle electrochemical processes can proceed at room temperature and atmospheric pressure, that is, under mild experimental conditions. As already discussed above, operating on p and especially on T helps to improve and optimize the performance of an electrolytic cell, but it is not an absolute experimental necessity.

According to the Butler–Volmer equation, the dependence of η on I is linear in a range of few millivolts around the reversible electrode potential, whereas it becomes logarithmic at $\eta > 50$–100 mV away from equilibrium conditions, depending on the degree of reversibility of the specific electrode reaction:

$$\eta = a + b \ln I \tag{7.17}$$

Equation (7.17) is the Tafel equation and expresses the way in which the applied potential difference operates to enhance the reaction rate [22]. Since the unit of η is volts, the units of a and b are also volts; a is called the Tafel intercept, that is, the overpotential at $I = 1$ (which depends on the units of I, A or mA or μA); b is known as the Tafel slope, that is, the variation of η per decade of current.

The current I is an extensive quantity, in that it depends on the size of the electrode. For this reason, the reaction rate is conveniently referred to the unit surface area ($I/S = j$, current density). Even so, the current density continues to be an extensive quantity if referred to the geometric (projected) surface area since electrodes are as a rule rough and the real surface does not coincide with the geometric surface [23]. Conversely, b is an *intensive* quantity, in that it depends only on the reaction mechanism and not on the size of the electrode. The term b is the most important kinetic parameter in electrochemistry also because of the easy and straightforward procedure for its experimental determination. Most electrode mechanisms can be resolved on the basis of Tafel lines only.

7.3.5
Electrocatalysis

Activation energies depend on the strength of chemical bonds formed and/or broken during a given reaction. In particular, electrode reactions are heterogeneous reactions since they occur at the boundary between two immiscible phases. This implies that bonds are formed between the electrode surface and reactants, products or, most often, intermediates. It is thus self-evident that if the material of the electrode is modified, the activation energy and/or the reaction mechanism will also be modified. This scenario is typical of heterogeneous catalysis and in fact it is referred to as heterogeneous catalysis at electrodes, which in a single word has become the well-known term *electrocatalysis* [24]. Incidentally, it is convenient to use this occasion to warn that in many cases the term *electrocatalysis* is meant to indicate that a given reaction is catalyzed electrochemically, that is, by electrons formed electrochemically to be transferred from the electrode to the reactant. This is a completely wrong concept, definitely to be rejected.

According to Equation (7.17) (Tafel line), an electrode reaction can be accelerated by simply increasing the overpotential (i.e. either making E more positive for anodic reactions or more negative for cathodic reactions). Nevertheless, the basic overpotential (a) cannot be changed, unless the electrode surface area is modified (geometric factors). However, if at constant I (or, better, j) the electrode material is changed, the overpotentials will become different because the activation energies are different (electronic factors). This is the very target of electrocatalysis: a search for new materials and/or new operating conditions in order to: (i) improve activity, efficiency and selectivity, (ii) reduce investment and/or operational costs, (iii) increase lifetime (stability) and (iv) avoid pollution.

The targets of electrocatalysis are at the basis of recent developments in the field of water electrolysis. First, it is necessary to distinguish between materials evaluation and materials selection. The former is the search for materials with better and better properties for the wanted electrode process. The latter implies global considerations of applicability. This is probably what makes academic research differ from R&D. The former is favored by scientifically exciting performance, in the latter it is necessary to find a compromise between, for instance, activity and stability or between efficiency and economic convenience.

Point (i) above implies the search for new procedures to prepare a given electrode material or new composite materials to look for possible synergetic effects. The latter usually are electronic factors, whereas the former often result in geometric effects. It is therefore necessary to be able to distinguish between the two effects to establish with certainty some predictive basis for the design and optimization of electrocatalysts.

Point (ii) above leads in general to the replacement of expensive with cheaper materials. This explain the great efforts in looking for less noble compounds that are usually less active and must therefore be improved with specific approaches. Also, expensive materials can be retained but employed in much lower amounts. This concept has led to the development of the so-called activated electrodes. These consist in an inert (and inexpensive) support on which the expensive (and active) material is deposited as a film of a few micrometers thickness. This minimizes the cost of the valuable component without losing its high activity. The configuration of activated electrodes is nowadays very general. It entails problems of reproducibility of preparation, especially of adhesion of the overlayer to the support. Hence, in this respect, point (ii) converges with point (iii) above to find a compromise between activity and stability.

Finally, point (iv) has become a more rigid discriminant than other criteria. Materials with even a small probability of pollution are no longer tolerated. For this reason, some of the most common electrode materials during the last century have been definitely banned from applications, for example, lead and mercury.

7.3.5.1 Theory of Electrocatalysis

Electrocatalytic activity is always assessed on a relative basis by comparing the current density at constant potential or the potential at constant current density. If current density is normalized to the apparent (geometric) surface area, apparent electrocatalytic effects will also include geometric effects. If current density is referred to the real surface area, true electronic effects will emerge. The determination of the real surface area is not at all an obvious operation and only in a very few cases have reliable approaches been developed [23]. In practice, fairly sound methods are available only for a few pure metal electrodes (Pt, Au, Ag, Hg) that have no practical impact in water electrolyzers. For materials of industrial interest, only approximate approaches are still available.

Comparison of Tafel lines for different materials is the most straightforward way to assess the relative activity. Since the slope of a Tafel line is determined by the mechanism of the electrode reaction, it follows that the Tafel lines of two materials can intersect, so that one of the materials turns out more active at low current density, and the other at high current density (Figure 7.4). Whereas the relative position of two Tafel lines can be modified by modifying both geometric and electronic factors, the Tafel slope changes only if the mechanism changes. In this respect, the search for new materials or the optimization of existing materials is oriented to obtain as low Tafel slopes as possible. Since the latter expresses the rate of increase in overpotential with increasing current, lower Tafel slopes allow lower energy consumption for the same production rate.

The theory of electrocatalysis is still in its infancy. It was developed first for the hydrogen evolution reaction in the second half of the 1900s. The grounds can be traced back in a seminal paper by Horiuti and Polanyi [25]. Accordingly, for a simple one-electron electrode reaction:

$$A + e \rightarrow B_{ads} \tag{7.18}$$

producing an adsorbed intermediate, the activation energy will be a function of the strength of interaction of the intermediate with the electrode material. The stronger the adsorption bond, the lower is the activation energy. Simple kinetic considerations show that if reaction (7.18) is the rate-determining step, the reaction rate will be an exponential function of $D(M-B)$, where D is the bond strength and M is an electrode surface site. Therefore, it turns out that

$$\log j \propto D(M-B) \tag{7.19}$$

that is, the reaction is facilitated by materials adsorbing B more strongly.

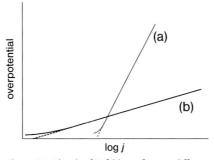

Figure 7.4 Sketch of Tafel lines for two different materials.

The situation is more complex, however, than it appears at first sight. As $D(M–B)$ increases, the surface of M will be more and more occupied by adsorbed B, which in turn will subtract part of the electrode surface to further discharge of A. In other words, B_{ads} acts as a sort of inhibitor for the electrode reaction. Since the amount of B_{ads} (ϑ_B = coverage) is governed by an adsorption isotherm, that is, ϑ_B is a function of $D(M–H)$, $\log j$ turns out also to be a linear function of $-D(M–B)$, that is, it is depressed by an increase in coverage with the intermediate [24].

Since $\log j$ increases linearly with $D(M–B)$ for energetic reasons, but decreases linearly with $D(M–B)$ for reasons of steric inhibition, at low $D(M–B)$ the former will prevail over the latter, whereas at high $D(M–B)$ the latter will prevail over the former. The outcome (Figure 7.5) is a so-called *volcano* curve, where poor catalysts are located on the extreme left-hand side, strong catalysts are located in the extreme right-hand side and moderate catalysts are placed around the apex of the curve. The curve is a graphical representation of the Sabatier principle according to which the best catalysts are those adsorbing relevant species neither too weakly nor too strongly. Volcano curves are known also for catalytic reactions (on the other hand the principles are precisely the same), the only difference being that they are called Balandin curves.

Volcano curves lay down the grounds for a predictive tool in electrocatalysis. One definite problem in establishing a volcano curve is the actual values of the quantity $D(M–B)$, which is seldom known for conditions resembling those relevant to an electrode reaction. As illustrated later, this issue is approached case by case, depending on the nature of the electrode reaction.

A complete theory of electrocatalysis leading to volcano curves has been developed only for the process of hydrogen evolution and can be found in a seminal paper by Parsons in 1958 [26]. The approach has shown that a volcano curve results irrespective of the nature of the rate-determining step, although the slope of the branches of the volcano may depend on the details of the reaction mechanism.

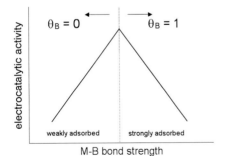

Figure 7.5 Sketch of the dependence of the electrocatalytic activity on the intermediate adsorption bond strength (volcano curve).

7.4
The Hydrogen Evolution Reaction

The discharging particles producing H_2 evolution at a cathode are H_2O molecules in alkaline solutions and protons in acidic solutions [27]. Thus, at intermediate pH there may be a transition from H^+ to H_2O with increasing current density, which manifests itself as a maximum in overpotential. Usually H_2 evolution is carried out in strongly alkaline or strongly acidic environments. Since H_2 evolution produces alkalinization of the catholyte:

$$H_2O + e \rightarrow \frac{1}{2}H_2 + OH^- \qquad (7.20)$$

a strongly alkaline medium always develops close to the cathode surface, which would make the pH control of solutions of weak acidity or alkalinity very difficult.

7.4.1
Reaction Mechanisms

The first step in the mechanism of H_2 evolution is always the formation of an adsorbed H intermediate, in both acids and alkalis:

$$H_2O + e \rightarrow H_{ads} + OH^- \qquad (7.21a)$$

or

$$H^+ + e \rightarrow H_{ads} \qquad (7.21b)$$

followed by either *electrochemical* desorption:

$$H_{ads} + H_2O + e \rightarrow H_2 + OH^- \qquad (7.22a)$$

or

$$H_{ads} + H^+ + e \rightarrow H_2 \qquad (7.22b)$$

or by *chemical* desorption:

$$H_{ads} + H_{ads} \rightarrow H_2 \qquad (7.23)$$

The choice of the route (7.21) → (7.22) or (7.21) → (7.23) depends on the electrocatalytic properties of the material. As a rule, electrode materials with weak $D(M-H)$ develop H_2 via (7.21) → (7.22) with step (7.21) being rate determining, whereas materials with strong $D(M-H)$ liberate H_2 via (7.21) → (7.22) with step (7.22) being rate determining. Accordingly, H_2 is evolved via (7.21) → (7.23) with step (7.23) being rate determining by materials with $D(M-H)$ values close to the apex of the volcano curve.

The above mechanism can be indicated synthetically by a sequence of letters ($E \equiv$ electrochemical step, $C \equiv$ chemical step). Thus, (7.21) → (7.22) is EE whereas (7.21) → (7.23) is EEC – two H_{ad} need be formed in step (7.21) to produce one H_2 in step (7.23). In general, in the sequence of letters the last step is rate determining. Since the Tafel slope is determined by the reaction mechanism, it can be shown that if reaction (7.21) is rate determining (E), the Tafel slope (at 25 °C) is 120 mV per decade of current (density). In the case of the EE mechanism, the Tafel slope is predicted to be 40 mV, and for the EEC mechanism only 30 mV [22]. These values of b are valid for low ϑ_H. For high ϑ_H the value of b may depend on ϑ_H.

7.4.2
Electrocatalysis

According to the fundamental theory of electrocatalysis, adsorption of atomic hydrogen on the electrode surface enhances proton discharge while slowing H adsorption. Hence for metals, activity is predicted to vary with heat of adsorption as a volcano-shaped curve. The predictions of the theory have been confirmed experimentally [28], although collecting data under perfectly comparable conditions is difficult. Nevertheless, as Figure 7.6 shows, a plot of $\ln j$ (activity) at constant potential (viz. overpotential) against the M–H bond strength reproduces a fairly symmetric volcano-shaped curve as expected. A problem is the assessment of the value of $D(M–H)$ under operating conditions, especially because $D(M–H)$ depends on the coverage ϑ_H. It has been reported, however, that $D(M–H)$ can be approached, to a first approximation, by the quantity determined experimentally in gas-phase adsorption, neglecting solvent displacement effects in solution.

In Figure 7.6, the sp-metals are located on the left-hand branch of the volcano curve that also includes the metals of group 1B (Cu, Ag, Au). These metals adsorb H rather weakly and they evolve H_2 via a primary discharge as a rate-determining step ($b = 120$ mV). Cu, Ag and Au show higher activity since they, particularly Cu, adsorb H more strongly than sp-metals. The equilibrium coverage is always small and the value of $D(M–H)$ is thus the governing factor. Nevertheless, $D(M–H)$ is $< 1/2 D(H–H)$, which implies a positive $\Delta H(H_2)_{ads}$.

Most transition metals (for Fe, Co and Ni see the following paragraphs) are located on the descending branch of the volcano curve. For these metals, the adsorption of H is strong and ϑ_H close to saturation, making H_2 desorption from the surface difficult. For these metals, the inhibiting effect of ϑ_H prevails over the

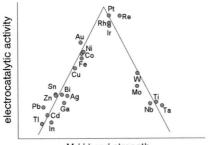

Figure 7.6 Experimental volcano curve for H_2 evolution on metals. M–H bond strength from overpotential data. Adapted from Ref. [28].

enhancing effect of $D(M–H)$. The *electrochemical* desorption with step (7.22) being rate determining governs the mechanism of H_2 evolution. Since ϑ_H is close to saturation, a Tafel slope b between 40 and 120 mV is as a rule observed, depending on $D(M–H)$. For transition metals, $D(M–H)$ is generally so strong as to be $> {}^1/_2 D(M–H)$, which entails a high negative $\Delta H(H_2)_{ads}$.

It is intriguing that analysis of the volcano curve predicts that the apex of the curve occurs at $\Delta H(H_2)_{ads} = 0$ (formally, $\Delta G = 0$) [26]. This value corresponds to the condition $D(M–H) = 1/2 D(H–H)$, that is, forming an M–H bond has the same energetic probability as forming an H_2 molecule. This condition is that expressed qualitatively by the Sabatier principle of catalysis and corresponds to the situation of maximum electrocatalytic activity. Interestingly, the experimental picture shows that the group of precious transition metals lies close to the apex of the curve, with Pt in a dominant position. It is a fact that Pt is the best catalyst for electrochemical H_2 evolution; however, its use is made impractical by its cost. On the other hand, Pt is the best electrocatalyst on the basis of electronic factors only, other conditions being the same.

The position of Fe, Co and Ni in Figure 7.6 is intriguing. As transition metals, they should appear on the descending branch of the curve and this is the case if values of $D(M–H)$ derived from gas-phase experiments are used. If, however, $D(M–H)$ is obtained under operational conditions (during H_2 evolution), the group appears on the ascending branch as shown in the figure [28]. This apparent contradiction can be reconciled if the phenomenon of H *absorption* is considered [27]. In some metals, H adsorbed on the surface shows a trend to penetrate into the lattice depending on the particular conditions. H penetration is tantamount to the formation of a metal hydride. H absorption affects the properties of the metal surface. In particular, subsurface H decreases the M–H bond strength of the metal surface since it saturates part of the bonding capacity of the atoms on the external surface.

The apparently contradictory behavior of Fe, Co and Ni is in fact explained by their ability to absorb atomic hydrogen, which reduces the strength of H adsorption. The shift of Fe, Co and Ni from the descending to the ascending branch of the volcano curve indicates that $D(M–H) > {}^1/_2 D(H–H)$ in the gas phase but $< {}^1/_2 D(H–H)$ in solution under H_2 evolution. These phenomena of H absorption are presumably responsible for the time-dependent properties of some electrodes during H_2 discharge.

While Figure 7.6 shows pure metals only, a volcano curve suggests that modulation of $D(M–H)$ would lead to tailoring of the activity of electrodes. It has therefore been argued that combination of a metal on the descending branch with a metal on the ascending branch should result in a compound with an intermediate value of $D(M–H)$, and hence higher activity. This argument has later been standardized as combination of hyper-d-electron metals with hypo-d-electron metals, in other words, a metal at the end with a metal at the beginning of a transition period [29–31].

Experiments have confirmed the soundness of such an argument and this has led to the search for and development of H_2 cathodes based on alloys and intermetallic compounds. Thus, for instance, combination of Ni with Mo shows activity improvement [32]. The same is the case for PtMo$_x$ [33–35]. Since in these cases the d-electron rule is not so obvious, the activity enhancement is generally defined as *synergetic effects*. These might also result from geometric factors, that is, a morphology depending on composition.

7.4.3
Materials for Cathodes [27, 36–38]

As discussed above, Pt is the reference electrode material for H$_2$ evolution since it is the most active elemental cathode. H$_2$ is formed on Pt with a Tafel slope of 30–40 mV, the lowest ever observed for this reaction. Its cost makes this metal unsuitable for routine applications. In fact, cathode materials traditionally used in technology have long been iron or mild steel in acidic solution and Ni in (strongly) alkaline solution. Steel can also be used in moderately basic solution.

In alkaline electrolyzers, Ni is the only elemental cathode that can be used. It is generally considered as a fairly good electrocatalyst, but in facts it exhibits two shortcomings: (i) its activity decreases with time [cf. the ΔV_t term in Equation (7.16)] especially under conditions of intermittent electrolysis and (ii) shutdown of industrial cells (for maintenance) leads to Ni dissolution at the cathode since this electrode is driven to more positive potentials by short-circuit with the anode. These shortcomings can be alleviated if Ni cathodes are *activated*, that is, if they are coated with a thin layer of more active and more stable materials. Activation has been attempted with a variety of materials from sulfides to oxides, from alloys to intermetallic compounds.

It has been mentioned above that intermetallics are compounds in which the electronic structure of one of the component is sizably modified by the combination with the other. Several intermetallic compounds where Ni is the main component have been investigated, such as LaNi$_5$, CeNi$_3$ and Ni$_3$Ti. If the activity of compounds of Ni with, for example, Ta, Ti, Nb, Hf and Zr is plotted against the calculated enthalpy for the formation of the related hydride – proportional to D(M–H) – a sort of volcano-shaped curve is again obtained, as shown in Figure 7.7 [39, 40]. Carbides also belong to this category of cathode materials.

Among the metals, precious metals have been shown to be among the most active. Nevertheless, their use is impeded by their cost. In the 1960s, it was discovered that the oxides of precious metals, besides being more resistant to anodic corrosion, were also much more active for Cl$_2$ evolution than the metals themselves [41]. That was the beginning of the dimensionally stable anode (DSA) era, a breakthrough from the industrial point of view [42]. Later, it was discovered that these oxides are in fact most versatile electrocatalysts, being active not only for anodic reactions but even for cathodic processes, including H$_2$ evolution [43].

In principle, precious metal oxides such as RuO$_2$, IrO$_2$ and Rh$_2$O$_3$ are thermodynamically unstable under the conditions of H$_2$ evolution since they should be reduced to the metals. In fact, this is not the case, with the exception of Rh$_2$O$_3$, whose stability depends on the details of the preparation [44, 45]. The reason for this apparent thermodynamic contradiction lies in the electronic conductivity of

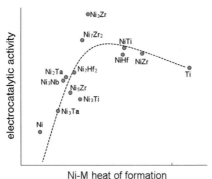

Figure 7.7 Experimental volcano curve for H_2 evolution on Ni intermetallics. M–H bond strength assumed proportional to the heat of formation of intermetallics. Adapted from [27].

(especially) RuO_2 and IrO_2. These oxides are indeed metallic conductors. Actually, as H_2 is evolved on the oxides, the surface sites adsorb H atoms so that the molecular entities lying at the surface are formally reduced. However, bulk reduction would be possible only by penetration of H^+ into the lattice, which does not occur since no electric field can be built up towards the interior of the solid due to the presence of metallic conduction.

7.4.4
Factors of Electrocatalysis

Two are the main factors governing the activity of materials: (i) electronic factors, related to chemical composition and structure of materials influencing primarily the M–H bond strength and the reaction mechanism, and (ii) geometric factors, related to the extension of the real surface area influencing primarily the reaction rate at constant electronic factors. Only the former result in true electrocatalytic effects, whereas the latter give rise to apparent electrocatalysis.

The two factors are seldom completely independent; most times they are interdependent. A typical example is the effect of particle size. A decrease in particle size produces an increase in surface area at constant amount of material. At the same time, as the particle size decreases the surface-to-volume ratio increases, which may lead to modifications of the electronic properties of surface atoms.

From a practical point of view, which effect is responsible for the performance of an electrode material is in principle uninteresting in that both converge to improve the electrode behavior. However, from a fundamental point of view, such a distinction is essential to be able to improve and optimize the experimental situation. In terms of reaction rate (current density), only knowledge of the real surface area can allow one to separate experimentally the two factors. If a plot of j against S (real surface area) at constant potential gives a straight line, the effects are more likely to be geometric only. On the other hand, if the correlation deviates from linearity, electronic factors are most likely to operate. This approach assumes that the

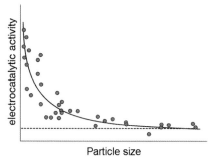

Figure 7.8 Electrocatalytic activity of Pt for H_2 evolution as a function of Pt particle size. Adapted from [46].

methods of surface area measurement are significantly reliable, which is seldom the case. Therefore, in most cases specific approaches have to be derived.

If we consider Ni^{2+} as an active site for Ni-based materials, changing the environment in which the ion is immersed is expected to influence its electronic properties. This is in principle the reason for testing a series of alloys or intermetallic compounds of Ni. On the other hand, on changing the environment, bond lengths will also be modified and this will modify the actual concentration of active sites, in turn determining the active surface area. A few examples can better illustrate these concepts.

Figure 7.8 shows the current density of H_2 evolution on Pt microcrystals [46]. It is intriguing that the activity increases as the particle size decreases, although the current is referred to unit real surface area. The excess increase in activity is definitely to be attributed to especially active surface atoms emerging in very small particles.

Figure 7.9 shows the activity for H_2 evolution of three samples of Ni [27]. Smooth and sandblasted Ni exhibit the same reaction mechanism (same Tafel slope, b), but a higher current for the latter. This is clearly due to the rougher surface of sandblasted Ni, that is, to purely geometric effects. The third sample is Raney Ni. This is obtained from an alloy of Ni with Zn or Al that are then leached away in alkaline solution [47–49]. This leaves a very porous solid with intrinsically very small particle size. The figure shows that Raney Ni, in addition to a much lower overpotential for H_2 evolution, also exhibits a lower Tafel slope. This is clear evidence for the occurrence of electronic effects (different mechanism) together presumably with important geometric effects.

A coating of NiS appears to activate smooth Ni considerably. This has led to intense investigations of sulfides as possible electrocatalysts for H_2 evolution. In principle, Ni^{2+} in the presence of S^{2-} can possess different electronic properties. However, specific studies have revealed that in fact NiS is hydrogenated:

$$NiS + H_2 \rightarrow Ni + H_2S \qquad (7.24)$$

with liberation of H_2S. The Ni surface becomes covered with a very fine powder of Ni that is responsible for the apparent activation of the underlying metal.

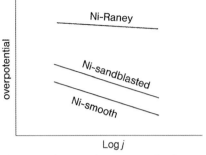

Figure 7.9 Electrocatalytic activity of Ni for H_2 evolution depending on the procedure of surface treatment.

Hence the final effect is close to that of Raney Ni, although the stability is unsatisfactory since the resulting powder is not stabilized mechanically [27].

A similar situation is that of oxides. Metal ions in an oxide are definitely in a different environment, which can be regarded as the origin of electronic effects. However, deposition of a thin film of Ru [50] or of RuO_2 [51] on Ni activates the electrode surface dramatically. It is difficult to separate the two factors but they are certainly both operating, especially with RuO_2 that, because of the method of preparation, as a rule is formed with a very high roughness factor. In these cases the reference surface could be that of a single crystal whose surface is usually regarded as ideally smooth. These examples introduce a discussion of the effect of the topography of electrode surfaces which joins electronic and geometric effects together but with some possibility of disentangling their effects since single-crystal faces are well-defined surfaces.

Another factor that has been claimed to influence the electrocatalytic properties of materials is the degree of crystallinity [52]. In particular, metals and metal alloys in an amorphous state have attracted interest as electrocatalysts for H_2 evolution. In practice, amorphization is promoted by adding a non-metal such as B or P. However, evident effects are minor and there is no definitive proof that such an approach is worth practical consideration. Moreover, recrystallization of materials may take place under operating conditions [27].

The widespread use of supported materials in catalysis and also in fuel cells has recently suggested a new line of research for H_2 evolution materials [53, 54]. The well-known concept of strong metal–support interaction (SMSI) in catalysis is being tentatively transposed to the field of electrocatalysis. This implies that very thin films of active electrocatalysts are deposited (supported) on materials whose interaction with the electrocatalysts is expected to generate synergetic effects. At the basis of this approach there is still the idea of combination of hyper- with hypo-d-electron metals. Thus, metals could be supported, for example, on TiO_2 powder giving high surface area electrodes with operating SMSI. In practice, this would be a change in the experimental configuration of electrodes rather than a really new conceptual approach. The convenience of this idea is still to be demonstrated convincingly, especially in relation to long-term performance.

7.5
The Oxygen Evolution Reaction

The reaction at the anode of water electrolyzers is a demanding reaction. This is because the mechanism involves three or more steps and the intermediates are species of high energy, thus involving high activation energies. For these reasons, O_2 evolution absorbs most of the overpotential to drive water electrolysis, thus bearing negatively on the economy of the process more than H_2 evolution.

As a demanding reaction, it is very sensitive to the structural and compositional details of the anode materials. For this reason, research on anodes for O_2 evolution calls for close characterization of electrocatalysts, especially from the point of view of materials chemistry and physics.

The species that are oxidized in the first step of O_2 formation are H_2O molecules in acidic solution:

$$H_2O \rightarrow OH_{ads} + H^+ + e \tag{7.25a}$$

and hydroxyl ions in alkaline solution:

$$OH^- \rightarrow OH_{ads} + e \tag{7.25b}$$

The intermediate energy depends on the interaction with the electrode surface. Materials binding OH_{ads} weakly as a rule show high overpotentials for O_2 evolution. Their mechanism is dominated by step (7.25) as the rate-determining step and the Tafel slope is close to 120 mV.

If OH_{ads} is adsorbed strongly, its formation is fast and successive steps of OH_{ads} desorption become rate determining. Under similar conditions, the Tafel slope is lower and electrode materials are electrocatalytically more active. Along the same conceptual lines of volcano curves, if the adsorption of OH_{ads} is too strong, its removal becomes very difficult and the overpotential increases again. Thus, in principle, a volcano curve should be expected as in the case of H_2 evolution.

7.5.1
Reaction Mechanisms

What makes the approach to electrocatalysis much more complex is the fact that after step (7.25) there is never just one more step, but at least two more. As a consequence of the complexity of the surface chemistry of OH species, several mechanisms have been proposed for O_2 evolution, differing in small and sometimes speculative details. However, most of the experimental observations can be interpreted on the basis of three simplified schemes.

Chemical oxide path (only the scheme for acidic solution is shown):

$$H_2O \rightarrow OH_{ads} + H^+ + e \tag{7.25a}$$

$$OH_{ads} + OH_{ads} \rightarrow O_{ads} + H_2O \tag{7.26}$$

$$O_{ads} + O_{ads} \rightarrow O_2 \tag{7.27}$$

The name of the mechanism comes from the chemical formation of a surface oxide. Thus step (7.26) is a chemical step and the mechanism can be written as EEC if step (7.26) is rate determining, thus predicting a Tafel slope of 30 mV for low OH_{ads} coverage (i.e. on the ascending branch of the volcano curve). Should step (7.27) be rate determining, the mechanism would be EEEEC with a predicted Tafel slope of 15 mV. Unfortunately, an electrode exhibiting such a low Tafel slope has never been observed, so that so an active material still remains a dream.

Electrochemical oxide path:

$$H_2O \to OH_{ads} + H^+ + e \qquad (7.25a)$$

$$OH_{ads} \to O_{ads} + H^+ + e \qquad (7.28)$$

$$O_{ads} + O_{ads} \to O_2 \qquad (7.27)$$

In this case, the formation of a surface oxide (O_{ads}) occurs electrochemically with two successive electron transfers. Therefore, if step (7.28) is rate determining, the mechanism is EE with a predicted Tafel slope of 40 mV at low OH_{ads} coverage.

Although the most active O_2 evolution electrocatalysts exhibit Tafel slopes in the range 30–40 mV, in some cases slopes close to 60 mV have been reported. For these cases, a third mechanism has been proposed in which the formation of O_{ads} is preceded by the acid–base dissociation of OH_{ads}. This mechanism is known for the name of its proposer, but it could be defined as follows.

Chemical acid–base equilibrium path:

$$H_2O \to OH_{ads} + H^+ + e \qquad (7.25a)$$

$$OH_{ads} \to O^-_{ads} + H^+ \qquad (7.29)$$

$$O^-_{ad} \to O_{ad} + e \qquad (7.30)$$

$$O_{ads} + O_{ads} \to O_2 \qquad (7.27)$$

The observed slope of 60 mV can only be explained with step (7.29) as rate determining. In this case, the mechanism is EC and the predicted Tafel slope 60 mV. However, if the rate-determining step shifts to step (7.30), the mechanism becomes ECE, which is kinetically equivalent to EE, and the Tafel slope turns again to 40 mV.

7.5.2
Anodic Oxides

As implied in the schemes of the mechanisms above, a surface oxide is formed during O_2 evolution. Since M–O is a much stronger bond than M–H, absorption of O is much more probable than that of H. Thus, whereas H_2 evolution can be treated as occurring on bare metal surfaces, O_2 evolution cannot. In the end, after O_2 evolution, a metal surface turns out coated by an oxide layer electrolytically grown during the anodic process. Often this leads to the observation of time-dependent performances of anodes and a way to allow for that is to keep the electrode under intense O_2 evolution for some time and then to carry out measurements on the stabilized oxide surface [55].

Electrolytic oxides are responsible for the passivity of corroding metals, for example TiO_2 and NiO. However, this is not generally the case under O_2 evolution conditions. If an oxide passivates a surface, it is not a good electrocatalyst for O_2 evolution. On the other hand, oxides that are good catalysts for O_2 evolution very often are unstable under O_2 evolution and dissolve. For instance, NiO passivates Ni in alkali and is also a good electrocatalyst for O_2 evolution. However, it dissolves in acids and the metal cannot be used for water electrolysis at low pH. Similarly, Ru is easily oxidized anodically but the oxide is not stable and dissolution occurs under O_2 evolution both in acid and in base [43, 56]. Nevertheless, if Ru oxide is electrodeposited during anodic polarization of aqueous solutions of $RuCl_3$, the electrodeposited Ru oxide is catalytically active for O_2 evolution, as shown by the decrease in anodic overpotential. However, such a configuration is impractical for water electrolysis since the liquid phase should contain $RuCl_3$, which would be deposited everywhere in the cell circuit.

Anodic oxides are often referred to as *hydrous* oxides in that they are naturally hydrated, that is, they are more similar to hydroxides. In these phases, metal ions are surrounded by an environment which resembles that of metal ions in solution. For this reason, the dissolution of hydrous oxides does not require a high energy of activation. If hydrous oxides are dehydrated, they become *dry* oxides, which therefore acquire higher resistance to anodic dissolution. The most straightforward way to obtain dry oxides is to subject hydrous oxides to thermal treatments or better to prepare them as thin surface films by a non-electrochemical technique (thermal decomposition, chemical vapor deposition, reactive sputtering, etc.).

7.5.3
Thermal Oxides (DSA)

For a long time, conventional alkaline electrolyzers used Ni as an anode. This metal is relatively inexpensive and a satisfactory electrocatalyst for O_2 evolution. With the advent of DSA (a Trade Name for dimensionally stable anodes) in the chlor-alkali industry [41, 42], it became clear that thermal oxides deposited on Ni were much better electrocatalysts than Ni itself with reduction in overpotential and increased stability. This led to the development of activated anodes. In general, Ni is a support for alkaline solutions and Ti for acidic solutions. The latter, however, poses problems of passivation at the Ti/overlayer interface that can reduce the stability of these anodes [43]. On the other hand, in acid electrolysis, the catalyst is directly pressed against the membrane, which eliminates the problem of support passivation. In addition to improving stability and activity, the way in which dry oxides are prepared (particularly thermal decomposition) develops especially large surface areas that contribute to the optimization of their performance.

Oxide surfaces are high-energy surfaces that interact with water molecules becoming covered by a carpet of OH groups. The latter, in contact with aqueous solutions, behave as weak acids or weak bases, giving rise to dissociation that is the main origin of surface charging:

$$M-OH + H^+ \rightarrow M-OH_2^+ \tag{7.31}$$

$$M-OH + OH^- \rightarrow M-O^- + H_2O \tag{7.32}$$

Accordingly, the magnitude and sign of the surface charge will be a function of solution pH. The value of pH at which the surface is chemically and electrically neutral identifies the point of zero charge (pzc) [57, 58]. Since the acidity or the basicity depends on the nature of the M–OH surface interaction, Figure 7.10 shows that a linear correlation exists between the pzc and the electronegativity of the oxide surface, empirically defined as the geometric mean of the elemental components of the oxide, for example,

$$x(MnO_2) = [x(Mn)x^2(O)]^{\frac{1}{3}} \tag{7.33}$$

Figure 7.10 shows that oxides with strong M–OH interaction possesses low values of pzc (very acidic surfaces), whereas oxides with weak M–OH interaction possess high values of pzc (basic surfaces). A corollary is that Figure 7.10 also predicts the anodic stability of oxides used as electrodes. Concepts of surface chemistry indicate that the pzc coincides with the pH of minimum solubility [57]. Therefore, acid oxides are expected to be stable in strongly acidic media, whereas basic oxides are stable in alkaline environments, as is in fact the case. IrO_2 (pzc < 2) is the most stable oxide electrode ever known in strongly acidic solution under O_2 evolution, whereas NiO (pzc > 9) is among the most stable oxide electrodes in strongly alkaline solution. Other (thermal) oxides are also stable but as a rule below a certain critical potential (e.g. RuO_2), which makes these materials interesting for many applications but not for extreme conditions [42, 43].

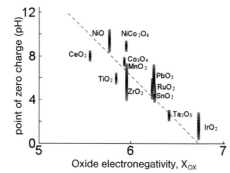

Figure 7.10 Dependence of the pzc of thermally prepared oxides on the oxide electronegativity calculated by means of Equation (7.33) [58].

Whereas Pt in an acidic solution saturated with H_2 acquires the reversible potential of the hydrogen electrode, this is not the case for the same Pt electrode in an acidic solution saturated with O_2. This is related to the high activation energies involved in breaking and forming chemical bonds. Thus the O_2 reaction is known to be highly irreversible. In particular, a Pt electrode in O_2-saturated solution acquires a potential $E \approx 0.9$ V (SHE) rather than 1.23 V. Hence an overpotential of >0.3 V can already be expected from an analysis of the equilibrium conditions.

7.5.4
Electrocatalysis

Pt is, of course, not a good electrocatalyst for the O_2 evolution reaction, although it is the best for the O_2 reduction reaction. However, also with especially active oxides of extended surface area, the theoretical value of $E°$ has never been observed. For this reason, the search for new or optimized materials is a scientific challenge but also an industrial need. A theoretical approach to O_2 electrocatalysis can only be more empirical than in the case of hydrogen in view of the complexity of the mechanisms. However, a chemical concept that can be derived from scrutiny of the mechanisms mentioned above is that oxygen evolution on an oxide can be schematized as follows [59]:

$$MO + xH_2O \rightarrow MO_{1+x} + 2xH^+ + 2xe \qquad (7.34)$$

$$MO_{1+x} \rightarrow MO + \frac{x}{2}O_2 \qquad (7.35)$$

As an oxide is polarized anodically, it is first oxidized to an unstable higher oxide whose rapid decomposition gives back the original oxide with liberation of O_2. This is a chemical view of a cycle whose feasibility is governed by the $\Delta H_t°$ of transition from the lower to the higher oxide. Such a ΔH plays the role of $\Delta H_{ads}(OH)$ [the analogous of $\Delta H_{ads}(H)$ for the H_2 reaction].

A plot of overpotential (at constant apparent current density) against $\Delta H_t°$ generates a sort of volcano-shaped curve, as Figure 7.11 shows [59]. As in the case of the H_2 reaction, oxides on the ascending branch are weak electrocatalysts (e.g. PbO_2 is good for O_3 generation but not for O_2 evolution), whereas those on the descending branch are too easily oxidized to higher compounds that do not decompose easily. As usual, the best electrocatalysts are found at intermediate values of $\Delta H_t°$. Here lie RuO_2 and IrO_2, the prototype of DSA. It is also interesting that in acids or in alkalis the rank of electrocatalytic properties does not change appreciably. What changes is the stability, as mentioned above in terms of pzc.

The performance of oxide electrodes depends on both factors, electronic and geometric. The latter is especially important since the preparation of oxide layers as a rule produces very high surface areas. A way to disentangle the two factors is to scrutinize the behavior of an *intensive* property. In electrochemical kinetics, the Tafel slope is the most appropriate, since it depends closely on the reaction mechanism and not on the extension of the surface area.

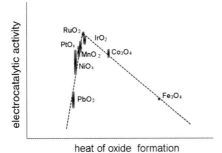

Figure 7.11 Electrocatalytic activity of thermal oxides for O_2 evolution as a function of heat of oxide formation [59].

7.5.5
Factors of Electrocatalysis

Figure 7.12 shows how the mechanism (Tafel slope) of O_2 evolution on RuO_2 varies with particle size [60]. The latter is represented by the surface charge at constant potential as measured by cyclic voltammetry. A high surface charge is associated with a high surface area. Hence the particle size increases as the surface area decreases. The figure shows that provided that the particle size is small, the mechanism does not vary. This does not rule out electronic effects from operating, but this cannot emerge from such a plot. However, as the particle size increases, there appears a progressive change in mechanism that for particle size $\to \infty$ coincides with the mechanism observed with the (110) face of a single crystal [61]. These data indicate that on small particles, edge effects presumably operate, which makes the material become particularly active (lower Tafel slope).

Another typical example is given in Figure 7.13. In this case, Ru + Ir oxides are prepared using three different procedures [62]. It is readily evident that the mechanism of O_2 evolution varies with composition differently depending on the preparation method. Scrutiny of the data reveals that thermal decomposition

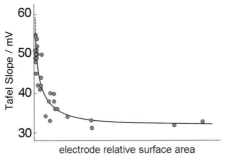

Figure 7.12 Dependence of the mechanism (Tafel slope) of O_2 evolution on the electrode surface area ($\propto$ 1/particle size) [60].

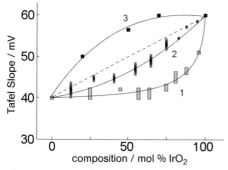

Figure 7.13 Dependence of the mechanism (Tafel slope) of O_2 evolution on the composition of $IrO_2 + RuO_2$ mixtures depending on the procedure of mixture preparation. (1) Thermal decomposition, precursors dissolved in water; (2) thermal decomposition, precursors dissolved in nonaqueous solvents; (3) reactive sputtering [62].

of precursors dissolved in aqueous solution results in the formation of mixed phases (RuO_2 and IrO_2 behave independently) whereas reactive sputtering produces solid solutions where mixing is at the atomic level.

Composite materials allow modulation of the electrocatalytic properties and possibly synergetic effects. However, the possibility of surface segregation is often overlooked in dealing with mixed oxides. More seriously, it is often not realized that surface segregation depends on the preparation procedure. Figure 7.14 shows data for O_2 evolution on $SnO_2 + IrO_2$ mixed oxides [63]. The activity is seen to reach a maximum for about 20% IrO_2. This is not, however, a synergetic effect but a consequence of surface segregation of IrO_2. Such an effect is not rare but rather a constant. Figure 7.15 shows the surface concentration of RuO_2 against the nominal concentration in $RuO_2 + Co_3O_4$ mixed oxides [64]. Table 7.1 summarizes the case of several mixed oxides prepared by thermal decomposition.

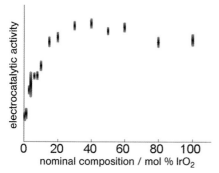

Figure 7.14 Electrocatalytic activity of $SnO_2 + IrO_2$ mixtures for O_2 evolution as a function of nominal composition [63].

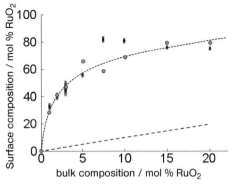

Figure 7.15 Surface enrichment with RuO_2 of $RuO_2 + Co_3O_4$ mixtures as determined by XPS (●) and electrochemical techniques [64].

Mixing two oxides produces several effects. In addition to surface segregation, the surface area is usually not monotonically related to composition. As a rule, it passes through a maximum at some intermediate compositions, which can be interpreted as being related to decreasing particle size. Also, stability is affected by mixing. Thus, active components (e.g. RuO_2) are mixed with inactive but chemically inert components (e.g. TiO_2) both to 'dilute' the precious metal component (for instance, in Figure 9.14 only 20% of IrO_2 is sufficient to obtain the same behavior of pure IrO_2), and to increase its chemical and electrochemical stability. It has been exhaustively proven experimentally, that TiO_2 stabilizes RuO_2 but also that IrO_2 stabilizes RuO_2 [56]. A suitable combination of components can thus result in the optimization of both the activity and the stability.

The two factors governing the activity of electrode materials, the surface area and the electronic structure, inspired the present trends in research for more efficient electrocatalysts. Thus, in order to maximize the surface area, materials are more and more dispersed so as to increase the surface-to-volume

Table 7.1 Surface enrichment in mixed thermal oxides.

Mixed oxides	Surface-enriched component
$RuO_2 + IrO_2$	Ir
$Co_3O_4 + RuO_2$	Ru
$IrO_2 + SnO_2$	Ir
$RuO_2 + ZrO_2$	Zr
$IrO_2 + Ta_2O_5$	Ta
$RuO_2 + TiO_2$	Ti
$RuO_2 + PtO_x$	Pt
$NiO_x + FeO_x$	Ni
$Co_3O_4 + NiO_x$	Ni
$RuO_2 + RhO_x$	Rh

ratio. On the other hand, new and more specific technologies of preparation are being tested in the search for conditions resulting in more and more enhanced surface area. Since the improvement of the activity of single-compound materials is doomed to reach a limit, the target is the exploitation of possible synergetic effects. This results in the testing of myriads of combinations (sometimes using combinatorial approaches) in the search for yet untested composite materials or in an attempt to achieve better mixing of already tested electrocatalysts. Figure 7.13 shows how much the degree of mixing in composite materials can affect the performance of mixed electrocatalysts.

7.5.6
Intermittent Electrolysis

Water electrolysis is usually carried out at constant current so that the performance of electrocatalysts should be constantly stable. Nevertheless, it has been shown that shutdown of cells in alkaline electrolysis results in the temporary dissolution of Ni cathodes. This implies that any variation of the electrolysis regime results in weakening of the resistance of materials to aggressive conditions. In other words, perturbations in the conditions of electrolysis are responsible for the appearance and the growth of the instability [cf. ΔV_t in Equation (7.16)].

The above problems arise in particular under the conditions of intermittent electrolysis [65–67]. If fossil fuels are to be replaced to avoid pollution, electrolysis cells should be driven by electric power generated by an alternative renewable energy source. In addition to nuclear energy that is already available, technologies that are closer to implementation are solar and eolic energy conversion. However, both are naturally discontinuous, wind because it does not flow constantly and not everywhere and sun because of the unavoidable alternation of day and night. For these reasons, both energy sources need energy vectors to be transmitted and, more importantly, both need storage to be accumulated. Storage can only be electrochemical since there exists no other way to store electrical energy.

As an electrolytic cell is powered by voltaic cells, the current reaches a maximum in the middle of the day and drops to practically short-circuit in the middle of the night [65]. These conditions are known as intermittent electrolysis. The point is that under intermittent electrolysis, electrode materials are subjected to much more severe conditions that reduce their stability. Materials good for constant electrolysis may therefore give poor performance in intermittent electrolysis. This is particularly the case for Ni as both anode and cathode, whose overpotential is seen to increase as the intermittent regime starts. This is illustrated in Figure 7.16, where the behavior of Ni and NiO is shown in comparison with activated Ni, covered with a thin layer of thermal oxides. Co_3O_4 is seen to be more stable, whereas Fe-doped NiO exhibits a surprising improvement under conditions of intermittent electrolysis. A conclusion drawn from these results is that the search for good materials for intermittent electrolysis should follow different and more specific criteria of choice.

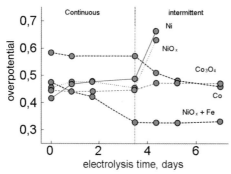

Figure 7.16 Variation of activity for O_2 evolution as a function of electrolysis time for continuous and intermittent electrolysis [65].

7.6
Electrocatalysts: State-of-the-Art

Myriads of electrocatalysts of the most varied composition have been tested thus far [36–38, 67]. It would be impractical to mention all of them and almost impossible to attempt to establish a rank of activity. Most of them exhibit electrocatalytic properties sufficiently close to each other. Under similar circumstances, it is only possible to summarize the situation briefly.

It is a definite fact that conventional electrolysis (alkaline cells) can be systematically improved by using activated electrodes for both anodes and cathodes. In the case of anodes, it appears that Fe-doped Ni-based oxides exhibit the best performance, for example $NiO_x + FeO_x$ and especially $NiCo_2O_4 + FeO_x$. For cathodes, Raney Ni is still a good choice, but it can be activated with coatings of different metal alloy or intermetallics, for example Ni + Ru, Co + Mo, Pt + Mo and Ni + Fe. It has also been found that Re shows good properties for H_2 evolution and partial oxidation of metal surfaces appears to improve their electrocatalytic activity. On the other hand, as anticipated above, some metallic oxides such as RuO_2 and IrO_2 are excellent electrocatalysts for H_2 evolution, [68] with IrO_2 slightly better than RuO_2 (but IrO_2 is much more expensive).

In the case of SPE cells (acid electrolysis), a much younger technology, the extreme anodic conditions restrict the choice to mixtures of IrO_2 with Ta_2O_5 or SnO_2, although $IrO_2 + RuO_2$ have been found to be more active [69–71]. This is not surprising in view of the activity of RuO_2 whereas Ta_2O_5 and SnO_2 are inactive, but the stability of such a composition can give more problems. The situation with cathodes is even more serious since most materials do not withstand the contact with the acidic membrane. However, since the MEA allows the use of gas diffusion conditions, powdered materials can easily be used. Under similar circumstances, a dispersion of Pt on a support of carbon material will perform excellently from the electrocatalytic point of view without severe investment costs.

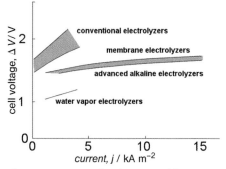

Figure 7.17 Ranges of performance of the various types of water electrolysis. Redrawn from Ref. [72], with permission of Springer-Verlag.

7.7
Water Electrolysis: State-of-the-Art

Figure 7.17 shows a summary of the available conditions of water electrolysis [72]. For each configuration there exists a range of performance. Conventional electrolyzers, which nevertheless are still the most common in the current production of H_2 on the intermediate and small scale, show high overpotential and a relatively small production rate. Membrane (SPE) and advanced alkaline electrolyzers show very similar performance, with somewhat lower overpotential but a much higher production rate. Definite improvements in energy consumption would come from high temperature (steam) electrolysis, which is, however, still far from optimization because of a low production rate and problems of material stability.

The average conditions extracted from Figure 7.17 are summarized in Table 7.2. Overall, actual alkaline electrolysis requires an energy consumption of 4.0–4.9 kW h m^{-3} of H_2. Consumption somewhat lower than 4.0 kW h m^{-3} has recently been claimed for SPE cells [73]. The current yield and H_2 purity are seen to be close to 100% in alkaline electrolysis.

7.8
Beyond Oxygen Evolution

The high overpotential for O_2 evolution could be avoided if the reaction were replaced with a different anodic reaction. This replacement could in turn reduce ΔE, the minimum cell potential difference, which depends on the nature of the electrode reactions. Such a strategy has already been applied with success in the chlor-alkali industry, where the Cl_2–H_2 couple ($\Delta E = 1.35$ V) has been replaced with Cl_2–O_2 ($\Delta E \approx 0.90$ V) (O_2 is reduced at the so-called air cathode).

A few attempts to apply the same strategy have been made recently. In one case, anodic O_2 evolution was replaced with oxidation of carbonaceous materials dispersed in the solution (slurry) [74]. This would correspond to an average practical $\Delta V = 0.5$ V. However, in a scenario of H_2 economy to reduce the production of greenhouse gases, an anode where C is oxidized to CO_2 would constitute a striking contradiction.

In another case, the electrolysis of an aqueous solution of NH_3 was proposed [75–77]. The net reaction would be

$$2NH_3 \rightarrow N_2 + 3H_2 \tag{7.36}$$

with oxidation of NH_3 at the anode. The theoretical ΔE would be 0.06 V with a decrease of almost 1.2 V with respect to conventional electrolysis. The experimental conditions are summarized in Table 7.3. The amount of energy consumption is impressively small, only about 30% of the conventional. However, the production rate is also strikingly small (about 1% of the conventional). This is certainly related to the fact that NH_3 is a solute and as such its oxidation at the phase boundary is subject to mass transfer limitations. Moreover, NH_3 oxidation calls for very sophisticated electrocatalysts (precious metals) that are subject to poisoning phenomena. In the end, such a proposed change in technology of H_2 production does not appear very close to implementation.

Further examples of recent attempts to reduce the consumption of electrical energy are the electrolysis of aqueous solutions of methanol (but CO_2 is still produced at the anode) [78, 79] and water electrolysis using ionic liquids as electrolytes [80]. In the latter case, the authors claimed the possibility of obtaining high hydrogen production efficiencies using an inexpensive material such as low-carbon steel.

Table 7.2 Alkaline water electrolysis: summary of ranges of experimental parameters.[a]

$H_2O \rightarrow H_2 + {}^1\!/_2 O_2$
Electrolyte: 25–30% KOH
$\Delta V = 1.65\text{–}2.00$ V, $j = 1\text{–}10$ kA m^{-2}
Energy consumption: 4.0–4.9 kW h m^{-3}
Current yield: 98–99.9%
H_2 purity: >99.8%

[a] Data from Ref. [72].

Table 7.3 Electrolysis of aqueous solutions of ammonia: experimental parameters.[a]

$2NH_3 \rightarrow N_2 + 3H_2$
Electrolyte: 1 M NH_3 in 5 M KOH
$\Delta V = 0.55$ V, $j = 25$ A m^{-2}
Anode: Rh + Pt
Cathode: Ni/C
Energy consumption: 1.3 kW h m^{-3}

[a] Data from Ref. [77].

Acknowledgment

The financial support of MIUR (Rome), PRIN Projects, is gratefully acknowledged.

References

1 Trasatti, S. (1999) *Journal of Electroanalytical Chemistry*, **460**, 1.
2 Trasatti, S. (1999) *Journal of Electroanalytical Chemistry*, **476**, 90.
3 Marban, G. and Vales-Solis, T. (2007) *International Journal of Hydrogen Energy*, **32**, 1625.
4 Imarisio, G. (1981) *International Journal of Hydrogen Energy*, **6**, 153.
5 Commission of the European Communities (1983) *Hydrogen as An Energy Carrier*, Reidel, Dordrecht.
6 Wendt, H. and Imarisio, G. (1988) *Journal of Applied Electrochemistry*, **18**, 1.
7 Pletcher, D. and Walsh, F.C. (1990) *Industrial Electrochemistry*, Chapman and Hall, London.
8 Trasatti, S. (1990) Electrode kinetics and electrocatalysis of hydrogen and oxygen electrode reactions. 1. Introduction, in *Electrochemical Hydrogen Technologies* (ed. H. Wendt), Elsevier, Amsterdam, Chapter 1.1.
9 Divisek, J. (1990) Water electrolysis in low- and medium temperature regime, in *Electrochemical Hydrogen Technologies* (ed. H. Wendt), Elsevier, Amsterdam, Chapter 3.
10 Lu, S., Zhuang, L. and Lu, J. (2007) *Journal of Membrane Science*, **300**, 205.
11 Rosa, V.M., Santos, M.B.F. and Dasilva, E.P. (1995) *International Journal of Hydrogen Energy*, **20**, 697.
12 Kikuchi, K., Tanaka, Y., Saihara, Y., Maeda, M., Kawamura, M. and Ogumi, Z. (2006) *Journal of Colloid and Interface Science*, **298**, 914.
13 Vogt, H. and Balzer, R.J. (2005) *Electrochimica Acta*, **50**, 2073.
14 Balzer, R.J. and Vogt, H. (2003) *Journal of the Electrochemical Society*, **150**, E11.
15 Stojić, D.L., Marceta, M.P., Sovilj, S.P. and Miljanić, S.S. (2003) *Journal of Power Sources*, **118**, 315.
16 Rasten, E., Hagen, G. and Tunold, R. (2003) *Electrochimica Acta*, **48**, 3945.
17 Song, S.D., Zhang, H.M., Liu, B., Zhao, P., Zhang, Y.N. and Yi, B.L. (2007) *Electrochemical and Solid-State Letters*, **10**, B122.
18 Fateev, V.N., Archakov, O.V., Lyutikova, E.K., Kulikova, L.N. and Porembskii, V.I. (1993) *Elektrokhimiya*, **29**, 551.
19 Borup, R., Meyers, J., Pivovar, B., Kim, Y.S., Mukundan, R., Garland, N., Myers, D., Wilson, M., Garzon, F., Wood, D., Zelenay, P., More, K., Stroh, K., Zawodzinski, T., Boncella, J., McGrath, J.E., Inaba, M., Miyatake, K., Hori, M., Ota, K., Ogumi, Z., Miyata, S., Nishikata, A., Siroma, Z., Uchimoto, Y., Yasuda, K., Kimijima, K.I. and Iwashita, N. (2007) *Chemical Reviews*, **107**, 3904.
20 Stucki, S. (1981) *Europhysics News*, **12**, 9.
21 Divisek, J., Mergel, J. and Schmitz, H. (1990) *International Journal of Hydrogen Energy*, **15**, 105.
22 Trasatti, S. (2003) Reaction mechanism and rate determining steps, in *Handbook of Fuel Cells – Fundamentals, Technology and Applications*, Vol. **2**, Part 2 (eds W. Vielstich, A. Lamm and H.A. Gasteiger), John Wiley & Sons, Ltd, Chichester.
23 Trasatti, S. and Petrii, O.A. (1991) *Pure and Applied Chemistry*, **63**, 711.
24 Trasatti, S. (2003) Adsorption – volcano curves, in *Handbook of Fuel Cells – Fundamentals, Technology and Applications*, Vol. **2**, Part 2 (eds W. Vielstich, A. Lamm and H.A. Gasteiger), John Wiley & Sons, Ltd, Chichester.
25 Horiuti, J. and Polanyi, M. (1935) *Acta Physicochimica URSS*, **2**, 505.

26 Parsons, R. (1958) *Transactions of the Faraday Society*, **54**, 1053.
27 Trasatti, S. (1992) Electrocatalysis of hydrogen evolution: progress in cathode activation, in *Advances in Electrochemical Science and Engineering* (eds H. Gerischer and C.W. Tobias), VCH Verlag GmbH, Weinheim.
28 Trasatti, S. (1972) *Journal of Elaectroanalytical Chemistry*, **39**, 163.
29 Jaksic, M.M. (1984) *Electrochimica Acta*, **29**, 1539.
30 Jaksic, M.M. (2001) *International Journal of Hydrogen Energy*, **26**, 559.
31 Jaksic, J.M., Krstajić, N.V., Grgur, B.N. and Jaksić, M.M. (1998) *International Journal of Hydrogen Energy*, **23**, 667.
32 Jaksic, J.M., Vojnović, M.V. and Krstajić, N.V. (2000) *Electrochimica Acta*, **45**, 4151.
33 Marceta Kaninski, M.P., Nikolić, V.M., Potkonjak, T.N., Simonović, B.R. and Potkonjak, N.I. (2007) *Applied Catalysis A–General*, **321**, 93.
34 Stojić, D.L., Grodzić, T.D., Marčeta Kaninski, M.P., Maksić, A.D. and Simić, N.D. (2006) *International Journal of Hydrogen Energy*, **31**, 841.
35 Stojić, D.L., Maksić, A.D., Marčeta Kaninski, M.P., Cekić, B.D. and Mijanić, S.S. (2005) *Journal of Power Sources*, **145**, 278.
36 Suffredini, H.B., Cerne, J.L., Crnkovic, F.C., Machado, S.A.S. and Avaca, L.A. (2000) *International Journal of Hydrogen Energy*, **25**, 415.
37 Lessing, P.A. (2007) *Journal of Materials Science*, **42**, 3477.
38 Hu, W.K. (2000) *International Journal of Hydrogen Energy*, **25**, 111.
39 Jaksic, M.M. (2000) *Electrochimica Acta*, **45**, 4085.
40 Yeager, E. and Tryk, D. (1984) in *Hydrogen Energy Progress V* (eds T.N. Veziroğlu and J.B. Taylor), Pergamon Press, Oxford, Vol. 2, p. 827.
41 Trasatti, S. (2000) *Electrochimica Acta*, **45**, 2377.
42 Trasatti, S. (ed)., (1980/81) *Electrodes of Conductive Metallic Oxides Parts A and B*, Elsevier, Amsterdam.

43 Trasatti, S. (1994) Transition metal oxides: versatile materials for electrocatalysis, in *The Electrochemistry of Novel Materials* (eds J. Lipkowski and P.N. Ross), VCH Publishers, Inc., New York.
44 Campari, M., Tavares, A.C. and Trasatti, S. (2002) *Hemijska Industrija (Chemical Industry Belgrade)*, **56**, 230.
45 Laplante, F., Bernard, N., Tavares, A., Trasatti, S. and Guay, D. (2006) X-ray photoelectron spectroscopy and X-ray diffraction characterization of rhodium oxides in reductive conditions, in *Electrocatalysis*, Vol. **2005-11** (eds G.M. Brisard, R. Adzic, V. Birss and A. Wieckowski), The Electrochemical Society, Pennington, NJ.
46 Bagotzky, V.S. and Skundin, A.M. (1984) *Electrochimica Acta*, **29**, 757.
47 Wendt, H. and Plzak, V. (1990) Electrode kinetics and electrocatalysis of hydrogen and oxygen electrode reactions. 2. Electrocatalysis and electrocatalysts for cathodic evolution and anodic oxidation of hydrogen, in *Electrochemical Hydrogen Technologies* (ed. H. Wendt), Elsevier, Amsterdam, Chapter 1. 2.
48 Borucinsky, T., Rausch, S. and Wendt, H. (1997) *Journal of Applied Electrochemistry*, **27**, 762.
49 Birry, L. and Lasia, A. (2004) *Journal of Applied Electrochemistry*, **34**, 735.
50 Bianchi, I., Guerrini, E. and Trasatti, S. (2005) *Chemical Physics*, **319**, 192.
51 Tavares, A.C. and Trasatti, S. (2000) *Electrochimica Acta*, **45**, 4195.
52 Kirk, D.W., Thorpe, S.J. and Suzuki, H. (1997) *International Journal of Hydrogen Energy*, **22**, 493.
53 Neophytides, S.G., Murase, K., Zafeiratos, S., Papakonstantinou, G., Paloukis, F.E., Krstajić, N.V. and Jaksić, M.M. (2006) *Journal of Physical Chemistry. B*, **110**, 3030.
54 Paunović, P., Popovski, O., Dimitrov, A., Slavkov, D., Lefterova, E. and Jordanov, S.H. (2006) *Electrochimica Acta*, **52**, 1810.
55 Trasatti, S. (1990) Electrode kinetics and electrocatalysis of hydrogen and oxygen electrode reactions. 4. The oxygen

evolution reaction, in *Electrochemical Hydrogen Technologies* (ed. H. Wendt), Elsevier, Amsterdam, Chapter 1. 4.

56 Trasatti, S. (1999) Interfacial electrochemistry of conductive metal oxides for electrocatalysis, in *Interfacial Electrochemistry: Theory, Practice, Applications* (ed. A. Wieckowski), Marcel Dekker, New York.

57 Daghetti, A., Lodi, G. and Trasatti, S. (1983) *Materials Chemistry and Physics*, **8**, 1.

58 Ardizzone, S. and Trasatti, S. (1996) *Advances in Colloid and Interface Science*, **64**, 173.

59 Trasatti, S. (1980) *Journal of Elaectroanalytical Chemistry*, **111**, 125.

60 Lodi, G., Sivieri, E., De Battisti, A. and Trasatti, S. (1978) *Journal of Applied Electrochemistry*, **8**, 135.

61 Castelli, P., Trasatti, S., Pollak, F.H. and O'Grady, W.E. (1986) *Journal of Elaectroanalytical Chemistry*, **210**, 189.

62 Angelinetta, C., Trasatti, S., Atanasoska, Lj.D., Minevski, Z.S. and Atanasoski, R.T. (1989) *Materials Chemistry and Physics*, **22**, 231.

63 De Pauli, C.P. and Trasatti, S. (2002) *Journal of Electroanalytical Chemistry*, **538–539**, 145.

64 Krstajić, N. and Trasatti, S. (1995) *Journal of the Electrochemical Society*, **142**, 2675.

65 Trasatti, S. (1995) *International Journal of Hydrogen Energy*, **20**, 835.

66 Schiller, G., Henne, R., Mohr, P. and Peinecke, V. (1998) *International Journal of Hydrogen Energy*, **23**, 761.

67 Trasatti, S. (2001) *Portugaliae Electrochimica Acta*, **19**, 197.

68 Kodintsev, I.M. and Trasatti, S. (1994) *Electrochimica Acta*, **39**, 1803.

69 Marshall, A., Borresen, B., Hagen, G., Sunde, S., Tsypkin, M. and Tunold, R. (2006) *Russian Journal of Electrochemistry*, **42**, 1134.

70 Marshall, A., Borresen, B., Hagen, G., Tsypkin, M. and Tunold, R. (2006) *Electrochimica Acta*, **51**, 3161.

71 Slavcheva, E., Radev, I., Bliznakov, S., Topalov, G., Andreev, P. and Budevski, E. (2007) *Electrochimica Acta*, **52**, 3889.

72 Wendt, H. and Kreysa, G. (1999) *Electrochemical Engineering*, Springer, Berlin.

73 Marshall, A., Børresen, B., Hagen, G., Tsypkin, M. and Tunold, R. (2007) *Energy*, **32**, 431.

74 Seehra, M.S., Ranganathan, S. and Manivannan, A. (2007) *Applied Physics Letters*, **90**, 044104.

75 Patil, P., De Abreu, Y. and Botte, G.G. (2006) *Journal of Power Sources*, **158**, 368.

76 Vitse, F., Cooper, M. and Botte, G.G. (2005) *Journal of Power Sources*, **142**, 18.

77 Cooper, M. and Botte, G.G. (2006) *Journal of the Electrochemical Society*, **153**, A1894.

78 Take, T., Tsurutani, K. and Umeda, M. (2007) *Journal of Power Sources*, **164**, 9.

79 Hu, Z.Y., Wu, M., Wei, Z.D., Song, S.Q. and Shen, P.K. (2007) *Journal of Power Sources*, **166**, 458.

80 de Souza, R.R., Padilha, J.C., Gonçalves, R.S., de Souza, M.O. and Rault-Berthelot, J. (2007) *Journal of Power Sources*, **164**, 792.

8
Energy from Organic Waste: Influence of the Process Parameters on the Production of Methane and Hydrogen

Michele Aresta and Angela Dibenedetto

8.1
Introduction

The utilization of organic waste [both fresh vegetables (FVGs) not rich in cellulose and residual organic compounds] for the generation of energy products such as methane and dihydrogen or other biofuels (alcohol, oil, biodiesel) is practiced more and more worldwide. The conversion of waste contributes to avoiding landfilling (which is under strict limitation in many countries), reduces water and soil pollution and produces usable energy that would otherwise be lost. Commonly, FVGs, and also industrial organic residual compounds and sludges, are treated by aerobic or anaerobic digestion. In the former process, the cost of which depends on the aeration frequency [1], the degradable fraction is usually converted without energy recovery. Conversely, the latter converts organic carbon into methane and CO_2 (biogas) with energy production, even if with moderate efficiency (30–50%) due to the low biodegradability of part of the solid fraction and long retention times (20–30 days) [2]. Biogas production still needs optimization of the process, also by integration in better engineered waste treatment plants. Various technological solutions (depending on temperature, solid content, type of reactor) are used, which in some cases are advantageously integrated with other waste treatments [3]. Energy is recovered as methane that can be separated from CO_2 and other minority gaseous compounds and used for thermal or electric energy production. The FVG biomass undergoes several steps, such as depolymerization, acidogenesis, acetate formation, methanogenesis and methanation of CO_2, which require different bacterial communities and a complex metabolic food chain [4–6]. In the whole process, H_2 and organic carboxylic acids, such as acetic acid, are key intermediates: it is important to maintain a low H_2 partial pressure as key biological reactions may occur that for thermodynamic reasons do not take place at higher H_2 pressures [4].

The anaerobic digestion of fatty acids, alcohols and organic compounds is accomplished by a syntrophy between H_2-producing and H_2-consuming methanogenic archaea [6] that favors the better use of the energy content of primary substrates [7].

Table 8.1 Metal enzymes involved in the conversion of CO_2 or H_2.

Enzyme/coenzyme	Metal in the active site	Reaction catalyzed
Conversion of CO_2		
Formate dehydrogenase	W	$CO_2 \rightarrow HCOO^-$
Tetrahydrofolate (THF)	Ni (in F-430 factor of CH_3–S–CoM)	$CO_2 \rightarrow {}^-CH_3/CH_4$
Methanofuran (MFR)		
Tetrahydromethanopterin (H_4-MPT)		
CH_3–S–CoM methyl reductase		
Methyl transferase (cobalamin)	Co	Methyl transfer
Carbon monoxide dehydrogenase (CODH)	Ni, Fe	$CO_2 \rightarrow CO$ or CH_3COOH
Dihydrogen formation/consumption		
Hydrogenases	Fe	$H^+ \rightarrow H_2$ (and $H_2 \rightarrow H^+$)
Hydrogenases	Ni, Fe	$H_2 \rightarrow H^+$, mainly Ni
Hydrogenases	Ni, Fe, Se	

The aim of this work was to investigate the role of iron, nickel and cobalt in the production of biogas during the anaerobic digestion of a sludge. These metals were chosen considering their role in anaerobic metabolism during methanogenic fermentation. In fact, they constitute the active center in several enzymes which play a key role in the complex methanation process (Table 8.1). In particular, nickel is the active center of the methyl-coenzyme M reductase (known as F430) and several H_2-consuming hydrogenases [8, 9] and acetate-forming enzymes [10–13]. Iron is present in several hydrogenases (H_2 uptake or evolution) and, as Fe_4S_4 protein, in carbon monoxide dehydrogenase (CODH), which has a key role in the anaerobic formation of acetic acid [10, 13, 14]. Cobalt is part of cobalamin, a catalyst of the transfer of methyl groups [15a]. All the above enzymes work together for the production of methane and carbon dioxide during the anaerobic digestion of sludges [15b] (Scheme 8.1).

Another interesting issue is the utilization of organic compounds present in process effluents for the recovery of carbon and energy. An example is the use of bioglycerol for the synthesis of added value products. In particular, bioglycerol has been used for the synthesis of 1,3-propanediol [16] that may find use in the polymer industry, or 2,3-butanediol [17] using selected microorganisms. Dihydrogen can also be obtained from the same organic source [18] using selected bacterial strains.

In this chapter, we present part of our work on the conversion of bioglycerol into 1,3-propanediol, 2,3-butanediol and H_2 using different microorganisms and discuss the use of a single bacterial strain isolated from anaerobic fermentation media, which is able to produce either diols or H_2 according to the conditions in which it is grown.

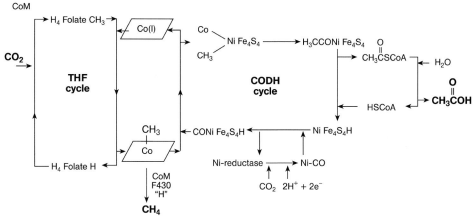

Scheme 8.1 Acetic acid formation from CO_2: the role of CODH and tetrahydrofolate (THF).

8.2 Experimental

8.2.1 Methanation of Residual Biomass

The anaerobic digestion of FVG organics was carried out under anaerobic conditions by using two spherical base Pyrex batch reactors, each of 2 L capacity, one being used as a control. Each reactor was double-walled (for thermostatic liquid circulation). The suspension was stirred at 250–500 rpm at 310 K. The gas produced in each of the reactors was collected in a separate gas -meter, from which samples were withdrawn under controlled conditions for gas chromatography (GC) analysis [Chromosorb SII column, 3 m × 2.1 mm i.d., thermal conductivity detection (TCD)]. The digestion of the fermentable biomass (3% by weight of a homogenized mixture whose composition was analogous to that of a domestic FVG waste) was initiated by using a sludge (3.5% solid material) taken from an anaerobic water treatment plant of a beer factory. The influence of metals was studied by adding 1 mL of either a 1 M $NiCl_2 \cdot 6H_2O$, $CoCl_2 \cdot 6H_2O$ or $FeCl_2 \cdot 6H_2O$ to the reactor in independent tests, unless specified otherwise. The total amount of metals in the reactor never reached the toxic concentration. The metal concentration in the liquid or solid phase was measured on a sample withdrawn at fixed times under controlled conditions from the reactor and treated according to the IRSA analytical method [19] using atomic absorption spectrometry (Shimadzu AA 6200). The solid was separated from the liquid phase and the metal distribution between the two phases was estimated on the basis of the results of the metal analysis on the liquid and solid, respectively. The metals were shown to accumulate slowly in the solid phase as they are adsorbed by extracellular polysaccharides or other solids. The addition of Ni and Fe was done 1 week after the anaerobic fermentation was started; cobalt was added to the sludge after 18 days. During the first 10 days of biogas production, the acidogenesis phase was prevalent over methanogenesis, so it was possible to monitor the H_2 production. The effect of each metal was monitored at two different time intervals during 8 h (the analysis of biogas was done every 30 min during 8 h) and 5 days (the analysis was made every 24 h during 5 days). The first test was useful for monitoring the effect of each addition (of metal or feed) during the acidogenesis phase of the fermentation and, thus, the production of H_2. The second test gave information about the effect of each metal addition on CH_4 and CO_2 production. All data in the figures are the average of three measurements with standard deviation represented as error bars.

8.2.2
Bioconversion of Glycerol

8.2.2.1 Characterization of Strains K1–K4

Bacterial strains were isolated from anaerobic sludge via membrane filtration and selective growth on culture media such as ADA for *Aeromonas*, Pseudomonas medium for *Pseudomonas*, TBCS for *Vibrio* and MacConkey for *Klebsiella*. Six strains were isolated from the MacConkey medium and supposed to be *Klebsiella* and four strains were isolated from the Pseudomonas medium and considered to be *Pseudomonas*. All the isolated strains were tested for their ability to grow on glycerol. Three strains were isolated showing the best performances among the six supposed *Klebsiella* (strains K1, K2 and K3) and one among the four supposed *Pseudomonas* (strain P4). Strains K1, K2 and K3 were characterized as *Klebsiella*. Strain P4 (renamed K4), although isolated using a Pseudomonas medium from a waste anaerobic fermentation sludge, was characterized as a *Klebsiella pneumoniae* subsp. *pneumoniae* by the NCCB Institute in Utrecht. K4 is a Gram-negative bacterium able to growth aerobically and anaerobically.

8.2.2.2 Use of Strains K1, K2 and K3

Strains K1, K2 and K3 were pretreated aerobically for 24 h at 310 K in 100 mL sterile flasks using a preculture medium [glycerol, 20; $KHPO_4$, 3.4; KH_2PO_4, 1.3; $(NH_4)_2SO_4$, 2.0; $MgSO_4 \cdot 7H_2O$ 0.2; yeast extract, 1.0; $CaCO_3$, 2; $FeSO_4 \cdot 7H_2O$, 0.005; $CaCl_2$, 0.002 g L^{-1}; 2 mL L^{-1} of a trace element solution ($ZnCl_2$ 0.07; $MnCl_2 \cdot 4H_2O$, 0.1; H_3BO_3, 0.06; $CoCl_2 \cdot 6H_2O$, 0.2; $CuCl_2 \cdot 2H_2O$, 0.02; $NiCl_2 \cdot 6H_2O$, 0.025; $NaMoO_4 \cdot 2H_2O$, 0.035 g L^{-1}). All subsequent experiments were carried out in a 250 mL bottle using 200 mL of culture medium (glycerol, 30; KCl, 1.6; NH_4Cl, 6.7; $CaCl_2$, 0.28; yeast extract, 2.8 g L^{-1}).

8.2.2.3 Use of Strain K4

Strain K4 was incubated for 24 h at 310 K in a 250 mL bottle containing 100 mL of sterile preculture medium (peptone 16.0; hydrolyzed casein 10; K_2SO_4 10; $MgCl_2$ 1.4 g L^{-1}). All experiments were carried in a 200 mL bottle filed with 100 mL of medium with a culture medium with the following composition: glycerol, 20; peptone, 3.0; KH_2PO_4, 1.5; $MgSO_4 \cdot 7H_2O$, 1.5 g L^{-1}).

8.2.2.4 Tests Under Aerobic Conditions

The K1, K2 and K3 bacterial growth was followed spectrophotometrically, monitoring the absorbance at 578 nm. Glycerol, 1,3-propanediol and 2,3-butanediol were quantified by GC (Agilent 6850). The organic carboxylic acids formed were determined by high-performance liquid chromatography using a Perkin-Elmer Series 4 liquid chromatograph connected with an LC 290 UV/Vis spectrophotometric detector. Strain K4 was also tested in the same conditions and produced 1,3-propanediol and 2,3-butanediol.

8.2.2.5 Tests Under Microaerobic or Anaerobic Conditions

Strain K4 was tested in the absence of oxygen at various concentrations of glycerol (10, 20, 30, 40 g L^{-1}). Two serum bottles were used with a modified Hungate technique: one of the bottles contained resazurin and a reducing agent, namely cysteine–HCl, and the other did not contain resazurin. The gas was analyzed by GC with TCD.

In a separate experiment, various concentrations of glycerol were used in the modified Hungate technique (see above). The bacterial strain K4 was also used in a reactor attached to a gas meter in order to evaluate the maximum amount of dihydrogen produced. The same strain was used in a reactor with periodic withdrawal of gas in order to evaluate the effect of pressure. The same strain was also used in a high-pressure autoclave equipped with a pressure gauge and the maximum pressure of dihydrogen produced was monitored.

8.3
Results and Discussion

8.3.1
Biogas from Waste

The production of biogas (CH$_4$ and CO$_2$) from residual vegetables or organics such as proteins or polysaccharides usually follows three phases, depolymerization, acidogenesis and methanogenesis [4–6]. In the first phase, the depolymerization of complex structures takes place under the action of hydrolytic bacteria or fungi. Monomers are formed, such as sugars, amino acids, peptides and acids. Such monomers are converted by fermentative bacteria into the so-called volatile fatty acids (VFAs), that is, acetic, propionic and butyric acid, and H$_2$. Ammonia and CO$_2$ are also formed. In a subsequent step, VFAs are converted into CO$_2$ and H$_2$. In the last step, acetic acid, CO$_2$ and H$_2$ are converted by methanogens into CH$_4$ and CO$_2$. In such a complex process, metal enzymes play a key role as they drive important reactions such as H$_2$ formation and conversion, CO$_2$ reduction to CO and the formation of acetic acid from CO, among others. Table 8.1 lists the enzymes and the relevant active centers involved in H$_2$ formation.

Iron-only hydrogenases have been isolated from several microorganisms [20, 21] and shown to be able both to produce and to consume dihydrogen.

The active site in *Clostridium pasteurianum* (Figure 8.1) contains a large unit characterized by an unusual arrangement of two moieties, an Fe$_4$S$_4$ iron protein

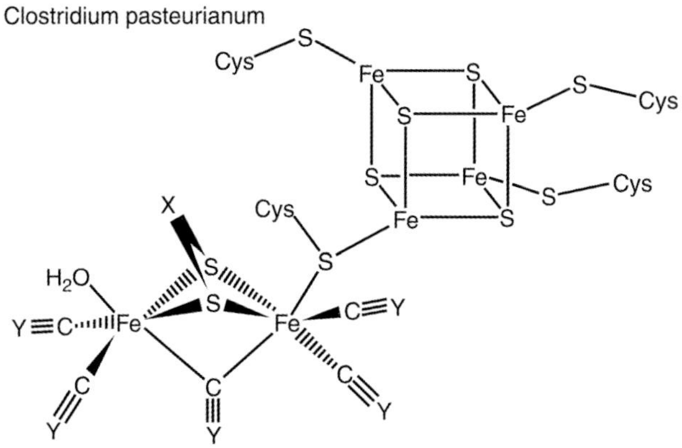

Clostridium pasteurianum

[Fe$_2$(S$_2$X)CY](CY)$_4$(H$_2$O)[Fe$_4$S$_4$](S$^\gamma$Cys)$_4$

Figure 8.1 The active center of *Clostridium pasteurianum* hydrogenase.

linked through a cysteine-S to an Fe$_2$ cluster in which two octahedral irons bear five C–X groups (CO or CN), one water molecule and three bridging S. Such a 'large domain' is accompanied by four 'small domains' containing either the Fe$_4$S$_4$ protein or the Fe$_2$ non-proteic cluster.

Nickel–iron hydrogenases [NiFe] (Figure 8.2) are present in several bacteria. Their structure is known [22, 23] to be a heterodimeric protein formed by four subunits, three of which are small [Fe] and one contains the bimetallic active center consisting of a dimeric cluster formed by a six coordinated Fe linked to a pentacoordinated Ni (III) through two cysteine-S and a third ligand whose nature changes with the oxidation state of the metals: in the reduced state it is a hydride, H$^-$, whereas in the oxidized state it may be either an oxo, O^{2-}, or a sulfide, S^{2-}.

It has also been shown that in some microorganisms, such as *Desulfomicrobium baculatum*, a cycteine-S has been replaced by a cysteine-SE [24], giving rise to a trinuclear [FeNiSe] hydrogenase.

The mechanism of action has been studied by several authors [20, 25] and the involvement of an Ni(III)–Ni(II)–Ni(0) system has been proposed. Ni and Fe enzymes apparently have a different role in H$_2$ production and consumption. In

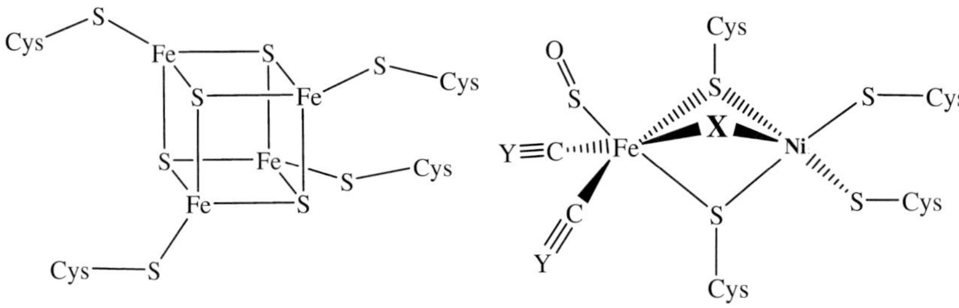

[Fe$_4$S$_4$](S$^\gamma$Cys)$_4$ [Ni-X-Fe](S$^\gamma$Cys)$_4$(CY)$_2$(SO)
X=S or O; Y=O or N

Figure 8.2 Two different subunits present in FeNi-hydrogenases.

fact, Ni enzymes are more specifically involved in H_2 consumption, whereas Fe enzymes are more involved in H_2 production.

Ni and Fe are equally involved in the synthesis of the acetyl moiety from CO_2 [26–28], whereas Co (as cobalamin) is well known to act as a carrier of methyl groups.

We tested the influence of the Fe, Ni and Co concentration on the composition of biogas by monitoring the H_2, CO_2 and CH_4 production during 8 h after the addition of feed and over 5 days. Each metal produced a different effect, as expected, on the basis of the role of the enzymes in which they are present. Nickel, iron and cobalt were added to the sludge either separately or in combination, always at a sub-toxic concentration. In a typical case, the starting concentration of metal ions in the sludge was 6, 13.6 and 4.46 ppm for Ni(II), Fe(II) and Co(II), respectively, in the liquid phase. Their levels in the solid phase were 14.6, 54.12 and 13.43 ppm under the same conditions. The addition of 0.5 mL of 1 M $NiCl_2$ to the reactor filled with 1.7 L of sludge produced an instantaneous increase in the metal concentration in the liquid phase from an initial 6 ppm up to 22.4 ppm. The concentration in the solid phase rose from 14.6 to 16.08 ppm. The metal concentration in solution decreased very slowly, remaining at the level of 90% of the value read soon after the addition after 5 days and decreasing to 70% after 2 weeks. Therefore, during the test during 8 h (which starts soon after the addition of metal and feeding) the bacteria really feel the increase in the concentration of the metal available in solution. The response to such an increase in metal concentration is, therefore, a real cause–effect process. An increase in the concentration of Ni in solution caused a decrease in dihydrogen in the gas phase (Figure 8.3a).

This observation agrees well with the role of nickel as part of the active site in H_2-consuming hydrogenases [11]. Therefore, the addition of Ni(II) promotes the activity of such H_2-consuming hydrogenases. Conversely, when Fe(II) was added, the increase in H_2 production in the batch reactor was observed within 2.5 h after feeding (Figure 8.3b), in comparison with the control, which had a lower H_2 production. These data also agree with the role of iron in H_2-evolving hydrogenases, which promote the formation of H_2 [14]. Noteworthily, the addition of cobalt did not cause any variation in H_2 production at any time after the start of the fermentation. Such metal-dependent effects were evident only when fresh feed was added and the system had not reached the optimum performance. Conversely, once a sludge was aged, the

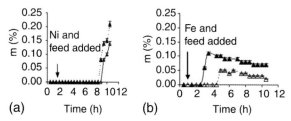

Figure 8.3 Influence of Ni and Fe on dihydrogen production during the acidogenesis phase in the anaerobic fermentation of FVG. The broken line represents the control.

evolution and accumulation of dihydrogen was not so evident when Ni or Fe was added.

Nevertheless, the added metal ions under these conditions showed a positive influence on methane formation, increasing the rate of formation and the amount of methane. Figure 8.4 shows the composition of biogas produced after the addition of (a) iron, (b) nickel and (c) cobalt: in all cases, an increase in the percentage of methane in the gas phase was observed.

All the results show that before the addition of the metals, both the reactor and control have the same or very close $CH_4:CO_2$ ratio, which varies with the

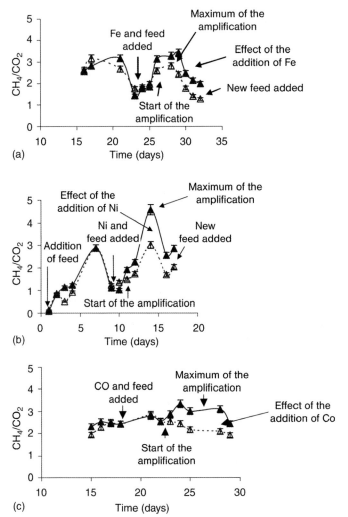

Figure 8.4 Effect of the addition of Fe, Ni and Co on the production of biogas: variation of the $CH_4:CO_2$ ratio. The broken line is the control.

addition of feed. The increase in the ratio after the addition of feed can be explained by considering the enhanced production of dihydrogen that is made available also for direct CO_2 methanation. After the metal had been added, a net increase in the CH_4:CO_2 ratio was observed in the reactor with respect to the control. Typically, the amplification started within 2 days after the addition of the metal and reached a maximum around the fifth or sixth day. The difference remained evident for 7–10 days also during the phase of decrease in methanation. If more feed was added, the amplification was again evident. After 20–30 days, the added metal seemed still able to affect the CH_4:CO_2 ratio. It is important to note that during such a period the concentration of the metal in the liquid phase in the reactor was always higher than that in the control. The increase in the CH_4:CO_2 ratio is produced by a more effective conversion of CO_2 to methane: Ni(II) increases the ratio to 4.5 compared with 3 found in the biogas in the control (Figure 8.4).

The observed effect of the increased concentration of the metals matches the role of metals in enzymes. Among the three metal ions, nickel has the greatest effect (compare Figure 8.4a–c). These results suggest that the controlled addition of Fe, Ni and Co could be beneficial for improving the methanation process of waste.

8.3.2
Dihydrogen from Bioglycerol

Bioglycerol is predicted to become a large-volume organic waste subsequent to the increased production of biodiesel. Therefore, in recent years, there has been increasing interest in its use as a raw material for the production of chemicals and energy products. The key point with the use of glycerol is that it is produced in the biodiesel industry as an aqueous solution containing salts and the most desirable use would be the direct utilization of such a process liquid stream. In this context, biotechnologies and chemical processes may find different applications, as the latter may require water-free samples for most conversions. We used aqueous glycerol as a starting material for the production of 1,3-propanediol and 2,3-butanediol or dihydrogen. The bacterial strains used were isolated from the sludge of the anaerobic fermentation of organics obtained when the maximum production of dihydrogen was observed (see above) Three strains, putatively assigned to the *Klebsiella* strain, were used as reported above, the *Klebsiella* K1, K2 and K3 and the *Klebsiella pneumoniae* subsp. *pneumoniae* K4 [29].

Klebsiella is known to be able to use glycerol under either aerobic [30] or anaerobic conditions. In particular, *Klebsiella pneumoniae* has been shown to produce 1,3-propanediol anaerobically [31–35], but there have been no reports about its use for the production of dihydrogen.

Strains K1, K2 and K3 were used for the production of 1,3-propanediol under aerobic conditions. The conditions described in the Experimental section were used: the starting concentration of glycerol was 30% by mass. Table 8.2 summarizes the results obtained. Strain K2 is the most active, showing a 47% conversion of glycerol after 144 h and the best selectivity towards 1,3-propanediol.

Table 8.2 Production 1,3-propanediol and 2,3-butanediol from glycerol with strains K1–K3.

Microorganism	Glycerol consumption (%)	Time (h)	1,3-Propanediol	2,3-Butanediol	Products of conversion (mol mol^{-1} glycerol)				
					Acetic acid	Lactic acid	Ethanol	Formic acid	Succinic acid	Butyric acid
K1	14	24	0.070	0.000	0.020	0.021	Trace	Trace	Trace	Trace
K2	47	24	0.330	0.098	0.026	0.018	Trace	Trace	Trace	Trace
K3	42	120	0.055	0.059	0.039	0.019	Trace	Trace	Trace	Trace

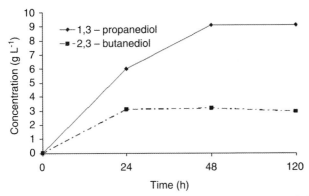

Figure 8.5 Kinetics of formation of 2,3-butanediol and 1,3-propanediol.

The other products formed in a significant amounts are 2,3-butanediol, acetic acid and lactic acid. Ethanol, formic acid, succinic acid and butyric acid were detected in trace amounts. K2 affords a 33% molar conversion of glycerol into 1,3-propanediol with 9.6% molar conversion into 2,3-butanediol. Figure 8.5 shows that K2 has fast kinetics during the first 48 h. 2,3-Butanediol is formed mainly in the first 24 h, whereas 1,3-propanediol accumulates for up to 48 h, by which time the formation of both 1,3-propanediol and 2,3-butanediol has ceased. This may be due to the accumulation of acids that may inhibit this metabolic path. A UV/Vis analysis of the solution for bacterial growth shows that the optical density continues to increase after 48 h and until 144 h. In the meantime, the concentration of glycerol in the interval 48–144 h decreases only slightly compared with the consumption observed during the first 24 h. Under the best conditions, the glycerol concentration is decreased to 60% of the original concentration after 24 h and to 53% after 144 h. Compared with other *Klebsiella* strains reported in the literature, K2 shows an interesting performance that needs to be optimized for application.

An interesting result was obtained with strain K4. It has been reported that *Klebsiella pneumoniae* is able to convert glycerol into 1,3-propanediol [34]. In fact, when the K4 strain was grown aerobically on the same culture medium as K1–K3, 1,3-propanediol was produced. Conversely, when it was grown anaerobically as reported in the Experimental section in presence of peptone and hydrolyzed casein, a new behavior was observed: strain K4 produced dihydrogen as the main metabolite of glycerol rather than 1,3-propanediol. This finding is new as H_2 has never been reported to be the main metabolite of glycerol with other *Klebsiella pneumoniae* strains. Table 8.3 shows the dependence of the formation of dihydrogen on the conditions.

The data show that there is an effect of the initial glycerol concentration on the conversion into dihydrogen. A concentration of $0.1 \, \text{mol L}^{-1}$ of glycerol seems to represent the best condition for its conversion.

An interesting result was obtained when strain K4 was grown in a high-pressure vessel that allowed the accumulation of dihydrogen. A pressure gauge allowed the pressure generated by dihydrogen to be measured (Figure 8.6).

Table 8.3 Production of dihydrogen from glycerol using strain K4 under anaerobic conditions[a].

Concentration of glycerol, (mmol L^{-1})	Consumed glycerol (%)	H$_2$ produced (mmol)	Molar Ratio H$_2$/glycerol	Purity of H$_2$ (%)	Formic acid (mol)	Lactic acid (mol)	Acetic acid (mol)
110	95	8.1	0.76	97	0.0013	nd[b]	0.0035
220	91	8.2	0.41				
330	79	6.96	0.27				
440	63	6.0	0.24				
330 with resazurin	68	4.50	0.20				

[a] In all cases 100 mL of solution were used. All systems were monitored for 120 h.
[b] nd, Not detected.

Figure 8.6 Production of dihydrogen by strain K4 grown on glycerol in a high-pressure vessel.

Table 8.4 Production of high-pressure dihydrogen via glycerol metabolism by strain K4.

Concentration of glycerol at $t=0$ (mol L^{-1})	Buffer	Glycerol consumed (%)	Gas produced (mol mol^{-1} glycerol)	Dihydrogen production rate, (mol L^{-1} h^{-1})	Max. H$_2$ pressure (MPa)
0.22	No	88	0.58	0.014	0.32
0.22	Yes	100	1.0	0.016	0.53
0.44[a]	Yes	74	0.44[a]	0.013	0.32
0.71[a]	Yes	64	0.34	0.014	0.36

[a] Dihydrogen was withdrawn at fixed times from the high-pressure reactor.

Table 8.4 reports the analytical data and main features relevant to such a system. Interestingly, dihydrogen was collected at a pressure of 0.53 MPa, which allows its distribution in external utilization circuits.

This is the first report to describe the collection of gas under pressure in a biosystem. This finding is of great interest as it would permit the distribution of H$_2$ to some utilities without any use of pumping systems. For example, H$_2$ could be conveniently distributed to fuel cells. This bacterial strain is still under investigation with a view to better exploitation of its properties.

The complete conversion of glycerol with formation of dihydrogen and volatile organic carboxylic acids is an interesting feature of the system. The carboxylic acids could be isolated, if present in a significant amount, or further converted into H$_2$ and CO$_2$ by using an *ad hoc* bacterial strain. The photosynthetic bacterium *Rhodospirillum rubrum* seems to be the most appropriate for such an application. A pH adjustment to ~6 helps to improve the acid conversion with a very low residual organic acid content [36].

An interesting comparison can be made with chemical techniques for glycerol treatment and dihydrogen production. Table 8.5 compares the performance of the above strain K4 with that of a catalytic conversion of aqueous glycerol [37].

Catalytic dihydrogen production form biomass is known as aqueous phase reforming (APR) [38]. It is interesting to note the absence of CO in the dihydrogen produced by the biosystem, which makes it immediately suitable for use in fuel-cells

Table 8.5 Comparison of the biological versus catalytic production of dihydrogen from glycerol.

Parameter	Biological: 2–6% glycerol	Catalytic: 1–20% glycerol
Conversion of glycerol (%)	100 at 2% feed	100 at 1% feed
Purity of H$_2$ (%)	>99	90
Presence of CO	Absent	Present
Presence of CO$_2$	Traces	Present
Temperature (K)	Ambient	500–600
Pressure (MPa)	0.6 MPa	2.0–3.0
Lifetime of the catalyst	>7 days	2 weeks
Co-products	Organic acids, ethanol	Organic acids and others

without preliminary treatment for cleaning. The bioconversion of glycerol also has the advantage of low-temperature operation. For full exploitation, improvements are necessary, such as:

- optimizing the H_2 production;
- increasing the pressure;
- developing a new buffer system for better control of the reaction system;
- increasing the lifetime of bacteria;
- working at higher concentration of the substrate (direct use of industrial waters);
- converting the organic acid co-products using photosynthetic bacteria in one pot;
- maximizing the use of carbon;
- scaling-up for checking the exploitation potential.

Acknowledgment

Financial support by the EU Project TOPCOMBI is gratefully acknowledged (H_2 production from bioglycerol).

References

1 Bernard, S. and Gray, F. (2000) *Water Research*, **34** (3), 725–734.
2 Weemaes, M., Grootaerd, H., Simoens, F. and Verstraete, W. (2000) *Water Research*, **34** (8), 2330–2336.
3 International Energy Agency (IEA) (1994).
4 Zehnder, A.J.B. (1978) in *Water Pollution Microbiology*, **Vol. 2** (ed. R. Mitchell), John Wiley & Sons, Inc., New York, pp. 349–376.
5 Zeikus, J.G. (1983) *Microbes and their Natural Environments* (eds J.H. Slater and J.W. Whittenbury), Cambridge University Press, Cambridge, pp. 423–462.
6 Schink, B. (1983) *Environmental Microbiology of Anaerobes* (ed. A.J.B. Zehnder), John Wiley & Sons, Inc., New York, pp. 1466–1473.
7 Thauer, R.K., Jungermann, K. and Decker, K. (1997) *Bacteriological Reviews*, **41**, 100–180.
8 Walsh, C.T. and Orme-Johnson, W.H. (1987) *Biochemistry*, **26**, 4901–4906.
9 Aresta, M., Quaranta, E. and Tommasi, I. (1988) *Journal of the Chemical Society, Chemical Communications*, 450–451.
10 Aresta, M., Giannoccaro, P., Quaranta, E., Tommasi, I. and Fragale, C. (1998) *Inorganica Chimica Acta*, **272**, 38–42.
11 Albracht, S.P.J. (1994) *Biochimica et Biophysica Acta*, **1188**, 167–204.
12 Ellefson, W.L. and Wolfe, R.S. (1981) *Journal of Biological Chemistry*, **256**, 4259–4262.
13 Rouviere, P.E., Escalante-Semerena, J.C. and Wolfe, R.S. (1985) *Journal of Bacteriology*, **162**, 61–66.
14 Adams, M.W.W. (1990) *Biochimica et Biophysica Acta*, **1020**, 115–145.
15 Hippler, B. and Thauer, R.K. (1999) *FEBS Letters*, **449**, 165–168; Schonheit, P., Mool, J. and Thauer, R.K. (1979) *Archives of Microbiology*, **123**, 105–107.
16 Ito, T., Nakashimada, Y., Senba, K., Matsui, T. and Nishio, N. (2005) *Journal of Bioscience and Bioengineering*, **100** (3), 260–265.

17 Biebl, H., Menzel, K., Zeng, A..-P. and Deckwer, W..-D. (1999) *Applied Microbiology and Biotechnology*, **52**, 297–298.
18 Kapdan, I.K. and Kargi, F. (2006) *Enzyme and Microbial Technology*, **38**, 569–582.
19 Istituto di Ricerca Sulle Acque (IRSA) Publications Vol. 3 (1985).
20 Dance, I. (1999) *Chemical Communications*, 1655–1656.
21 Nicolet, Y., Piras, C., Legrand, P., Hatchikian, E.C. and Fontecilla-Camps, J.C. (1999) *Structure*, **7**, 13–23.
22 Volbeda, A., Charon, M.H., Piras, C., Hatchikian, E.C., Frey, M. and Fontecilla-Camps, J.C. (1995) *Nature*, **373**, 580–587.
23 Volbeda, A., Garcin, E., Piras, C., de Lacey, A.L., Fernandez, V.M., Hatchikian, E.C., Frey, M. and Fontecilla-Camps, J.C. (1996) *Journal of the American Chemical Society*, **118**, 12989–12996.
24 Garcin, E., Vernede, X., Hatchikian, E.C., Volbeda, A., Frey, M. and Fontecilla-Camps, J.C. (1999) *Structure*, **7**, 557–566.
25 Higuchi, Y., Ogata, H., Miki, K., Yasuoka, N. and Yagi, T. (1999) *Structure*, **7**, 549–556.
26 Hu, Z., Spangler, N.J., Anderson, M.E., Xia, J., Ludden, P.W., Lindahl, P.A. and Munck, E. (1996) *Journal of the American Chemical Society*, **118**, 830–845.
27 Tommasi, I., Aresta, M., Giannoccaro, P., Quaranta, E. and Fragale, C. (1998) *Inorganica Chimica Acta*, **272**, 38–42.
28 Aresta, M., Narracci, M., Dibenedetto, A. and Tommasi, I. (2002) 223rd ACS Meeting, – Inorganic Chemistry Division, Boston, 18–22 August 2002, Abstract No. 61.
29 Dibenedetto, A. and Lonoce, R. (2006) TOPCOMBI Project Report. Aresta, M., Dibenedetto, A., Narracci, M. and Lonoce, R. (2008) TOPCOMBI Project Report.
30 Forage, R.G. and Lin, E.C.C. (1982) *Journal of Bacteriology*, **151** (2), 591–599.
31 Johnson, E.A., Burke, S.K., Forage, R.G. and Lin, E.C.C. (1984) *Journal of Bacteriology*, **160**, 55–60.
32 Solomon, B.O., Zeng, A.P., Biebl, H., Ejiofor, O.A., Posten, C. and Deckwer, W.D. (1994) *Applied Microbiology and Biotechnology*, **42**, 222–226.
33 Chen, X., Xiu, Z., Wang, J., Zhang, D. and Xu, P. (2003) *Enzyme and Microbial Technology*, **33**, 386–394.
34 Xiu, Z.L., Song, B.H., Wang, Z.T., Sun, L.H., Feng, E.M. and Zeng, A.P. (2004) *Biochemical Engineering Journal*, **19**, 189–197.
35 Cheng, K.K., Liu, H.J. and Liu, D.H. (2005) *Biotechnology Letters*, **27**, 19–22.
36 Levin, D.B. and Love, M. (2004) *International Journal of Hydrogen Energy*, **29**, 173–185.
37 Boonyanuwat, A., Jentis, A. and Lercher, J.A. (2006) International Conference on Catalysis for Sustainable Energy Production, Florence, 29 November–1 December 2006.
38 Czernik, S., French, R.J., Feik, C.J. and Chornet, E. (2002) *Industrial and Engineering Chemistry Research*, **41**, 4209–4215.

9
Natural Gas Autothermal Reforming: an Effective Option for a Sustainable Distributed Production of Hydrogen

Paolo Ciambelli, Vincenzo Palma, Emma Palo, and Gaetano Iaquaniello

9.1
Introduction

This chapter deals with the catalytic autothermal reforming of methane for hydrogen production. The goal of producing hydrogen in small-scale plants is attracting increasing attention with the continued interest in the potential widespread application of fuel cells, on-site hydrogen generators and the establishment of distributed hydrogen infrastructures.

After a short overview of the technologies available today for hydrogen production from fossil fuels, attention is focused on the autothermal reforming process, which represents an effective technological option for a distributed energy system owing to greater reactor simplicity and compactness and lower construction costs compared with steam reforming.

In the first part of the chapter, a state-of-the-art review and also a thermodynamic analysis of the autothermal reforming reaction are reported. The former, relevant to both chemical and engineering aspects, refers to the reaction system and the relevant catalysts investigated. The latter discusses the effect of the operating conditions on methane conversion and hydrogen yield.

In the second part of the chapter, attention is focused on research activities at the University of Salerno on the autothermal reforming of methane. In particular, the design, assembly, setup and optimization of the performance of a compact reactor are reported. It is demonstrated that such an autothermal reforming reactor can assure very fast start-up, high flexibility and stable operation over a wide range of conditions. Moreover, when coupled with innovative catalytic formulations and geometries, it may represent an effective solution for the main component of a distributed hydrogen production. Finally, this option is also investigated from the economic point of view.

To realize the concept of sustainable development, it is important to consider both pollution and energy issues, because environmental degradation is already threatening human health and a reduction in energy resources will hinder development

Catalysis for Sustainable Energy Production. Edited by P. Barbaro and C. Bianchini
Copyright © 2009 WILEY-VCH Verlag GmbH & Co. KGaA, Weinheim
ISBN: 978-3-527-32095-0

for future generations. It seems worthwhile to consider how we can preserve such resources and to construct a new energy system in order to solve this problem [1]. The pursuit of sustainable development requires the improvement and modification of each country's political, economic and social systems, and, in addition, novel technological solutions [1].

A distributed energy system is an efficient, reliable and environmentally friendly alternative to traditional energy systems. In a distributed power generation system, the units for energy conversion are situated close to energy consumers and large units are replaced by smaller ones, thus making it possible for single buildings to be completely self-supporting in terms of electricity, heat and cooling energy [2]. In particular, hydrogen-based energy systems offer several advantages over traditional energy systems and would contribute to partially or wholly resolving the concerns regarding national security, emissions of greenhouse gases, finite sources of fossil fuels and environmental quality [3]. Indeed, hydrogen-based distributed power generation could combine the greatest potential benefits of hydrogen in terms of reduced emissions of pollutants and greenhouse gases with safety and security requirements, the high production capacity, storage and far-and-wide transportation being inherently more dangerous in terms of security and safety hazards. End-use technologies that employ hydrogen, such as fuel cells and combustion engines, have a safety record that advances protection of the environment and public health [3].

The versatility of hydrogen as a fuel is represented by the multiple process technologies and feedstocks from which it can be derived, including fossil fuels, biomass and water [4]. Each technology, in a different stage of development, offers different opportunities, benefits and challenges. The choice and timing of the various options for hydrogen production are closely linked to the local availability of feedstock, maturity of the technology, market applications and demand, policy issues and costs. When hydrogen is produced from fossil fuels, CO_2 capture and sequestration, allowing the decoupling of fossil fuels from CO_2 emissions, make this feedstock comparable, from an environmental point of view, with renewable energy sources [5].

Despite these benefits, there are substantial obstacles that need to be overcome in order to realize a hydrogen economy. In fact, although a wide choice of technologies for hydrogen production, storage and distribution is available, they need to be adapted for use in each specific energy system. Building a new hydrogen energy infrastructure would require considerable financial investment over multiple decades and it would involve logistical problems in matching supply and demand during the transition period [3, 6].

Distributed production of hydrogen from natural gas is based on small-scale steam reforming technology. The production unit can be located at the consumer refueling site and the unit capacity can be tailored to the site's fuelling requirements, thus eliminating the need for an extensive hydrogen delivery infrastructure. This process may be the most viable for introducing hydrogen as an energy carrier since it requires less capital investment for the smaller hydrogen volumes needed initially in the transition phase to the hydrogen economy. Small-scale reformers will permit the use of existing natural gas pipelines for the production of hydrogen at the site of the consumer, thus representing an important technology for the transition to a larger hydrogen supply.

Given the intrinsic advantages of fossil fuels, such as their availability, relatively low cost and existing infrastructures for delivery and distribution, they are likely to play a major role in energy and hydrogen production in the near- to medium-term future [7]. Considering availability and economic performance, energy consumption and reforming efficiency are important factors in hydrogen production [1]. In particular, among the potential hydrogen sources, natural gas offers many advantages: it is relatively abundant around the world and can be transported in large pipeline systems in many of the developed countries, resulting in the most readily available primary fuel [8].

The general term usually applied to the process of converting liquid or gaseous light hydrocarbon fuels to hydrogen and carbon monoxide is *reforming*.

A typical fuel processor for hydrogen production involves a series of steps. Generally, the reforming step is followed by CO abatement stages through water gas shift (WGS) and preferential oxidation (PROX) reactions and further final product purification [pressure swing adsorption (PSA) or membrane separation] to obtain a pure hydrogen stream (Figure 9.1).

Natural gas feedstock is very dependent of the source location: in some cases it has high levels of H_2S, CO_2 and hydrocarbons. Organic sulfur compounds must be removed because they will irreversibly deactivate both reforming and WGS catalysts. Hence a preliminary feed desulfurization step is necessary. This process consists in a medium-pressure hydrogenation (usually on a cobalt–molybdenum catalyst at 290–370 °C), which reduces sulfur compounds to H_2S, followed by H_2S separation through ZnO adsorption (at 340–390 °C) or amine absorption [9].

The CO abatement stage is generally carried out through the WGS reaction, which allows the conversion of CO to CO_2 and H_2. This step is essential when the stream is fed to the Pt anode of a polymer electrolyte membrane (PEM) fuel cell, since even a small amount of carbon monoxide (>10 ppm) leads to poisoning of the electrocatalyst. Generally, the process consists of a high-temperature WGS step on an Fe–Cr oxides catalyst (350–400 °C) followed by a low-temperature WGS step on a Cu–Zn–Al catalyst (~200 °C), where a higher CO conversion is realized [10].

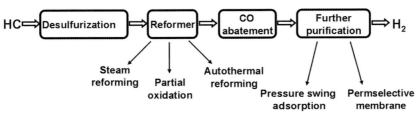

Figure 9.1 Schematic of a typical fuel processor.

A further purification step is carried out through the CO preferential oxidation reaction, generally employing precious metal-based catalysts, mostly Pt [10–13].

Finally, it is possible to obtain a pure hydrogen stream through several techniques, such as PSA, cryogenic distillation or membrane separation. PSA and cryogenic distillation processes are commercially available separation techniques [14].

In the PSA process, the reformer product gas is passed through an adsorption column, where impurities are preferentially adsorbed. After a set time, the feed gas to the reactor column is diverted to a parallel column. At this stage, the first column is depressurized, allowing the bed regeneration. It is a cyclic process, the two columns being alternately pressurized and depressurized [15].

On the other hand, membranes, because of their physical nature, allow only selected materials to permeate across them. An advantage deriving from the integration of the membrane inside a reformer is the possibility of overcoming the thermodynamic limitations of the reforming reaction, giving the possibility of attaining a high methane conversion at lower temperature with respect to a traditional reactor [16]. The membrane reactor can become a novel reforming technology [17] and promises economic small-scale hydrogen production combined with inexpensive CO_2 capture because of the high concentration and pressure of the exiting gas stream [18, 19]. This could avoid a dedicated hydrogen infrastructure, facilitate CO_2 capture at small scale and thus contribute to a more rapid cutting of greenhouse gas emissions.

With regard to the reforming step, there are several ways to convert the fuel:

- steam reforming (SR)
- partial oxidation (POX)
- autothermal reforming (ATR).

SR provides the highest concentration of hydrogen and gives a high fuel processing conversion efficiency. POX is a very fast process, characterized by a small reactor size with a rapid response to the load changes. Non-catalytic POX operates at temperatures of approximately 1400 °C, but on adding a catalyst [catalytic partial oxidation (CPOX)] this temperature is lowered to about 700 °C. The combination of SR with CPOX is termed ATR.

SR is the established process for converting natural gas and other hydrocarbons into syngas. The SR process has been practiced since 1930. The first plant using light alkanes as feed began operation in 1930 at Standard Oil in the USA and 6 years later at ICI at Billingham, UK [19, 20].

The SR process involves two reactions, the conversion of hydrocarbon with steam to form hydrogen and carbon oxides [reaction (9.1)] and the WGS reaction for the conversion of carbon monoxide into carbon dioxide [reaction (9.2)]:

$$CH_4 + H_2O \rightleftarrows CO + 3H_2 \qquad \Delta H^\circ_{298K} = +206 \text{ kJ mol}^{-1} \qquad (9.1)$$

$$CO + H_2O \rightleftarrows CO_2 + H_2 \qquad \Delta H^0_{298K} = -41 \text{ kJ mol}^{-1} \qquad (9.2)$$

The higher hydrocarbons in natural gas also react with steam in a similar way according to the reaction

$$C_nH_m + nH_2O \rightleftarrows nCO + (n + m/2)H_2 \quad (9.3)$$

Natural gas is reacted with steam on an Ni-based catalyst in a primary reformer to produce syngas at a residence time of several seconds, with an H_2:CO ratio of 3 according to reaction (9.1). Reformed gas is obtained at about 930 °C and pressures of 15–30 bar. The CH_4 conversion is typically 90–92% and the composition of the primary reformer outlet stream approaches that predicted by thermodynamic equilibrium for a CH_4:H_2O = 1:3 feed. A secondary autothermal reformer is placed just at the exit of the primary reformer in which the unconverted CH_4 is reacted with O_2 at the top of a refractory lined tube. The mixture is then equilibrated on an Ni catalyst located below the oxidation zone [21]. The main limit of the SR reaction is thermodynamics, which determines very high conversions only at temperatures above 900 °C. The catalyst activity is important but not decisive, with the heat transfer coefficient of the internal tube wall being the rate-limiting parameter [19, 20].

The SR reactors are well suited for long periods of steady-state operation. The high endothermicity of the primary SR reaction results in reactor performances limited by heat transfer more than by reaction kinetics. Consequently, the reactors are designed to promote heat exchange and tend to be large and heavy [22].

The catalyst typically employed in this process is a high Ni content catalyst ($\sim$12–20% Ni as NiO) supported on a refractory material such as α-alumina containing a variety of promoters. Key additives are potassium and/or calcium ions, which mainly serve to suppress excessive carbon deposition on the catalyst [9].

Although the stoichiometry for reaction (9.1) suggests that one only needs 1 mol of water per mole of methane, excess steam must be used to favor the chemical equilibrium and reduce the formation of coke. Steam-to-carbon ratios of 2.5–3 are typical for natural gas feed. Carbon and soot formation in the combustion zone is an undesired reaction which leads to coke deposition on downstream tubes, causing equipment damage, pressure losses and heat transfer problems [21].

Carbon formation can occur by thermal cracking of hydrocarbons according to the reaction

$$C_nH_{2n+2} \rightleftarrows nC + (n+1)H_2 \quad (9.4)$$

Since higher hydrocarbons tend to decompose more easily than methane, therefore the risk of carbon formation is higher with vaporized liquid petroleum fuels than with natural gas.

Other sources of carbon formation are the Boudouard reaction (9.5) and the reverse of steam gasification, reaction (9.6):

$$2CO \rightleftarrows C + CO_2 \quad (9.5)$$

$$CO + H_2 \rightleftarrows C + H_2O \quad (9.6)$$

Higher hydrocarbons are much more reactive than methane and, over the usual SR nickel catalysts, they readily lead to the generation of coke, which can rapidly deactivate the catalyst. The problem of carbon formation may be solved effectively by the insertion of a low-temperature, fixed-bed adiabatic prereformer prior to the primary steam reformer [23]. In the adiabatic prereformer, the higher hydrocarbons are completely converted into C_1 fragments (CH_4, CO and CO_2). This is similar to the process taking place in conventional reforming, but the relatively low temperatures (350–550 °C) eliminate the potential for carbon formation. Moreover, the prereformer allows for higher inlet temperatures in the primary reformer, thereby reducing its size [24, 25].

As SR is a very intensive energy process, extensive work has been done over the years in order to develop more attractive alternative options.

As the name implies, *partial oxidation* is the partial or incomplete combustion of a fuel according to the reaction

$$CH_4 + \frac{1}{2}O_2 \rightarrow CO + 2H_2 \tag{9.7}$$

A substoichiometric amount of air or oxygen is used. The reaction can be carried out in the presence or the absence of a catalyst. In the non-catalytic process, a mixture of oxygen and natural gas is preheated, mixed and ignited in a burner; the reactor temperature must be high enough to reach complete CH_4 conversion (typically 1200–1500 °C). Combustion products such as CO_2 and H_2O are also formed to a certain extent.

The advantage over catalytic processes is that fuel components such as sulfur compounds do not need to be removed. High-temperature POX can also handle much heavier petroleum fractions than catalytic processes and is therefore attractive for processing diesel, logistic fuels and residual fractions cuts. Some carbon is formed by the thermal cracking of methane and has to be removed by washing. Texaco and Shell have commercialized this conversion process. The use of a catalyst can substantially reduce the operating temperature, allowing the use of less expensive materials for the reactor such as steel. Generally, Ni and noble metals such as Rh, Pt and Pd are used in the POX reaction [26]. It should be noted that the POX reaction produces less hydrogen per mole of methane than the SR reaction. This means that POX (either non-catalytic or catalyzed) is usually less efficient than SR for fuel cell applications.

ATR is a stand-alone process which combines POX and SR in a single reactor. The ATR process was first developed in the late 1950s by Topsøe, mainly for industrial synthesis gas production in ammonia and methanol plants [27].

The hydrocarbon feedstock is reacted with a mixture of oxygen or air and steam in a sub-stoichiometric flame. In the fixed catalyst bed the synthesis gas is further equilibrated. The composition of the product gas will be determined by the thermodynamic equilibrium at the exit pressure and temperature, which is determined through the adiabatic heat balance based on the composition and flows of the feed, steam and oxygen added to the reactor. The synthesis gas produced is completely soot-free [28].

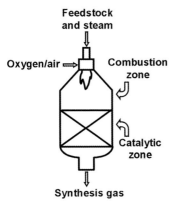

Figure 9.2 Schematic representation of an ATR reactor.

A schematic representation of an ATR reactor is shown in Figure 9.2.

The autothermal reformer consists of a thermal and a catalytic zone. The feed is injected into a burner and mixed intensively with steam and a substoichiometric amount of oxygen or air. In the combustion or thermal zone part of the feed reacts essentially according to the reaction

$$CH_4 + \frac{3}{2}O_2 \rightarrow CO + 2H_2O \qquad \Delta H^{\circ}_{298K} = -519 \text{ kJ mol}^{-1} \qquad (9.8)$$

By proper adjustment of the oxygen-to-carbon and steam-to-carbon ratios, the partial combustion in the thermal zone [reaction (9.8)] supplies the heat for the subsequent endothermic SR reaction (9.1) [24]. The CO shift reaction (9.2) also takes place in the catalytic zone.

Unlike the methane steam reformer, the autothermal reformer requires no external heat source and no indirect heat exchangers. This makes autothermal reformers simpler and more compact than steam reformers, resulting in lower capital cost. In an autothermal reformer, the heat generated by the POX reaction is fully utilized to drive the SR reaction. Thus, autothermal reformers typically offer higher system efficiency than POX systems, where excess heat is not easily recovered.

The ATR is carried out in the presence of a catalyst, which controls the reaction pathways and thereby determines the relative extents of the oxidation and SR reactions. The SR reaction absorbs part of the heat generated by the oxidation reaction, limiting the maximum temperature in the reactor. The net result is a slightly exothermic process. However, in order to achieve the desired conversion and product selectivity, an appropriate catalyst is essential. The lower temperature provides many benefits such as less thermal integration, less fuel consumed during the start-up phase, wider choice of materials, which reduces the manufacturing costs, and reduced reactor size and cost due to a minor need for insulation [22].

For small-scale units providing hydrogen for fuel cells, the choice of the optimal technology may be dictated by parameters such as simplicity and fast transients response, in particular for automotive applications. For these reasons, even if the SR process may be the most energy efficient in industrial processing, CPOX or ATR using air may be the preferred choice for small-scale or distributed hydrogen production [29].

9.2
Autothermal Reforming: from Chemistry to Engineering

The industrial processes for hydrogen production are well established [9, 30, 31], but may not be appropriate for small-scale stationary applications such as residential fuel cells or for unattended operation such as on-site hydrogen generation. These new applications allow for new process designs based on catalyst and engineering improvements [10, 32]. Hence a study of the ATR process has to involve both chemical and engineering aspects.

9.2.1
The Catalyst

Catalyst formulations for ATR fuel processors mainly depend on the fuel and the operating temperature. ATR catalysts are required to be active simultaneously for hydrocarbon oxidation and SR reactions, be robust at high temperatures for extended periods and be resistant to sulfur poisoning and carbon deposition, especially in the catalytic zone that runs oxygen limited [33]. Moreover, they must be resistant to intermittent operation and cycles, especially in start-up and shut-down steps.

As ATR is a combination of POX and SR reactions, the active species to be employed for the preparation of a good ATR catalyst are the same as those for these two processes, especially Ni, Pt, Pd, Rh, Ru and Ir.

Ni-based catalysts have been widely employed in reforming reactions and they are used extensively in industry because they offer appreciable catalyst activity, good stability and low cost. The commercial catalysts contain NiO supported on ceramic materials such as α-Al_2O_3, MgO and $MgAl_2O_4$ spinel. The average content of NiO is around 15 wt% [34]. The magnesium aluminate-supported catalyst has a higher melting point and in general a higher thermal strength and stability than the alumina-based catalyst [19].

Small amounts of other compounds can be added to Ni-based catalysts to improve the functional characteristics of the final catalyst. Typically, they are: calcium aluminate to enhance the mechanical resistance of the catalyst pellets, potassium oxide to improve the resistance to coke formation and silica to form a stable silicate with potassium oxide [34]. Promotion with rare earth oxides such as La_2O_3 also results in enhanced resistance to coking.

A series of nickel/(rare earth phosphate) catalysts were investigated by Nagaoka et al. [35] for methane ATR. Among them, Ni/(Gd, Ce or Er phosphate) showed good activity, maintaining a stable CH_4 conversion during time on-stream tests and exhibiting an extremely high resistance against coking. Therefore, they could be considered as novel candidate catalytic systems for ATR applications.

In order to improve the resistance of Ni/Al_2O_3-based catalysts to sintering and coke formation, some workers have proposed the use of cerium compounds [36]. Ceria, a stable fluorite-type oxide, has been studied for various reactions due to its redox properties [37]. Zhu and Flytzani-Stephanopoulos [38] studied Ni/ceria catalysts for the POX of methane, finding that the presence of ceria, coupled with a high nickel dispersion, allows more stability and resistance to coke deposition. The synergistic effect of the highly dispersed nickel/ceria system is attributed to the facile transfer of oxygen from ceria to the nickel interface with oxidation of any carbon species produced from methane dissociation on nickel. The effect of adding zirconia to Ni/ceria-based catalysts in reforming and POX reactions has also been widely investigated [39–44]. Monoclinic zirconia has high thermal stability at high temperature in addition to a considerable surface area. Furthermore, because it has both basic and weakly acidic sites, zirconia could be resistant to coke formation. The remarkable catalytic performance of Ni/CeO_2-ZrO_2 is attributed to the combination of several effects, namely the high oxygen storage capacity of ceria in ceria–zirconia solid solutions [45–48], the strong interaction between Ni and ceria–zirconia and the fairly high capability for H_2 uptake.

Dong and co-workers studied the effect of adding a transition metal (Cu, Co and Fe) to $Ni/Ce_{0.2}Zr_{0.1}Al_{0.65}O_\delta$ catalyst on the activity in the ATR of methane [49, 50]. It was observed that the presence of Cu and Co allows a significant improvement of the catalytic activity at lower temperature, whereas the addition of Fe leads to a remarkable decrease in CH_4 conversion. Results of catalyst characterization showed that the improvement in the catalytic activity when Cu is added as a promoter is due to the improvement of NiO dispersion, with inhibition of $NiAl_2O_4$ formation.

The noble metals seem to be more active than Ni for the ATR reaction, as proposed by Ashcroft and co-workers [51, 52], but they are more expensive [34].

Ayabe et al. [53] studied the catalytic ATR of methane and propane over Al_2O_3-supported metal catalysts. In the case of methane they found the following scale of activity: Rh > Pd > Ni > Pt > Co, and limited carbon deposition, whereas in the case of propane a large amount of carbon deposition was observed even under steam-rich conditions.

Souza and Schmal [54] reported the activity of a 1 wt% $Pt/ZrO_2-Al_2O_3$ catalyst for the ATR of methane. Higher activity and stability were found in time on-stream testing at 800 °C compared with Pt/Al_2O_3 and Pt/ZrO_2 catalysts. The higher stability was related to a higher coking resistance due to $Pt-Zr^{n+}$ interactions at the metal–support interface.

Villegas et al. [55] studied the ATR of isooctane over a 2 wt% $Pt/Ce_{0.67}Zr_{0.33}O_2$ catalyst. The catalyst performed well at high space velocities, showing a stable activity during time on-stream testing under the experimental conditions suggested as optimum by thermodynamic evaluation. The catalyst did not show any apparent poisoning by carbon deposition, suggesting that the ceria–zirconia support would play an active role in the gasification of the carbonaceous species originating from isooctane decomposition.

A 1% Pt/CeO$_2$ catalyst was investigated for the ATR of synthetic diesel by Cheekatamarla and Lane [56] with specific attention to the effect of sulfur poisoning on the catalytic activity. The catalyst exhibited good stability in spite of its tendency for poisoning by sulfur-containing fuels. A further study by the same authors [57] elucidated the overall deactivation mechanism for this catalyst. Temperature programmed reduction (TPR) experiments suggested that sulfur poisoning prevents ceria from transferring oxygen to the metal, and temperature programmed desorption (TPD) and X-ray photoelectron spectroscopy (XPS) analysis revealed the formation of chemisorbed sulfur entities.

A similar Pt/CeO$_2$ catalyst, studied by Recupero *et al.* [58] in the ATR of propane, was found to give high H$_2$ and CO selectivity values in the temperature range 600–700 °C. The catalyst did not show any carbon deposition after time on-stream testing carried out with sequential start-up and shut-down cycles.

Alternative catalyst formulations for methane ATR based on bimetallic catalysts have been studied, aiming at increasing the activity of nickel catalysts by the addition of low contents of noble metals.

Nickel–platinum bimetallic catalysts showed higher activity during ATR than nickel and platinum catalysts blended in the same bed. It was hypothesized that nickel catalyzes SR, whereas platinum catalyzes POX and, when they are added to the same support, the heat transfer between the two sites is enhanced [59, 60]. Advanced explanations were reported by Dias and Assaf [60] in a study on ATR of methane catalyzed by Ni/γ-Al$_2$O$_3$ with the addition of small amounts of Pd, Pt or Ir. An increase in methane conversion was observed, ascribed to the increase in exposed Ni surface area favored by the noble metal under the reaction conditions.

In addition to the activity, other important requirements for the catalyst are the capability to start the reaction rapidly without the necessity for previous reduction with hydrogen and to perform effectively with intermittent operation; these are essential properties for the catalyst in reformers, especially for portable and small-scale stationary fuel cell applications. In this respect, Dias and Assaf [61] focused on the potential of Pd, Pt and Ir to promote fast and intermittent ignition of methane ATR in Ni/γ-Al$_2$O$_3$. They concluded that the three metals are very good promoters of the reduction of the nickel catalyst with methane, but the lower cost of palladium makes this metal more suitable than Pt and Ir for small fuel cells.

Cheekatamarla and Lane [62, 63] studied the effect of the presence of Ni or Pd in addition to Pt in the formulation of catalysts for the ATR of synthetic diesel. For both metals, a promotional effect with respect to catalytic activity and sulfur poisoning resistance was found when either alumina or ceria was used as the support. Surface analysis of these formulations suggests that the enhanced stability is due to strong metal–metal and metal–support interactions in the catalyst.

In recent years, research on catalysts for the ATR of hydrocarbons has paid considerable attention to perovskite systems of general formula ABO$_3$. In the perovskite structure, both A and B ions can be partially substituted, leading to a wide variety of mixed oxides, characterized by structural and electronic defects. The oxidation activity of perovskites has been ascribed to ionic conductivity, oxygen mobility within the lattice [64], reducibility and oxygen sorption properties [65, 66].

High activity of perovskite-type metal oxides with B sites partially exchanged with ruthenium was reported by Liu and Krumpelt [67] for hydrogen production by ATR of diesel fuel. The catalysts exhibited excellent reforming efficiency and CO_x selectivity in addition to sulfur tolerance during an aging test with diesel surrogate fuel. The ruthenium was found to be highly dispersed in the perovskite lattice.

A series of cerium- and nickel-substituted $LaFeO_3$ perovskites were investigated by Erri et al. [68] in the ATR of JP-8 fuel surrogate. The catalytic activity tests carried out at high space velocity showed that partial substitution with cerium leads to a slight decrease in activity but improves hydrogen selectivity for the nickel-rich catalysts. Moreover, Ce doping had a beneficial effect on coking inhibition. However, a phase segregation was detected after reaction.

The optimization of the catalyst formulation is relevant not only to the active species but also to the structure of the support. Indeed, structured catalysts in the form of monolith or foam offer great advantages over pellet catalysts, the most important one being the low pressure drop associated with the high flow rates that are common in environmental applications.

Since structured catalysts can operate at significantly higher space velocities than pellet bed reactors, the consequent reduction in reactor size saves on both cost and weight. A secondary benefit of having a smaller reactor to be heated is that a rapid thermal response to transient behavior can be achieved. Whether the application is a load-following stationary fuel cell or an on-site hydrogen generator stepping up from standby mode to full operation, the reactors need to be able to respond quickly to changes in temperatures and flow rates [32].

Due to the severe operating conditions (high temperature and high flow rates), typical of short contact time reactions, heat and mass transfer properties are expected to play a decisive role in the behavior of the reactor. Thus, in principle, the choice of catalyst support can greatly affect the reactor performance [69].

A structured ruthenium catalyst (metal monolith supported) was investigated by Rabe et al. [70] in the ATR of methane using pure oxygen as oxidant. The catalytic activity tests were carried out at low temperature (<800 °C) and high steam-to-carbon ratios (between 1.3 and 4). It was found that the lower operating temperature reduced the overall methane conversion and thus the reforming efficiency. However, the catalyst was stable during time on-stream tests without apparent carbon formation.

Structured catalysts were also employed at Argonne National Laboratory in the ATR of methane and heavy hydrocarbons [71, 72].

Qi et al. studied perovskite systems obtained by partially replacing La with Ce at the A-site of $LaNiO_3$ for ATR of gasoline and its subrogate n-octane with or without thiophene additive. They found similar activities and mechanical and thermal stabilities with both pellet and monolith $La_{0.8}Ce_{0.2}NiO_3$ [73].

Generally, the preparation of washcoated structured catalysts is governed by several parameters, such as the nature and particle size of the precursor powder, loading of powder, nature and concentration of dispersants, temperature of the slurry, use of binders in the slurry and deposition of a primer layer on the monolith.

An extensive review by Avila *et al.* takes into account the influence of these variables [74].

With regard to the ATR of isooctane, Villegas *et al.* reported a study dealing with the influence of the drying procedure on the Ni distribution on alumina-washcoated cordierite monoliths [75]. The wet impregnated monoliths were dried by three different methods: in a ventilated oven, at room temperature and in a microwave oven. The microwave drying allows a more homogeneous deposition of Ni, although irrespective of the drying method, at the microscopic scale, a surface enrichment of Ni in the alumina top layer was evidenced.

9.2.2
Kilowatt-scale ATR Fuel Processors

Experimental results concerning the development of a small-scale 1 kW autothermal reformer of propane were reported by Rampe *et al.* [76]. In the proposed reactor, two reactions occur on a metal honeycomb structure coated with platinum. Air and water are mixed before they are fed to the reactor in counterflow to the product gas outside the reactor wall, where the water is vaporized and the steam and air are heated up. Then, they are mixed with propane at the bottom of the reactor. It was verified that the preheating operation mode led to about a 4% higher efficiency, since the higher inlet air temperature causes a higher temperature level in the reaction zone, resulting in improved kinetics of the reforming reaction.

Heinzel *et al.* [77] compared the performance of a natural gas autothermal reformer with that of a steam reformer. The ATR reactor was loaded with a Pt catalyst on a metallic substrate followed by a fixed bed of Pt catalyst. In the start-up phase, the metallic substrate was electrically heated until the catalytic combustion of a stoichiometric methane–air mixture occurred. The reactor temperature was increased by the heat of the combustion reaction and later water was added to limit the temperature rise in the catalyst, while the air flow was reduced to sub-stoichiometric settings. With respect to the steam reformer, the behavior of the ATR reactor was more flexible regarding the start-up time and the load change, thus being more suitable for small-scale stationary applications.

Liu *et al.* [71] reported the experimental results obtained with a kilowatt-scale adiabatic reformer fed with diesel surrogates, developed at the Argonne National Laboratory. An electric heater coil was used as igniter, and the reactor catalytic bed was constituted of four monolith catalysts packed in series at the core of the reactor. The ATR reactor achieved stable performance with a smooth radial temperature distribution. The reforming efficiency, evaluated by changing the gas hourly space velocity (GHSV) up to $100\,000\,h^{-1}$, was influenced to a major extent by the oxygen-to-carbon ratio and to a minor extent by the steam-to-carbon ratio.

Lenz and Aicher reported the experimental results obtained with an autothermal reformer fed with desulfurized kerosene employing a metallic monolith coated with alumina washcoat supporting precious metal catalysts (Pt and Rh) [78]. The experiments were performed at steam-to-carbon ratios $S/C = 1.5$–2.5 and air-to-fuel ratios $\lambda = 0.24$–0.32. The GHSV was varied between 50 000 and 300 000 h^{-1}. The best efficiency was achieved at S/C = 1.5 and $\lambda = 0.28$ at a GHSV of 50 000 h^{-1}.

We reported on the design, development and optimization of a compact self-sustained ATR reactor for the production of about 5 m^3 (STP) h^{-1} of hydrogen [79, 80]. The reactor was thermally integrated by two heat exchangers for the preheating of the air and liquid water fed to the reactor at the expense of the hot reactor outlet stream. The reactor operated in a stable manner over a wide range of operating conditions with both pellets and structured catalysts.

A conceptual design and selection of an ATR biodiesel processor for a vehicle fuel cell auxiliary power unit were reported by Specchia *et al.* [81]. Three processor options were compared for H_2 production with respect to efficiency, complexity, compactness, safety, controllability and emissions. The ATR with both high-temperature shift (HTS) and low-temperature shift (LTS) reactors showed the most promising results.

Other simulation studies reported on the differences between ATR and SR fuel processors for liquid hydrocarbons [82]. The results showed that a fuel processor based on the SR technology gives a higher power than an ATR-based fuel processor. However, this higher performance is counterbalanced by a much higher plant complexity, resulting in increased cost and an impact on system controllability and start-up time.

An experimental study by Lee *et al.* [72] reported the development and testing of a natural gas fuel processor, which incorporates a catalytic autothermal reformer, a sulfur trap and a WGS reactor. The fuel processor was successfully run over 2300 h of continuous operation. The ATR reactor gave over 40% H_2 (dry basis) in the ATR reformate and 96–99.9% methane conversion over the entire test duration.

A compact design for a gasoline fuel processor for auxiliary power unit (APU) applications, including an autothermal reformer followed by WGS and selective oxidation stages, was reported by Severin *et al.* [83]. The overall fuel processor efficiency was about 77% with a start-up time of 30 min.

A 2 kW hydrogen generation unit, developed and tested by Cipitì *et al.* [84], included an ATR unit followed by intermediate WGS and preferential oxidation stages. Preliminary experimental results showed that propane was completely converted and that the CO content in the outlet stream was 0.2% (dry basis). However, a start-up time of about 50 min was observed.

A stand-alone 1 kW integrated fuel processor for gasoline, incorporating an autothermal reformer followed by high- and low-temperature WGS reactors, was reported by Qi *et al.* [85]. The start-up of the ATR reformer lasted less than 5 min and stabilized in around 50 min for the whole system.

9.3
Thermodynamic Analysis

Thermodynamic analysis allows one to predict the influence of the operating parameters (preheating of reactants and O_2:CH_4 and H_2O:CH_4 molar feed ratios) on the equilibrium chemical composition. The aim of this section is to identify the most favorable thermodynamic operating conditions at which methane is converted to hydrogen in the ATR process. Moreover, the results also provide a valuable indication for programming the catalytic experiments.

The chemical equilibrium can be determined by two approaches, one based on the equilibrium constants and the other on the minimization of the free energy of reaction. One of the disadvantages of using the former approach is that it is difficult to take into account the solid carbon which can be generated during the reforming process. Therefore, the method of minimizing the Gibbs free energy is normally preferred in fuel reforming analysis [86, 87]. We chose this method and carried out the thermodynamic calculations by assuming the gas mixture to be a perfect gas.

The general reaction of methane conversion for the thermodynamic analysis of the ATR reactor can be written as follows:

$$CH_4 + xO_2 + yH_2O + (3.76x)N_2 \rightarrow products \quad (9.9)$$

where the main products are CH_4, CO, CO_2, $C(s)$, O_2, N_2, H_2 and H_2O. The $O_2:CH_4$ molar ratio was varied from 0.1 to 0.9. For each value of this ratio, the $H_2O:CH_4$ molar ratio was varied from 0 to 1.2. The evaluations were carried out by assuming a constant reactor pressure (1 atm) and adiabatic conditions for the ATR reactor, that is, no heat transfer to or from the reactor.

9.3.1
Effect of Preheating the Reactants

In Figure 9.3, the effect of preheating the reactants on (a) the H_2 yield (calculated as moles of H_2 formed/moles of CH_4 fed) and (b) the CH_4 conversion is reported as a function of the $O_2:CH_4$ and $H_2O:CH_4$ molar feed ratios. The preheating temperature was varied from 25 to 400 °C.

It was observed that at any $H_2O:CH_4$ molar ratio investigated, the increase in the preheating temperature of reactants leads to an increase in H_2 yield. In particular, this effect is more pronounced at higher $H_2O:CH_4$ values and at $O_2:CH_4$ molar ratios ranging from 0 to 0.7. Similar results were reported for CH_4 conversion, for which the preheating effect, resulting in increased conversion, is more pronounced at higher $H_2O:CH_4$ values and at $O_2:CH_4$ molar ratios ranging from 0 to 0.8.

9.3.2
Effect of $O_2:CH_4$ and $H_2O:CH_4$ Molar Feed Ratios

The results of thermodynamic analysis are reported in Figure 9.4 in terms of adiabatic temperature, CH_4 conversion (a) and $C(s)$ moles produced (b) as function of $O_2:CH_4$ and $H_2O:CH_4$ molar ratios. The calculations were carried out for a preheating temperature of reactants of 400 °C. The same results are also reported by Seo et al. [86].

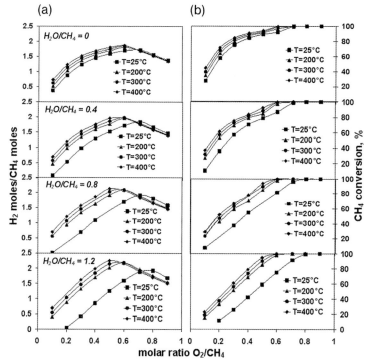

Figure 9.3 Effect of the reactant preheating temperature on H_2 yield (a) and CH_4 conversion (b).

Both CH_4 conversion and adiabatic temperature are significantly affected by the $O_2:CH_4$ and $H_2O:CH_4$ feed ratios. The CH_4 conversion increases with the $O_2:CH_4$ ratio up to the value 0.6, at which total conversion is reached. For values greater than 0.6, the adiabatic temperature continues to increase although the CH_4 conversion remains at 100%. This is due to the oxidation reaction of H_2 and CO to H_2O and CO_2 favored by the excess O_2 supply. On the other hand, on increasing the $H_2O:CH_4$ ratio at a fixed $O_2:CH_4$ value, the adiabatic temperature decreases,

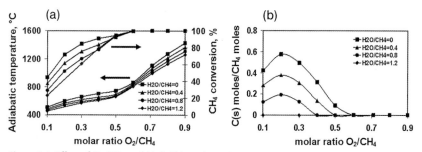

Figure 9.4 Effect of $O_2:CH_4$ and $H_2O:CH_4$ molar ratios on adiabatic temperature and CH_4 conversion (a) and on C(s) formation (b).

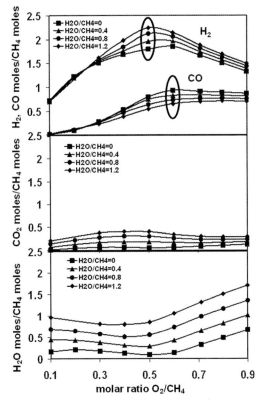

Figure 9.5 Effect of O_2:CH_4 and H_2O:CH_4 molar ratios on H_2, CO, CO_2, H_2O yields.

due to the strongly endothermic SR reaction. As a result of the lower reactor temperature, the CH_4 conversion is also reduced. By increasing the H_2O:CH_4 ratio, coke formation is predicted at lower O_2:CH_4 ratios. Moreover, fixing the O_2:CH_4 feed ratio, higher H_2O:CH_4 ratios allow the reduction of coke formation. For an H_2O:CH_4 ratio of 1.2, no coke is generated at any value of the O_2:CH_4 ratio. Moreover, at fixed H_2O:CH_4 values, coke formation is depressed by increasing the O_2:CH_4 ratio.

The effect of O_2:CH_4 and H_2O:CH_4 ratios on the equilibrium composition has also been evaluated (Figure 9.5). The H_2 and CO yields reach a maximum at an O_2:CH_4 molar ratio of 0.5 and 0.6, respectively. As the H_2O:CH_4 ratio increases, the H_2 yield also increases, but conversely, the CO yield decreases, leading to increase in the H_2:CO ratio. On the other hand, if the O_2:CH_4 ratio is increased above 0.5, the H_2 yield drops more steeply compared with the decrease in the CO yield, due to the more favored oxidation of H_2 than CO at high O_2:CH_4 ratios.

In conclusion, thermodynamic analysis can provide the optimal operating conditions to achieve the best ATR performance. It has been found that preheating the reactants promotes H_2 production. This effect is more pronounced at higher H_2O:CH_4 ratios and at O_2:CH_4 molar ratios ranging from 0.1 to 0.7. Employing a preheating temperature of 400 °C, thermodynamic evaluation showed that an $O_2:CH_4$ molar ratio of 0.6 maximizes the CH_4 conversion, limits the effect of $H_2O:CH_4$ molar ratio on the adiabatic temperature and avoids coke formation. Moreover, with an $O_2:H_4$ feed ratio of 0.6, the $H_2O:CH_4$ molar ratio which allows the maximum H_2 production is 1.2.

9.4
A Case Study

This section focuses on a case study analyzing all the operations to be carried out during the setting-up of a kilowatt-scale ATR reactor, such as the optimization of the start-up phase and of thermal integration. The ATR reactor described here was developed at the University of Salerno [79, 80] in the frame of the Italian FISR Project 'Idrogeno puro da gas naturale mediante reforming a conversione totale ottenuta integrando reazione chimica e separazione a membrana'.

The aim of this section is to demonstrate that the feasibility of a distributed hydrogen system based on methane ATR is the result of a combination between optimized engineering (design and operation) and catalyst (formulation and support structures and geometry).

A fast-start capability is a key requirement for a compact fuel processor, especially crucial for specific cases such as on-board automotive application. A possible means for the fast start-up of an ATR reactor is starting by feeding to the reactor a mixture with an $O_2:C$ molar ratio typical of a rich combustion. The goal of this mode is to raise the reactor temperature quickly and simultaneously avoid catalyst oxidation. Then the reactor will move to the ATR mode through the addition of steam and by decreasing the $O_2:C$ feed ratio to the desired value.

9.4.1
Laboratory Apparatus and ATR Reactor

The main features of the experimental apparatus are reported below.

Mass flow controllers (Brooks) are used for each cylinder gas (99.999% purity, SOL SpA).

A pressurized tank (10 bar) is used for the storage of distilled water fed to the reactor. To feed liquid water, a specific mass flow controller (Quantim, Brooks) is used.

A stainless-steel reactor (36 mm i.d.), specifically designed and realized, is employed to test the ATR reaction. At the reactor outlet, a fixed fraction of the exhaust stream is sent to the analysis section after removing water in a cold trap.

In the analysis section, CH_4, CO and CO_2 concentrations are monitored with an online NDIR multiple analyzer (Advance Optima, Uras 14). For the analysis of O_2 a continuous paramagnetic analyzer (Advance Optima, Magnos 106) is employed; H_2 is analyzed with a thermal conductivity detector (Advance Optima, Caldos 17). An analog–digital board (National Instruments, AT-M IO 64E) allows

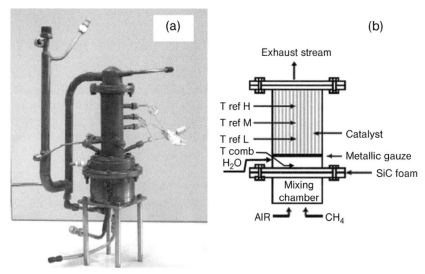

Figure 9.6 ATR reactor as realized (a) and as a schematic drawing (b).

for the computer acquisition of concentration data from the analyzers. Software for data acquisition and products composition was specifically developed in the LabView environment.

The ATR reactor is shown in Figure 9.6, (a) as realized and (b) as a schematic drawing.

It consists of two main parts: (i) a lower section where only during the start-up phase does methane react with air at a fixed $O_2:CH_4$ ratio and (ii) an upper section where the reforming reactions occur in the presence of the catalyst. During the reactor start-up, due to the high $O_2:CH_4$ ratio and the related exothermicity of the hydrocarbon oxidation, heat is generated in the lower section and transferred to the reforming section to heat the catalytic bed until the catalyst threshold temperature is reached. Methane and air are fed at the bottom of the reactor and premixed in a mixing chamber; a commercially available SiC foam (Figure 9.7a) is placed at the exit of the mixing chamber in order to obtain a well-distributed and homogeneous flame during the start-up phase and a very well-mixed gas flow in the stationary conditions.

In the start-up phase, the ignition of a methane–air mixture is induced by a voltaic arc between two spark plugs placed on the surface of the SiC foam (Figure 9.7b).

The catalytic bed (70 cm^3), supported by a metallic gauze, is located in the reforming section. Water is fed to the reactor at the bottom of the metallic gauze. The temperature inside the reactor is monitored by four thermocouples: one (Tcomb) is located on the SiC foam and the other three (T ref L, T ref M, T ref H) are located at 25, 50 and 75%, respectively, of the catalytic bed height to provide the reactor temperature axial profile. Moreover, additional thermocouples monitor

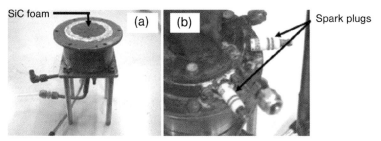

Figure 9.7 SiC foam (a) and spark plugs (b).

the temperature of (i) the preheated water and air, (ii) the exhaust stream before and after thermal exchange and (iii) the stream in the premixing chamber.

A differential pressure sensor monitors the pressure drop across the reactor, giving also an indication of the coke formation. The outside shell of the reactor is thermally insulated to limit heat loss.

The ATR reactor is integrated with two heat exchangers to preheat the air (Figure 9.8a) and the water (Figure 9.8b) fed to the reactor by using the sensible heat of the reactor outlet stream.

As shown in Figure 9.8a, cold air inflowing at the top of the reactor is heated by the hot reaction products exiting the reactor and then enters at the bottom of the reactor.

After preheating the air, the reactor outlet stream passes through the second heat exchanger, where cold liquid water is heated and vaporized before being fed to the reactor (Figure 9.8b).

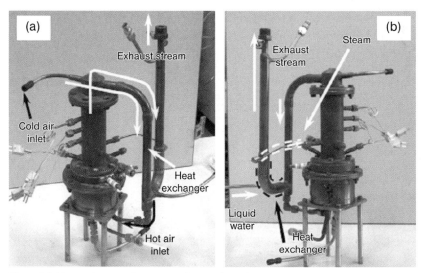

Figure 9.8 Heat exchangers for air (a) and water (b) preheating.

9.4.2
ATR Reactor Setup: Operating Conditions

In order to set up the ATR reactor, a commercial pelletized catalyst, 0.3%Pt/Al$_2$O$_3$ (Engelhard), was used.

The catalytic activity tests were carried out under different operating conditions, by changing the GHSV and the feed ratios O$_2$:CH$_4$ and H$_2$O:CH$_4$ over the ranges of values $0.59 \leq O_2{:}CH_4 \leq 1.36$, $0 \leq H_2O{:}CH_4 \leq 1.33$ and $10\,000\,h^{-1} \leq GHSV \leq 24\,400\,h^{-1}$.

In the following, the O$_2$:CH$_4$ and H$_2$O:CH$_4$ feed ratios will be denoted x and y, respectively. The value of GHSV is calculated as the ratio between the total gas volume flow rate (STP) and the catalyst bed volume, and the percentage conversion of methane is calculated as

$$\text{methane conversion (\%)} = (C_{CO} + C_{CO_2})/(C_{CH_4} + C_{CO} + C_{CO_2}) \times 100$$

where C is the dry basis concentration of each gas in the exhaust stream [88]. The amount of coke deposited during the reaction was so small that it could be neglected in all the results presented below.

9.4.3
ATR Reactor Setup: Start-up Phase

The results relevant to the optimized conditions for the start-up phase, as defined for specific experiments, are reported in Figure 9.9 in terms of H$_2$, CO and CO$_2$

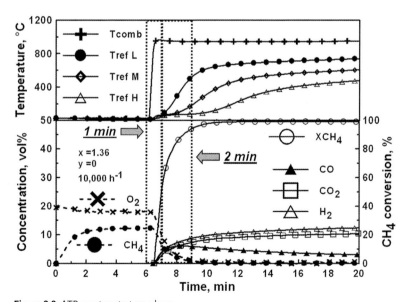

Figure 9.9 ATR reactor start-up phase.

concentrations (vol.%, dry basis) and of CH_4 conversion (%) as a function of time (lower part of the plot). Moreover, temperature values measured inside the reactor are reported against the reaction time (upper part of the plot). O_2:CH_4 (x) and H_2O:CH_4 (y) ratios and GHSV values are also reported.

The reactor start-up was performed by feeding a water-free mixture of methane and air with an O_2/CH_4 molar ratio of 1.36 and by inducing for few seconds the voltaic arc between the spark plugs. When the mixture is ignited, the temperature on the SiC foam suddenly (1 min) reaches around 1000 °C. Furthermore, due to the heat transfer, the temperature in the catalytic zone reaches in about 2 min the light-off value with full reactants conversion. The whole start-up phase is no longer than 3 min.

After the ignition phase, water is fed to the reactor and the O_2:CH_4 and H_2O:CH_4 ratios are varied until the desired operating conditions are reached. It must be noted that after the ignition phase, due to the lower values of the O_2:CH_4 ratio, the homogeneous combustion of methane is inhibited and consequently POX, SR and WGS reactions occur simultaneously in the catalytic bed, while the temperature on the SiC foam decreases to values lower than 400 °C.

9.4.4
ATR Reactor Setup: Influence of Preheating the Reactants

In order to establish the influence of preheating of the reactants on the ATR reactor performance, catalytic activity tests were carried out with and without preheating the reactants. For the tests performed with air and water preheating, two heat exchangers were integrated in the previous version of ATR reactor.

The results of catalytic activity tests carried out without preheating are shown in Figure 9.10.

The first step is relevant to the start-up phase, which in this particular case we chose to extend for up to 1 h in order to verify the reactor stability also in these conditions, where water is not present and while there is a higher oxygen concentration in the feed gas with respect to the ATR conditions. By lowering the O_2:CH_4 ratio, the H_2 concentration at the reactor outlet increases, approaching the value expected by thermodynamic evaluation and CH_4 conversion is still complete. A further decrease in the O_2:CH_4 feed ratio to values lower than 1.16 corresponds to an abrupt decrease in temperature in the lower section and a simultaneous temperature increase in the catalytic reforming section.

These results can be explained by considering that at this O_2:CH_4 ratio the homogeneous combustion reaction is not favored and consequently the feed mixture reacts in the catalytic section where the heat developed by the exothermic reactions is responsible for the remarkable temperature increase.

By lowering the x value at 0.84, the average temperature in the catalytic section is lowered by about 100 °C whereas the H_2 concentration increases up to about 30 vol.%. In the next step, by lowering the x value to 0.71, a dramatic effect on CH_4 conversion is observed. In fact, the new value is lower than 90%. The observed temperature profile along the catalytic bed (a higher value at the catalyst inlet and a decrease along the

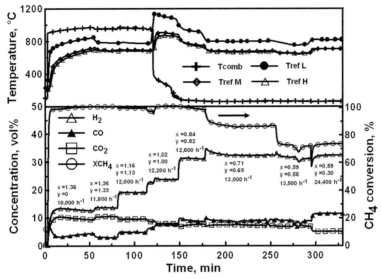

Figure 9.10 Results of CH_4 ATR catalytic activity test on a Pt/Al_2O_3 catalyst without preheating of reactants.

catalytic bed) is the typical profile of an ATR reactor, in agreement with both experimental [71, 72, 78] and modeling studies [89]. This temperature profile is likely a result of a reaction sequence involving the initial exothermic oxidation of methane followed by the strongly endothermic SR reactions and the mildly exothermic WGS reaction [90, 91].

It must be evidenced that the reactor performance is stable under all investigated operating conditions.

In order to compare the different preheating operations, and thus evaluating this effect on the reactor performance, temperature values measured in the reactor are reported in Table 9.1 for the following operating conditions: $O_2:CH_4 = 0.71$, $H_2O:CH_4 = 0.69$, GHSV = 13 000 h^{-1}.

It was observed that higher temperatures were reached in the reactor on changing the preheating method, the effect being more pronounced at $O_2:CH_4$ feed ratios lower than 0.7, due to the minor extent of the oxidation reaction. However, the trend of the temperature profile along the reactor was unchanged irrespective of the means of reactant preheating.

Table 9.1 Reactor temperature profiles as a function of preheating method.

Preheating method	T comb (°C)	T ref L (°C)	T ref M (°C)	T ref H (°C)
Without preheating	60	798	678	668
Air preheating	260	915	797	716
Air and water preheating	389	922	814	740

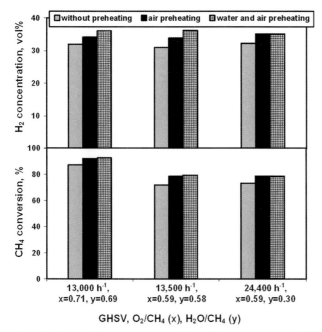

Figure 9.11 H_2 outlet concentration and overall CH_4 conversion in different preheating methods.

The H_2 concentration and CH_4 conversion at the reactor outlet are reported in Figure 9.11 for different preheating operations.

The influence of preheating on H_2 concentration is more pronounced at lower $O_2:CH_4$ ratios. Up to an $O_2:CH_4$ value of 0.84 (not reported), H_2 concentrations were almost the same, and very high CH_4 conversions were reached due to the oxidation reaction, which proceeds to a major extent. When the $O_2:CH_4$ ratio reaches 0.7, preheating of the reactants generally leads to an enhancement of ATR reactor performance with respect to both CH_4 conversion and reactor outlet H_2 concentration.

This result is in good agreement with the thermodynamic analysis previously reported, according to which the influence of preheating of reactants is more pronounced for $O_2:CH_4$ values in the range 0–0.7. Moreover, the preheating of the reactants by heat exchange with the hot outlet products improves the reformer performance, because less fuel needs to be consumed to reach the necessary reactor temperature, resulting in more fuel being left for reforming [72, 92].

9.4.5
Catalytic Activity Test Results

For a sustainable production of H_2, the role of the reforming catalyst is very significant. Even though there are several commercially developed reforming catalysts, their use is still limited to typical operating conditions such as high temperature (above 800 °C for CH_4) low space velocity (GHSV = 3000–8000 h^{-1}) owing to slow kinetics [31] and high steam-to-carbon ratio (>3). However, a distributed hydrogen production system requires improvements in catalyst durability, processor size reduction and inexpensive catalysts. Furthermore, as the system has to be installed in space-restricted areas, it needs to operate with a small quantity of catalyst. For the purpose of wide-range installation of the system, not only reductions in catalyst volume but also cheaper catalysts are necessary. These requirements drive researchers to develop more active and stable innovative catalysts [93]. Moreover, it must be taken into account that a very high steam-to-carbon ratio, although preventing coke formation, may represent a kinetic limitation to the SR reaction and also an additional heat supply for steam generation. Therefore, we decided to investigate the catalytic activity of both commercial and non-commercial catalysts under more severe operating conditions with respect to the traditional conditions employed for this kind of reaction. In particular, the catalytic activity tests described here were carried out at higher space velocities and lower steam- to-carbon ratios.

The catalysts investigated were the following:

- Pt/Al_2O_3 pellets (Engelhard)
- Ni/Al_2O_3 spheres (our laboratory, Figure 9.12a)
- Ni/Al_2O_3–aluminate spheres
- noble metal-based ceramic monolith (ATR7B, Engelhard) (Figure 9.12b).

Nickel-based catalysts supported on Al_2O_3 spheres (12 wt% as NiO) were prepared through wet impregnation of support with aqueous solutions of nickel acetate tetrahydrate, $Ni(CH_3COO)_2 \cdot 4H_2O$ (Carlo Erba). The resulting compound was dried at 120 °C for 12 h and then calcined in air at 950 °C for 1 h.

With regard to the operating conditions, catalytic activity tests were carried out at fixed $O_2:CH_4$ and $H_2O:CH_4$ feed ratios of 0.56 and 0.49, respectively, while the space velocity was varied from GHSV = 45 000–90 000 h^{-1}.

The results of catalytic activity tests are reported in Figure 9.13a in terms of the temperature profile as a function of the catalyst bed length and in Figure 9.13b in

Figure 9.12 Ni/Al_2O_3 (a) and noble metal-based monolith (b) catalysts.

terms of CH_4 conversion, H_2 concentration and H_2 throughput per unit catalyst volume as a function of the space velocity.

The results show the typical ATR reaction temperature profile for all the catalysts at any space velocity investigated. The only exception is represented by the temperature profile obtained with Ni/Al_2O_3 spheres, for which a change occurs at $68\,000\,h^{-1}$. In particular, the temperature profile is characterized by a higher temperature in the medium zone of the catalyst bed. The temperature profile seems to follow the order of activity; however, looking at the catalyst performance in terms of product distribution, it can be observed that the best performance was shown by the ATR7B monolith catalyst, which gives a CH_4 conversion very close to the equilibrium value and higher with respect to the other catalysts. Moreover, its activity is not affected by the increase in space velocity since the CH_4 conversion and H_2 concentration are unchanged [94].

The lower catalytic activity shown by the Pt/Al_2O_3 catalyst is probably due to the greater tendency of this catalyst to promote the hydrocarbon oxidation reaction [59, 95].

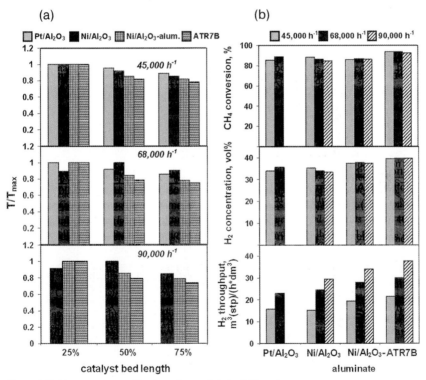

Figure 9.13 Temperature profile along the catalytic bed (a) and CH_4 conversion, H_2 concentration and H_2 throughput per unit catalyst volume (b) in CH_4 ATR catalytic activity tests carried out at high space velocity.

The enhanced activity of the Ni/Al$_2$O$_3$–aluminate catalyst with respect to the Ni/Al$_2$O$_3$ catalyst could be due to the higher Ni content and presence of CaAl$_2$O$_4$. Indeed, it has been reported that Ni supported on spinel compounds such as CaAl$_2$O$_4$ leads to improved activity and stability of these catalysts with respect to Ni supported on Al$_2$O$_3$ catalysts [96]. These structures, also employed for the carbon dioxide reforming of methane [97], allow carbon formation to be limited. The best performance in terms of H$_2$ throughput per unit catalytic bed volume [about 40 m^3 (STP) h^{-1} dm^{-3}] was obtained with the ATR7B catalyst at the highest space velocity employed in the catalytic tests.

In order to investigate the performance of the ATR7B monolith at lower temperatures, other catalytic activity tests were carried out [98], the results of which are reported in Figure 9.14. The experimental conditions were changed in order to achieve stable low-temperature reactor behavior.

The typical ATR reactor temperature profile was obtained for all the tests except for that carried out at O$_2$:CH$_4$ = 0.25. It can be observed that a decrease in GHSV and an increase in the y value (with x = 0.6) allows a lower temperature to be attained

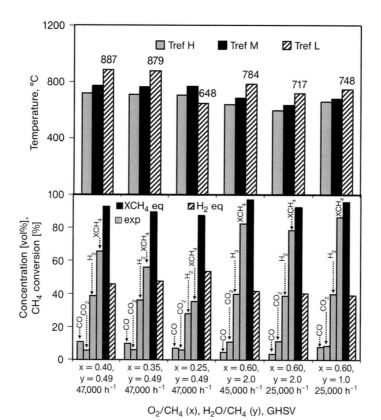

Figure 9.14 Results of CH$_4$ ATR catalytic activity tests on noble metal-based monolith at lower temperature.

in the catalytic bed and a more uniform temperature profile. Therefore, the activity of this catalyst is high enough to reach CH_4 conversion and H_2 concentration very close to the equilibrium values.

9.5 Economic Aspects

An economic evaluation was carried out for a traditional ATR + PSA scheme for H_2 production (Figure 9.15).

The overall plant could be supplied in three skids: the process skid of $2 \times 3 \times 5.0$ m (Figure 9.16), the PSA unit and the combustor unit where PSA purge gas is disposed and which acts also as a flare.

The variable operating costs, the investment costs and the rate of depreciation were taken into account. The cost analysis was kept simple because the main target was simply to evaluate the overall production costs.

In Table 9.2, the basic economic assumptions and parameters are reported.

The variable operating costs include the consumption of feed + fuel, cooling water and electricity. For their evaluation, it was taken into account that in the actual economic scenario the costs of such a scheme are mainly related to those of the natural gas and of the plant thermal efficiency. The evaluation is reported in Table 9.3.

The investments are estimated on the basis of long-term experience in building hydrogen production plants and on the assumption of building 10 identical units in order to optimize the construction costs. Table 9.4 gives the plant cost estimate as a percentage of the delivered equipment costs [99]. Such an estimate does not

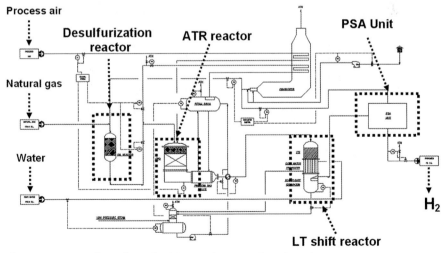

Figure 9.15 Process flow scheme for ATR + PSA H_2 manufacturing unit.

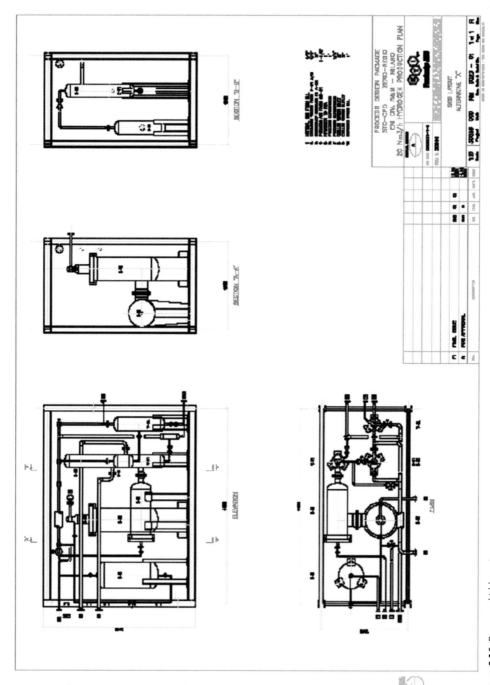

Figure 9.16 Process skid layout.

Table 9.2 Basic economic assumptions and parameters.

Parameter	Basic assumption
Natural gas price	€0.18 m^{-3} (STP) [LHV = 8700 kcal m^{-3} (STP)]
Electricity price	€0.085 (kW h)$^{-1}$
Cooling water price	€0.07 m^{-3}
Capacity factor	80% at design capacity
H$_2$ plant capacity	50 m^3 h^{-1} (STP)
Depreciation	10% per year of investment
Maintenance materials and labor (H$_2$ section)	2.5% of the investment

include the costs of the hydrogen storage vessel and the high-pressure hydrogen compressor.

Finally, in Table 9.5 the various components (variable operating costs, operating and maintenance costs and depreciation) and the overall results expressed in terms of euros per cubic meter (STP) of hydrogen produced through ATR and SR are given.

Although the costs are almost the same, the advantage of the ATR technology with respect to the SR approach is represented by the higher reactor compactness and flexibility, thus being more suitable for small-scale applications.

Table 9.3 Variable costs estimation.

	Feed + fuel	Water	Electricity	Overall cost
Specific consumption	4850 kcal m^{-3} (STP)	0.03	0.35 kW h m^{-3} (STP)	
Specific costs	0.10	0.002	0.03	0.132

Table 9.4 Investment costs estimation.

	10^3 €	%
Equipment cost (delivered including PLC[a])	450	100
Bulk materials (piping, instrumentation, electrical, etc.)	150	33
Equipment and bulk materials outside battery limit	200	44
Total direct costs	800	177
Engineering	50	11[b]
Construction	300	67
Total direct + indirect costs	590	255
Contractor profit and project contingency	100	22
Total investment cost	1250	277

[a]PLC = programmable logic controllers
[b]Engineering costs have been shared on 10 identical units.

Table 9.5 Hydrogen production costs [€ m^{-3} (STP)].

Case	Variable operating costs	Operating and maintenance costs	Depreciation rate	Total
ATR technology	0.13	0.04	0.36	0.53
SR technology	0.11	0.05	0.36	0.52

9.6
Conclusions and Perspectives

The feasibility of a hydrogen-based distributed power generation has been examined from both technological and economic points of view. The natural gas catalytic ATR technology was chosen for H_2 production. The reason of this choice was due to the particular attributes of the ATR reactor, which is a very compact and integrated system and thus suitable for decentralized small-scale units. The state of the art reported in the first part of this chapter was aimed to demonstrate that an effective hydrogen production system through the ATR reaction is a compromise between innovative catalyst formulations with regard to active species and support geometries and innovative engineering devices.

The experimental data provided in the last section are a contribution to support the feesibility of developing a compact and stand-alone kilowatt-scale stable ATR reactor capable to work in a stable manner over a wide range of operating conditions. Moreover, in order to improve its performance and increase its compactness, it has been integrated with two heat exchangers for the preheating of the air and liquid water fed to the reactor by the hot exhaust stream.

The high performance of this ATR reactor was also investigated under very severe operating conditions (low S/C ratio and high GHSV values). Using an innovative structured noble metals based catalyst allows both high CH_4 conversion and H_2 throughput per unit catalyst bed volume of about $40 \, m^3 \, (STP) \, h^{-1} \, dm^{-3}$ to be achieved.

From an economic point of view, the cost comparison between SR and ATR technologies for a plant with a capacity of $50 \, m^3 \, (STP) \, h^{-1}$ of H_2, show that in spite of the fairly similar costs of the two technologies, the greater reactor compactness, flexibility and ease of operation of the ATR reactor mean that the natural gas ATR could be an effective option for distributed small-scale hydrogen production.

Acknowledgments

This work was financed by the FISR Project DM 17/12/2002 'Idrogeno puro da gas naturale mediante reforming a conversione totale ottenuta integrando reazione chimica e separazione a membrana', http://www.fisrproject.com.

References

1 Kikuchi, R. (2004) *Environment, Development and Sustainability*, **6**, 453–471.
2 Alanne, K. and Saari, A. (2006) *Renewable and Sustainable Energy Reviews*, **10**, 539–558.
3 Dixon, R.K. (2007) *Mitigation and Adaptation Strategies for Global Change*, **12**, 325–341.
4 Turner, J.A. (2004) *Science*, **305**, 972–974.
5 Conte, M., Iacobazzi, A., Ronchetti, M. and Vellone, R. (2001) *Journal of Power Sources*, **100**, 171–187.
6 Friedmann, S.J. and Homer-Dixon, T. (2004) *Foreign Affairs*, **83**, 72–83.
7 Muradov, N.Z. and Veziroğlu, T.N. (2005) *International Journal of Hydrogen Energy*, **30**, 225–237.
8 Dicks, A.L. (1996) *Journal of Power Sources*, **61**, 113–124.
9 Armor, J.N. (1999) *Applied Catalysis A–General*, **176**, 159–176.
10 Farrauto, R., Hwang, S., Shore, L., Ruettinger, W., Lampert, J., Giroux, T., Liu, Y. and Ilinich, O. (2003) *Annual Review of Materials Research*, **33**, 1–27.
11 Kahlich, M.J., Gasteiger, H.A. and Behm, R.J. (1997) *Journal of Catalysis*, **171**, 93–105.
12 Kahlich, M.J., Gasteiger, H.A. and Behm, R.J. (1999) *Journal of Catalysis*, **182**, 430–440.
13 Kahlich, M.J., Gasteiger, H.A. and Behm, R.J. (1998) *Journal of New Materials for Electrochemical Systems*, **1**, 39–46.
14 Adhikari, S. and Fernando, S. (2006) *Industrial and Engineering Chemistry Research*, **45**, 875–881.
15 Park, J.-H., Kim, J.-N. and Cho, S.-H., (2000) *Arche Journal*, **46**, 790–802.
16 Kikuchi, E. (2000) *Catalysis Today*, **56**, 97–101.
17 Sjardin, M., Damen, K.J. and Faaij, A.P.C. (2006) *Energy*, **31**, 2523–2555.
18 Jordal, K., Bredesen, R., Kvamsdal, H.M. and Bolland, O. (2003) Proceedings of the 6th International Conference on Greenhouse Gas Control Technologies, Kyoto (eds J. Gale and Y. Kaya), Pergamon Press, p. 135.
19 Navarro, R.M., Peña, M.A. and Fierro, J.L.G. (2007) *Chemical Reviews*, **107**, 3952–3991.
20 Rostrup-Nielsen, J.R. (1984) *Catalysis Science and Technology* (eds J.R. Anderson and M. Boudart), Springer, Berlin, Chapter 1.
21 Peña, M.A., Gómez, J.P. and Fierro, J.L.G. (1996) *Applied Catalysis A–General*, **144**, 7–57.
22 Ahmed, S. and Krumpelt, M. (2001) *International Journal of Hydrogen Energy*, **26**, 291–301.
23 Rostrup-Nielsen, J.R., Dybkjaer, I. and Christensen, T.S. (1998) *Studies in Surface Science and Catalysis*, **113**, 81.
24 Joensen, F. and Rostrup-Nielsen, J.R. (2002) *Journal of Power Sources*, **105**, 195–201.
25 Christensen, T.S. (1996) *Applied Catalysis A–General*, **138**, 285–309.
26 York, A.P.E., Xiao, T. and Green, M.L.H. (2003) *Topics in Catalysis*, **22**, 345–358.
27 Haldor Topsoe A/S. (1988) *Hydrocarbon Processing*, **67**, 77.
28 Aasberg-Petersen, K., Bak Hansen, J.-H., Christensen, T.S., Dybkjaer, I., Christensen, P., Seier Stub Nielsen, C., Winter Madsen, S.E.L. and Rostrup-Nielsen, J.R. (2001) *Applied Catalysis A–General*, **221**, 379–387.
29 Rostrup-Nielsen, J.R. (2000) *Catalysis Today*, **63**, 159–164.
30 Kondratenko, E.V. and Baerns, M. (2003) *Encyclopedia of Catalysis*, John Wiley & Sons, Inc., New York.
31 Farrauto, R.J. and Bartholomew, C.H. (1997) *Fundamentals of Industrial Catalytic Processes*, Blackie Academic and Professional, London.
32 Giroux, T., Hwang, S., Liu, Y., Ruettinger, W. and Shore, L. (2005) *Applied Catalysis B–Environmental*, **55**, 185–200.

33 Faur Ghenciu, A. (2002) *Current Opinion in Solid State and Materials Science*, **6**, 389–399.
34 Freni, S., Calogero, G. and Cavallaro, S. (2000) *Journal of Power Sources*, **87**, 28–38.
35 Nagaoka, K., Eiraku, T., Nishiguchi, H. and Takita, Y. (2006) *Chemistry Letters*, **35**, 580–581.
36 Chu, W., Yan, Q., Liu, X., Li, Q., Yu, Z. and Xiong, G. (1998) *Studies in Surface Science and Catalysis*, **119**, 849–854.
37 Trovarelli, A., de Leitenburg, C., Boaro, M. and Dolcetti, G. (1999) *Catalysis Today*, **50**, 353–367.
38 Zhu, T. and Flytzani-Stephanopoulos, M. (2001) *Applied Catalysis A–General*, **208**, 403–417.
39 Roh, H.-S., Dong, W.-S., Jun, K.-W. and Park, S.-E. (2001) *Chemistry Letters*, **1**, 88.
40 Roh, H.-S., Jun, K.-W., Dong, W.-S., Park, S.-E. and Baek, Y.-S. (2001) *Catalysis Letters*, **74**, 31–36.
41 Roh, H.-S., Jun, K.-W., Dong, W.-S., Chang, J.-S., Park, S.-E. and Joe, Y.-I. (2002) *Journal of Molecular Catalysis A–Chemical*, **181**, 137–142.
42 Dong, W.-S., Roh, H.-S., Jun, K.-W., Park, S.-E. and Oh, Y.-S. (2002) *Applied Catalysis A–General*, **226**, 63–72.
43 Pengpanich, S., Meeyoo, V. and Rirksomboon, T. (2004) *Catalysis Today*, **93–95**, 95–105.
44 Roh, H.-S., Potdar, H.S. and Jun, K.-W. (2004) *Catalysis Today*, **93–95**, 39–44.
45 Fornasiero, P., Di Monte, R., Rao, G., Ranga Kaspar, J., Meriani, S., Trovarelli, A. and Graziani, A.M. (1995) *Journal of Catalysis*, **151**, 168–177.
46 Trovarelli, A., Zamar, F., Llorca, J., de Leitenburg, C., Dolcetti, G. and Kiss, J.T. (1997) *Journal of Catalysis*, **169**, 490–502.
47 Hori, C.E., Permana, H., Ng, K.Y.S., Brenner, A., More, K., Rahmoeller, K.M. and Belton, D. (1998) *Applied Catalysis B–Environmental*, **16**, 105–117.
48 Vidal, H., Kaspar, J., Pijolat, M., Colon, G., Bernal, S., Cordón, A., Perrichon, V. and Fally, F. (2000) *Applied Catalysis B–Environmental*, **27**, 49–63.
49 Dong, X., Cai, X., Song, Y. and Lin, W. (2007) *Journal of Natural Gas Chemistry*, **16**, 31–36.
50 Cai, X., Dong, X. and Lin, W. (2006) *Journal of Natural Gas Chemistry*, **15**, 122–126.
51 Ashcroft, A.T., Cheetham, A.K., Foord, J.S., Green, M.L.H., Grey, C.P., Murrel, A.J. and Vernon, P.D.F. (1990) *Nature*, **344**, 319–321.
52 Vernon, P.D.F., Green, M.L.H., Cheetham, A.K. and Ashcroft, A.T. (1990) *Catalysis Letters*, **6**, 181–186.
53 Ayabe, S., Omoto, H., Utaka, T., Kikuchi, R., Sasaki, K., Teraoka, Y. and Eguchi, K. (2003) *Applied Catalysis A–General*, **241**, 261–269.
54 Souza, M.M.V.M. and Schmal, M. (2005) *Applied Catalysis A–General*, **281**, 19–24.
55 Villegas, L., Guilhaume, N., Provendier, H., Daniel, C., Masset, F. and Mirodatos, C. (2005) *Applied Catalysis A–General*, **281**, 75–83.
56 Cheekatamarla, P.K. and Lane, A.M. (2005) *Journal of Power Sources*, **152**, 256–263.
57 Cheekatamarla, P.K. and Lane, A.M. (2006) *Journal of Power Sources*, **154**, 223–231.
58 Recupero, V., Pino, L., Vita, A., Cipitì, F., Cordaro, M. and Laganà, M. (2005) *International Journal of Hydrogen Energy*, **30**, 963–971.
59 Ma, L. and Trimm, D.L. (1996) *Applied Catalysis A–General*, **138**, 265–273.
60 Dias, J.A.C. and Assaf, J.M. (2004) *Journal of Power Sources*, **130**, 106–110.
61 Dias, J.A.C. and Assaf, J.M. (2005) *Journal of Power Sources*, **139**, 176–181.
62 Cheekatamarla, P.K. and Lane, A.M. (2005) *International Journal of Hydrogen Energy*, **30**, 1277–1285.
63 Cheekatamarla, P.K. and Lane, A.M. (2006) *Journal of Power Sources*, **153**, 157–164.
64 Islam, M.S., Cherry, M. and Catlow, C.R.A. (1996) *Journal of Solid State Chemistry*, **124**, 230–237.
65 Tejuca, L.G. and Fierro, J.L.G. (eds) (1993) *Properties and Applications of Perovskite-type Oxides*, Marcel Dekker, New York.

66 Forni, L. and Rossetti, I. (2002) *Applied Catalysis B–Environmental*, **38**, 29–37.
67 Liu, D.-J. and Krumpelt, M. (2005) *International Journal of Applied Ceramic Technology*, **2**, 301–307.
68 Erri, P., Dinka, P. and Varma, A. (2006) *Chemical Engineering Science*, **61**, 5328–5333.
69 Maestri, M., Beretta, A., Groppi, G., Tronconi, E. and Forzatti, P. (2005) *Catalysis Today*, **105**, 709–717.
70 Rabe, S., Truong, T.-B. and Vogel, F. (2007) *Applied Catalysis A–General*, **318**, 54–62.
71 Liu, D.-J., Kaun, T.D., Liao, H.-K. and Ahmed, S. (2004) *International Journal of Hydrogen Energy*, **29**, 1035–1046.
72 Lee, S.H.D., Applegate, D.V., Ahmed, S., Calderone, S.G. and Harvey, T.L. (2005) *International Journal of Hydrogen Energy*, **30**, 829–842.
73 Qi, A., Wang, S., Fu, G., Ni, C. and Wu, D. (2005) *Applied Catalysis A–General*, **281**, 233–246.
74 Avila, P., Montes, M. and Miró, E.E. (2005) *Chemical Engineering Journal*, **109**, 11–36.
75 Villegas, L., Masset, F. and Guilhaume, N. (2007) *Applied Catalysis A–General*, **320**, 43–55.
76 Rampe, T., Heinzel, A. and Vogel, B. (2000) *Journal of Power Sources*, **86**, 536–541.
77 Heinzel, A., Vogel, B. and Hübner, P. (2002) *Journal of Power Sources*, **105**, 202–207.
78 Lenz, B. and Aicher, T. (2005) *Journal of Power Sources*, **149**, 44–52.
79 Ciambelli, P., Palma, V., Palo, E. and Sannino, D. (2005) Proceedings of 7th World Congress of Chemical Engineering, Glasgow, 10–14 July 2005.
80 Palo, E. (2007) PhD thesis, University of Salerno, Department of Chemical and Food Engineering.
81 Specchia, S., Tillemans, F.W.A., van den Oosterkamp, P.F. and Saracco, G. (2005) *Journal of Power Sources*, **145**, 683–690.
82 Specchia, S., Cutillo, A., Saracco, G. and Specchia, V. (2006) *Industrial and Engineering Chemistry Research*, **45**, 5298–5307.
83 Severin, C., Pischinger, S. and Ogrzewalla, J. (2005) *Journal of Power Sources*, **145**, 675–682.
84 Cipitì, F., Recupero, V., Pino, L., Vita, A. and Laganà, M. (2006) *Journal of Power Sources*, **157**, 914–920.
85 Qi, A., Wang, S., Fu, G. and Wu, D. (2006) *Journal of Power Sources*, **162**, 1254–1264.
86 Seo, Y.S., Shirley, A. and Kolaczkowski, S.T. (2002) *Journal of Power Sources*, **108**, 213–225.
87 Chan, S.H. and Wang, H.M. (2000) *Fuel Processing Technology*, **64**, 221–239.
88 Mukainakano, Y., Li, B., Kado, S., Miyazawa, T., Okumura, K., Miyao, T., Naito, S., Kunimori, K. and Tomishige, K. (2007) *Applied Catalysis A–General*, **318**, 252–264.
89 Springmann, S., Bohnet, M., Docter, A., Lamm, A. and Eigenberger, G. (2004) *Journal of Power Sources*, **128**, 13–24.
90 Vermeiren, W.J.M., Blomsma, E. and Jacobs, P.A. (1992) *Catalysis Today*, **13**, 427–436.
91 Hochmuth, J.K. (1992) *Applied Catalysis B–Environmental*, **1**, 89–100.
92 Docter, A. and Lamm, A. (1999) *Journal of Power Sources*, **84**, 194–200.
93 Roh, H.S., Jun, K.-W. and Park, S.-E. (2003) *Applied Catalysis A–General*, **251**, 275–283.
94 Iaquaniello, G., Mangiapane, A., Ciambelli, P., Palma, V. and Palo, E. (2005) *Chemical Engineering Transactions*, **7**, 261–266.
95 Veser, G., Ziauddin, M. and Schmidt, L.D. (1999) *Catalysis Today*, **47**, 219–228.
96 Lu, Y., Liu, Y. and Shen, S. (1998) *Journal of Catalysis*, **177**, 386–388.
97 Lemonidou, A.A., Goula, M.A. and Vasalos, I.A. (1998) *Catalysis Today*, **46**, 175–183.
98 Ciambelli, P., Palma, V., Palo, E., Iaquaniello, G., Mangiapane, A. and Cavallero, P. (2007) *Chemical Engineering Transactions*, **11**, 437–442.
99 Peters, M.S. and Timmerhaus, K.D. (1980) *Plant Design and Economics for Chemical Engineers*, 3rd edn, McGraw-Hill.

Part Four
Industrial Catalysis for Sustainable Energy

10
The Use of Catalysis in the Production of High-quality Biodiesel

Nicoletta Ravasio, Federica Zaccheria, and Rinaldo Psaro

10.1
Introduction

Biodiesel may be chemically represented as a mixture of fatty acid methyl esters (FAMEs). It is a naturally derived liquid fuel, produced from renewable sources which, in compliance with appropriate specification parameters, may be used in place of diesel fuel both for internal combustion engines and for producing heat in boilers.

Plant oils were already used as a fuel by Rudolf Diesel in 1912, but more recent testing revealed that pure oils, even of fully refined quality, do not fit the modern fast-running diesel engines with their high efficiency and with low emission profile. Methyl esters are the derivatives of choice, simple to produce and come very close to diesel in their fuel properties (Table 10.1).

There are slight but acceptable differences in density and viscosity; the higher flash point is a beneficial safety feature and the absence of sulfur in plant oils is the reason for the excellent SO_x emission profile of biodiesel.

Moreover, European regulations in 2005 restricted the sulfur content in diesel fuel to $50\,\text{mg}\,\text{kg}^{-1}$. Sulfur organic compounds are known to provide diesel fuel with a lubricity that will disappear as the regulations take effect. Addition of biodiesel at a level of 1–2% to diesel blends has the effect of restoring lubricity through an antiwear action on engine injection systems, which is specific for polar molecules.

An important characteristic of diesel fuels is their ability to autoignite, quantified by the cetane value. Generally the cetane value is higher for biodiesel, resulting in smoother running of the engine with less noise.

Blends of biodiesel and petroleum diesel are designated B followed by the volume percentage of biodiesel present. B5 and B20, the most common blends, can be used in unmodified diesel engines. Biodiesel is by nature an oxygenated fuel with an oxygen content of about 10%. This improves combustion and reduces CO, soot and unburnt hydrocarbon emissions (Table 10.2, Figure 10.1) while slightly increasing the NO_x emissions and causing a 7% lower calorific value.

Catalysis for Sustainable Energy Production. Edited by P. Barbaro and C. Bianchini
Copyright © 2009 WILEY-VCH Verlag GmbH & Co. KGaA, Weinheim
ISBN: 978-3-527-32095-0

Table 10.1 Physical–chemical properties of diesel compared with biodiesel.

Standardized properties	Unit	Diesel EN 590	Biodiesel (FAME) EN 14214
Density at 15 °C	$kg\,m^{-3}$	820–845	860–900
Viscosity at 40 °C	$mm^2\,s^{-1}$	2.00–4.50	3.5–5.0
Flash point	°C	>55	>120
Sulfur	% (m/m)	<0.20	<0.01
Cetane No.		>49	>51
Oxygen content	% (m/m)	0.0	10.9
Caloric value	$MJ\,dm^{-3}$	35.6	32.9
Efficiency degree	%	38.2	40.7

The advantages, especially of environmental nature, which can potentially result from the widespread use of biodiesel are manifold:

- First, being a naturally derived material and thus produced from photosynthesis, its combustion does not contribute towards increasing the net atmospheric carbon dioxide concentration, one of the major factors responsible for the greenhouse effect. For this reason, biodiesel is one of the fuels whose use should allow the objectives envisaged in the Kyoto Agreement to be achieved. Greenhouse gas emissions of B20 and B100 are compared with those of other alternative fuels in Figure 10.2.

- Biodiesel (FAMEs) is the only alternative fuel currently available that have an overall positive life cycle energy balance. Thus, it yields as much as 3.2 units of fuel product energy for every unit of fossil energy consumed in its life cycle, compared with only 0.83 units for petroleum diesel, while considerably more energy (about 29%), including high-grade fossil fuel, is required to produce ethanol from corn than is available in the energy ethanol output [1].

- Considering that the triglyceride oils used for the production of biodiesel are sulfur free and that sulfur is not added to the end product in any way, the use of biodiesel does not contribute towards the phenomenon of acid rain.

Table 10.2 Average biodiesel emissions (%) compared with conventional diesel.

Emission type	B20	B100
Total unburned hydrocarbons	−20	−67
CO	−12	−48
CO_2	−16	−79
Particulate matter	−12	−47
NO_x	+2	+10
SO_x	−20	−100
Polycyclic aromatic hydrocarbons (PAHs)	−13	−80
Nitrated PAHs	−50	−90

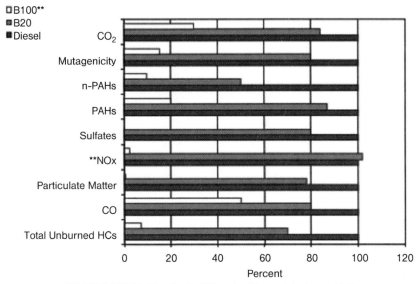

Figure 10.1 Relative emissions: diesel versus biodiesel.

- Due to its particular composition, biodiesel is biodegradable.
- For economies such as the European economy, which has to sustain the demand for energy with massive imports of fossil fuels, this is an excellent opportunity to have available an autonomous and also renewable energy source. The European Community strongly encourages this initiative, which allows the use of marginal land, not dedicated to food production, with undoubted advantages for safeguarding and increasing the work force and protecting and safeguarding the environment.

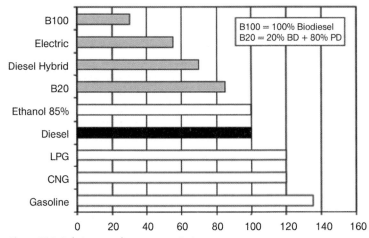

Figure 10.2 Relative greenhouse gas emissions.

The first biodiesel initiatives were reported in 1981 in South Africa and in 1982 in Austria, Germany and New Zealand. Since then, the production of this alternative fuel has seen enormous developments, particularly in Europe, where it reached 5.7 millions tons in 2007. It is expected to increase further to fulfill the recent decision of the European Parliament to substitute 10% of transport fuels with biofuels by 2020. According to assessments of the European Community, to reach this target, the production of bioethanol, biodiesel and second-generation biofuels should reach 36 Mtep (tep = tonnes equivalents petrol) in 2020.

Most biodiesel is produced today through transesterification of triglycerides of refined edible oils (Scheme 10.1).

Both acid and base catalysts can be used, but base catalysts are 4000 times more active and cause fewer corrosion problems than acid catalysts. Methanol, due to its low cost, is the alcohol most commonly used up to now, with the exception of Brazil, where ethanol is preferred. Homogeneous alkaline catalysts, such as NaOH, KOH and sodium methoxide (NaOMe), are generally used. NaOMe is the most active, most

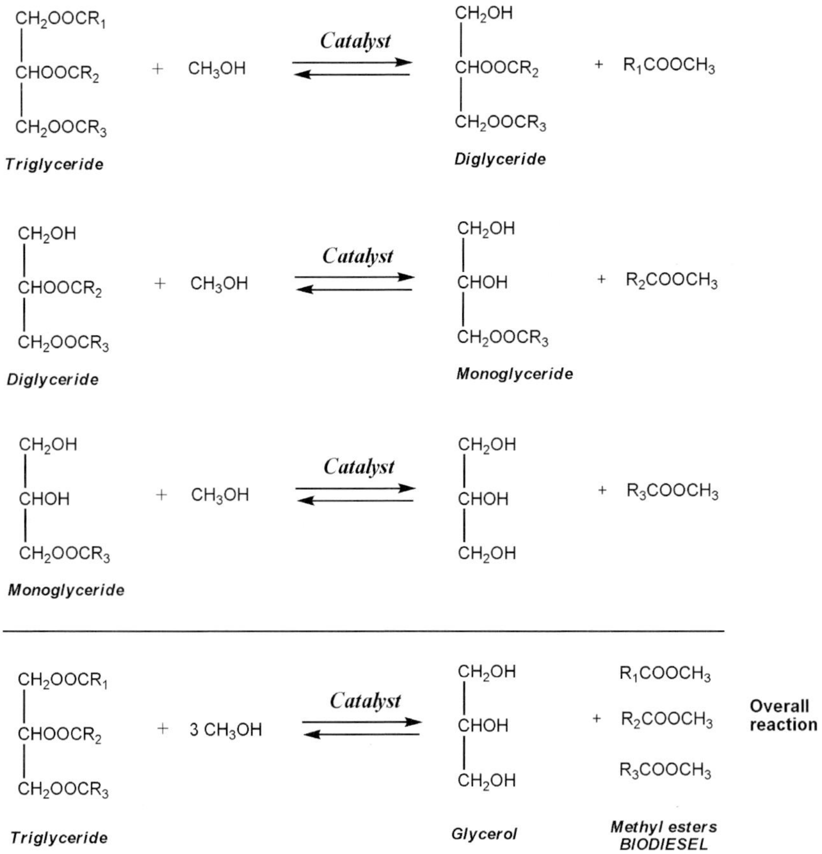

$$\begin{array}{c}CH_2OOCR_1\\|\\CHOOCR_2\\|\\CH_2OOCR_3\end{array} + CH_3OH \underset{}{\overset{Catalyst}{\rightleftarrows}} \begin{array}{c}CH_2OH\\|\\CHOOCR_2\\|\\CH_2OOCR_3\end{array} + R_1COOCH_3$$

Triglyceride → Diglyceride

$$\begin{array}{c}CH_2OH\\|\\CHOOCR_2\\|\\CH_2OOCR_3\end{array} + CH_3OH \underset{}{\overset{Catalyst}{\rightleftarrows}} \begin{array}{c}CH_2OH\\|\\CHOH\\|\\CH_2OOCR_3\end{array} + R_2COOCH_3$$

Diglyceride → Monoglyceride

$$\begin{array}{c}CH_2OH\\|\\CHOH\\|\\CH_2OOCR_3\end{array} + CH_3OH \underset{}{\overset{Catalyst}{\rightleftarrows}} \begin{array}{c}CH_2OH\\|\\CHOH\\|\\CH_2OH\end{array} + R_3COOCH_3$$

Monoglyceride

$$\begin{array}{c}CH_2OOCR_1\\|\\CHOOCR_2\\|\\CH_2OOCR_3\end{array} + 3\,CH_3OH \underset{}{\overset{Catalyst}{\rightleftarrows}} \begin{array}{c}CH_2OH\\|\\CHOH\\|\\CH_2OH\end{array} + \begin{array}{c}R_1COOCH_3\\R_2COOCH_3\\R_3COOCH_3\end{array}$$

Triglyceride — Glycerol — Methyl esters BIODIESEL — Overall reaction

Scheme 10.1 Transesterification reactions of glycerides with methanol.

expensive and most widely used compound, with over 60% of industrial plants using alkaline catalysts.

Sodium is recovered after the transesterification reaction as sodium glycerate, sodium methylate and sodium soaps in the glycerol phase. An acidic neutralization step (e.g. with aqueous hydrochloric, acetic or sulfuric acid) is required to neutralize these catalysts. In that case glycerol is obtained as an aqueous solution containing sodium chloride, acetate or sulfate. Depending on the process, the final glycerol purity is about 80–95%. When NaOH is used as catalyst, side-reactions forming sodium soaps generally occur. This type of reaction is also observed when NaOMe is employed and traces of water are present. Sodium soaps are soluble in the glycerol phase and must be isolated after neutralization by decantation as fatty acids. The loss of esters converted to fatty acids can be as high as 1%.

Strict feedstock and reagent specifications are therefore the main issue with this process. In particular, the total free fatty acid (FFA) content associated with the lipid feedstock must not exceed 0.5 wt%, otherwise soap formation will seriously hinder the production of fuel-grade biodiesel. Soap production gives rise to the formation of gels, increased viscosity and greatly increases product separation costs [2]. The alcohol and catalyst must also comply with rigorous specifications. They have to be essentially anhydrous since the presence of water in the feedstock promotes hydrolysis of the alkyl ester to FFAs, which in turn give soaps. To conform to such demanding feedstock specifications, only highly refined vegetable oils can be used.

Production costs are rather high. However, it is extremely important to note that the main factor affecting the cost of biodiesel is not the process but the cost of raw material, accounting for up to 85% of the final product cost. It is therefore not surprising that the skyrocketing price of vegetable oils during 2007–08 has caused half of the biodiesel plants in Europe to stop production.

Moreover, the use of edible oils such as rapeseed, soybean and palm oils is the subject of continuous attack from both the media and some political parties, while concerns have been raised also by the Food and Agriculture Organization of the United Nations [3]. From the environmental point of view, an increasing demand for palm oil is spurring the destruction of Asia's forests, and rapeseed, widely grown in Europe, is said to lower biodiversity.

For these reasons, research all over the world, and in our group, is focused on the use of alternative feedstocks, including waste oils, rendering fats, oils from plants growing in marginal areas and polyunsaturated oils of both vegetable and animal origin. These alternative feedstocks belong to two main families, namely highly acidic oils and polyunsaturated oils. The use of these fats and oils gives excellent opportunities to those involved in catalytic process development.

A comprehensive review on heterogeneous catalysts for the production of biodiesel has been recently published by Santacesaria and co-workers [4].

In this chapter, we report just a few selected examples of heterogeneous catalytic systems for the esterification of fatty acids and for the simultaneous esterification and transesterification of acidic oils and fats, and we discuss the use of selective hydrogenation as a tool for the production of high-quality biodiesel from non-edible raw materials.

10.2
Heterogeneous Transesterification and Esterification Catalysts

10.2.1
Heterogeneous Basic Catalysts

The use of heterogeneous basic catalysts for the transesterification of triglycerides has long been considered the main tool to reduce processing costs in the production of biodiesel, as it would lead to simplified operations and eliminate waste streams.

The patent and academic literature lists a number of basic heterogeneous catalysts. They are not as active as homogeneous catalysts, requiring higher reaction temperatures (200–250 °C) and pressures.

The only commercial plant exploiting a heterogeneous catalyst is the Diester Industrie 160 000 t yr^{-1} plant at Sète, France, based on the Esterfip-H technology developed by the Institut Français du Pétrole [5]. This continuous plant operates at 230 °C and 50 atm with a catalyst consisting of mixed Zn and Al oxides. The desired chemical conversion required to produce biodiesel according to the European specifications is reached with two successive stages of reaction and glycerol separation in order to shift the equilibrium of methanolysis (Table 10.3).

The catalyst section includes two fixed-bed reactors, fed with vegetable oil and methanol at a given ratio. Excess of methanol is removed after each reactor by partial evaporation. Then, esters and glycerol are separated in a settler. The glycerol outputs are collected and the residual methanol removed by evaporation. In order to obtain biodiesel according to the European specifications, the last traces of methanol and glycerol have to be removed. The purification section for methyl esters coming from decanter 2 consists of a finishing methanol vaporization under vacuum followed by a final purification step in an adsorber to remove the soluble glycerol (Figure 10.3).

Table 10.3 Main features of biodiesel obtained with the Esterfip-H technology (starting material rapeseed oil) (Reproduced with permission of Elsevier from ref. [56]).

	Biodiesel from reactor 1	Biodiesel from reactor 2	European specification
Methyl esters (wt%)	94.1	98.3	>96.5
Monoglycerides (wt%)	2.0	0.5	<0.8
Diglycerides (wt%)	1.1	0.1	<0.2
Triglycerides (wt%)	1.6	0.1	<0.02
Free glycerol (wt%)	—	—	<0.02
Metal content (mg kg^{-1})	<2	<2	<5
Group I (Na + K) (mg kg^{-1})	<2	<2	<5
Group II (Ca + Mg) (mg kg^{-1})	<2	<2	<5
Zn (mg kg^{-1})	<1	<1	—
Phosphorus content (mg kg^{-1})	<10	<10	<10
Acid number (mg KOH kg^{-1})	<0.3	<0.3	<0.5

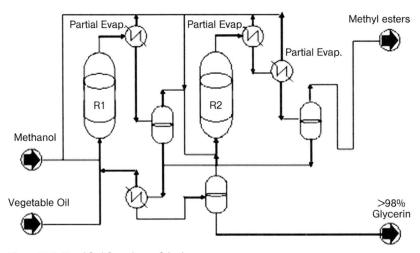

Figure 10.3 Simplified flow sheet of the heterogeneous Esterfip-H based process (Reproduced with permission of Elsevier from ref. [56]).

The main advantage of this process is the high purity of glycerol co-produced. It is limpid and colorless. The glycerol content is at least >98%. Neither ash nor inorganic compounds are detected, the main impurities being water, methanol and MONG ('matter organic non-glycerol'), such as methyl esters. The high quality of the glycerol by-product is a very important economic parameter.

As already mentioned, when simple alkaline derivatives, such as sodium alcoholates, soda or potash, are used as catalysts, one arrives at a pure product usable as fuel and a glycerol at the required standard only after many stages.

Thus, these alkaline compounds are found in both glycerol and esters. They should be eliminated by washing and/or neutralization of the ester fraction, followed by drying of the latter. In the glycerol phase, it is necessary to neutralize the soaps and alcoholates present, and sometimes to eliminate any salts formed. The glycerol thus obtained contains water, generally between 5 and 40 mass%. It also contains salts resulting from the neutralization of alkaline catalyst, for example sodium chloride when the catalyst is soda or sodium methylate and when neutralization is carried out with hydrochloric acid. The salt concentration in glycerol resulting from these processes is generally 3–6 mass%. Obtaining glycerol of high purity starting from crude glycerol resulting from these processes thus imposes stages of purification such as distillation under reduced pressure, sometimes associated with treatments on ion-exchange resins. All these treatments have a great influence on the price of the final glycerol. In every case, the semi-purified glycerol coming from the heterogeneous catalytic process also needs to be purified further to fulfill the international specification for pure/FU/USP glycerol.

It is therefore apparent that the main advantage of this process is the possibility of obtaining high-purity glycerol, with important savings in glycerol refining, whereas the low activity of the catalytic system requires a two-step process and high investment costs, mainly based on the necessity to carry out the reaction at high pressure.

It is also worth noting that although the IFP group reported on different mixed oxide catalytic systems, namely Ti and Al mixed oxides [6], Bi, Ti and Al mixed oxides [7], Sb and Al [8] and Zn, Ti and Al [9], none of them is able to convert the FFAs eventually present in the oil. Feedstock specifications are therefore demanding also for these basic catalytic systems.

10.2.2
Heterogeneous Acid Catalysts

Methodologies based on acid-catalyzed reactions have the potential to lower production costs since acid catalysts do not show a measurable susceptibility to FFAs.

The homogeneous acid-catalyzed transesterification process does not enjoy the same popularity in commercial application as its counterpart, the base-catalyzed process, one of the main reasons being that it is about 4000 times slower, due to the different mechanism [10]. Thus, in the reaction sequence triglyceride is converted stepwise to diglyceride, monoglyceride and finally glycerol with formation of one molecule of methyl ester at each step (Scheme 10.1).

The transesterification pathway for an acid-catalyzed reaction is shown in Scheme 10.2. The key step is the protonation of the carbonyl oxygen that increases the electrophilicity of the adjoining carbon atom, making it more susceptible to nucleophilic attack. The formation of a more electrophilic species also occurs with homogeneous and heterogeneous Lewis acid catalysts through coordination of the carbonyl oxygen atom to the Lewis acid center. In this case, the rate-determining step depends on the acid site strength, strong acid sites inhibiting desorption of the product [11].

In contrast, base catalysis takes a more direct route, creating first an alkoxide ion, which directly acts as a strong nucleophile (Scheme 10.3).

This crucial difference, that is, the formation of a more electrophilic species versus that of a strong nucleophile, is ultimately responsible for the observed difference in activity.

From the kinetic point of view, three regimes can categorize the overall reaction process of soybean oil transesterification using sulfuric acid as the catalyst [12]. Initially the reaction is characterized by a mass transfer-controlled regime resulting from the low miscibility of the catalyst and reagents, nonpolar oil phase and polar methanol–acid phase. As the reaction proceeds, the partial glycerol ester product acts as an emulsifier and a kinetically controlled regime emerges, characterized by a sudden surge in product formation. Finally, the last regime is reached once equilibrium is approached, near reaction completion. In this case, a very large 30:1 molar ratio of alcohol to oil is needed to have acceptable reaction rates. In this way, the forward reactions follow pseudo-first-order kinetics, whereas the reverse reactions exhibit second-order kinetics (see Scheme 10.1).

Such a large excess of methanol increases both the alcohol recovery and separation costs. However, ester formation increases sharply, going from 3:1 (77%) to 6:1 (87.8%), and ultimately reaches a plateau value of 30:1 (98.4%) [13].

180 °C with a methanol-to-substrate ratio of 12 [35]. Maximum activity was observed for titanium loadings of 3–11%: on the surface of these materials, dispersed surface TiO_x species, in both tetrahedral and octahedral coordination, and small TiO_2 crystallites are present. It is interesting that an optimal range of strength for Lewis acid sites responsible for the catalytic activity exists, the strongest, similar to those found on crystalline anatase, being almost inactive. Although a direct comparison has not been reported, the grafted catalysts appear more active than both TS-1 and TiO_2/SiO_2 prepared by impregnation [36].

12-Tungstophosphoric acid (TPA) impregnated on four different supports, hydrous zirconia, silica, alumina and activated carbon, was also tested for biodiesel production from low-quality canola oil containing up to 20% FFAs. The hydrous zirconia catalyst was found to be the most active, giving a 90% yield at 200 °C and a 1:9 oil-to-methanol stoichiometric ratio. Lewis acid sites generated by the interaction of TPAs with surface hydroxyl group of zirconia, forming Zr–O–W bonds exerting an electron-withdrawing effect on surface Zr^{4+} cations, account for the high activity observed [37].

A correlation between the transesterification activity and the concentration of Lewis acid sites has also been observed when studying the use of the hydrophobic Fe–Zn double metal cyanide (DMC) $K_4Zn_4[Fe(CN)_6]_3 \cdot 6H_2O \cdot 2(tert\text{-BuOH})$, prepared by adding a solution of a tri-block polymer in tert-butanol to the mixture obtained by mixing a $ZnCl_2$ in water–tert-butanol and a solution of $K_4Fe(CN)_6$ in water. These complexes possess a zeolite-like cage structure and are insoluble in most organic solvents and even in aqua regia. Pyridine adsorption followed by DRIFT spectroscopy revealed the presence of strong Lewis acid sites whereas Brønsted sites were absent. This catalyst showed good activity in the esterification–transesterification of a number of unrefined vegetable oils (Table 10.4), used oils and non-edible oils at 150 °C with an oil-to-methanol ratio of 1:15. The catalyst performance is not influenced by the water content and it can be reused after separation by centrifugation without any further purification [38, 39].

The role of surfactant molecules in the synthesis of Fe–Zn DMC is probably to increase the surface area and acid sites density, thereby enhancing the catalytic

Table 10.4 Esterification–transesterification of different oils in the presence of Fe–Zn double metal cyanide.

Vegetable oil	Conversion (%)	Acid (wt% oil)	Acid (wt% FAMEs)
Coconut	99.8	1.3	0.6
Sesame	98.2	2.2	0.7
Peanut	97.0	1.8	0.7
Used margarine	98.0	2.7	0.6
Rubber seed	97.1	15.6	1.1
Jatropha	86.9	2.4	0.7
Pinnai	84.2	17.1	2.4
Karanja	88.3	3.0	0.2

activity in both esterification and transesterification reactions. Coordinatively unsaturated Zn^{2+} ions are the probable active sites for both reactions.

10.3
Selective Hydrogenation in Biodiesel Production

Another pool of raw materials that up to now has not been considered for the manufacture of biodiesel is highly unsaturated vegetable oils, such as those obtained from linseed or camelina, which give a yield of oil per hectare higher than soybean.

The degree of unsaturation of the fatty acids is normally expressed as the iodine value (IV), that is, the number of grams of iodine that have reacted with 100 g of product analyzed, under controlled experimental conditions. The higher is IV, the greater is the degree of unsaturation. For example, for biodiesel intended for haulage use, the most remunerative use, a maximum IV limit of 120 g I_2 per 100 g is envisaged.

Biodiesel is normally produced by starting from vegetable oils having an IV of ≤ 130, that is, having a low unsaturation index, such as rapeseed oil (IV = 110–115), sunflower oil (IV = 120–130) and soybean oil (IV = 125–135).

The major fatty acids present in plant-derived fatty substances are oleic acid (9-octadecenoic acid, C18:1), linoleic acid (9,12-octadecadienoic acid, C18:2) and the conjugated isomers thereof and linolenic acid (9,12,15-octadecatrienoic acid, C 18:3) (Scheme 10.4). Their rates of oxygen absorption are 1:40:100, respectively [40], hence partial hydrogenation with consequent lowering of the IV would lead to a significant increase in oxidative stability, particularly when C18:3 is reduced. The dependence of oxidative stability on the content of oleic acid (C18:1) in vegetable oils has long been known (Figure 10.5).

Also, the ability to autoignite, quantified by the cetane value, increases with the same order of oxidative stability (Figure 10.6), that is, with lowering of the IV.

On the other hand, in order to preserve the cold properties of the fuel (cloud point, pour point and low-temperature filterability), it is essential not to increase the melting point, which in turn depends on both the saturated compound (stearic acid, C18:0,

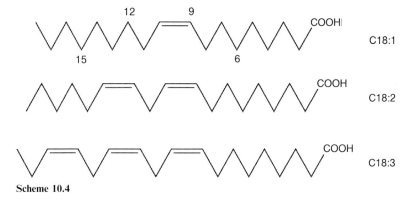

Scheme 10.4

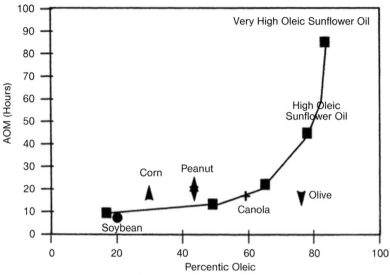

Figure 10.5 Dependence of oxidative stability on the percentage of oleic acid in different vegetable oils.

Figure 10.7) content and the extent of *cis–trans* and positional isomerization as the difference in melting point between the *cis* and *trans* isomers is at least 15 °C according to the double bond position, as shown in Table 10.5.

For all these reasons, the monounsaturated methyl oleate and methyl palmitoleate (C16:1) have been identified as the ideal components of biodiesel [41].

The use of copper hydrogenation catalysts in the fats and oils industry mainly relies on copper chromites for the production of fatty acid alcohols from esters [42]. However, copper-based systems have long been known in edible oil hydrogenation as being the most selective for the reduction of linolenate (C18:3) to oleate (C18:1) without affecting linoleate, the valuable component from the nutritional point of view. Monoenes are not reduced, hence the percentage of saturates is scarcely changed during the hydrogenation process [43, 44]. Moreover, it has already been

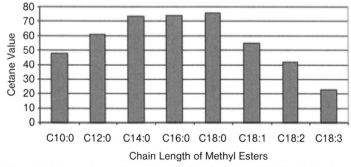

Figure 10.6 Dependence of cetane value on length and number of double bonds in different FAMEs.

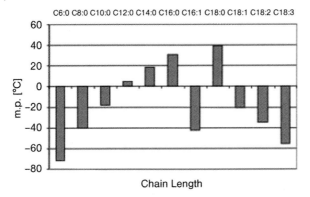

Figure 10.7 Melting points of different saturated and unsaturated methyl esters.

shown that Cu/SiO$_2$ catalysts prepared by chemisorption hydrolysis and prereduced *ex situ* are effective and very selective in the hydrogenation of rapeseed oil methyl esters, allowing up to 88% of C18:1 derivative to be obtained without modifying the amount of C18:0 and with a *trans* content of about 20% [45].

This prompted us to investigate the possibility of selectively hydrogenating highly unsaturated oils, unsuitable for the production of biodiesel, in order to improve their oxidative stability while retaining the cold properties.

We focused our attention on tall oil, a by-product of the paper industry, whenever this is prepared according to the Kraft process. This material consists of a mixture of highly unsaturated fatty acids (many of which have conjugated diene systems) and terpene-derived rosin acids. The rosin acids have the molecular formula $C_{20}H_{30}O_2$ and thus belong to the diterpenes (pimaric and abietic acids). Tall oil has an IV of ~170 g I$_2$ per 100 g.

The peculiarity of this material is that it consists of a mixture of acids and not triglycerides; therefore, its transformation into biodiesel requires only an esterification

Table 10.5 Influence of double bond number, position and geometry on the melting points of the 18-carbon atom fatty acids.

Fatty acid	No. of double bonds	Mp (°C)	
9,12,15 C18:3	3	−13	
9,12 C18:2	2	−7	
9 C18:1	1	+16	
C18:0	0	+70	
Monounsaturated fatty acid: influence of double bond position			
		Mp (*cis*) (°C)	Mp (*trans*) (°C)
Δ^6		28	53
Δ^9		16	45
Δ^{12}		27	52
Δ^{15}		40	58

reaction instead of a transesterification reaction and therefore does not produce glycerol, making the total economy lighter and independent of the critical marketing of this polyalcohol. However, reports on the use of tall oil to produce biodiesel are very rare. Liu et al. [46] developed a process for producing a diesel oil additive, not biodiesel, from pine oil. That process produced a diesel oil cut which was blended with a base diesel fuel.

Hydrotreating has been proposed by Arbokem in Canada [47] as a means of converting crude tall oil into biofuels and fuel additives. However, this process is a hydrogenation process that produces hydrocarbons rather than biodiesel.

Recently, two processes for biodiesel production from crude tall oil have been proposed [48, 49]. They rely on the use of a homogeneous acid catalyst or of an acyl halide for the esterification reaction, but no information was given on the properties of the fuel obtained, particularly concerning the oxidative stability and conformity with European specification EN 14214:2003 for IV.

In the process described here [50], the esterification reaction was carried out in the homogeneous phase and was followed by distillation under vacuum of the methyl esters obtained, allowing the removal of rosin acids as a bottom product. Hydrogenation of tall oil methyl esters obtained in this way gave the results reported in Table 10.6 and Figure 10.8.

The reaction was very fast and led to nearly complete suppression of conjugated dienes and a significant reduction in linoleic acid derivative, leaving the stearic acid content unaffected. Owing to this very high selectivity, not only can the IV be reduced to meet the European standard ($IV_{max} = 120$ g I_2 per 100 g) but also excellent cold properties can be obtained (Figure 10.1). The hydrogenated oil is completely colorless. Moreover, the treatment greatly improves the Conradson carbon residue (CCR), representing the tendency for the fuel to form carbon deposits when burnt with stoichiometric quantities of comburent, such as in internal combustion engines.

The IV, CCR and oxidation stability are three strictly co-related parameters. As a general rule, a reduction in IV (on the same feedstock) dramatically improves the oxidation stability. In contrast, the distillation step removes the main part of naturally occurring antioxidants. For this reason, even after hydrogenation, the Rancimat induction time, which measures the oxidative stability, of the hydrogenated sample does not fulfill the EN 14214 requirement (6 h at 110 °C), the measured induction

Table 10.6 Selective hydrogenation of tall oil methyl esters (Copyright with kind permission of Taylor & Francis).

Catalyst	P (atm)	C18:2	C18:2$_{conj}$	C18:1	C18:0	IV	Pour point (°C)	Oxidation stability (h)
Initial composition	—	44	11	37	0.4	145	—	<1
Cu/SiO$_2$	6	22	2	68	0.5	117	−18	4
Cu/Al$_2$O$_3$	6	24	4	63	1	112	−12	4
Cu/Al$_2$O$_3$	3	19	4	68	1	—	—	—
Ni/SiO$_2$	4	25	0.5	56	11	104	+3	—

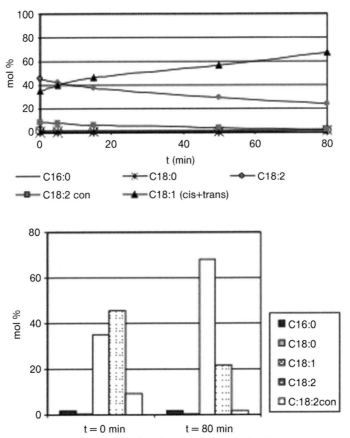

Figure 10.8 Reaction profile and product distribution obtained in the hydrogenation reaction carried out using Cu/SiO$_2$ (Copyright with kind permission of Taylor & Francis).

period being 4 h. Nevertheless, the non-hydrogenated, distilled tall oil methyl ester shows a Rancimat induction time shorter than 1 h, demonstrating an evident stabilization effect provided by partial selective hydrogenation. To fulfill the EN requirement, it is common practice to use synthetic antioxidants [51]. A further demonstration of the quality improvement can be obtained from the visual evaluation of tubes used for CCR evaluation. Tests carried out at Mercedes Benz suggest that fuels having IVs higher than 115 are not acceptable because of excessive carbon deposits.

For FAME fuels, CCR correlates with the respective residual amount of glycerides, FFAs, soaps and catalyst polyunsaturated residues [52]. Moreover, the parameter is also influenced by high concentrations of FAMEs and related polymers [53]. In our case, the first-mentioned factor has no impact, as the feedstock origin is the same in terms of major and minor constituents. In contrast, the important influence of selective hydrogenation reaction on CCR is clearly demonstrated in Figure 10.9, where CCR-containing tubes obtained from tall oil FAMEs non-hydrogenated and

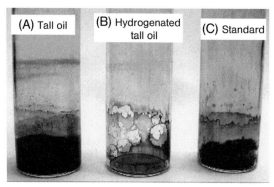

Figure 10.9 Conradson carbon residue of tall oil (A) and hydrogenated tall oil (B) compared with a standard oil (C) (Copyright with kind permission of Taylor & Francis).

selectively hydrogenated are shown. The third tube contains a reference product and it is shown to provide a visual idea of specification limit.

No significant differences were observed among the different catalysts used. On the contrary, a catalyst with the same Cu loading but prepared by the conventional incipient wetness technique did not show any activity even after a long reaction time. The sharp difference in activity between catalysts prepared by our technique and the incipient wetness method has been closely investigated in the case of Cu/TiO_2 systems and attributed to both very high dispersion of the metallic phase and the morphology of Cu crystallites [54].

A comparison test carried out over a commercial Ni catalyst showed that early formation of stearic acid (C18:0) results in a very high pour point, thus indicating that very high selectivity is required to improve oxidative stability while retaining cold properties.

10.4
Conclusions and Perspectives

According to one of the pioneers of biodiesel in Europe, Professor Martin Mittelbach, 'One main driving force for establishing a biofuels industry was to become less dependent on mineral oil, so only local feedstock and biofuel production can prevent a new dependency on vegetable oil producing countries. The future lies in the production of high quality biodiesel from all different kinds of feedstocks in multi-feedstock plants, leading to optimum biodiesel blends for all diesel engines, all climates and all kinds of applications' [55]. Thus, the use of vegetable oil for energy production poses several ethical problems as the yearly amount of natural oils available for this application is not infinite and the increase in demand for biodiesel production may raise the prices on the international market and, at the same time, may lower the availability of oils and fat for food purposes in developing countries. Hence the main challenge for biodiesel producers will be to single out fatty materials having no impact on the food chain. To do so will require the use of industrial crops such as *Jatropha curcas*, but also requalification of non-edible oils such as fish oils, linseed oil, camelina oil, hempseed oil and tall oil fatty acids by means of selective hydrogenation, which might make available important quantities of feedstocks that can be used for biodiesel production with reduced or no impact on the food chain. Also, the use of several streams coming from animal fat rendering can be reconsidered as a suitable source of biofuels. This offers enormous opportunities for workers involved in heterogeneous catalysis.

We have reported our work in the field of selective hydrogenation for the production of high-quality biodiesel. To the best of our knowledge, this is the only example of such a process.

Catalysts able to promote both transesterification of triglycerides and esterification of FFAs in an oil or fat are still rare and up to now none of them have found a commercial application. Thus far it seems that a moderate to high concentration of strongly acidic sites and a hydrophobic surface are mandatory to achieve good conversions in simultaneous esterification and transesterification reactions, but more research and creative thinking are still needed in this area.

Acknowledgment

The authors wish to thank Dr Paolo Bondioli, Stazione Sperimentale Oli e Grassi, Milan, for helpful discussions and critical reading of the manuscript.

References

1 Pimentel, D. (2003) *Natural Resources Research*, **12**, 127.
2 Ma, F.R. and Hanna, M.A. (1999) *Bioresource Technology*, **70**, 1.
3 Sustainable Bioenergy: a Framework for Decision Makers (2007), http://esa.un.org/un-energy/pdf/susdev.Biofuels.FAO.pdf.
4 Di Serio, M., Tesser, R., Pengmei, L. and Santacesaria, E. (2008) *Energy and Fuel*, **22**, 207.
5 Stern, R., Hillion, G., Rouxel, J. and Leporq, J. (1999) US Patent 5,908,946.
6 Lacome, T., Hillion, G., Delfort, B., Renaud, R., Leporq, S. and Acakpo, G. (2005) US Patent Application 2005266139 A1.
7 Delfort, B., Hillion, G., Le Pennec, D. and Lendresse, C. (2004) French Patent Application 2004 4731 FR 2869613 A1.
8 Lacome, T., Hillion, G., Delfort, B., Revel, R., Leporq, S. and Chaumonnot, A. (2003) French Patent 03 06338.
9 Hillion, G., Delfort, B., Lendresse, C. and Le Pennec, D. (2004) French Patent 04 04730.
10 Lotero, E., Liu, Y., Lopez, D.E., Suwannakarn, K., Bruce, D.A. and Goodwin, J.G. Jr, (2005) *Industrial and Engineering Chemistry Research*, **44**, 5353.
11 Bonelli, B., Cozzolino, M., Tesser, R., Di Serio, M., Piumetti, M., Garrone, E. and Santacesaria, E. (2007) *Journal of Catalysis*, **246**, 293.
12 Freedman, B., Butterfield, R.O. and Pryde, E.H. (1986) *Journal of the American Oil Chemists Society*, **63**, 1375.
13 Canacki, M. and Van Gerpen, J. (1999) *Transactions of the ASAE*, **42**, 1203. Rathore, V. and Madras, G. (2007) *Fuel*, **86**, 2650.

14 Sarin, R., Sharma, M., Sinharay, S. and Malhotra, R.K. (2007) *Fuel*, **86**, 1365.
15 Meher, L.C., Kulkarni, M.G., Dalai, A.K. and Naik, S.N. (2006) *European Journal of Lipid Science and Technology*, **108**, 389.
16 Konwer, D. and Taylor, S.E. (1989) *Journal of the American Oil Chemists Society*, **66**, 223.
17 Nabi, M.N., Akter, M.S. and Shahadat, M.M.Z. (2006) *Bioresource Technology*, **97**, 372.
18 Ghadge, S.V. and Raheman, H. (2005) *Biomass and Bioenergy*, **28**, 601.
19 Giannelos, P.N., Zannikos, F., Stournas, S., Lois, E. and Anastopoulos, G. (2002) *Industrial Crops and Products*, **16**, 1.
20 Usta, N. (2005) *Biomass and Bioenergy*, **28**, 77.
21 Veljkovic, V.B., Lakicevic, S.H., Stamenkovic, O.S., Todorovic, Z.B. and Lazic, M.l. (2006) *Fuel*, **85**, 2671.
22 Pasias, S., Barakos, N., Alexopoulos, C. and Papayannakos, N. (2006) *Chemical Engineering and Technology*, **29**, 1365.
23 Banavali, R.M. and Benderly, A. (2008) European Patent Application EP 1878716.
24 Mbaraka, I.K. and Shanks, B.H. (2005) *Journal of Catalysis*, **229**, 365.
25 Toda, M., Takagaki, A., Okamura, M., Kondo, J.N., Hayashi, S., Domen, K. and Hara, M. (2005) *Nature*, **438**, 178.
26 Zong, M.H., Duan, Z.Q., Lou, W.-Y., Smith, T.J. and Wu, H. (2007) *Green Chemistry*, **9**, 434.
27 Nebel, B.A., Auvinen, J. and Mittelbach, M. (2007) Book of Abstracts, International Congress on Biodiesel, Vienna, 5–7 November 2007, p. 42.
28 Omota, F., Dimian, A.C. and Bliek, A. (2003) *Chemical Engineering Science*, **58**, 3175.
29 Yadav, G.D. and Murkute, A.D. (2004) *Journal of Catalysis*, **224**, 218.
30 Furuta, S., Matsuhashi, H. and Arata, K. (2004) *Applied Catalysis A–General*, **269**, 187.
31 Furuta, S., Matsuhashi, H. and Arata, K. (2004) *Catalysis Communications*, **5**, 721.
32 Furuta, S., Matsuhashi, H. and Arata, K. (2006) *Biomass Bioenergy*, **30**, 870.
33 Hillion, G. and Le Pennec, D. (2004) European Patent Application, EP 1460124.
34 Hillion, G., Leporq, S., Le Pennec, D. and Delfort, B. (2004) US Patent Application, 2004234448 A1.
35 Cozzolino, M., Di Serio, M., Tesser, R. and Santacesaria, E. (2007) *Applied Catalysis A–General*, **325**, 256.
36 Oku, T., Nonoguchi, M. and Moriguchi, T. (2005) PCT International Patent Application WO 2005/021697.
37 Kulkarni, M., Gopinath, R., Meher, L.C. and Dalai, A.K. (2006) *Green Chemistry*, **8**, 1056.
38 Sreeprasanth, P.S., Srivastava, R., Srinavas, D. and Ratnasamy, P. (2006) *Applied Catalysis A–General*, **314**, 148.
39 Srinavas, D., Srivastava, R. and Ratnasamy, P. (2007) PCT International Patent Application WO 2007/043063.
40 Frankel, E.N. (ed.) (2005) *Lipid Oxidation*, 2nd edn, The Oily Press, Bridgewater.
41 Knothe, G. (2008) *Energy and Fuels*, **22**, 1358.
42 Noweck, K. and Ridder, H. (1987) *Ullmann's Encyclopedia of Industrial Chemistry*, Vol. A10 (ed. W. Gerhartz), 5th edn, VCH Verlag GmbH, Weinheim, p. 277.
43 Johansson, L.E. (1980) *Journal of the American Oil Chemists Society*, **57**, 16.
44 Koritala, S. and Dutton, H.J. (1969) *Journal of the American Oil Chemists Society*, **46**, 265.
45 Ravasio, N., Zaccheria, F., Gargano, M., Recchia, S., Fusi, A., Poli, N. and Psaro, R. (2002) *Applied Catalysis A–General*, **233**, 1.
46 Liu, D.D.S., Monnier, J. and Tourigny, G. (1998) *Petroleum Science and Technology*, **16**, 597.
47 http://www.arbokem.com/nat_r/cetane_2.html.
48 Chatterjee, S.G., Omori, S., Marda, S. and Shastri, S. (2007) US Patent Application 2007130820.

49 Logan, M.J., Richard, P.R. and Dick, D.G. (2008) US Patent Application 20080015373.
50 Bondioli, P.F., Ravasio, M.N. and Zaccheria, F. (2007) PCT International Patent Application WO 2006/111997 A1; European Patent Application 06745284.7.
51 Schober, S. and Mittelbach, M. (2004) *European Journal of Lipid Science and Technology*, **106**, 382.
52 Mittelbach, M. (1996) *Bioresource Technology*, **56**, 7.
53 Mittelbach, M. and Enzelsberger, H. (1999) *Journal of the American Oil Chemists Society*, **78**, 545.
54 Boccuzzi, F., Chiorino, A., Gargano, M. and Ravasio, N. (1997) *Journal of Catalysis*, **165**, 140.
55 Biorenewable Resources No.1, supplement to INFORM; International News on Fats Oils and Related Materials (2006) 7.
56 Bournay, L., Casanave, D., Delfort, B., Hillion, G. and Chodorge, J.A. (2005) *Catalysis Today*, **106**, 190–192.

11
Photovoltaics – Current Trends and Vision for the Future
Francesca Ferrazza

11.1
Introduction

The direct conversion of sunlight into electricity is a very elegant process for generating renewable (and sustainable) energy from the most abundant and promising source of energy – the Sun. This process, usually referred to as photovoltaic (PV) solar energy, is inherently modular and quiet and has a huge potential. Therefore, it has a broad range of applications and can contribute substantially to our future energy needs. Although the basic principles of PV were discovered in the nineteenth century, it took until the 1950s and 1960s before solar cells found practical use as electricity generators. This was mainly triggered by the early development of silicon semiconductor technology for electronic applications. Today, a range of PV conversion technologies are commercially available and under development on a laboratory scale.

Complete PV systems consist of modules (also referred to as panels), which contain solar cells and the so-called balance of system (BoS). The BoS mainly comprises electronic components, cabling, support structures and, if applicable, electricity storage or optics and Sun trackers (the latter for concentrator systems). The BoS costs also include labor costs for turn-key installation.

Although reliable PV systems are commercially available and widely applied, ambitious further development of PV technology is crucial to enable PV to become a major source of sustainable energy and to strengthen the position of the European PV industry sector. In fact, despite a decade of unprecedented growth at approximately 40% per year, photovoltaics represents only a marginal share of the world's energy mix. The current price level of PV systems, which has dropped substantially compared with its levels of only 10 years ago, allows solar electricity to compete with peak power in grid-connected applications and with alternatives such as diesel generators in stand-alone applications, but it does not yet allow direct competition with consumer or wholesale electricity prices. A drastic further reduction of turn-key system prices is therefore needed – and fortunately possible.

Further development is also required to enable the European PV industry to maintain and strengthen its position in the global market, which is highly competitive

Catalysis for Sustainable Energy Production. Edited by P. Barbaro and C. Bianchini
Copyright © 2009 WILEY-VCH Verlag GmbH & Co. KGaA, Weinheim
ISBN: 978-3-527-32095-0

and characterized by rapid innovation. Research and technology innovation in all steps of the manufacturing chain is crucial for the further development of PV.

11.2
Market: Present Situation and Challenges Ahead

The photovoltaic market has increased by an order of magnitude in the last decade and at least by 40% per year in the last 5 years. It has now reached a yearly production of about 2.5 GW and an installed capacity of about 6 GW world-wide, driven mainly by two large subsidized markets – Germany and Japan [1]. Other countries are now following, such as Spain, Italy and the USA (particularly California), and are expected to make larger and growing contributions in the coming years. At the same time, manufacturing costs and selling prices have decreased at about 5% per year and solar cell efficiency has greatly improved, with the best performers now producing full-size cells – not laboratory samples! – well into the 20% range. Other important steps ahead include a drastic reduction in the energy payback time. Once considered (wrongly) as almost infinite, it has been clearly shown in different studies, such as that published by IEA-PVPS [2], that complete systems installed in southern Europe have paid their energy duty back after less than 2 years. All this did not happen by chance: it is the result of the combination of market-assisting measures, research, development and demonstration activities, with both private and public support (Figure 11.1).

Wafer-based crystalline silicon has been the dominant technology since the birth of photovoltaics. It is abundant, reliable and scientifically well understood, as it has enjoyed the knowledge and technology originally developed for the microelectronics

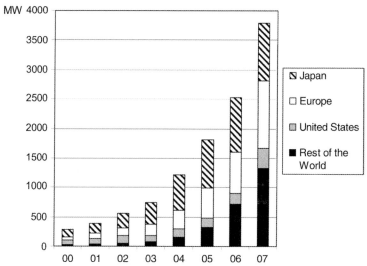

Figure 11.1 PV module production World-wide by region. Note totals may vary by 30% according to different soucrces.

industry. Progress in silicon wafer-based technology with time has determined the price-learning curve of PV modules – a decrease of about 20% for each doubling of capacity – driven by market size and technology improvement. Silicon wafer-based technology has represented 85–95% of the global PV market production in the last decade, with a growing trend until very resently. Between 2000 and 2006, the share was over 90%, broken into more than 50% multicrystalline silicon, around 35% single-crystalline and less than 5% ribbon technology. In 2007 however thin films gained over 10% of the market and are expected to grow even more as new capacity is added.

Amongst the largest areas of improvement, the thickness of silicon wafers has decreased from 400 µm in 1990 to 240 µm in 2005 and around 200 µm nowadays, cell surface area has increased from 100 to 240 cm^2 and modules have improved in efficiency from about 10% in 1990 to an average of about 13–15% today, with the best performers up to above 17%. Further, manufacturing facilities have increased from the typical 1–5 MW per year size of 1990 to several hundred megawatts per year today, with the expectation of gigawatt-sized factories to be seen soon.

In the meanwhile, other technologies are gaining larger volumes of the market – although not yet larger shares – specifically thin films and ribbon technologies, taking advantage of the growing demand for PV products and of a shortage of raw silicon, which has been creating some tension in the availability of silicon feedstock for wafer manufacturing. Many initiatives are under way with thin films – and other technologies too, such as concentrators and organic solar cells – as witnessed by the interest in equipment vendors to supply turn-key plants. It is possible, therefore, to say that besides market assistance measures such as the feed-in tariffs which have allowed demand to grow, the resulting silicon supply bottleneck has induced different technologies to begin industrial scale up.

The challenge for the whole PV sector is to reduce manufacturing costs and increase volumes in order to provide electricity at prices comparable to conventional sources (and therefore being capable of getting rid of incentives) and to represent a much larger share in the energy mix compared with the negligible numbers of today (well below 0.1% of the electricity supply world-wide).

The challenge has been represented in a price roadmap, developed by the European PV Technology Platform and shown in Figure 11.2 [3].

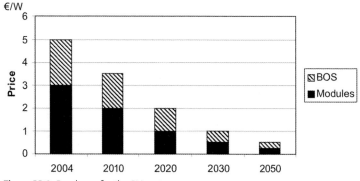

Figure 11.2 Roadmap for the PV sector.

The roadmap has been intentionally represented in terms of prices, because that is what is perceived by the public. This obviously means the expected costs are much lower. If this is the case, the period 2010–20 is when PV electricity will be competitive with retail prices from the grid (starting from the sunniest regions) and peak load generation. Later, the electricity can be competitive at the wholesale level, that is, with all conventional electricity production today.

From what we know today, crystalline silicon-based technology has the capability to continue to follow the established price experience curve, with direct production costs expected to achieve significant reduction to around €1.00 W^{-1} in 2013 and €0.75 W^{-1} in 2020 and even lower in the long term. This is true for other technologies also, which, however, are even less developed and in the case of thin films need to overcome limitations such as low efficiency (<10%), stability, the use of toxic or rare materials, complexity in some cases and the need for expensive equipment. If this can be done, thin films are well placed, for instance, to take advantage of low material usage implying potential low costs, manufacturing sequences with cell and module at the same time, high industrial potential and good esthetic aspects.

This will happen if R&D effort is directed to address the most critical issues and the technology areas most likely to allow continued progress of PV towards full sustainability.

11.3
Crystalline Silicon Technology

Crystalline silicon modules are typically produced in a complex and articulated manufacturing chain [4], all of which have greatly improved over the years – and which still have margins for further improvements.

11.3.1
From Feedstock to Wafers

In the first step, ingots are grown using pure silicon as the starting material for a melting and crystallization process which produces crystals of high purity and variable degrees of crystal perfection. The highest is monocrystalline silicon, typically grown by the Czochralski method, in which a preformed silicon seed is slowly pulled out of molten pure silicon, allowing a cylindrical silicon crystal to grow. The lowest is multicrystalline silicon, produced in square-based ingots with various similar methods by melting and solidifying the silicon in the same or in different containers (single container or casting), where solidification is achieved by careful control of the heat extraction. A third category is represented by thin ribbons or sheets, which already have approximately the desired thickness; different technologies for the realization of ribbons exist, which in turn have different degrees of crystal perfection and silicon purity. Ingot dimensions are typically 100 kg for monocrystals and 250–400 kg for multicrystals. Ribbons are generally in long sheets or hollow tubes. The size of ingots today is an improvement compared with the standard in industry

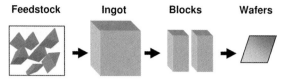

Figure 11.3 Process flow from silicon feedstock to wafers for the case of multicrystalline silicon.

of only 10 years ago, when crystals were typically about 30 and 150 kg for mono- and multicrystalline silicon, respectively. Improvements have been made from the point of view of equipment, more automated and better engineered for energy consumption and maintenance costs and therefore from the point of view of productivity. Quality has also improved as a result of improved homogeneity in larger ingots. In the case of multicrystalline ingots and ribbon technology, dimensions have also been optimized to produce several wafer sizes with acceptable kerf loss (Figure 11.3).

The starting material, pure polysilicon, is produced by a complex sequence of energy-consuming and waste-producing processes in large and expensive plants. Polysilicon is produced using part of the metallurgical silicon resulting from the carbothermic reduction of quartz sand in arc furnaces at very high temperatures (2000 °C) and making it react with HCl to produce silicon precursors such as trichlorosilane ($SiHCl_3$), which are then distilled to achieve the required purity. The precursors are then decomposed at high temperature in special reactors, where pure Si finally deposits on special rods (Siemens process, the most common) or on fine Si particles in a fluidized bed configuration.

After solidification, the ingots are shaped using diamond-coated band saws or wires in an abrasive medium in order to have square multicrystalline or pseudo-square rounded angle monocrystalline blocks. This process removes the outer parts of the ingots, generally out of specification for dimensions and with some contamination from the containers or the furnaces.

The square or pseudo-square blocks are cut into thin wafers using multi-wire saws, a technique which was developed specifically for PV and which is now common in the semiconductor industry also. A great deal of progress has been made with equipment and process control aspects and in introducing reduced diameter wires several hundred kilometers long for reduced kerf loss.

This step is evidently not required for ribbons, apart from edge trimming and cutting the wafers from the sheets. This is considered to be a major advantage over ingot technology in terms of costs.

11.3.2
From Wafers to Cells and Modules

The wafers are processed into solar cells, the majority of which have a diode structure, as sketched in Figure 11.4, characterized by a thin, diffused, doped emitter, screen-printed front and back contacts and a front-surface antireflective coating. Prior to the effective cell manufacturing step, a chemical treatment of the silicon wafers removes

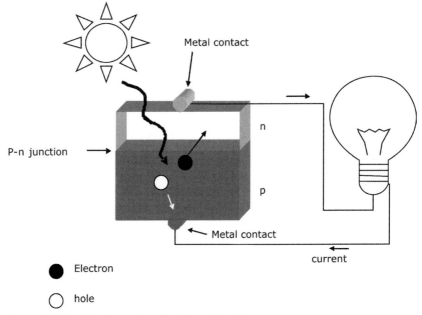

Figure 11.4 Principle of operation of a silicon solar cell.

dirt and damage from the surface coming from the wafering step and provides surface structuring which reduces reflectance losses.

The cell manufacturing is rather simple in itself. The major efforts devoted to improvement over the years have been on the one hand the introduction of the silicon nitride antireflective coating, capable also of passivating defects in multicrystalline silicon, and the ability to process thinner, larger wafers at increasing yield. This means that the automation of the processing lines has become much less invasive on the brittle wafers and handling has become much less troublesome. Also, cell processing has had to deal with wafer bowing because of the different thermal expansion coefficients of silicon and aluminum (the back contact).

Cells are then individually sorted and classified according to their illuminated electrical parameters and are electrically interconnected in strings, which are then encapsulated to form a module, which is designed to be weather proof and produce electrical output for more than 25 years.

The obvious critical aspect of wafered silicon technology lies in the complexity of the manufacturing and supply chain, which involves rather different industry sectors – metallurgy, electronic components, building and power conditioning. Most skeptical observers also point out the inherent difficulty of reducing the costs in a cell and module manufacturing process which need to handle globally a number of wafers in the billion pieces range. However, this is exactly what has happened so far – and those working in the silicon sector are confident that it can continue happening.

11.3.3
Where to Cut Costs

Modules are currently sold at around €3 W^{-1} irrespective of the technology on which they are based, with a slight slowing of the price learning curve compared with historical data. The reason for this deviation is a silicon feedstock shortage that started a few years ago, which has produced a slight reduction in the wafered silicon share in 2005 compared with 2004 and has caused and increase in silicon feedstock prices to \$45 kg^{-1} – although a few hundred dollars per kilogram were reported for spot-market purchases. As the demand for modules has greatly exceeded the supply, the module selling price has grown or at least it has not reduced. This situation is expected to change in terms of availability in the last part of the decade, as new capacity for silicon feedstock will be effectively in place. It is more uncertain as to what the feedstock prices could be, given the relatively modest degree of innovation of the expected new plants in terms of cost-cutting measures.

The total silicon consumption for the PV industry eventually exceeded that of the electronics industry for the first time in 2006, with an annual level of over 20 000 t.

The continued success of silicon technology needs to keep progressing in terms of both a reduction of material consumption and an increase in device efficiency. Other goals relate to reduced energy content, environmental aspects, standards and manufacturing aspects such as process automation and control.

The availability of abundant, low-price silicon feedstock, in the €10–20 kg^{-1} range for the long term, is a very important point for the possibility of silicon technology progressing as expected in terms of cost reduction.

The development of new, less energy-intensive techniques for silicon feedstock preparation dedicated to the solar sector is being carried out in many different ways. Different processes have been studied and are currently under development, all of them with the aim of reducing costs and complexity while increasing availability. Some processes dealing with the direct purification of metallurgical silicon, with no need for precursor formation and dissociation, have the potential to reduce costs substantially and have not so far reached technical maturity, despite encouraging laboratory results and long-lasting research activities. Other approaches target cost reduction potential in the existing processes and are in part already being utilized for commercial 'solar-grade silicon', which is cheaper than the semiconductor-grade polysilicon but without fundamental changes. Fluidized bed reactors of various kinds are also under development for the decomposition of monosilane precursors and vapor to liquid approaches.

On top of the actions needed to control feedstock costs and availability (and hence prices), a further point for improving cost figures is to consider internal recycling processes, as 60% or more of the starting material is lost during processing. In fact, about 15% of the crystallized silicon ingots do not reach the wafering stage, depending on the ingot technology considered. Part of it can be recycled, at some expense, and part is lost as silicon powder or unusable scrap. A great deal of work is going into this part of the process, because of the improvement margins for better silicon utilization through recycling and for reduced rejection.

A further roughly 40% of the material is lost as dust and other yield losses in the wafering process, which is obviously a very delicate point from the point of view of the processing, equipment and recycling possibilities. Besides improvements in current wafer technology with more automated, more controlled equipment and thinner and thinner wafers, a certain focus is being put into alternative wafering technologies, such as laser-assisted cutting. This issue does not apply to ribbon technology, which is in fact seen as one of the potential evolution areas of wafer-based silicon technology, especially for the long and medium term. Also, wafer substitutes or equivalents on low-cost substrates are seen as a possibility to overcome ingot/wafer manufacturing limitations, especially given the increasing challenges to be faced for wafer thickness below 150 µm, in terms of manufacturability and optimal cell processing.

Finally, about 10% is lost as broken wafers between wafer handling in the wafer/ribbon manufacturing stage and the subsequent cell process. Cell manufacturing needs to improve automation to reduce breakage in absolute numbers and in order to allow the introduction of even thinner wafers, necessary to achieve the cost reduction targets, with the achievement of manufacturing (mechanical) yield losses well below 5%. Recycling is already being used, with the effect of alleviating the very heavy losses indicated above.

The whole silicon production chain is expected, by virtue of R&D actions taken at the manufacturing stage, to move from the present $10\,t\,MW^{-1}$ consumption figure to below $3\,t\,MW^{-1}$ in the long term. It is clear, however, that although the short- and probably medium-term targets can be achieved by lowering the specific consumption even at moderately high feedstock prices, to fulfill the cost targets for the long term both feedstock price and silicon consumption will have to decrease substantially.

Another important aspect is cell (and module) efficiency. A great deal of progress is being made in this area. Just as an example, the laboratory world record on $1\,cm^2$, 24.7%, is now not so far away commercially, considering the best commercial cells are in the 22–23% range, on a surface of $2250\,cm^2$! Laboratory cells have traditionally been realized on small areas, in cleanroom facilities and using vacuum technologies for the deposition of metal contacts.

Most of the development work in recent years on commercially available devices, based on screen printing technology for contact formation, has been focused on improving the manufacturing sequences with higher performing equipment, by introducing an Al back layer which provides the electrical contact and some carrier confinement and, in the case of multicrystalline silicon, by introducing hydrogen passivation in the silicon nitride anti-reflective coating (ARC) formation step. This last step alone, which has allowed the transition from about 13% to >15% efficient cells t, had been demonstrated at the laboratory level long ago, but was only introduced into most manufacturing facilities when reliable processing plasma-enhanced chemical vapor deposition (PECVD) equipment became available, shortly after 2000.

Current state-of-the-art manufacturing processes are realized in non-cleanroom classified areas. Automation has been introduced in most manufacturing lines, but substantial progress is needed in view of the further decrease in wafer thickness.

Only three high-efficiency processes have so far been scaled up to production level, namely the laser grooved buried grid developed at the University of New South Wales,

Australia, and scaled up by BP Solar in Spain, the heterojunction with intrinsic thin layer (HIT) cells developed by Sanyo by replacing the diffused P-doped emitter with an amorphous silicon layer and the back contact cells developed by Stanford University for use in concentrator technology and now converted to a large area for flat plate use. All three use single-crystalline silicon, while the majority of screen-printed cells use multicrystalline silicon wafers.

Commercial module efficiency values (total outer dimensions) are in the 12–14% range for the screen-printed modules and 15–17.5% range for the best performers. It is expected that more device designs capable of achieving efficiency values in the 18+ % range will be transferred to the production scale. Promising candidates for such developments are all-rear contacted cells of different kinds, on both single- and multicrystalline substrates, which have the added advantage of reducing cell interconnection complexity in automated sequences. Different contacting schemes will also most likely be introduced, which will influence the cell architecture substantially. As wafers decrease in thickness, in fact, the present full aluminum rear contact of the cell will have to be replaced by local contacts, both for bow prevention and for reduced recombination at the metal–silicon interface. An example of different cell structures including enhanced passivation is the crystalline silicon–amorphous silicon (c-Si–a-Si) heterostructure cells, which are already at the commercial level (Sanyo HIT modules) with best cell efficiency exceeding 22%. Examples of novel promising structures include laser-fired contact cells. Also, more effective means of contact passivation will be introduced and material quality will be enhanced during processing for lower starting quality materials and kept as high as possible for high-quality materials. As the market size grows and the consumption of all materials becomes relevant, it will also be necessary to reduce or avoid totally the use of silver pastes, now consumed at an average of 80–90 kg MW^{-1} or some 130 t yr^{-1}.

In the long term, it is expected that silicon technology will still play an important role in the PV sector, although a higher degree of uncertainty exists in terms of all the numerical parameters identified, and, more importantly, in the cell and module architecture and component materials after 2020, when the market size is expected to be in the 30 GW yr^{-1} range. It is likely that silicon technology by this time will have incorporated aspects which are now related to novel or emerging technologies and that new materials will also be included in the processing sequences. Also, the distinction between cells and modules may no longer hold and the wafer or active layer may be so thin that true distinctions between wafer and thin-film technologies may no longer be appropriate. It is, in fact, clearly expected that module efficiency will be able to rise higher than the current laboratory record. This may only be possible by incorporating technologies at the periphery of the device such as up or down converters.

11.4
Thin Films

Thin films are suitable candidates for low-cost photovoltaic modules because of reduced material consumption and predicted manufacturing advantages due to a shorter supply chain. There are three principle inorganic thin-film technologies, amorphous silicon (in various configurations, including amorphous microcrystalline heterostructures), cadmium telluride and copper indium selenide (CIS) or copper indium gallium selenide (CIGS), all of which have demonstrated large area, reasonably efficient solar modules and which have a number of general principles in common. In each of the thin-film technologies, only very small amounts of semiconductor materials (typically 0.001 mm thick) are used and the materials used for encapsulation, such as glass and plastic, are relatively inexpensive. The availability of large-area deposition equipment developed in other industry sectors (such as the flat plate display) and related process technology and also strong synergies with the architectural glass industry and the flat panel display industry offer significant opportunities for high-volume, low-cost manufacturing. The monolithic series connection used for thin-film PV simplifies module assembly compared with series connection by tabbing and stringing commonly used with crystalline silicon solar cells. In addition, flexible lightweight modules can be produced using thin polymer or metal substrates and roll coating techniques. Currently, the energy payback time of thin-film modules is short, around 1.5 years in Central Europe, with a long term potential of below 6 months.

These features imply that the cost reduction potential for thin-film technology is very high and it is thus capable, in the longer term, of extending the PV learning curve beyond the point that can be reached by crystalline silicon technology.

At present, three competitive inorganic thin-film technologies, based on thin-film amorphous/microcrystalline silicon (TFSi), polycrystalline CdTe and $Cu(In,Ga)(S, Se)_2$, also known as CIS, are under widespread investigation. The high efficiency potential of each technology has already been proven in the laboratory. Each of these technologies has demonstrated its capabilities in pilot production lines and they are now being or has been transferred to large-area, high-volume production lines. This transfer is not straightforward since there is a lack of maturity of many of the required large-area production processes and the related production equipment.

The thin-film PV industry is taking off and the challenge is now to scale up fast enough and to establish a significant presence in the PV marketplace.

11.4.1
Technology and Improvement Requirements

Thin-film modules are generally produced by applying a thin (1–2 μm) layer of active semiconductor material directly on a glass substrate or superstrate and applying thin metal contacting layers in general directly on the glass – but also on patterns on the film itself. The deposition techniques for all materials are in general under vacuum by chemical or physical vapor deposition methods such as sputtering, evaporation, PECVD and many other variations. The beauty of such a technology is, in addition to a very low material content and therefore an inherent possible low cost, the possibility of simultaneously fabricating cells and modules, by depositing the film and patterning it via a laser. Handling therefore becomes the transport of large glass sheets – much less cumbersome than individual brittle wafers. However, the true advantage of thin films has not yet shown itself at the commercial scale, at least not so far, despite many decades of intense research work. The efficiency of thin-film modules, to start with, is nowhere close to that of crystalline silicon modules. At the commercial scale, on large areas, the average module efficiency is below 10% and manufacturing yields not up to the 90% + level of silicon wafer-based technology. Things may change rapidly in the future, however.

It is clear that production equipment plays a crucial role in the reduction of costs. Standard equipment with well-defined processes needs to be developed enabling higher throughput, improved up-times and improved yields. Equipment manufacturers will play a vital role in this development and it is important to exploit synergies with existing industries where possible. Productivity parameters such as process yield, throughput and availability of production equipment can be improved by the introduction of adapted process control and process optimization. To improve both production yield and efficiency and to ensure the quality of the final products, improved quality assurance procedures and in-line production monitoring techniques need to be further developed. Whenever possible, the integration of subsequent production and processing steps into one line and similar environmental conditions should allow for more efficient and cheaper production. A recent feature of the thin-film sector is the direct interest of some equipment vendors in stepping into the PV arena and selling turn-key plants – tools and process together. This is expected to have an impact.

The other main areas of development required to overcome present limitations are in the improvement of efficiency and reliability. At the basic research level, this implies a better understanding of the relation between the deposition processes, the electronic material properties and the resulting device properties and the improvement in the quality of all individual layers building up the module.

Finally, as with any new product, dedicated recycling processes need to be developed, both for recycling during production and for end-of-life recycling. Although the energy payback time is already favorable, a further reduction to less than 6 months is possible.

11.5
Other Technology-related Aspects

Cell and module efficiency contribute directly to the overall euros per watt cost (and price) of a PV module and are historically an area of great development. Increasing the efficiency of the solar cells and the power density of the modules is, together with a reduction in the specific consumption of silicon, the major cost-saving factor of PV technology. An increase of 1% in efficiency alone, keeping the other cost components fixed, is able to reduce overall costs per watt by about 5% (compound growth rate), as shown in the simulation in Figure 11.5, in which direct costs have been kept constant, starting from a €2 W^{-1} base case at 15% efficiency.

The consumption of material in general, which directly influences costs but also availability and environmental issues, needs to decrease substantially. In silicon

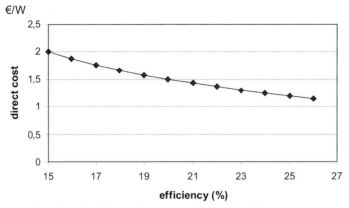

Figure 11.5 Sensitivity of cost per watt as a function of efficiency.

wafer or ribbon manufacturing, it is expected that recycling of silicon scrap and dust will be maximized, together with slurry recovery and reuse of silicon powder from the cutting phases. Further, all chemicals for cleaning and etching will be reduced in terms of specific consumption and eventually eliminated, and it is expected that equipment will reduce energy consumption. In the crystallization process, reusable crucibles may be an option to reduce a cost which will have an increasing weight in euros per kilogram, considering the single-use characteristics of present crucibles.

The energy intensity of the processes contributes to the energy payback time of modules. Current silicon feedstock production is energy intensive, around $150\,kW\,h\,kg^{-1}$. Together with the low silicon utilization, this leads to a module energy payback time between 1.5 and 4 years, which, although much shorter than the module lifetime, needs to decrease.

In the module itself, the aluminum frame represents the largest energy-containing component, €$30\,kg^{-1}$, so long-term module development will need to be frameless or with a non-aluminum frame.

Investment costs represent a non-negligible part of the cost breakdown of crystalline modules and should be reduced while equipment improves in terms of standardization, energy consumption and optimized production volume. It is expected that specific investment costs will reduce from the M€1–$2\,MW^{-1}$ of current manufacturing facilities to less than M€$0.5\,MW^{-1}$ in the long term, if the whole production chain from crystallization to module assembly is considered.

Scale factors are as important in achieving the required level of progress as all of the different steps discussed earlier. EU-funded studies such as MUSIC-FM and its recent recalculation [5] have shown the feasibility of large-scale plants of up to $500\,MW\,yr^{-1}$, already at a time when the total market was just over 100 MW globally, showing the associated volume benefits as well. It is expected that the current base case of a $100\,MW\,yr^{-1}$ plant will grow to $500\,MW\,yr^{-1}$ in the short term and probably an order of magnitude higher in the medium term. Equipment will then be completely different from the fragmented single-process tools of today, loosely connected by some custom-made automation. Instead, it is likely that each single line will have multiple processes integrated and that batch processes will tend to disappear. Module assembly, for instance, will turn into automated sequences in which the encapsulating materials/sheets will be fed in on reels and spools, as much as possible. Evidently, major efforts are needed to achieve such progress.

Another important aspect related to both process technology and equipment design is the need for yield management and process control techniques embedded in the tools, which will have to provide accurate and reliable means for avoiding yield losses in manufacturing and contributing to overall improvement. From this point of view, it is expected that standards in equipment and in characterization techniques will be developed or implemented if already available.

Despite the clear focus on cost reduction, due regard should also be given to product and process safety. For PV to become widely adopted at the Gigawatts (GWp) scale with a large production workforce and components distributed on millions of commercial and domestic roofs, then safe systems must be inherent in future products. Safety must start with the materials, equipment and the process of manufacture, and include the product safety for system installation and, most importantly, any long-term fire and health aspects of unattended PV systems.

11.6
Advanced and Emerging Technologies

In addition to silicon wafers and thin films, other photovoltaic technologies are gaining interest and importance such as concentrating PV technologies, in combination with very high-efficiency cells.

Concentrating photovoltaics (CPV) is not a new idea, as it has been under development since the beginning of research in the sector. However, recently a number of initiatives have moved towards commercial application. The progress in high-efficiency cells, with multijunction III–V (GaInP/GaInAs/Ge) material-based cells above 40% at the laboratory scale, makes the economics of concentrators much more attractive than in the past. Concentrator technology is in principle a simple option, because the expensive and complicated cell manufacturing only needs to address very small cells, the rest of the work being assigned to lenses which focus the sunlight on the cell. However, system design and integration are complicated because the Sun needs to be tracked and alignment aspects are crucial. System complication and reliability issues have so far hindered the widespread use of concentrating PV. However, this situation may change in the future, as more companies come to the market.

Organic solar cells have already been a subject of R&D efforts for a long time because of the potentially very low cost of the active layer material, the low-cost substrates, the ease of up-scaling and the low energy input. The breakthrough for solar cells incorporating an organic part in the active layer came with the advent of concepts which were radically different from the planar hetero- or homojunction solar cells. The generic idea behind these concepts is the existence of nanosized domains resulting in a bulk-distributed interface to increase the exciton dissociation rate and thereby the collection of photogenerated carriers. Within this class one can distinguish between the hybrid approach in which there is still an inorganic component (e.g. the Graetzel cell) and fully organic approaches (e.g. bulk donor–acceptor heterojunction solar cells). The main challenges in this field are related to increases in efficiency, stability improvement and the development of the roll-to-roll technology. Within this field, Europe has built up a world-leading position both in R&D and in first industrial up-scaling efforts. Moreover Europe is in an excellent position to remain at the cutting edge of these technologies because of its strong position in related fields such as organic electronics and organic memories.

Most of the other novel PV technologies suggested so far can be categorized as high-efficiency approaches, which can be divided between approaches that modify and tailor the properties of the active layer to match it better to the solar spectrum versus approaches that modify the incoming solar spectrum and are applied at the periphery of the active device (without fundamentally modifying the active layer properties).

In both cases, nanotechnology and nanomaterials are expected to provide the necessary toolbox to bring about these effects. Nanotechnology allows the introduction of features with reduced dimensionality (quantum wells, quantum wires, quantum dots) in the active layer. There are three basic ideas behind the use of structures with reduced dimensionality within the active layer of a photovoltaic device. The first aims at decoupling the basic relation between the output current and output voltage of the device. By introducing quantum wells or quantum dots consisting of a low-bandgap semiconductor within a host semiconductor with a wider bandgap, the current should increase while retaining (part of) the higher output voltage of the host semiconductor. A second approach aims at using the quantum confinement effect to obtain a material with a higher bandgap. The third approach aims at the collection of excited carriers before they thermalize to the bottom of the energy band concerned. The reduced dimensionality of the quantum dot material tends to reduce the allowable phonon modes by which this thermalization process takes place and increases the probability of harvesting the full energy of the excited carrier. Several groups in Europe have built up a strong position in the growth, characterization and application of these nanostructures in various structures (III–V, Si, Ge), and also at the conceptual level ground-breaking R&D is being performed (e.g. the metallic intermediate band solar cell).

Tailoring the incoming solar spectrum to the active semiconductor layer relies on up- and down-conversion layers and plasmonic effects. Again, nanotechnology might play an important role in the achievement of the required spectral modification. Surface plasmons have been proposed as a means to increase the photoconversion efficiency in solar cells by shifting energy in the incoming spectrum towards the wavelength region where the collection efficiency is maximum or by increasing the absorbance by enhancing the local field intensity. This application of such effects in photovoltaics is definitely still at a very early stage, but the fact that these effects can be tailored to shift the limits of existing solar cell technologies by merely introducing modifications outside the active layer represents an appreciable asset of these approaches which would reduce their time-to-market considerably.

It is evident that both modifications to the active layer and application of the peripheral structures could be combined eventually to obtain the highest beneficial effects.

11.7
System Aspects

Photovoltaic systems can be implemented in a wide range of applications, sizes and situations and to meet a wide range of power needs.

At the system level, requirements are for a reliable, cost-effective and attractive solution to energy supply needs.. This means pursuing solutions that can reduce costs at the component and/or system level, increase the overall performance of the system and improve the functionality of and the services provided by the system.

Traditionally, PV systems are divided into two major categories depending on whether they are connected to the electricity grid system or are operated in an off-grid configuration. In turn, the grid-connected systems can be divided into central systems, which feed all the electricity generated into the grid, and dispersed systems, where the electricity goes to meet local loads first with only the excess being fed into the grid. It should be noted, however, that whereas most large ground-based systems fall into the first category, building-related systems of all sizes can be operated in either mode, depending on the financial arrangements. The off-grid, or stand-alone, systems can also be divided into professional applications (e.g. telecommunications, remote sensing) and rural development applications (e.g. irrigation, lighting, electrification of health centers and schools). Consumer products are a special category in which the PV cells are integrated into a product to provide the required power supply. While this has been an important market for PV, especially in terms of public awareness, the developments in this area are driven by commercial needs for new product development. These systems are not considered in this strategic research agenda, which looks at the research needs for widespread use of PV as a general energy supply technology.

Because of the wide variations in system applications, it is not possible to give definitive values for system costs, but indicative values can be provided as examples. The module has traditionally been the most costly component in the system, typically accounting for 50–70% of the costs at the system level. However, this varies considerably with application and system size, since the relative impact of the BoS (power conditioning, mounting structures, cables, switches, etc.) and installation costs will vary substantially. In order to meet the cost targets required for a high penetration of PV technology into the energy supply market in the period 2020–30, substantial and consistent system-level cost reductions must be made alongside those for the PV module. The system-level costs can be broadly divided into those for BoS components (whether part of the energy generation and storage system or components used for control and monitoring) and installation (including labor). Although it can be generally stated that there is scope for cost reduction at the component level, it is also of major importance to address installation issues by harmonizing, simplifying and integrating components to reduce the site-specific overheads.

It is usual to express cost targets for PV systems in terms of Watt peak (Wp) or kWp levels. Ultimately, at the system level, the cost comparison must be made between the unit cost of the electricity generated from the PV system and from the alternative energy source(s) for that application. Clearly, both of these values will vary with application and system details and a full treatment of the relationship between the kW h cost and the Wp cost would not be feasible here. The target price for the non-module aspects of a typical PV system has been set in EU-related work [3] at €1 Wp^{-1} for 2010 and <€0.5 Wp^{-1} for 2030 and beyond. Taking a reasonable profit margin of 20%, this results in costs of 0.8 and <€0.4 Wp^{-1}, respectively (all excluding VAT). These numbers have to be interpreted with the utmost care since they may vary substantially with system type (roof top, building integrated or ground based), with module efficiency and with the country involved. Taking into account the targets for the module cost, this implies that the kW h cost of PV-generated electricity will be comparable to the consumer tariff in most European countries in the period 2015–25, depending on the local irradiation, so making PV a cost-effective alternative. These cost targets for grid-connected systems provide a useful reference and have been retained in determining this research agenda, but it should also be recognized that the condition of cost-effectiveness will be met at other Wp costs for different applications.

Consideration of typical BoS costs for current systems illustrates the challenge facing the research in this area. Studies in Germany, The Netherlands and the UK indicate BoS prices (including components and installation) of €1.6–2.5 Wp^{-1} for building mounted systems on domestic properties. Lower costs can be obtained for large ground-mounted systems, where the effects of component standardization can be seen in the reduction in costs for both mounting systems and labor. Recent (2006) information from Germany indicates that turn-key systems have been realized at low BoS prices between 0.85 and €1.2 Wp^{-1} (for large ground-based and small roof-top systems, respectively).

The BoS costs are made up of both power-related and area-related costs (e.g. the cost of the mounting structure is dependent on the area of the array). In the latter case, these costs are highly dependent on the efficiency of the module used and so the cost goals are more challenging for lower efficiency modules. Currently, area-related costs are significant and range from 0.6 (in the case of domestic PV systems with 12% efficient modules mounted on a frame) to 1.5–2 (in the case of ground-based large-area PV systems with 12% efficient modules) times the power-related costs. Increasing module efficiency can, therefore, help significantly in achieving the system-related goals.

The relationship between kW h cost and Wp cost at the system level is a function not only of the initial capital cost of components and installation, but also of the lifetime of all components, the sustained performance of the system over its lifetime and of any aspects of multifunctionality or added value that are realized. Moreover, it is dependent on the energy produced by the modules per Wp of installed module power (this is related to the module behavior under non-standard conditions, such as higher temperatures, lower light intensities, low angles of light incidence, spectral variations, etc.).

11.8
Conclusions

PV technology is improving in performance and reducing its costs. In turn, together with market-assisted measures, this allows for strong market growth. However, PV electricity costs need to fall even more for the technology to be deployed at the very large scale – which is not yet the case.

The main technologies in the field have been revised and the main issues for each have been addressed.

From the point of view of market leader, silicon wafer technology, technical improvements identified in the previous sections, in connection with appropriate factory scale-up and integration, are capable of reducing direct costs as much as shown in Figure 11.5. Wafer costs are expected to decrease by up to 50% in the short term and module costs to become a fraction of the entire cost in the long term.

However, it must be noted that this is a very aggressive forecast, which will be possible only if a large effort is put into the activities described previously to reach all the technical goals and if the cost of starting material can be brought down to the €10–20 kg^{-1} levels (Figure 11.6).

From the point of view of thin films, efficiency improvements and reliability issues need to be overcome for technology to be able to follow the same curve (with a different breakdown of components, however).

Novel and emerging technologies have the long-term potential to provide high-efficiency, low-cost and highly scalable products, but need substantial research effort to achieve important results.

The long-term solution, by 2030 and beyond, may be a combination of existing and novel technologies, such as spectrum shifting technologies in combination with advanced processing of very thin silicon wafers. In any event, intense R&D efforts are needed to provide the solutions for photovoltaic electricity to be an important actor in the future.

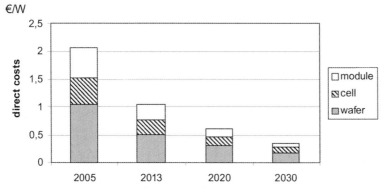

Figure 11.6 Possible evolution of direct costs for PV module components.

References

1 Photon International magazine (2007) www.photon-magazine.de.
2 IEA-PVPS T10 – 01:2006, www.eupvplatform.org.
3 Strategic Research Agenda, www.eupvplatform.org.
4 (2003) *Practical Handbook of Photovoltaics: Fundamentals and Applications*, T. Markvart and L. Castaner (eds.) Elsevier, Amsterdam.
5 www.c.europa.eu/research/energy/pdf/european_photovoltaics_en.pdf.

12
Catalytic Combustion for the Production of Energy

Gianpiero Groppi, Cinzia Cristiani, Alessandra Beretta, and Pio Forzatti

12.1
Introduction

In recent decades, catalytic combustion has been explored as a primary control method for the production of heat and energy with two main goals: (i) to achieve ultra-low emissions of NO_x, CO and unburned hydrocarbons (UHCs) and (ii) to obtain stable combustion under conditions not allowed by conventional methods.

In particular, catalytic combustion in gas turbines (GTs) has attracted specific interest; GTs nowadays represent the preferred energy conversion technology in medium- and large-scale power stations and strong opportunities are predicted in the small-scale distributed power generation sector. NO_x emissions are a major concern of GT systems. Several NO_x control technologies for GTs have been developed based on both primary and secondary methods. Dry low NO_x (DLN) burners based on lean premixed combustion guarantee NO_x emission levels of 20–25 ppm, but operation at 9–10 ppm emission is claimed by manufacturers. Further reductions may be precluded by flame stability problems. To meet the most stringent emission regulations, many installations include a selective catalytic reduction (SCR) unit, which allows a further reduction in NO_x emission levels.

Catalytic combustion has been commercially demonstrated to reduce NO_x emissions to below 3 ppm while keeping CO and UHC emissions below 10 ppm without the need for expensive exhaust clean-up systems. In addition, a catalytic combustor reduces typical DLN problems such as risk of blow-out and flame instability. Also, the economic advantage of primary methods including catalytic combustion as opposed to secondary clean-up measures (SCR and SCONOx) has recently been assessed [1].

The potential of catalytic combustion has been recognized for more than 30 years, but only recently has this technology been proven to be commercially viable and finally commercialized, although to a limited level.

New designs have also been investigated in recent years. These include fuel-rich combustion, a staged process configuration, in which an initial catalytic section converts natural gas into an H_2/CO-enriched stream, and a final homogeneous stage completes the lean combustion. Such a configuration exploits the stabilizing effect of

---

Catalysis for Sustainable Energy Production. Edited by P. Barbaro and C. Bianchini
Copyright © 2009 WILEY-VCH Verlag GmbH & Co. KGaA, Weinheim
ISBN: 978-3-527-32095-0

H_2 on flames and has been proven, at the demonstration scale, to yield extremely low NO_x emissions. Another novel concept is represented by oxycombustion for zero emission GTs, in which pure O_2 is used instead of air and the combustion is highly diluted by flue gases (H_2O and CO_2). In principle, the formation of NO_x is prevented and concentrated streams of CO_2 are easily produced, suitable for C-sequestration techniques.

In the following, a review of the traditional and novel concepts of catalytic combustion for GTs is addressed, with emphasis on the requirements and challenges that the different applications open to catalysis. The most relevant characteristics of PdO-supported catalysts and of transition metal-substituted hexaaluminates (which have been most extensively considered for lean combustion applications) are described, along with those of noble metal catalysts adopted in rich combustion systems.

Further, the development of miniaturized devices for the generation of power and/or heat is discussed here as it represents an emerging field of application of catalytic combustion. Due to the presence of the catalytic phase, the microcombustors have the potential to operate at significantly lower temperatures and higher surface-to-volume ratios than non-catalytic microcombustors. This makes them a viable solution for the development of miniaturized power devices as an alternative to batteries.

12.2
Lean Catalytic Combustion for Gas Turbines

12.2.1
Principles and System Requirements

In a conventional GT system, a fraction of the air delivered by the compressor and the fuel, typically natural gas, are mixed, then combusted in a flame and the hot gas expands and drives the turbine.

Flame stability requires adiabatic combustion temperatures as high as 1600–1800 °C, which must be reduced to 1100–1450 °C by means of cooling by-pass air before delivering the hot compressed gas to the turbine to avoid damaging the inlet blades. At such temperatures, within the tens milliseconds residence time required for complete burnout of fuel and CO, significant amounts of NO_x are produced, mostly by the Zeldovich thermal mechanism [2].

In a catalytic burner, the combustion is ignited and stabilized under ultra-lean conditions, which results in adiabatic temperatures close to those allowed for delivering the hot compressed gas to the turbine. Hence the need for by-pass air is minimized and the formation of thermal NO_x is almost prevented due to the absence of a hot combustion zone. Reduction of NO_x emission has been reported to be even larger than expected from the lower combustion temperature if a significant fraction of the fuel is oxidized on the catalyst surface [3]. This effect has been attributed either to the reduction in the formation of prompt NO_x in view of the

12.2 Lean Catalytic Combustion for Gas Turbines

Table 12.1 Design criteria and operating conditions of GT combustors.

Design criteria	
Emission targets	$NO_x < 5$ ppm
	$CO < 10$ ppm
	UHCs < 10 ppm
Pressure drops	$<5\%$
Catalyst durability	8000 h
Operating conditions	
Inlet temperature	300–450 °C
Outlet temperature	1100–1400 °C
Pressure	1–2 MPa
Mass flow rate	100–200 kg m^{-2} s^{-1}
Residence time	10–30 ms

decrease in CH radicals in the gas phase due to complete fuel oxidation on the catalyst surface [4] or to the reduction in the formation of NO_x due to release of H_2O produced at the catalyst surface. Indeed, a beneficial effect of an increase in H_2O concentration on the diminution of NO_x emission under typical GT operating conditions has been reported [5].

Operating constraints of GT systems (Table 12.1) pose severe requirements on the catalytic combustor. Air is delivered by the compressor at temperatures which typically range from 300 to 450 °C depending on the load conditions and nominal pressure ratio of the machine.

Upon accurate fuel–air mixing, ignition must readily occur and fuel conversion should proceed rapidly up to completion while the gas reactants heat up to the adiabatic combustion temperature. Due to high gas velocity (10–30 m s^{-1} at the combustor inlet) and size constraints of the combustion chamber, the process must be completed within a few tens of milliseconds and the overall pressure drop (including mixing) must be kept below 5% of the turbine inlet pressure to prevent significant energy efficiency losses.

These characteristics of GT operations result in the following requirements:

1. Highly active catalysts able to ignite the combustion of natural gas at temperatures as close as possible to those at the compressor discharge must be employed. Indeed, to fill the gap between the compressor discharge temperature and catalyst ignition temperature, a homogeneous preburner is needed which can produce significant amounts of NO_x.

2. Materials with high thermal stability able to hinder catalyst deactivation by sintering, phase transformation and volatilization and also to secure mechanical integrity upon thermal shocks. In fact, strong temperature excursions are experienced during start-ups and shut-downs and particularly during the load trip of the turbine. To prevent overspeeding and destruction of the turbine in this case, the fuel feed is immediately shut off while air continues to flow; this results in a temperature decrease of several hundred degrees in less than 1 s. Under typical

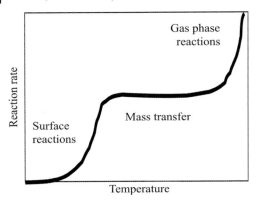

Figure 12.1 The kinetic regimes of a combustion process.

GT operation, a lifetime of the catalyst section of >8000 h must be guaranteed, corresponding to yearly replacement during scheduled inspection of hot combustor parts. Catalyst stability against poisoning by air- and fuel-borne contaminants must therefore also be considered.

3. Ignition of gas-phase reactions in order to secure complete fuel conversion and CO burnout within the imposed residence time constraints. As illustrated in Figure 12.1, the reaction rate is governed by different mechanisms on progressively increasing the temperature, that is, kinetics of surface reactions, gas–solid mass transfer and kinetics of gas-phase reactions. Accordingly, the onset of mass transfer limitation would prevent complete fuel conversion and CO burnout in the presence of catalytic reactions only, unless the reactor is greatly oversized.

4. Design of a structured catalyst configuration able to cope with mass transfer and pressure drop constraints.

12.2.2
Design Concepts and Performance

Different design concepts have been proposed to match the severe requirements of catalytic combustors. A main classification criterion is based on fuel/air stoichiometry in the catalyst section, which has a dominant effect on the selection of catalytic materials and on the operating characteristics of the combustor. In this section, only configurations based on lean catalytic combustion will be described. The peculiar characteristics of rich catalytic combustion will be described in a separate section.

12.2.2.1 Fully Catalytic Combustor
The first design concept tried to exploit fully the potential of catalytic combustion by completing the process in a single catalyst section. In such a configuration, a

preheated, premixed fuel–air stream is fed to the catalyst section. Ignition occurs at the catalyst walls, which rapidly reach the adiabatic reaction temperature, and the reaction rate is controlled by gas–solid mass transfer of the fuel. The heat released at the catalyst surface results in a progressive increase in the temperature in the gas phase and eventually causes ignition of homogeneous combustion, allowing for rapid fuel burnout within or immediately after the catalyst section. The major challenge in this concept is the development of catalytic materials able to withstand thermal stresses resulting from operation at temperatures slightly higher than those required at the turbine inlet. In addition, complete fuel conversion is difficult to achieve within catalyst sections of reasonable size at the high flow velocities characteristic of modern GTs [6]. Almost two decades of research effort have not yielded satisfactory results [6–10]. Accordingly, also in view of the trend towards higher turbine inlet temperatures, this concept was abandoned and novel design approaches have been pursued which try to keep the temperature of the catalyst section well below the exit combustor temperature.

12.2.2.2 Fuel Staging

A method to control the catalyst temperature is to reduce the fuel concentration and consequently the adiabatic reaction temperature. This can be achieved by splitting the fuel feed partly to the catalyst and partly to a downstream homogeneous section where combustion is completed, being stabilized by the hot gas stream from the catalyst section [11–14]. The fuel/air ratio fed to the catalyst was adjusted to keep the adiabatic reaction temperature below 1000 °C, considered a critical limit to prevent catalyst deterioration. Specific efforts were devoted to optimizing the mixing of fuel fed to the downstream homogeneous section with the hot stream coming from the catalyst; such mixing is critical to avoid NO_x formation according to thermal and prompt mechanisms. For this purpose, in addition to a specific design of the catalyst section and of the fuel distribution system, air staging was adopted. The concept was proven up to a scale equivalent to one combustor of a 10 MW multi-can type of GT. Despite good emission performance [15], the need for careful control of the complex fuel–air distribution pattern remains a major drawback of this design concept.

12.2.2.3 Partial Catalytic Hybrid Combustor

Catalytica Energy Systems and Tanaka Kikinzoku Kogyo have developed a configuration in which all the fuel–air mixture required to achieve the desired combustor outlet temperature is fed to the catalyst section. Here combustion proceeds only to partial fuel conversion (about 50%), being completed in a downstream homogeneous section [16]. The catalyst wall temperature is kept well below the adiabatic reaction level by means of a proprietary catalyst design based on (i) the temperature self-regulation characteristics of a PdO–Pd system in CH_4 combustion, (ii) the use of metal monolith supports with internal heat-exchange capabilities between nearby active and passive channels and (iii) the use of a diffusion barrier on top of the active catalyst layer.

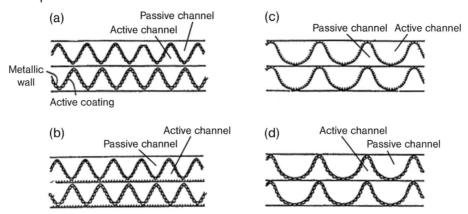

Figure 12.2 Examples of monoliths with internal heat exchange [18].

The peculiar features of the PdO–Pd catalyst will be discussed in a following section.

Metal monoliths with internal heat exchange capabilities are obtained by assembling single side-coated flat and corrugated sheets as illustrated in Figure 12.2 [17, 18].

Heat generated at the catalytic wall of the active channels is efficiently transmitted by conduction through the thin metal foil and is dissipated in the gas flow on both the catalytic and the non-catalytic side, allowing the wall temperature to be kept well below the adiabatic reaction temperature. The structure can be adjusted in order to tune the fraction of active channels in the monolith cross-section.

The deposition of an inert porous diffusion barrier on top of the catalyst layer can significantly hinder the rate of mass transfer of reactants to the catalyst surface, at the same time affecting only negligibly the rate of heat transfer to the gas phase. The effect is equivalent to that observed with fuels, like higher hydrocarbons, whose mass diffusivity in air is considerably lower than thermal diffusivity of the fuel–air mixture (Lewis number >1). Such unbalancing of heat and mass transfer rates results in a significant decrease in the catalyst wall temperature.

A scheme of the combustor configuration is shown in Figure 12.3.

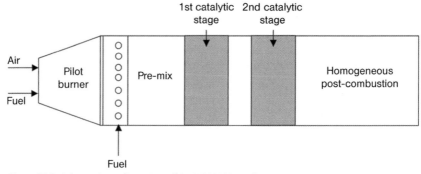

Figure 12.3 Schematic configuration of the XONON combustor.

The catalyst section consists of an inlet stage with high activity designed to minimize the ignition temperature and to operate at low to medium wall temperature and one or more subsequent stages less active than the inlet stage, designed to operate at medium to high temperature and to provide an outlet gas temperature high enough to guarantee fast ignition of gas-phase combustion in the downstream homogeneous section where fuel conversion is completed and the gas stream heats up to the adiabatic combustion temperature. The use of flame holders [19] or other means of hydrodynamic flame stabilization [20] of homogeneous combustion has also been proposed to secure a more efficient burnout of CO and UHC within the limited residence time (10–15 ms).

The partial catalytic hybrid combustor concept has been developed to a commercial stage for small-sized GTs (1.5 MW) which can be operated in a co-generative configuration adequate for the distributed generation of heat and power. Emission levels below 2.5 ppm NO_x (at 15% O_2) were certified by the US Environmental Protection Agency (EPA) [21]. In a recent campaign with a larger sized (10 MW) machine, fairly good emission performances were demonstrated under different load conditions (Figure 12.4).

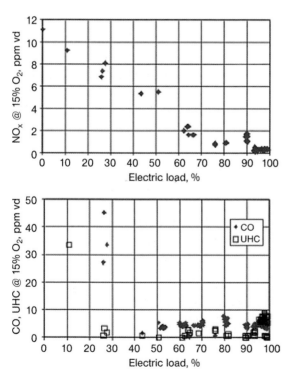

Figure 12.4 Emission performances of a GE10-1 (11 MWe) GT equipped with a catalytic combustor. Taken from [22].

12.3
Fuel-rich Catalytic Combustion

After having been explored since the pioneering years of catalytic combustion as a method to reduce fuel NO_x formation [3] and to process different liquid fuels [23], the fuel-rich catalytic combustion-based air staging design concept has recently attracted interest following the results obtained by Precision Combustion Inc. (PCI). The base configuration of the rich catalytic lean (RCL) burn technology developed by PCI is illustrated in Figure 12.5.

Air from the compressor is split into two streams: primary air is premixed with the fuel and then fed to the catalyst, which is operated under O_2 defect conditions; secondary air is used first as a catalyst cooling stream and then mixed with the partially converted stream from the catalyst in a downstream homogeneous section where ignition of gas-phase combustion occurs and complete fuel burnout is readily achieved. The control of the catalyst temperature below 1000 °C is achieved by means of O_2 starvation to the catalyst surface, which leads to the release of reaction heat controlled by the mass transfer rate of O_2 in the fuel-rich stream and of backside cooling of the catalyst with secondary air. To handle both processes, a catalyst/heat exchanger module has been developed, which consists of a bundle of small tubes externally coated with an active catalyst layer, with cooling air and fuel-rich stream flowing in the tube and in the shell side, respectively [24].

Stabilization of gas-phase combustion in the downstream homogeneous section can be obtained by a thermal effect, associated with heat released in the catalyst section and/or by a chemical effect associated with H_2 production under fuel-rich conditions. The lower combustion temperature allowed by stabilization effects results in a significant reduction of NO_x emissions while CO and UHC are kept at an ultra-low level. However, to minimize NO_x emission, the homogeneous ignition delay downstream of the catalyst section should be long enough to allow effective premixing of secondary combustion air with catalyst effluent so as to minimize local over-temperatures. The relative importance of thermal and chemical stabilization

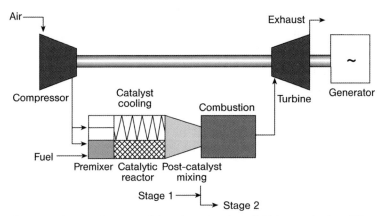

Figure 12.5 Base configuration of the rich catalytic lean (RCL) burn technology [25].

effects and also ignition delay are mainly controlled by the temperature and composition of the catalyst effluent, which in turn can be tuned by catalyst design. According to Smith et al. [25], in full pressure tests the operation of the fuel-rich catalyst was selective to full oxidation products (CO_2 and H_2O) since thermal stabilization was preferred.

Two configurations of the RCL burn technology have been designed: a catalytic pilot burner, which replaces the existing diffusion flame or partially premixed pilot of the DLN combustor [26], and a full catalytic burner [25].

The catalytic pilot burner processes only a fraction of the fuel and is targeted to retrofitting applications with minor combustor modifications. Test results indicate that to achieve effective stabilization of homogeneous combustion, 18–20% of the fuel–air must be processed in the catalytic pilot, which is a much higher fraction than the typical 2–5% processed in a conventional pilot burner. Under such conditions, test results demonstrated single-digit (<5 ppm at 15% O_2) emissions of NO_x and CO with low acoustics at 50 and 100% load conditions.

In the full catalytic burner, all the fuel is processed within an RCL burn module which replaces a conventional premixer–swirler arrangement in the DLN combustor. This configuration requires major design modification of the combustion with respect to the pilot burner, but has provided better emission performance: NO_x, CO and UHC emissions below 3, 10 and 2 ppm at 15% O_2, respectively, with negligible combustion acoustics (less than 0.15% peak-to-peak oscillation of mean combustor pressure).

Results obtained in full pressure tests demonstrated the following advantages of rich fuel over lean fuel catalytic combustion:

1. Low light-off temperature. Operation under a deficiency of O_2 exhibits a higher catalytic oxidation rate than under an excess of O_2, which in turn results in a lower ignition temperature. Further, much wider ignition/extinction hysteresis is observed and upon light-off the catalyst keeps ignited down to temperatures well below the practical range of relevance to GT operation (down to 215 °C in the case of the low-pressure Saturn T1200 engine). Accordingly, no homogeneous pre-burner is needed at steady state during partial and full load operation.

2. No risk of flash back in the catalyst section thanks to O_2 depletion in the fuel-rich stream and absence of fuel in backside cooling air.

3. Tolerance to high turbine firing temperatures and to fuel–air non-mixing. For a given geometry of the catalyst module and with fixed air flow rate and air staging ratio, the catalyst temperature was proven to be fairly insensitive to the overall fuel/air ratio (i.e. adiabatic combustion temperature) of the combustor [25]. The reason is that heat release and heat removal depend mainly on the gas–solid mass flow rate of O_2 in the catalyst side and the air flow rate in the back cooling side, respectively, and hardly depends on fuel type and concentration.

4. Fuel flexibility. Sub-scale tests at 1 MPa proved the viability of RCL technology to natural gas, landfill gas (70% CH_4, 30% CO_2), refinery gas (70% CH_4, 30%H_2), pre-vaporized diesel No. 2 and gasoline or blast furnace gas with a low heating

expected to yield high transformation efficiencies. However, the difficulties encountered in the machinery and operation of miniaturized moving parts [35] have subsequently oriented the research towards static conversion devices, such as thermoelectric systems based on the Seebeck effect [37, 38, 40], photovoltaic systems [41, 42] and thermionic generators [43]. Alternatively to the direct conversion of energy from the enthalpy of the combustion flue gases, processes are also being studied in which the heat recovery is obtained through endothermic reforming reactions with production of H_2 streams for the fuelling of H_2 fuel cells [44]. Methanol steam reforming [45, 46] and NH_3 decomposition [47–49] have been mostly studied for micro-scale applications.

Independently of the application of microcombustors (either for direct conversion of energy or heat recovery through endothermic H_2 production), several peculiarities characterize the combustion process in micro-spaces with respect to the process in conventional equipments [35, 36]. The decrease in the characteristic size of the combustor chamber amounts to two or even three orders of magnitude (from 10 cm to less than 1 mm), which is accompanied by (i) a dramatic increase in the surface-to-volume ratio and (ii) a significant decrease of the contact times within the reactor (a few milliseconds). Both factors are sources of instability for the combustion process. The increase in the surface-to-volume ratio favors both heat losses (thermal quenching) and termination of the radical species through adsorption and recombination into molecular species (chemical quenching), with consequent extinction of the combustion at insufficiently high flow rates and fuel concentration. Conversely, the extremely short residence times favor the occurrence of blow-out at high flow rates. Efficient insulation of the combustion chamber [47, 50] and the adoption of materials that do not cause radical termination [51] allow the combustor stability to be increased, which is still a key issue, unless a highly reactive fuel such as H_2 (unsuitable in miniaturized applications) is used [36].

To overcome such hurdles, special devices have been proposed and tested, which are based on internal heat recovery of waste heat through the adoption of 'Swiss roll' geometries, such as those reported in Figure 12.7 [52].

These microcombustors have demonstrated stable performance using propane as fuel [52, 53]; however, it must be observed that the presence of a central zone at very high temperature may be critical for the choice of materials and give rise to significant NO_x emissions [54].

Early studies in this field [35, 36] indicated that a high surface-to-volume ratio, which represents a hurdle for gas-phase combustion, is instead an advantage for catalytic combustion. In fact the small scale enhances considerably the rate of gas–solid mass transfer, which favors the kinetics of the combustion process and compensates for the short residence time. Also, as is well established for large-scale systems, the presence of a catalytic phase allows for stable combustion at significantly lower temperature than traditional homogeneous burners [55, 56]. This makes the design and operation of microcombustors more flexible. Several recent studies have explored the potential of catalytic microcombustors using H_2 [37, 38, 50], methane [37], propane [52, 53, 57] and mixtures of H_2 with propane [57], butane [38, 47, 52] and dimethyl ether [52].

12.6.2.1 PdO-based Catalysts

Supported palladium oxide is the catalyst of choice for GT combustors fuelled by natural gas in view of the following properties:

- maximum activity in CH_4 combustion, which results in minimum light-off temperature;
- unique temperature self-regulation features associated with the reversible PdO–Pd transformation;
- good thermal stability.

Such key features are strongly interconnected via the complex behavior of supported palladium. It is well known that CH_4 combustion activity depends markedly on the oxidation state of palladium. In Figure 12.8, a typical conversion curve obtained in temperature-programmed combustion (TPC) experiments during heating/cooling cycles is plotted.

As reported by many authors [62–66], a large conversion hysteresis is observed between heating and cooling branches which is associated with reversible transformation of highly active PdO to poorly active metallic Pd [62, 63]. Indeed, the temperature ranges where a negative apparent activation energy is observed are associated with PdO thermal decomposition in the heating branch and with PdO re-formation in the cooling branch. The detailed features of the PdO–Pd reversible transformation and its relation to CH_4 combustion activity are complex and still being debated. A thorough state-of-the-art discussion was given by Ciuparu et al. [67]. In the following, we summarize the main features of practical relevance. The position and amplitude of the activity hysteresis depends mainly on partial pressure of O_2 and on the nature of the support material. The threshold temperatures of both decomposition and re-formation of PdO increase markedly with O_2 partial pressure. Data on the effect of the support material are reported in Table 12.2.

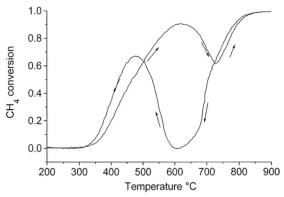

Figure 12.8 TPC test in the annular reactor at high GHSV. Feed composition: 0.5% CH_4, 2% O_2, 1% H_2O, balance He. GHSV = 10^6 N cm^3 g^{-1} h^{-1}; heating/cooling rate, 15 °C min^{-1}; catalyst, 10%Pd/La$_2$O$_3$(5%)–Al$_2$O$_3$.

Table 12.2 Temperature of onset of PdO decomposition (T_D) during the heating ramp and that of re-formation of PdO from Pd (T_R) during cooling in thermal cycling in air for PdO supported on different alumina-based materials.

Support	T_D (°C)	T_R (°C)
Al_2O_3	795	690
La_2O_3–Al_2O_3	800	690
CeO_2–Al_2O_3	795	755
La_2O_3–CeO_2–Al_2O_3	800	750
ZrO_2	800	725
YSZ	800	715

The threshold temperature of thermal decomposition of PdO does not depend significantly on the support and corresponds well with the value predicted by thermodynamics [10, 68]. On the other hand, the support markedly affects the threshold temperature of PdO re-formation during cooling, which is significantly lower than that of PdO decomposition. In particular, oxides with high oxygen mobility such as CeO_2, ZrO_2 and YSZ present markedly higher threshold temperatures of PdO re-formation and accordingly much lower PdO decomposition/re-formation hysteresis. The mechanism of such hysteresis has not yet been fully clarified. It has been proposed that passivation by chemisorbed O_2 on the Pd surface occurs at high temperature, which hinders the formation of bulk PdO [64]. It has also been suggested that supports with high oxygen mobility would likely promote nucleation of bulk PdO from the passivation layer [67, 69]. In a recent paper it was shown that in Al_2O_3-supported, CeO_2-doped catalysts, direct contact with CeO_2 particles is required to promote Pd reoxidation effectively (Figure 12.9) [70].

The variations of CH_4 combustion activity associated with the PdO–Pd reversible transformation are responsible for the unique thermostating ability of palladium-supported catalysts. Indeed, in the adiabatic combustion of CH_4, the

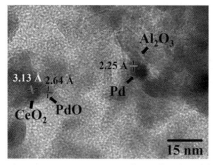

Figure 12.9 HRTEM image of sample Pd/Ce/Al_2O_3 after TPO cycle stopped at 833 K during cooling [70].

catalyst stabilizes at the temperature of PdO decomposition predicted by thermodynamics for the actual oxygen partial pressure (850–900 °C in typical GT conditions), that is, well below the adiabatic reaction temperature [16, 71]. Temperature oscillation problems have been reported by some authors [14, 72, 73], which are probably associated with the dynamics of the PdO–Pd transformation and of the hysteresis in the combustion activity. The use of ZrO_2-based supports, a preferred material for GT combustor catalysts [61], has been claimed to eliminate oscillation problems.

The light-off performances of palladium catalysts are determined by the kinetics of CH_4 combustion over the highly active PdO phase. There is general agreement in the literature [74–79] that this reaction exhibits a zero order dependence on O_2 concentration and a first-order dependence on CH_4 concentration. Strong H_2O inhibition has been reported typically with a negative order of −1 at low temperatures. A few studies have indicated that H_2O inhibition is still present up to 600 °C, although with a less negative reaction order, higher than −1 [66, 79, 80]. More controversial reports have appeared on the effect of CO_2, which has been indicated either to inhibit CH_4 combustion, particularly at high CO_2 concentration [76, 77, 81], or not to exert any effect at all [71, 78].

Apparent activation energies in the range 70–90 kJ mol^{-1} are typically obtained in dry (i.e. no H_2O in the feed) combustion experiments when assuming simple first-order kinetics in CH_4 combustion. However, it has been emphasized that such activation energies must be corrected to higher values by properly accounting for H_2O inhibition [78].

Several authors have proposed that CH_4 combustion over PdO occurs via a redox mechanism [82–85]. Methane activation through assisted hydrogen extraction is generally regarded as the rate-determining step, although there is not a general consensus on the nature of the adsorption sites. Further, desorption of H_2O by decomposition of surface hydroxyls has been reported to play a key role in reaction kinetics at temperatures below 450 °C [67, 86].

Structure sensitivity of CH_4 combustion over PdO is also a widely debated issue [67]. Ribeiro and co-workers [76, 87] carefully reviewed data reported in earlier studies and suggested that the wide scatter of turnover frequencies was mostly due to spurious factors such as neglect of H_2O inhibition, the presence of contaminants originating either from Pd precursors (e.g. Cl$^-$ [88]) or from the support (e.g. SiO_2 [89]) and poor control of the Pd oxidation state. The precise measurement of PdO surface area is another major problem in the accurate determination of turnover frequencies. Conventional chemisorption techniques require prereduction of PdO particles, which can result in major modification of morphology. Measurements of labeled ^{18}O exchange have been proposed for this purpose [90, 91]. By applying this technique to samples obtained from single crystals of Pd metal, Ribeiro and co-workers [87] confirmed that oxidation of Pd metal to PdO during catalyst pretreatment results in a marked increase in the active surface area due to surface roughening [90, 92]. In addition, CH_4 combustion was found to be insensitive on an amorphous PdO layer grown on different crystal faces of Pd metal. Unfortunately, this approach can hardly be applied to real supported catalysts since it has been observed that PdO facilitates the exchange of ^{18}O with the support, the exchange increasing with decreasing size of the PdO particles.

Finally, it is worth emphasizing that the collection of kinetic data on CH_4 combustion under conditions relevant to practical applications (high temperature and high reactant concentration) is a difficult task since the results are typically biased by the onset of temperature gradients associated with the strong exothermicity of combustion and by the impact of diffusion-limiting effects associated with the very high reaction rate. Novel structured catalytic reactors, obtained by deposition of thin catalyst layers on a support with well-defined geometry and assembled in a configuration capable of minimizing mass transfer effects and of dissipating the reaction heat efficiently, have been developed for this purpose. Examples of such structured reactors (annular reactor [64, 79], plate cell reactor [93]) are given in Figure 12.10.

Structured reactors equipped with measurement systems able to provide spatially resolved concentration and temperature profiles have also been developed [94].

Durability of catalyst performance under the harsh environment of a GT combustor is another key issue in the development of the technology. In addition to thermomechanical issues discussed in the previous section, volatility and sintering of the active catalytic species are major concerns in this respect. On the other hand, poisoning by sulfur and other contaminants has been recognized to have a minor effect on the catalyst performance due to the high temperature of this application [8].

Due to very high GHSV ($>10^7\,h^{-1}$), an extremely low limit of vapor pressure ($<10^{-4}$ Pa) has been fixed as a rough criterion to match the $8000\,h^{-1}$ constraint of catalyst life in GT combustors [95]. Estimates made considering all the relevant species (metals, oxides, hydroxides, oxyhydroxides) under the oxidizing and water-containing atmosphere of GT combustors showed that Pd is able to match such a constraint up to about 1000 °C, whereas most other components (including Pt) fail [95].

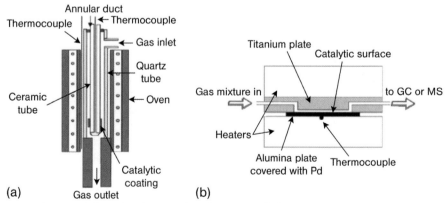

Figure 12.10 Examples of structured catalytic reactors for kinetic measurements: (a) annular reactor [47, 61]; (b) plate cell reactor [75].

The PdO–Pd transformation also plays a key role in the sintering behavior since coarsening of palladium particles occurs more readily upon thermal reduction [76, 96]. Sintering of Pd metal particles under conditions relevant to GT combustors has been investigated [96, 97]. The results showed that sintering occurs via Ostwald ripening up to an average particle size of hundreds of nanometers. It has been observed that reoxidation of large metal particles results in a significant decrease in crystal size [69, 92]. However, such fragmentation, which originates from the large differences in crystal structure and density between the tetragonal PdO phase and fcc Pd metal phase, occurs via the formation of multiple incoherent PdO domains [92] and does not result in a real redispersion of the palladium aggregates.

It is worth noting that, in order to match the combined requirements of high combustion activity and durability under harsh operating conditions, catalysts with a high Pd loading (about 10% w/w) must be adopted [16, 98]

12.6.2.2 Metal-substituted Hexaaluminate Catalysts

Hexaaluminate (HA) materials containing transition metal ions in the structure have been extensively investigated for GT applications, in view of their excellent thermal stability and catalytic activity [9, 99–101].

These materials can be represented with the general formula $ABM'_xM_{11-x}O_{19-x}$, where A and B are large cations such as Ba, Sr, Ca and La, M' is a transition metal ion such Cr, Mn, Fe Co, Ni and Cu and M stands for Al. They crystallize in two hexagonal structures, β-Al$_2$O$_3$ and magnetoplumbite, depending on the nature of the components (Figure 12.11).

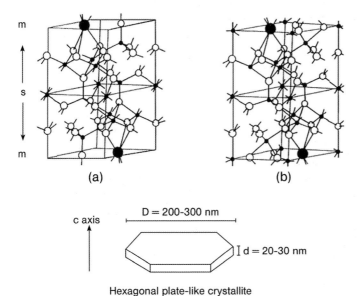

Figure 12.11 β-Al$_2$O$_3$ and magnetoplumbite structures (large closed circles = A and B cations; small closed circles = M and M' cations; medium open circles = oxygen ions).

The structures consist of spinel blocks, containing the M and M' ions, whereas the mirror planes contain the large cations. The alternate stacking of spinel blocks and mirror planes results in peculiar hexagonal plate-like crystallites, characterized by a strong anisotropy along the c-axis direction, that is responsible for the high thermal stability of these systems [99, 100]. Transition metal cations partially replacing Al^{3+} ions provide the significant methane combustion activity of the catalyst [100, 102].

Mn-substituted HA catalysts are the most active. Mn enters the structure at low concentration preferentially in tetrahedral Al sites with dominant oxidation state $+2$ and, at high concentrations, in octahedral Al sites with dominant oxidation state $+3$ [103]. The catalytic activity is also influenced by the composition of the mirror plane: high methane combustion activity has been reported for $Sr_{0.8}La_{0.2}Mn_1Al_{11}O_{19}$ [104]. This activity is comparable to that of $BaMn_2Al_{11}O_{19}$ and it is slightly lower than that measured over $LaMg_{0.5}Mn_{0.5}Al_{11}O_{19}$ [105]. It has been suggested that the incorporation of a divalent cation such as Mg^{2+} stabilizes the structure favoring the incorporation of Mn as Mn^{3+}, which is highly active [105].

These materials have been prepared both by hydrolysis of the alkoxides [100] and by coprecipitation from soluble salts of the constituents by using NH_4OH or $(NH_4)_2CO_3$ as precipitating agent [106]. Monophasic samples with surface areas in the range $10–15\,m^2\,g^{-1}$ have been obtained upon calcination at $1300\,°C$ [100, 106].

Recently, the synthesis of nano-sized HA has been proposed via reverse-microemulsion preparation, which is reported to be effective for controlling the hydrolysis and polycondensation of the alkoxides of the constituents. Using this preparation route, the nanoparticles crystallize directly to the desired phase at the relatively low temperature of $1050\,°C$ and maintain surface areas higher than $100\,m^2\,g^{-1}$ after calcination at $1300\,°C$ for $2\,h$ [107–109].

12.6.2.3 Rich Combustion Catalysts

Numerous studies have been published on catalyst material directly related to rich catalytic combustion for GT applications [73]. However, most data are available on the catalytic partial oxidation of methane and light paraffins, which has been widely investigated as a novel route to H_2 production for chemical and, mainly, energy-related applications (e.g. fuel cells). Two main types of catalysts have been studied and are reviewed below: supported nickel, cobalt and iron catalysts and supported noble metal catalysts.

Transition metal carbide catalysts have also been explored as methane partial oxidation catalysts [110]; promising results were obtained over Mo_2C systems and enhancements were reported with the addition of transition metal promoters.

Supported nickel and cobalt catalysts The catalytic partial oxidation of methane to synthesis gas (a mixture of CO and H_2) has been investigated since 1946, when the first study on Ni catalysts was published [111]. Since then, several investigations have confirmed the early studies; Ni is highly active for synthesis gas production, but it also catalyzes carbon formation. It was shown that stable operation can be obtained by feeding over-stoichiometric O_2–CH_4 mixtures (i.e. $O_2:CH_4 > 0.5$); different regions tend to establish along the catalytic bed: $NiAl_2O_4$, NiO/Al_2O_3 (active for methane oxidation to CO_2 and H_2O) and supported Ni metal particles (active for methane reforming with steam and CO_2 to synthesis gas) [112]. In order to improve the stability of the catalysts, much work was done on the modification of the support. Choudary and co-workers [113–121] studied nickel catalysts supported over CaO, SiO_2, TiO_2, ZrO_2, rare earth oxide-modified alumina and others. SiO_2 and TiO_2 were found not to be good supports, because of the formation of inactive binary metal oxide phases under high-temperature operating conditions. NiO containing MgO, CaO, rare earth oxides or alumina catalysts showed high catalytic activity at very short contact times. The Ni/MgO catalyst system has been extensively studied by Ruckenstein and Hu [122, 123] and Santos et al. [124]; it is believed that the observed high catalyst stability results from the formation of a solid solution between Ni and MgO (extending the catalyst lifetime and reducing sintering) and to the weak basicity of MgO (probably hindering carbon deposition) [125–127]. The use of rare earths oxides as supports or promoters has been suggested to reduce the impact of carbon deposition, due to the capability for oxygen storage (which can favor the gasification of deposited carbon), and enhance the morphological stability of the support, preventing sintering and loss of surface area during high-temperature operation.

Alternative approaches for stabilizing the Ni performance include (i) its incorporation in oxide matrices (for instance, controlled Ni metal crystallites were prepared by reduction of Ni-containing Mg/Al hydrotalcite-type precursors by Vaccari and co-workers [128, 129]) and (ii) the addition of an active component (Co, Fe and especially noble metals), reducing carbon deposition [130, 131].

Co and Fe catalysts have also been studied for the partial oxidation of methane to synthesis gas. Their potential relies on the fact that Co and Fe have higher melting and vaporizing points than Ni. Lower performances were mostly observed, however, which is probably related to the higher activity of CoO and Fe_2O_3 for the complete oxidation of methane [121, 132, 133]. The recognized order of reactivity for partial oxidation is in fact $Ni \gg Co > Fe$. However, it was observed that the performance of Co improves when a promoter is added. An extensive study of the catalytic partial oxidation of methane over Co/Al_2O_3 catalysts with different metals (0.1 wt% of Ni, Pt, Rh, Ru, Pd) and oxides (<4 wt% Fe_3O_4/Cr_2O_3, La_2O_3, SnO_2, K_2O) was recently performed by Lødeng et al. [134]. A comparison with Ni- and Fe-based catalysts was also addressed. It was found that addition of metal promoters, particularly Rh and Pt, enhanced the catalyst activity at low temperatures (which resulted in delayed extinction of the reaction during ramping at $-1\,°C\,min^{-1}$). However, addition of Ni promoted carbon formation. Addition of surface oxides typically promoted instability, deactivation and combustion (hence the formation of a stable Co metallic phase was hindered). It was found that Ni performed better than Co-based catalysts at all temperatures. However, Fe-based catalysts showed high combustion activity.

Noble metal catalysts High yields of synthesis gas were reported for the partial oxidation of methane over nearly all noble metal catalysts (Pd, Rh, Ru, Pt, Ir) [110]. The observed performances (degree of methane conversion and yields of CO and H_2) vary with the catalyst and the operating conditions but all reported data share these common features: (i) no carbon deposition is observed (Claridge and co-workers [135]

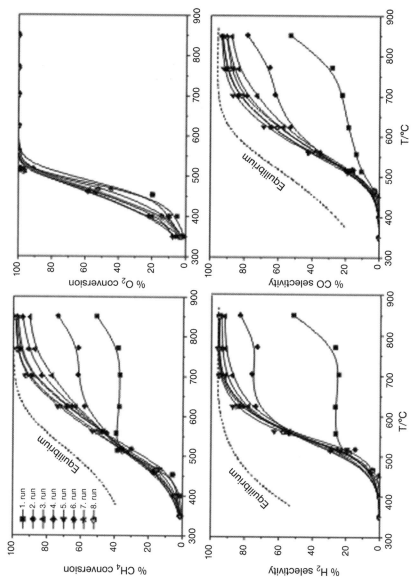

Figure 12.13 Conditioning of 0.5% Rh/Al$_2$O$_3$ during successive CH$_4$-CPO runs. Operating conditions: CH$_4$ = 4%, O$_2$:CH$_4$ = 0.56, balance N$_2$; GHSV = 800 000 Nl (kg cat.)$^{-1}$ h^{-1}. Dotted line = calculated thermodynamic equilibrium. Reprinted from [149].

Alternatively, specific preparation techniques (such as chemical vapor deposition [150] or the production of embedded Rh/Al$_2$O$_3$ catalysts [155]) have been developed to reduce surface heterogeneity and the related C-forming rate and guarantee stable and evenly distributed Rh particles.

Concerning the operation of catalysts under adiabatic conditions, Basini et al. [156] reported the results of methane partial oxidation runs in a pilot-scale reactor operating at high pressure and short contact times, showing stable activity (almost complete conversion of methane and over 90% selectivity to CO and H_2) during more than 500 h on-stream. In addition, operability for 20 000 h bench-scale testing has been claimed recently by the same group [157].

Still, it has been shown that operation of structured reactors under very severe conditions (e.g. extremely high space velocities and high preheating temperatures, which may result in surface temperatures exceeding 1000 °C) causes an activity loss (localized at the reactor inlet) which greatly affects the thermal behavior of the reactor, but hardly affects the overall reactor performance in terms of methane conversion and syngas selectivity. Indeed, model analyses have shown that a decrease in the catalyst activity (and in particular a decrease in the rate of endothermic reforming reactions) causes an increase in the inlet catalyst temperature, which partly compensates the same activity loss; as a consequence, outlet conversion and selectivity remain fairly constant and are hardly sensitive to catalyst aging [158].

12.7 Conclusions

Following wide R&D efforts in the 1990s, catalytic combustors for GTs have finally reached the commercialization stage. Such an achievement required a combined effort of designing novel engineering solutions and developing catalytic materials with high activity and stability performance. Both technical and fundamental issues are still open. The former are mainly the increase in catalyst robustness and durability, which should be extended towards a target of a 20 000 h life, the improvement of fuel flexibility and the integration of the catalytic combustor in a larger machine, whereas the latter are mainly related to the understanding of the complex behavior of Pd-based catalysts and the development of high-temperature combustion catalysts. However, the real potential of catalytic combustion for GTs will mainly depend on future environmental drivers towards ultra-low NO_x limits enforced by severe emission regulations.

Rich catalytic combustion will offer wide opportunities with respect to most of the above issues, including: flexible integration in different machines, low-temperature ignition ability, tolerance to fuel concentration and temperature non-uniformities and fuel flexibility. Further, the production of syngas in short contact time catalytic reactors could be exploited in several energy-related applications such as fuel cell and oxy-fuel combustion.

Finally, the development of microcombustors for energy conversion in small devices appears to be a stimulating field for catalytic combustion research, which once more will require the combined study of engineering solutions with advanced catalytic materials.

References

1. Major, B. and Powers, B. (1999) *Cost Analysis of NO_x Control Alternatives for Stationary Gas Turbines*, DOE Contract DE-FC02-97CHO877.
2. Miller, J.A. and Bowman, C.T. (1989) *Progress in Energy and Combustion Science*, **15**, 287.
3. Krill, V.W. and Kesselring, J.P. (1978) Proceedings of the 3rd Workshop on Catalytic Combustion, Asheville, NC, 3–4 October 1978, p. 261.
4. Schlegel, A., Buser, S., Benz, P., Bockhorn, H. and Mauss, F. (1994) Proceedings of Twenty-Fifth Symposium (International) on Combustion, Irvine, CA, August 1994, The Combustion Institute, Pittsburgh, PA, p. 1019.
5. Dalla Betta, R.A., Nickolas, S.G., Caron, T.J., McCarty, J.G. and Spencer, M.J., Corr, R.A. II (2003) US Patent 6,595,003.
6. Sadamori, H. (1999) *Catalysis Today*, **47**, 325.
7. Trimm, D.L. (1983) *Applied Catalysis*, **7**, 249.
8. Pfefferle, L.D. and Pfefferle, W.C. (1987) *Cataysis Reviews Science and Engineering*, **29**, 219.
9. Zwinkels, M., Jaras, S. and Menon, P.G. (1993) *Catalysis Reviews Science and Engineering*, **35**, 319.
10. Forzatti, P. and Groppi, G. (1999) *Catalysis Today*, **54**, 165.
11. Furuya, T., Hayata, T., Yamanaka, S., Koezuka, J., Yoshine, T. and Ohkoshi, A. (1987) ASME Paper No. 87-GT-99.
12. Furuya, T., Yamanaka, S., Hayata, T. and Koezuka, J. (1988) US Patent 4,731,989, assigned to Toshiba.
13. Ozawa, Y., Fujii, T., Kikumoto, S., Sato, M., Fukuzawa, H., Saiga, M. and Watanabe, S. (1995) *Catalysis Today*, **6**, 351.
14. Ozawa, Y., Fujii, T., Sato, M., Kanazawa, T. and Inoue, H. (1999) *Catalysis Today*, **47**, 399.
15. Ozawa, Y., Tochihara, Y., Mori, N., Yuri, I., Sato, J. and Kagawa, K. (2003) *Catalysis Today*, **83**, 247.
16. Dalla Betta, R.A., Ezawa, N., Tsurumi, K., Schlatter, J.C. and Nickolas, S.G. (1993) US Patent 5,183,401, assigned to Catalytica and Tanaka Kikinzoku.
17. Retallick, W.B. and Alcorn, W.R. (1993) US Patent 5,202,303, assigned to W.R. Grace.
18. Dalla Betta, R.A., Ribeiro, F.H., Shoji, T., Tsurumi, K., Ezawa, N. and Nickolas, S.G. (1993) US Patent 5,250,489, assigned to Catalytica and Tanaka Kikinzoku.
19. Dalla Betta, R.A., Yee, D.K., Magno, S.A. and Shoji, T. (1996) US Patent 5,518,697, assigned to Catalytica and Tanaka Kikinzoku.
20. Carroni, R., Griffin, T. and Kelsall, G. (2004) *Applied Thermal Engineering*, **24**, 1665.
21. Clapsaddle, C. and Van Osdell, D. (2000) Environmental Technology Verification Report (ETV) – NO_x Control Technologies – Catalytica Combustion System, Inc. XONON™ Flameless Combustion System, EPA Cooperative Agreement CR826152-01-2.
22. Cocchi, S., Modi, R. and Nutini, G. (2006) *Catalysis Today*, **117**, 419.
23. Brabbs, T.A. and Merrit, S.A. (1993) NASA Technical Paper 3281.
24. Smith, L.L., Etemad, S., Ulkarim, H., Castaldi, M.J. and Pfefferle, W.C. (2001) US Patent 6,174,159, assigned to Precision Combustion.
25. Smith, L.L., Karim, H., Castaldi, M.J., Etemad, S., Pfefferle, W.C., Khanna, V.K. and Smith, K.O. (2003) ASME Paper GT-2003-38129.
26. Karim, H., Lyle, K., Etemad, S., Smith, L.L., Pfefferle, W.C. and Smith, K.O. (2002) ASME Paper GT-2002-30083.
27. Smith, L.L., Karim, H., Castaldi, M.J., Etemad, S. and Pfefferle, W.C. (2006) *Catalysis Today*, **117**, 438.
28. Griffin, T., Sundkvist, S.G., Åsen, K.I. and Bruun, T. (2005) *Journal of Engineering for Gas Turbines and Power*, **127**, 81.

29 Linder, U., Eriksen, E.H. and Åsen, K.I. (2000) Swedish Patent Application 0002037.
30 Bruun, T., Werswick, B., Gronstad, L., Kristiansen, K. and Linder, U. (2000) Norwegian Patent Application 20006690.
31 Åsen, K.I. and Julsrud, S. (1997) Norwegian Patent Application 19972630.
32 Eriksson, S., Schnider, A., Mantzaras, J., Wolf, M. and Jaras, S. (2007) *Chemical Engineering Science*, **62**, 3991.
33 Jensen, K.F. (2001) Microreaction engineering – is small better? *Chemical Engineering Science*, **56**, 293.
34 Ehrfeld, W., Hessel, V. and Love, H. (2000) *Microreactors: New Technology for Modern Chemistry*, Wiley-VCH Verlag GmbH, Weinheim.
35 Carlos Fernandez-Pello, A. (2002) *Proceedings of the Combustion Institute*, **29**, 883.
36 Waitz, I.A., Gauba, G. and Tzeng, Y. (1998) *Journal of Fluids Engineering*, **120**, 109.
37 Vican, J., Gajdegzko, B.F., Dryer, F.L., Milius, D.L., Aksay, I.A. and Yetter, R.A. (2002) *Proceedings of the Combustion Institute*, **29**, 909.
38 Yoshida, K., Tanaka, S., Tomonari, S., Satoh, D. and Esashi, M. (2006) *Journal of Microelectromechanical Systems*, **15**, 195.
39 Epstein, A.H., Senturia, J.D., Al-Midani, O., Anathasuresh, G., Esteve, E., Frechette, L., Gauba, G., Ghodssi, R., Groshenry, C., Jacobson, S.A., Kerrebrock, J.L., Lang, J.H., Lin, C.C., London, A., Lopata, J., Mehra, A., Mur Miranda, J.O., Nagle, S., Orr, D.J., Piekos, E., Schmidt, M.A., Shirley, G., Spearing, S.M., Tan, C.S., Tzeng, Y.S. and Waitz, I.A. (1997) Micro-heat engines, gas turbines and rocket engines – The MIT Microengine Project, paper 97-1773, 28th AIAA Fluid Dynamics Conference.
40 Weinberg, F.J., Rowe, D.M., Min, G. and Ronney, P.D. (2002) *Proceedings of the Combustion Institute*, **29**, 941.
41 Yang, W.M., Chou, S.K., Shu, C. and Li, Z.W. (2002) *Applied Physics Letters*, **81**, 5255.
42 Yang, W.M., Chou, S.K., Shu, C., Xue, H., Li, Z.W., Li, D.T. and Pan, J.F. (2003) *Energy Conversion and Management*, **44**, 2625.
43 Zvezdin, A., Brignone, M., Repetto, P., Innocenti, G., Pizzi, M., Li Pira, N., Lambertini, V., Sgroi, M. and Bollito, G., (2005) US Patent Application 2005/0140341.
44 Kelley, S.C., Deluga, G.A. and Smyrl, W.H. (2002) *AIChE Journal*, **48**. 1071.
45 Cao, C., Wang, Y., Holladay, J.D., Jones, E.O. and Palo, D.R. (2005) *AIChE Journal*, **51**, 982.
46 Hollady, J.D., Jones, E.O., Phelps, M. and Hu, J. (2002) *Journal of Power Sources*, **108**, 21.
47 Arana, L., Schaevitz, S., Franz, A.J., Schmidt, M.A. and Jensen, K.F. (2003) *Journal of Microelectromechanical Systems*, **12**, 600.
48 Deshmuth, S.R. and Vlachos, D.G. (2005) *Industrial and Engineering Chemistry Research*, **44**, 4982.
49 Deshmuth, S.R. and Vlachos, D.G. (2005) *Chemical Engineering Science*, **60**, 5718.
50 Norton, D.G., Wetzel, E.D. and Vlachos, D.G. (2004) *Industrial and Engineering Chemistry Research*, **43**, 4833.
51 Miesse, C.M., Masel, R.L., Jensen, C.D., Shannon, M.A. and Short, M. (2004) *AIChE Journal*, **50**, 3206.
52 Sitzki, L., Borer, K., Schuster, E., Ronney, P.D. and Wussow, S. (2001) Third Asia–Pacific Conference on Combustion, Seoul, 24–27 June 2001.
53 Ahn, J., Eastwood, C., Sitzki, L. and Ronney, P.D. (2005) *Proceedings of the Combustion Institute*, **30**, 2463.
54 Kim, N., Kato, S., Kataoka, T., Yokomori, T., Maruyama, S., Fujimori, T. and Maruta, K. (2005) *Combustion and Flame*, **141**, 229.
55 Groppi, G., Cristiani, C. and Forzatti, P. (2008) Catalytic combustion, in *Handbook of Heterogeneous Catalysis*, 2nd edn (eds G. Ertl, H. Knözinger, F. Schüth and J. Weitkamp), Wiley-VCH Verlag GmbH, Weinheim, Chapter 11. 6.
56 Maruta, K., Takeda, K., Ahn, J., Borer, K., Sitzki, L., Ronney, P.D. and Deutschmann,

O. (2002) *Proceedings of the Combustion Institute*, **29**, 957.
57 Norton, D.G. and Vlachos, D.G. (2005) *Proceedings of the Combustion Institute*, **30**, 2473.
58 Chapman, L.R., Vigor, C.W. and Watton, J.F. (1982) US Patent 4,331,631, assigned to General Motors.
59 Chapman, L.R. and Watton, J.F. (1981) US Patent, 4,279,782 assigned to General Motors.
60 Dalla Betta, R.A., Tsurumi, K. and Ezawa, N. (1994) US Patent 5,281,128, assigned to Catalytica and Tanaka Kikinzoku.
61 Dalla Betta, R.A., Tsurumi, K., Shoji, T. and Garten, R.L. (1995) US Patent 5,405,260, assigned to Catalytica and Tanaka Kikinzoku.
62 Farrauto, R.J., Hobson, M.C., Kennelly, T. and Waterman, E.M. (1992) *Applied Catalysis A: General*, **81**, 227.
63 Farrauto, R.J., Lampert, J.K., Hobson, M.C. and Waterman, E.M. (1995) *Applied Catalysis B: Environmental*, **6**, 263.
64 McCarty, J.G. (1995) *Catalysis Today*, **26**, 283.
65 Groppi, G., Artioli, G., Cristiani, C., Lietti, L. and Forzatti, P. (2001) in *Natural Gas Conversion VI. Studies in Surface Science and Catalysis, Vol. 136* (eds E. Iglesia, J.J. Spivey and J.H. Fleisch), Elsevier, Amsterdam, p. 345.
66 Ciuparu, D. and Pfefferle, L. (2001) *Applied Catalysis A: General*, **209**, 415.
67 Ciuparu, D., Lyubovsky, M.R., Altman, E., Pfefferle, L.D. and Datye, A. (2002) *Catalysis Reviews*, **44**, 593.
68 Kleykamp, H. (1970) *Zeitschrift für Physikalische Chemie*, **71**, 142.
69 Forzatti, P. (2003) *Catalysis Today*, **83**, 3.
70 Colussi, S., Trovarelli, A., Groppi, G. and Llorca, J. (2007) *Catalysis Communications*, **8**, 1263.
71 Farrauto, R.J., Kennelly, T. and Waterman, E.M. (1990) US Patent 4,893,465, assigned to Engelhard.
72 Griffin, T., Weisenstein, W., Scherer, V. and Fowles, M. (1995) *Combustion and Flames*, **101**, 81.

73 Lyubovsky, M., Smith, L.L., Castaldi, M., Karim, H., Nentwick, B., Etemad, S., LaPierre, R. and Pfefferle, W.C. (2003) *Catalysis Today*, **83**, 71.
74 Cullis, C.F., Trimm, D.L. and Nevell, T.G. (1972) *Journal of the Chemical Society, Faraday Transaction 1*, **68**, 1406.
75 Yao, Y.Y. (1980) *Industrial and Engineering Chemistry Product Research and Development*, **19**, 293.
76 Ribeiro, F.H., Chow, M. and Dalla Betta, R.A. (1994) *Journal of Catalysis*, **146**, 537.
77 Fujimoto, K., Ribeiro, F.H., Avalos-Borja, M. and Iglesia, E. (1998) *Journal of Catalysis*, **179**, 431.
78 van Giezen, C., van den Berg, F.R., Kleinen, J.L., vanDillen, A.J. and Geus, J.W. (1999) *Catalysis Today*, **47**, 287.
79 Ibashi, W., Groppi, G. and Forzatti, P. (2003) *Catalysis Today*, **83**, 115.
80 Zhu, G., Han, J., Zemlyanov, D.Yu and Ribeiro, F.H. (2005) *Journal of Physical Chemistry B*, **109**, 2231.
81 Burch, R., Urbano, F.J. and Loader, P.K. (1995) *Applied Catalysis A: General*, **123**, 173.
82 Muller, C., Maciejewski, M., Koeppel, R., Tschan, R. and Baiker, A. (1996) *Journal of Physical Chemistry*, **100**, 2006.
83 Muller, C., Maciejewski, M., Koeppel, R. and Baiker, A. (1999) *Catalysis Today*, **47**, 245.
84 Au-Yeung, J., Bell, A.T. and Iglesia, E. (1999) *Journal of Catalysis*, **185**, 213.
85 Ciuparu, D., Altman, E. and Pfefferle, L. (2001) *Journal of Catalysis*, **203**, 64.
86 Burch, R., Crittle, D.J. and Hayes, M.J. (1999) *Catalysis Today*, **47**, 229.
87 Zhu, G., Han, J., Zemlyanov, D.Y. and Ribeiro, F.H. (2004) *Journal of the American Chemical Society*, **126**, 9896.
88 Simone, D.O., Kennelly, T., Brungard, N.L. and Farrauto, R.J. (1991) *Applied Catalysis*, **70**, 87.
89 Zhu, G., Fujimoto, K., Yu Zemlyanov, D., Datye, A.K. and Ribeiro, F.H. (2004) *Journal of Catalysis*, **225**, 170.
90 Monteiro, R.S., Zemlyanov, D., Storey, J.M. and Ribeiro, F.H. (2001) *Journal of Catalysis*, **199**, 37.

91 Monteiro, R.S., Zemlyanov, D., Storey, J.M. and Ribeiro, F.H. (2001) *Journal of Catalysis*, **199**, 291.

92 Datye, A.K., Bravo, J., Nelson, T.R., Atanasova, P., Lyubovsky, M. and Pfefferle, L. (2000) *Applied Catalysis A: General*, **198**, 179.

93 Lyubowsky, M. and Pfefferle, L. (1999) *Catalysis Today*, **47**, 29.

94 Reinke, M., Mantzaras, J., Bombach, R., Schenker, S. and Inauen, A. (2005) *Combustion and Flames*, **141**, 448.

95 McCarty, J.G., Gusman, M., Lowe, D.M., Hildebrand, D.L. and Lau, K.N. (1999) *Catalysis Today*, **47**, 5.

96 McCarty, J.G., Malukhin, G., Poojary, D.M., Datye, A.K. and Xu, Q. (2005) *Journal of Physical Chemistry B*, **199**, 2387.

97 Datye, A.K., Xu, Q., Kharas, K.C. and McCarty, J.G. (2006) *Catalysis Today*, **111**, 59.

98 Carstens, Jason N., Su, Stephen C. and Bell Alexis T. (1998) *Journal of Catalysis*, **176**, 136.

99 Machida, M., Eguchi, H.K. and Arai, H. (1987) *Journal of Catalysis*, **103**, 385.

100 Arai, H., Eguchi, K. and Machida, M. (1989) MRS International Meeting on Advanced Materials, Vol. 2, Materials Research Society, Tokyo, p. 243.

101 Groppi, G., Cristiani, C. and Forzatti, P. (1997) *Catalysis*, **13**, 85.

102 Machida, M., Kawasaki, H., Eguchi, H. and Arai, H. (1988) *Chemistry Letters*, 1461.

103 Bellotto, M., Artioli, G., Cristiani, C., Forzatti, P. and Groppi, G. (1998) *Journal of Catalysis*, **179**, 597.

104 Machida, M., Eguchi, K. and Arai, H. (1990) *Journal of Catalysis*, **123**, 477.

105 Groppi, G., Cristiani, C. and Forzatti, P. (2001) *Applied Catalysis B: Environmental*, **35**, 137.

106 Groppi, G., Bellotto, M., Cristiani, C., Forzatti, P. and Villa, P.L. (1993) *Applied Catalysis*, **104**, 101.

107 Zarur, A.J. and Ying, J.Y. (2000) *Nature*, **403**, 65.

108 Zarur, A.J., Hwu, H.H. and Ying, J.Y. (2000) *Langmuir*, **16**, 3042.

109 Teng, F., Tian, Z., Xu, J., Xiong, G. and Lin, L. (2004) in *Natural Gas Conversion VII, Studies in Surface Science and Catalysis, Vol. 147* (eds X. Bao and Y. Xu), Elsevier, Amsterdam, p. 493.

110 York, P.E., Xiao, T. and Green, M.L.H. (2003) *Topics in Catalysis*, **22**, 345.

111 Prettre, M., Eichner, Ch. and Perrin, M. (1946) *Transactions of the Faraday Society*, **42**, 335.

112 Dissanyake, D., Rosynek, M.P., Kharas, K.C.C. and Lunsford, J.H. (1991) *Journal of Catalysis*, **132**, 117.

113 Choudary, V.R., Mamman, A.S. and Sansare, S.D. (1992) *Angewandte Chemie (International Edition in English)*, **31**, 1189.

114 Choudary, V.R., Ramarjeet, R.M. and Rane, V.H. (1992) *Journal of Physical Chemistry*, **96**, 8686.

115 Choudary, V.R., Rajput, A.M. and Prabhakar, B. (1993) *Journal of Catalysis*, **139**, 326.

116 Choudary, V.R., Rane, V.H. and Rajput, A.M. (1993) *Catalysis Letters*, **22**, 289.

117 Choudary, V.R., Rajput, A.M. and Prabhakar, B. (1992) *Catalysis Letters*, **15**, 363.

118 Choudary, V.R., Sansare, S.D. and Mamman, A.S. (1992) *Applied Catalysis*, **90**, 1.

119 Choudary, V.R., Rajput, A.M. and Rane, V.H. (1992) *Catalysis Letters*, **16**, 296.

120 Choudary, V.R., Rane, V.H. and Rajput, A.M. (1997) *Applied Catalysis A – General*, **162**, 235.

121 Choudary, V.R., Rajput, A.M., Prabhakar, B. and Mamman, A.S. (1998), *Fuel*, **77**, 1803.

122 Ruckestein, E. and Hu, Y.H. (1999) *Applied Catalysis A – General*, **183**, 85.

123 Hu, Y.H. and Ruckestein, E. (1998) *Industrial and Engineering Chemistry Research*, **37**, 2333.

124 Santos, A., Menendez, M., Monzon, A., Santamaria, J., Miro, E.E. and Lombardo, E.A. (1996) *Journal of Catalysis*, **158**, 83.

125 Choudary, V.R. and Mamman, A.S. (2000) *Applied Energy*, **66**, 161.
126 Choudary, V.R. and Mamman, A.S. (1999) *Fuel Processing Technology*, **60**, 203.
127 Tang, S., Lin, J. and Tan, K.L. (1998) *Catalysis Letters*, **51**, 169.
128 Basini, L., D'Amore, M., Fornasari, G., Guarinoni, A., Matteuzzi, D., Del Piero, G., Trifirò, F. and Vaccari, A. (1998) *Journal of Catalysis*, **173**, 247.
129 Albertazzi, S., Basile, F., Benito, P., Del Gallo, P., Fornasari, G., Gary, D., Rosetti, V. and Vaccari, A. (2007) *Catalysis Today*, **128**, 258.
130 Provendier, H., Petit, H.C., Estournes, C., Libs, S. and Kiennemann, A. (1999) *Applied Catalysis A – General*, **180**, 163.
131 Arpentinier, F., Basile, F., Del Gallo, P., Fornasari, G., Gary, D., Rosetti, V. and Vaccari, A. (2006) *Catalysis Today*, **117**, 462.
132 Slagtern, A. and Olsbye, U. (1994) *Applied Catalysis A – General*, **110**, 99.
133 Slagtern, A., Swaan, H.M., Olsbye, U., Dahl, I.M. and Mirodatos, C. (1998) *Catalysis Today*, **46**, 107.
134 Lødeng, R., Biørgum, E., Enger, B.C., Eilertsen, J.L., Holmen, A., Krogh, B., Rønnekleiv, M. and Rytter, E. (2007) *Applied Catalysis A – General*, **333**, 11.
135 Tsang, S.C., Claridge, J.B. and Green, M.L.H. (1995) *Catalysis Today*, **23**, 3.
136 Veser, G., Ziauddin, M. and Schmidt, L.D. (1999) *Catalysis Today*, **47**, 219.
137 Aghalam, P., Park, Y.K. and Vlachos, D.G. (2000) *Catalysis*, **15**, 98.
138 Trimm, D.L. and Onsan, Z.I. (2001) *Catalysis Reviews*, **43**, 31.
139 Younes-Metzler, O., Johansen, J., Thorsteinsson, S., Jensen, S., Hansen, O. and Quaade, U.J. (2006) *Journal of Catalysis*, **241**, 74.
140 Burch, R. and Hayes, M.J. (1995) *Journal of Molecular Catalysis A: Chemical*, **100**, 13.
141 Wei, J. and Iglesia, E. (2004) *Journal of Catalysis*, **225**, 116.

142 Rostrup-Nilesen, J.R. (1973) *Journal of Catalysis*, **31**, 173.
143 Rostrup-Nielsen, J.R. and Bak Hansen, J.H. (1993) *Journal of Catalysis*, **144**, 38.
144 Hickman, D.A. and Schmidt, L.D. (1993) *Science*, **259**, 343.
145 Witt, P.M. and Schmidt, L.D. (1996) *Journal of Catalysis*, **163**, 465.
146 Bodke, A.S., Bharadway, S.S. and Schmidt, L.D. (1998) *Journal of Catalysis*, **194**, 138.
147 Horn, R., Williams, K.A., Degenstein, N.J., Bitsch-Larsen, A., Dalle Nogare, D., Tupy, S.A. and Schmidt, L.D. (2007) *Journal of Catalysis*, **249**, 380.
148 Bruno, T., Beretta, A., Groppi, G. and Forzatti, P. (2005) *Catalysis Today*, **99**, 89.
149 Beretta, A., Bruno, T., Groppi, G., Tavazzi, I. and Forzatti, P. (2007) *Applied Catalysis B: Environmental*, **70**, 515.
150 Beretta, A., Donazzi, A., Groppi, G., Forzatti, P., Dal Santo, V., Sordelli, L., De Grandi, V. and Psaro, R. (2008) *Applied Catalysis B: Environmental*, **83**, 96.
151 Wang, D., Dewaele, O. and Froment, G. (1998) *Journal of Molecular Catalysis A: Chemical*, **136**, 301.
152 Wei, J. and Iglesia, E. (2004) *Journal of Catalysis*, **224**, 370.
153 Liu, Z.-P. and Hu, P. (2003), *Journal of the American Chemical Society*, **125**, 1958.
154 Rostrup-Nielsen, J. and Norskov, J.K. (2006) *Topics in Catalysis*, **40** (1–4), 45.
155 Montini, T., Condò, A.M., Hickey, N., Lovey, F.C., De Rogatis, L., Fornasiero, P. and Graziani, M. (2007) *Applied Catalysis B: Environmental*, **73**, 84.
156 Basini, L., Aasberg-Petersen, K., Guarinoni, A. and Ostberg, M. (2001) *Catalysis Today*, **64**, 9.
157 Basini, L. (2006) *Catalysis Today*, **117**, 384.
158 Tavazzi, I., Beretta, A., Groppi, G., Maestri, M., Tronconi, E. and Forzatti, P. (2007) *Catalysis Today*, **129**, 372.

13
Catalytic Removal of NO_x Under Lean Conditions from Stationary and Mobile Sources

Pio Forzatti, Luca Lietti, and Enrico Tronconi

13.1
Introduction

Nitrogen oxides (NO_x) are formed during the combustion at high temperature of fossil fuels and of biomasses and are blamed for the production of acid rain, the formation of ozone in the troposphere and of secondary particulate matter and for causing a reduction in breathing functionality and damage to the cardio-circulatory system in humans.

Table 13.1 summarizes representative NO_x emission limits for coal-, oil- and natural gas-fired thermal power plants, gas turbines and incinerators in Europe.

The control of NO_x from stationary sources includes techniques of modification of the combustion stage (primary measures) and treatment of the effluent gases (secondary measures). The use of low-temperature NO_x burners, over fire air (OFA), flue gas recirculation, fuel reburning, staged combustion and water or steam injection are examples of primary measures; they are preliminarily attempted, extensively applied and guarantee NO_x reduction levels of the order of 50% and more. However, they typically do not fit the most stringent emission standards so that secondary measures or flue gas treatment methods must also be applied.

Tables 13.2 and 13.3 show the European emission standards for passenger cars and for heavy-duty diesel engines.

With the introduction of Euro V emission standards for passenger cars in 2009, all diesel engines will be equipped with a common rail (CR) fuel injection system and with a diesel particulate filter (DPF) to solve the problem of particulate emissions and of the smoke in transient operation. CO emissions in passenger cars produced by diesel engines will be lower than gasoline engines (0.5 versus $1\,g\,km^{-1}$). However, gasoline engines will still maintain about 60% lower NO_x emission than diesel engines (0.06 versus $0.18\,g\,km^{-1}$). This is the gap to be closed to reach the almost fuel-neutral Euro VI standards (0.06 $g\,km^{-1}$ in gasoline engines versus $0.08\,g\,km^{-1}$ in diesel engines).

Catalysis for Sustainable Energy Production. Edited by P. Barbaro and C. Bianchini
Copyright © 2009 WILEY-VCH Verlag GmbH & Co. KGaA, Weinheim
ISBN: 978-3-527-32095-0

Table 13.1 Representative NO_x emission limits for large thermal power plants (>50 MWth), gas turbines and incinerators in Europe.

Stationary units	NO_x emission limit (ppm)
Coal-fired power plants (at 6% O_2)	100
Oil-fired power plants (at 3% O_2)	75
Natural gas-fired power plants (at 3% O_2)	50
Gas turbines (at 15% O_2)	25
Incinerators (11% O_2)	35

For this purpose, in addition to the continuous evolution of CR and exhaust gas recirculation (EGR), novel primary measures are under study, including the long route EGR to cool the recirculated exhaust gas, the use of premixed combustion [which implies, however, higher CO and unburned hydrocarbon (UHC) emissions], the reduction of the compression ratio, the shaping of the injection rate and so on. Still, the after-treatment catalytic technologies for NO_x removal and for CO/hydrocarbon (HC) and particulate matter (PM) reduction in passenger cars must be improved significantly.

Also for the case of heavy-duty diesel engines, a substantial reduction in the emissions of NO_x and PM is required in the near future.

At present the most effective available after-treatment techniques for NO_x removal under lean conditions are ammonia selective catalytic reduction (SCR) [1–3] and NO_x storage reduction (NSR) [4–6]. Indeed, three-way catalysts (TWCs) are not able to reduce NO_x in the presence of excess oxygen, because they must be operated at air/fuel ratios close to the stoichiometric value. Also, non-thermal plasma (NTP) and hydrocarbon-selective catalytic reduction (HC-SCR) are considered, although they are still far from practical applications.

Table 13.2 EU emission standards for passenger cars (g km^{-1}).

Tier	Date	CO	HCs	HCs + NO_x	NO_x	PM
Diesel						
Euro III	January 2000	0.64	—	0.56	0.50	0.05
Euro IV	January 2005	0.50	—	0.30	0.25	0.025
Euro V	September 2009	0.50	—	0.23	0.18	0.005
Euro VI	September 2014	0.50	—	0.17	0.08	0.005
Gasoline						
Euro III	January 2000	2.30	0.20	—	0.15	—
Euro IV	January 2005	1.0	0.10	—	0.08	—
Euro V	September 2009	1.0	0.10	—	0.06	0.005
Euro VI	September 2014	1.0	0.10	—	0.06	0.005

Table 13.3 EU emission standards for heavy-duty diesel engines (g kW^{-1} h^{-1}) (smoke in m^{-1}).

Tier	Date	CO	HCs	NO$_x$	PM	Smoke
Euro III	October 2000	2.1	0.66	5.0	0.10	0.8
Euro IV	October 2005	1.5	0.46	3.5	0.02	0.5
Euro V	October 2008	1.5	0.46	2.0	0.02	0.5

In the following, ammonia SCR and NSR technologies are illustrated and discussed with a view to novel stationary and mobile applications.

13.2
Selective Catalytic Reduction

13.2.1
Standard SCR Process

The standard SCR process is based on the reduction of NO with ammonia to water and nitrogen according to the following main reaction:

$$4NO + 4NH_3 + O_2 \rightarrow 4N_2 + 6H_2O \tag{13.1}$$

In fact, NO accounts for 90–95% of NO$_x$ in exhausts. Aqueous ammonia or urea, as ammonia precursor, are typically employed. The term 'selective' refers to the unique ability of ammonia to react selectively with NO$_x$ instead of being oxidized by oxygen.

In the case of sulfur-containing fuels (e.g. coal or oil), SO$_2$ is produced during combustion in the boiler along with minor percentages of SO$_3$; SO$_2$ can further be oxidized to SO$_3$ over the catalyst:

$$SO_2 + \frac{1}{2}O_2 \rightarrow SO_3 \tag{13.2}$$

Reaction (13.2) is highly undesired because SO$_3$ reacts with water present in the flue gas in large excess and with ammonia to form sulfuric acid and ammonium sulfate salts. The ammonium sulfate salts deposit and accumulate on the catalyst if the temperature is not high enough, leading to catalyst deactivation, and on the cold equipment downstream of the catalytic reactor, causing corrosion and pressure drop problems. The catalyst deactivation by deposition of ammonium sulfate salts can be reversed upon heating.

Commercial SCR catalysts are made of homogeneous mixtures of titania, tungsta and vanadia (or molybdena). Titania in the anatase form is used as a high surface and sulfur-resistant carrier to disperse the active components. Tungsta or molybdena is employed in large amounts (10 and 6% w/w, respectively) to increase the surface acidity and the thermal stability of the catalyst and to limit the oxidation of SO$_2$. Vanadia is responsible for the activity in the reduction of NO$_x$, but it is also active in the oxidation of SO$_2$. Accordingly, its content is kept low, usually below 1–2% w/w.

Commercial catalysts also contain silicoaluminates and glass fibers as additives to improve the catalyst strength.

The study of V–Ti–O, W–Ti–O, Mo–Ti–O, V–W–Ti–O and V–Mo–Ti–O catalysts has proved that [7–25]:

1. The commercial catalysts are constituted by anatase TiO_2 with V_2O_5 and WO_3 (or MoO_3) well dispersed over the surface of the support due to the strong interaction with titania.

2. Vanadium oxide is present in the form of isolated vanadyls (at low V loading) and polymeric metavanadate species (at high V loading) over dry surfaces. Monomeric wolframyls and polymeric W_xO_y species are present over dehydrated W–Ti–O catalyst. The presence of molybdenyl species has been observed in the case of Mo–Ti–O catalysts, while the existence of Mo_xO_y species was not evidenced but cannot be excluded. The structures of the supported V and W/Mo surface oxide species are not significantly affected by the presence of each other, suggesting that the structural features of the surface oxide species are primarily controlled by the interaction with the TiO_2 surface.

3. Evidence for the occurrence of an electronic interaction between V and W has been reported, possibly involving the semiconductor character of the TiO_2 support and leading to a different reducibility of the catalysts.

4. The high activity of V–W–Ti–O and of V–Mo–Ti–O is due to a synergistic effect between V and W (Mo) oxide species and is related to the superior redox properties of the ternary catalysts.

The SCR catalysts are used in the form of honeycomb monoliths or plates to guarantee low pressure drops in view of large frontal area with parallel channels, high external surface area per unit volume of catalyst, high attrition resistance and low tendency for fly ash plugging. The SCR monoliths and plates are assembled into standard modules and inserted into the reactor to form catalyst layers.

The SCR process is highly sensitive to inhomogeneities in the ammonia distribution at the inlet of the catalyst layer [26, 27].

Figure 13.1 shows the increment of the catalyst volume required to keep the ammonia slip below 5 ppm with NH_3:NO_x inlet ratio α equal to 0.8 when the standard deviation of α [$\sigma(\alpha)$] increases; for $\sigma(\alpha) > 0.2$ an asymptotic behavior is apparent because the flow inside the channel is segregated and the local NH_3:NO_x ratio is higher than 1 at the entrance of a significant number of monolith channels. Non-uniform distribution of NH_3 and NO_x over the entire cross-section of the catalytic converter is minimized by proper design of the ammonia distribution grid and its precise tuning during plant startup, position of guide vanes, use of a dummy layer before the catalyst layers and optimization of the flow distribution on the basis of cold models or computational fluid dynamics (CFD) calculations. These problems typically lead to NO_x removal efficiencies of 80–85% in high-dust arrangements (catalytic reactor located immediately after the boiler and economizer) and of 90–95% in tail-end arrangements (catalytic reactor located after the electrostatic precipitator and the flue gas desulfurization unit).

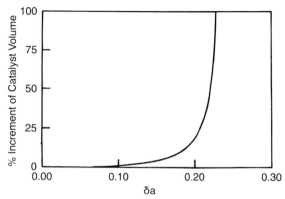

Figure 13.1 Influence of the maldistribution of the $NH_3:NO_x$ ratio (α) over the cross-section at the inlet of the catalytic converter on the percentage of extra catalyst volume requires to keep the ammonia slip below 5 ppm for $\alpha = 0.8$. Adapted from ref. [27].

An important requirement for SCR catalysts, especially for stationary applications and power stations, is to combine high activity in the de-NO_x reaction and very low (almost negligible) activity in the oxidation of SO_2 to SO_3.

As shown in Figure 13.2, the intrinsic first-order rate constant of NO_x reduction increases linearly with the vanadium content whereas the intrinsic first-order rate constant for the oxidation of SO_2 increases more than linearly with the vanadia content. This is consistent with the identification of the active sites for NO_x reduction with isolated V sites as opposite to the active sites for SO_2 oxidation that are probably associated with dimeric (or polymeric) sulfated vanadyls, in line with the consolidated picture of the active sites in commercial sulfuric acid catalysts [29]. Accordingly, high

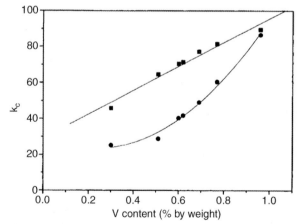

Figure 13.2 Dependence of the intrinsic kinetic constant k_C of de-NO_x and SO_2 oxidation reactions (full squares and circles, respectively) on the V content of the catalyst. $T = 350\,°C$, $k_{\text{de-}NO_x}$ (N m h^{-1}); $k_{SO_2-SO_3}$ (s^{-1}). Adapted from ref. [28].

dispersion of surface V oxide must be realized during the impregnation of the tungsta–titania or molybdena–titania powders with the solution of the vanadium salt, as indeed is attained in commercial catalysts, and typically the vanadia content must be low (<1.5% w/w).

In addition, the reduction of NO_x is a very fast reaction and is controlled by external and internal diffusion [27, 30]. In contrast, the oxidation of SO_2 is very slow and is controlled by the chemical kinetics [31]. Accordingly, the SCR activity is increased by increasing the catalyst external surface area (i.e. the cell density) to favor gas–solid mass transfer while the activity in the oxidation of SO_2 is reduced by decreasing the volume of the catalyst (i.e. the wall thickness); this does not affect negatively the activity in NO_x removal because significant ammonia concentrations are confined near the external geometric surface of the catalyst.

Different proposals have been advanced for the mechanism of the standard SCR reaction, which have been reviewed by Busca et al. [32]. Inomata et al. proposed that ammonia is first adsorbed as NH_4^+ at a V–OH Brønsted site adjacent to a $V^{5+}=O$ site and then reacts with gas-phase NO to form nitrogen and water while V^{4+}–OH groups are reoxidized to $V^{5+}=O$ by gaseous oxygen [33]. Janssen et al. demonstrated that one N atom of the N_2 product comes from NH_3 and the other from NO [34]. Ramis et al., based on Fourier transform infrared (FT-IR) spectroscopic evidence, proposed that ammonia is adsorbed over a Lewis acid site and is activated to form an amide species, which reacts with gas-phase NO to give a nitrosamide intermediate NH_2NO, which is known to decompose easily to N_2 and H_2O; the catalytic cycle is closed by reoxidation of the reduced catalyst by gaseous oxygen [35]. Topsøe et al. proposed that ammonia is adsorbed at a V^{5+}–OH site and is activated by a nearby $V^{5+}=O$ group, which is then reduced to V^{4+}–OH; again, the catalytic cycle is closed by reoxidation of V^{4+}–OH by gaseous oxygen [36]. It is worth noting that (i) both mechanisms proposed by Inomata et al. and by Topsøe et al. require the participation of dimeric or polymeric vanadyl species; (ii) in contrast, the mechanism proposed by Ramis et al. requires the participation of isolated vanadyls and this is consistent with the linear dependence of the rate constant of NO_x reduction on the vanadia loading shown in Figure 13.2; (iii) whatever the reaction mechanism is, a key step is represented by the formation of a reaction intermediate that decomposes rapidly and selectively to nitrogen and water (this is the case with the nitrosamide intermediate or of NH_4NO_2, the hydrated form of NH_2NO); and (iv) it has been recognized that other catalyst components, such as W or Mo and Ti surface sites, adsorb ammonia and participate in the reaction by providing a reservoir of ammonia adsorbed species [37].

In spite of the redox mechanism discussed above, the following Eley–Rideal (ER) kinetic expression has been proposed in the literature:

$$r_{NO} = \frac{k_c K_{NH_3} P_{NH_3} P_{NO}}{1 + K_{NH_3} P_{NH_3} + K_W P_{H_2O}} \quad (13.3)$$

where k_c is the intrinsic chemical rate constant and K_{NH_3} and K_W are the adsorption equilibrium constants of ammonia and water, respectively.

Equation 13.3 implies that the reaction between adsorbed NH_3 and gas-phase NO is rate determining and is also consistent with the fact that the reaction is virtually independent on the oxygen partial pressure above 1–2% v/v and is slightly depressed by water up to 5% v/v due to competition with ammonia for adsorption on the active sites. This equation is appropriate in the case of typical stationary SCR applications where a sub-stoichiometric $NH_3:NO_x$ feed ratio is employed to minimize the slip of unconverted ammonia. However, considering that water does not affect the NO_x removal in the concentration range of industrial interest (>5% v/v), the term $K_W P_{H_2O}$ can be incorporated in the kinetic constant k'_c, so that the following simplified rate equation has been successfully applied for the modeling of industrial reactors [38, 39, 27]:

$$r_{NO} = \frac{k'_c K_{NH_3} P_{NH_3} P_{NO}}{1 + K_{NH_3} P_{NH_3}} \tag{13.4}$$

If the $NH_3:NO$ ratio (α) is >1 and $K_{NH_3} P_{NH_3} > 1$, Equation 13.4 reduces to $r_{NO} = k'' P_{NO}$.

13.2.2
SCR Applications: Past and Future

The SCR process is best proven and is used worldwide due to its efficiency, selectivity and economics in coal-fired, oil-fired and gas-fired power stations, industrial heaters, chemical plants (e.g. NHO_3 tail gases, fluid catalytic cracking (FCC) regenerators, explosives manufacture plants) and in the steel industry. The introduction of stricter emission limits along with the high cost of alternative very efficient primary measures will favor in the future the more extensive application of the NO_x SCR control system in large gas turbines, incinerators and stationary diesel engines; novel applications are also predicted in the cement and glass industries.

In some of these applications, the reaction is preferably operated at low temperature and under transient conditions. For example, in the most energy-efficient arrangement of incinerators (Figure 13.3), the exhaust gases are first freed from dust in electrostatic precipitators, then dry adsorption agents are added to remove sulfur dioxide, halogens and dioxins, and finally they are passed through highly efficient bag filters. At the exit of the bag filters, the dust content is very low, the

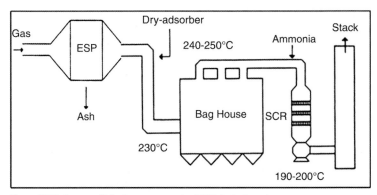

Figure 13.3 Energy-efficient arrangement of SCR in incinerators.

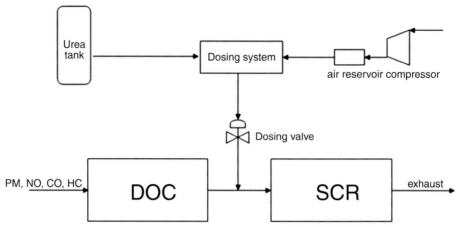

Figure 13.4 Layout of DOC and SCR in vehicles.

exhaust gas temperature is around 170–200 °C and the content of SO_2/SO_3 and other pollutant gases is almost zero. Then the exhaust gases enter the SCR reactor without any reheating or are reheated to a very limited extent. This eventually improves markedly the overall energy efficiency of the incineration process. In the case of stationary diesel engines, fast load changes are frequently experienced.

The SCR technology is also considered for the control of NO_x emission in diesel vehicles. Here the SCR catalyst is typically placed after the diesel oxidation catalyst (DOC), which is used to oxidize CO and UHCs and to convert part of the NO to NO_2. In this way, the SCR catalysts can take advantage of the fast SCR reaction to enhance significantly the de-NO_x efficiency at low temperature (Figure 13.4). The fast SCR reaction is based on the following stoichiometry:

$$NO + NO_2 + 2NH_3 \rightarrow 2N_2 + 3H_2O \quad (13.5)$$

The SCR catalyst in vehicles must operate under fast transients, be effective at low temperature where most NO_x emissions are produced and perform adequately in a wide temperature range (from 200 up to 500 °C).

Therefore, there is a strong motivation to develop a dynamic model of the SCR monolithic reactor suitable for extended temperature operation and to study the fast SCR reaction in view of future possible applications. In the following, we will focus on these two issues.

13.2.3
Modeling of the SCR Reactor

13.2.3.1 Steady-state Modeling of the SCR Reactor
A simple isothermal pseudo-homogeneous, single-channel, 1D model is typically adopted to model a monolith SCR reactor [27, 30, 38, 40–50], which implies uniform conditions over the entire cross-section of the monolith catalysts and accounts only

for the axial concentration gradients inside the channel, whereas the effects of interphase and intra-phase mass transfer are lumped into the overall effective pseudo-first-order rate constant k_{NO_x}:

$$k_{NO_x} = -AV \ln(1-x_{NO_x}) \tag{13.6}$$

where AV (area velocity) is the ratio of the volumetric flow rate to the overall geometric surface area of the monolith catalyst (m h^{-1}), x_{NO_x} is the conversion of NO_x ($0 < x_{NO_x} < 1$) and k_{NO_x} is given by

$$\frac{1}{k_{NO_x}} = \frac{1}{k_C} + \frac{1}{k_g} \tag{13.7}$$

where $k_C = \eta k_{SCR}$ (η = catalyst internal efficiency and k_{SCR} = intrinsic kinetic constant of the SCR reaction) is the effective rate constant of the chemical reaction that incorporates the effect of mass transfer in catalyst pores, and k_g is the gas–solid mass transfer coefficient. In fact, the SCR reaction operates under combined intra-particle and external diffusion control.

It has been demonstrated that k_g can be estimated by analogy with the Graetz–Nusselt problem governing heat transfer to a fluid in a duct with constant wall temperature ($Sh = Nu_T$) [30] and that the axial concentration profiles of NO and of NH_3 provided by the 1D model are equivalent and almost superimposed with those of a rigorous multidimensional model of the SCR monolith reactor in the case of square channels and of ER kinetics, which must be introduced to comply with industrial conditions for steady-state applications characterized by substoichiometric NH_3:NO feed ratio, that is, $\alpha < 1$ (see Section 13.2.1) [30].

The study of the intra-phase mass transfer in SCR reactors has been addressed by combining the equations for the external field with the differential equations for diffusion and reaction of NO and NH_3 in the intra-porous region and by adopting the Wakao–Smith random pore model to describe the diffusion of NO and NH_3 inside the pores [30, 44]. The solution of the model equations confirmed that steep reactant concentration gradients are present near the external catalyst surface under typical industrial conditions so that the internal catalyst effectiveness factor is low [27].

Laboratory data collected over honeycomb catalyst samples of various lengths and under a variety of experimental conditions were described satisfactorily by the model on a purely predictive basis. Indeed, the effective diffusivities of NO and NH_3 were estimated from the pore size distribution measurements and the intrinsic rate parameters were obtained from independent kinetic data collected over the same catalyst ground to very fine particles [27], so that the model did not include any adaptive parameters.

13.2.3.2 Unsteady-state Kinetics of the Standard SCR Reaction

The dynamic study of the ammonia SCR reactions was typically addressed by investigating the adsorption–desorption kinetics of ammonia and then the surface reactions.

Langmuir kinetics have usually been considered for ammonia adsorption–desorption kinetics [48, 51, 52]. On the other hand, Andersson *et al.* ruled out that the kinetics of ammonia adsorption–desorption can be described by a simple Langmuir approach because the complex shape of their ammonia temperature-programmed desorption (TPD) curves clearly indicated that the surface of the $V_2O_5/\gamma\text{-}Al_2O_3$ catalyst employed is not homogeneous in nature [51]. Still, a strong correlation between the estimates of adsorption and desorption parameters was apparent, so extrapolation of their kinetics to different experimental conditions is questionable [3].

Lietti and co-workers studied the kinetics of ammonia adsorption–desorption over V–Ti–O and V–W–Ti–O model catalysts in powder form by transient response methods [37, 52, 53]. Perturbations both in the ammonia concentration at constant temperature in the range 220–400 °C and in the catalyst temperature were imposed. A typical result obtained at 280 °C with a rectangular step feed of ammonia in flowing He over a $V_2O_5\text{–}WO_3/TiO_2$ model catalyst followed by its shut off is presented in Figure 13.5. Eventually the catalyst temperature was increased according to a linear schedule in order to complete the desorption of ammonia.

The results confirm that the adsorption of ammonia is very fast and that ammonia is strongly adsorbed on the catalyst surface. The data were analyzed by a dynamic isothermal plug flow reactor model and estimates of the relevant kinetic parameters were obtained by global nonlinear regression over the entire set of runs. The influences of both intra-particle and external mass transfer limitations were estimated to be negligible, on the basis of theoretical diagnostic criteria.

The activation energy for ammonia desorption was found to be close to zero and accordingly a non-activated ammonia adsorption process was considered ($E_a = 0$):

$$r_a = k_a^\circ C_{NH_3}(1-\theta_{NH_3}) \qquad (13.8)$$

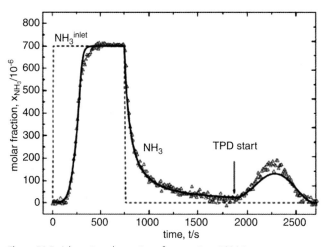

Figure 13.5 Adsorption–desorption of ammonia at 280 °C on a model $V_2O_5\text{–}WO_3/TiO_2$ catalyst. Dashed lines, inlet NH_3 concentration; triangles, outlet NH_3 concentration; solid lines, model fit with Temkin-type coverage dependence. Adapted from ref. [3].

This is consistent with the spontaneity of adsorption of a basic molecule, such as ammonia, over the acid catalyst surface.

In addition, a Temkin-type coverage dependence of the desorption energy had to be used to describe the data:

$$r_d = k_d^\circ \exp\left[-E_d^\circ(1-\gamma\theta_{NH_3})/RT\right]\theta_{NH_3} \tag{13.9}$$

Indeed, a simple Langmuir approach, which considers a constant value of the adsorption energy, was not appropriate. This is consistent with the presence of distinct types of acid sites (Lewis and/or Brønsted) characterized by different acid strength and associated with V, W and Ti surface sites [54].

The solid line in Figure 13.5 represents the model prediction: good agreement is apparent both in the case of the rectangular step feed change and shut off and the case of the subsequent TPD run. It is worth noting that the estimated kinetics account fairly satisfactorily for large variations in the ammonia surface coverage ($\theta_{NH_3} = 0$–0.8) and in the catalyst temperature ($T = 220$–$400\,°C$).

Similar experiments were performed for NO; they indicated that NO does not adsorb appreciably on the catalyst surface, in line with previous literature indications.

The transient response method has also been applied in our laboratories to investigate the kinetics of the surface reaction of strongly adsorbed NH_3 with gaseous or weakly adsorbed NO. Figure 13.6 shows typical results obtained over a V_2O_5–WO_3/TiO_2 model catalyst upon imposing a step feed change of ammonia in flowing He + 700 ppm NO 1% v/v O_2 followed by shut-off.

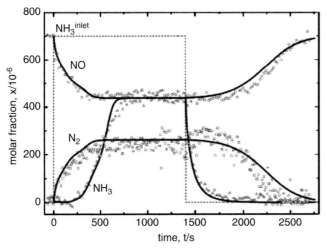

Figure 13.6 Step feed and shut-off of 700 ppm NH_3 in He + 700 ppm NO + 1% v/v O_2 over V_2O_5–WO_3/TiO_2 model catalyst at 220 °C. Dashed lines, inlet NH_3 concentration; solid lines, model fit with Temkin-type coverage dependence and modified Langmuir kinetics, Equation (13.10). (Adapted from ref. [52]).

The data demonstrate that the rate of the SCR reaction is unaffected by changes in the ammonia surface coverage at high coverage, since upon shut-off of ammonia the NH_3 concentration at the reactor outlet rapidly dropped to zero, but the NO and N_2 concentrations were not affected for several minutes. The results could be described fairly satisfactorily (solid line in Figure 13.6) by the adsorption–desorption kinetics of ammonia discussed above together with modified Langmuir kinetics for the de-NO_x surface reaction, which assume that the reaction is virtually independent of the ammonia surface coverage above a characteristic 'critical value' $\theta^*_{NH_3}$:

$$r_{NO} = k_C \, C_{NO} \, \theta^*_{NH_3} \left[1 - \exp\left(\theta_{NH_3}/\theta^*_{NH_3}\right)\right] \tag{13.10}$$

This empirical rate expression considers the active sites of the catalyst as only a fraction of the total adsorption sites for ammonia and is consistent with the presence of a 'reservoir' of ammonia adsorbed species which can take part in the reaction. The ammonia 'reservoir' is likely associated with poorly active but abundant W and Ti surface sites, which can strongly adsorb ammonia; in fact, $\theta^*_{NH_3}$ roughly corresponds to the surface coverage of V. Once the ammonia gas-phase concentration is decreased, the desorption of ammonia species originally adsorbed at the W and Ti sites can occur followed by fast readsorption. When readsorption occurs at the reactive V sites, ammonia takes part in the reaction. Also, the analysis of the rate parameter estimates indicates that at steady state the rate of ammonia adsorption is comparable to the rate of its surface reaction with NO, whereas NH_3 desorption is much slower. Accordingly, the assumption of equilibrated ammonia adsorption, which is customarily assumed in steady-state kinetics, may be incorrect, as also suggested by other authors [55].

The complete unsteady-state model for adsorption–desorption and surface reactions of the plug flow laboratory reactor was based on the following equations:

NH_3 mass balance on the catalyst surface:

$$\frac{\partial \theta_{NH_3}}{\partial t} = r_a - r_d - r_{NO} - r_{ox} \tag{13.11}$$

NH_3, NO and N_2 mass balances on the gas stream:

$$\frac{\partial C_{NH_3}}{\partial t} = -v \frac{\partial C_{NH_3}}{\partial z} + \Omega_{NH_3}(r_a - r_d - r_{NO} - r_{ox}) \tag{13.12}$$

$$\frac{\partial C_{NO}}{\partial t} = -v \frac{\partial C_{NO}}{\partial z} + \Omega_{NH_3} r_{NO} \tag{13.13}$$

$$\frac{\partial C_{N_2}}{\partial t} = -v \frac{\partial C_{N_2}}{\partial z} - \Omega_{NH_3}(r_{NO} + 0.5 r_{ox}) \tag{13.14}$$

where Ω_{NH_3} represents the catalyst NH_3 adsorption capacity (mol m_{cat}^{-3}).

In addition to adsorption and desorption of ammonia (r_a and r_d, respectively, with rate Equations 13.8 and 13.9) and to its surface reaction with NO to give nitrogen and

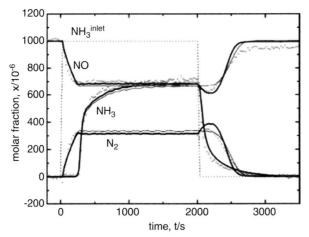

Figure 13.7 Step feed and shut-off of ammonia in He + 2% v/v O_2 + 1% v/v H_2O + 1000 ppm NO over V_2O_5–WO_3/TiO_2 model catalyst at 175 °C. Dashed lines, inlet NH_3 concentration; solid lines, model fit with MR rate law, Equation (13.16).

water (r_{NO}, with rate Equation 13.10), ammonia oxidation to N_2 (r_{ox}, with rate Equation 13.15) was also introduced in order to account for the direct oxidation of ammonia occurring at high temperatures:

$$r_{ox} = k_{ox}^{\circ} \exp(-E_{ox}/RT)\theta_{NH_3} \qquad (13.15)$$

More recently, specific attention was paid to transient experiments in the low-temperature region; accordingly, dedicated dynamic experiments have been performed on a commercial V_2O_5–WO_3/TiO_2 catalyst in a wide temperature range [56]. Figure 13.7 shows data recorded at 175 °C feeding NH_3 step pulses in flowing He + NO (1000 ppm) + O_2 (2% v/v) + H_2O (1% v/v).

It is worth noting that when the ammonia feed is shut off, the NO concentration trace decreases, passes through a minimum and then begins to increase, recovering the inlet value when the reaction is depleted. The symmetrical evolution of N_2 confirmed the occurrence of the SCR reaction. These effects were always observed in transient experiments performed at T < 250 °C and have also been reported in the past over different commercial SCR catalysts (see Figure 13.6 and refs [57] and [58]). They have been attributed to the inhibiting effect of excess NH_3, possibly caused by a competition between NO and ammonia in adsorbing on the catalyst. This effect is of limited interest for stationary applications of the standard SCR technology, but it may play a role in mobile applications, since it becomes more important at low temperature and significantly affects the dynamic response of the SCR systems.

This dynamic effect attributed to the inhibition of ammonia has recently been accounted for by a dual-site modified redox (MR) rate law, that assumes competition between ammonia and NO in adsorbing on the catalyst) [56]:

$$r_{NO} = \frac{k^{\circ\prime}_{NO}\exp[-(E_{NO}/RT)]C_{NO}\theta_{NH_3}}{[1+K'_{NH_3}(\theta_{NH_3}/1-\theta_{NH_3})]\left[1+K_{O_2}\left(C_{NO}\theta_{NH_3}/P_{O_2}^{\frac{1}{4}}\right)\right]} \quad (13.16)$$

In Figure 13.7, the model predictions obtained by using either the MR rate expression, Equation 13.16 (solid thick lines), or the modified ER kinetics, Equation 13.13 (solid thin lines), are compared: it clearly appears that the redox kinetics account better for the transient behavior upon shut-down of the ammonia feed.

Also, the MR kinetics provided a much better description than the modified ER kinetics of fast SCR transients originated by high-frequency NH_3 feed pulses in a stream of 1000 ppm NO, 2% v/v O_2 and 1% v/v H_2O and similar to those associated with the operation of SCR after treatment devices for vehicles [56]. Indeed, the MR model is definitively more chemically consistent than the modified ER model in view of the redox character of the standard SCR reaction.

13.2.3.3 Unsteady-state Models of the Monolith SCR Reactor

Andersson and co-workers developed a single-channel, transient, 1D non-isothermal model of the monolith SCR reactor to describe the removal of NO_x from the exhausts of diesel vehicles [51]. This model was developed for a coated monolith catalyst so that no account was taken of intra-porous diffusion effects. The model consisted of two energy balances for the gas and the solid phases (in order to follow thermal transients associated with changes in the exhaust gas temperature), two mass balances for NO and NH_3 in the gas bulk, two inter-phase continuity equations for NO and NH_3 and one mass balances for adsorbed ammonia. Accumulation terms were included only in the equations for the catalyst temperature and for the ammonia surface coverage; axial diffusion was neglected and gas–solid heat and mass transfer coefficients were estimated according to ref. [30]. Rate expression for ammonia adsorption–desorption determined by TPD runs were included and rate expression for the surface SCR reaction were derived from pilot-plant data. The model simulations were compared with results of dynamic tests collected with a 12 L engine equipped with a honeycomb SCR catalyst and operating with a stoichiometric injection of ammonia. A good match between measured and calculated NO conversions was achieved, but some adjustment to the rate parameters for ammonia desorption were required to account properly for temperature and ammonia slip.

In recent years, several proposals have been made for the numerical simulation of SCR catalysts, primarily of coated monoliths [59, 60], and for control-oriented purposes [61].

Tronconi *et al.* [46] developed a fully transient two-phase 1D + 1D mathematical model of an SCR honeycomb monolith reactor, where the intrinsic kinetics determined over the powdered SCR catalyst were incorporated, and which also accounts for intra-porous diffusion within the catalyst substrate. Accordingly, the model is able to simulate both coated and bulk extruded catalysts. The model was validated successfully against laboratory data obtained over SCR monolith catalyst samples during transients associated with start-up (ammonia injection), shut-down (ammonia

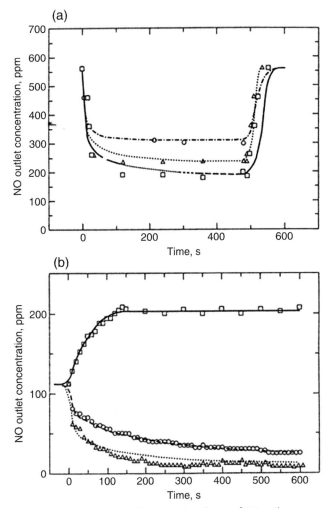

Figure 13.8 Experimental and simulated evolution of NO outlet concentration during start-up and shut-down of an SCR honeycomb reactor (open circles, $\alpha = 0.6$; triangles, $\alpha = 0.8$; squares, $\alpha = 1$) (a) and following step variation of NO inlet concentration (b). Adapted from ref. [46].

shut-down) and step changes of the feed composition at different temperatures and $NH_3:NO$ molar feed ratios, as shown in Figure 13.8.

The same model was applied to the simulation of typical transients occurring during the operation of industrial SCR monolith reactors in large power plants. In all cases it was found that the change in NO outlet concentration is considerably delayed with respect to the variation of the inlet NH_3 concentration. This is unfavorable for a feedback control system using the ammonia feed as the control variable and makes the adoption of a predictive dynamic model attractive. The dynamic model was also completed with account being taken of the SO$_2$ oxidation reaction in ref. [47].

Recently, the dynamic model of ref. [46] was specifically adapted for mobile SCR applications by Tronconi and co-workers [62]. The model consists of unsteady mass and enthalpy balances in the gas phase and in the solid phase:

Gas phase:

$$\frac{\partial C_j}{\partial t} = -\frac{v}{L}\frac{\partial C_j}{\partial z} - \frac{4}{d_h}k_{mt,j}\left(C_j - C_j^W\right) \quad j = NH_3, NO \tag{13.17}$$

$$\frac{\partial T_g}{\partial t} = -\frac{v}{L}\frac{\partial T_g}{\partial z} - \frac{4}{d_h}\frac{h(T_g - T_s)}{\rho_g C_p} \tag{13.18}$$

Solid phase:

$$0 = k_{mt,j}\left(C_j - C_j^W\right) + R_{eff,j} \quad j = NH_3, NO \tag{13.19}$$

$$\frac{\partial T_s}{\partial t} = \frac{h(T_g - T_s) - \sum_{j=1}^{NCG} \Delta H_j R_{eff,j}}{\rho_s C_{p,s} S_W (1 + S_W/d_h)} \tag{13.20}$$

where NCG is the number of gaseous species. Gas–solid mass and heat transfer coefficients are evaluated by analogy with the Graetz–Nusselt problem for developing laminar flow in square ducts [30], as in the case of steady-state models.

The strong intra-phase diffusion limitations are accounted for by the following equations for diffusion–reaction of the reactants in the catalytic monolith wall (Equation 13.21) with the appropriate boundary conditions (Equation 13.22):

$$0 = D_{eff,j}\frac{\partial^2 C_j^*}{\partial x^2} + S_W^2 R_j \quad j = NH_3, NO \tag{13.21}$$

$$R_{eff,j} = -\frac{D_{eff,j}}{S_W}\frac{\partial C_j^*}{\partial x}\bigg|_W \quad j = NH_3, NO \tag{13.22}$$

where R_j represents the volumetric intrinsic rate of formation of species j. Build-up of reactants in the gas phase within the catalyst pores is neglected, whereas accumulation–depletion of NH$_3$ adsorbed on the catalyst surface, which is the controlling factor in the dynamics of SCR monolith reactors, is described by

$$\Omega_{NH_3}\frac{\partial \vartheta_{NH_3}}{\partial t} = R_{NH_3*} \tag{13.23}$$

Accordingly, in addition to rate parameters and reaction conditions, the model requires the physicochemical, geometric and morphological characteristics (porosity, pore size distribution) of the monolith catalyst as input data. Effective diffusivities, $D_{eff,j}$, are then evaluated from the morphological data according to a modified Wakao–Smith random pore model, as specifically recommended in ref. [63].

To describe the $NH_3 + NO + O_2$ (standard SCR) reacting system, NH_3 adsorption–desorption, ammonia oxidation to nitrogen and standard SCR have been considered with the kinetics already presented in the previous section.

The resulting set of model partial differential equations (PDEs) were solved numerically according to the method of lines, applying orthogonal collocation techniques to the discretization of the unknown variables along both the z and x coordinates and integrating the resulting ordinary differential equation (ODE) system in time.

The model was validated at Daimler against a large number of transient experiments with step pulses of NH_3 in a continuous flow of NO over full-scale monolith catalyst samples during test-bench runs with diesel engine exhausts. Validation maps were generated using the steady-state measured engine operating points and the deviation between simulated NO_x conversion and experimental data was typically below 4%. The reactor model was also validated by performing engine test-bench experiments under real conditions in the temperature range 150–450 °C with ETC (European transient cycle) and ESC (European stationary cycle) measurements [62].

Figure 13.9 shows the model simulations (gray lines) and experimental results (solid black lines) for NO_x concentrations downstream of the catalyst along with the inlet NO_x concentration values (dotted black lines). The comparison between measurement and simulation for the NO_x reveals that excellent agreement was obtained for both cycles. The mean error of the model predictions in the de-NO_x efficiency was not higher than 3–4%. These results confirm that the model is able to predict accurately the dynamic behavior of SCR monolith catalysts under fast transient conditions. Such results were achieved only by using the MR rate expression (Equation 13.16); indeed, the adoption of a classical ER rate law or of modified ER kinetics (Equation 13.10) resulted in a significantly less accurate prediction of the NO_x conversion during the fast transient parts of the test cycles.

The ammonia SCR for heavy-duty diesel engines has been commercialized by Daimler since 2005 in Europe and the USA under the trade name Bluetec as an after-treatment system able to satisfy the Euro IV and Euro V emission standards.

13.2.4
Fast SCR

13.2.4.1 Mechanism of Fast SCR

To reduce the emissions during cold start of the engine and extended operation at low loads in vehicles, it has been proposed to use the so-called fast SCR reaction in which NH_3 reacts with NO and NO_2 according to

$$4NH_3 + 2NO + 2NO_2 \rightarrow 4N_2 + 6H_2O \tag{13.24}$$

This reaction has been known since the 1980s to be faster by one order of magnitude than the standard SCR (Equation 13.1) at low temperatures [64].

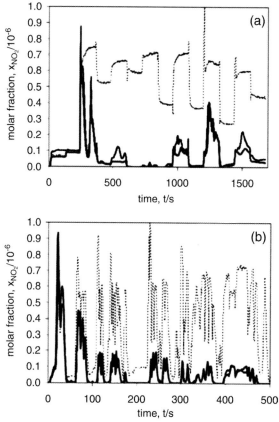

Figure 13.9 Validation of the dynamic model of the monolith SCR reactor during ESC and ETC tests. All concentrations are normalized by the respective maximum inlet value during the test cycle. Dotted black lines, inlet values; solid black lines, outlet measurements; gray lines, outlet simulations. Adapted from ref. [62].

Indeed, it is possible to install on vehicles a preoxidizing catalyst upstream of the SCR catalyst that converts part of the NO to NO_2; the DOC also oxidizes CO and UHCs to CO_2.

The study of the mechanism of the fast SCR over V–W–Ti–O catalysts was addressed first by Koebel and co-workers [65–68]. They suggested that (i) the reoxidation of the catalyst is rate determining at low temperature in the redox cycle of standard SCR catalyst, (ii) NO_2 reoxidizes the catalyst faster than O_2; the NO_2-enhanced reoxidation of the catalyst was demonstrated by *in situ* Raman experiments, (iii) the reaction occurs via the nitrosamide intermediate in both standard and fast SCR and (iv) ammonium nitrate is considered an undesired side-product.

The mechanism of fast SCR over a zeolite-based catalyst has also been addressed by Sachtler and co-workers using an IR technique [69, 70]. They concluded that nitrogen is produced through fast decomposition of ammonium nitrite (the hydrated form of nitrosamide), which is formed from equimolar NO/NO_2 feeds via N_2O_3 and its reaction with water and ammonia:

$$NO + NO_2 \rightarrow N_2O_3 \tag{13.25}$$

$$N_2O_3 + H_2O + 2NH_3 \rightarrow 2NH_4NO_2 \rightarrow 2N_2 + 4H_2O \tag{13.26}$$

Ammonium nitrite is unstable above 100 °C and the sum of reactions (13.25) and (13.26) results in the fast SCR reaction (13.24). This reaction scheme can explain the optimal 1:1 $NO:NO_2$ feed ratio of the fast SCR on the basis of well-known chemistry; however, it cannot explain all of the several products (N_2, NH_4NO_3, N_2O) observed in experiments covering the full range of $NO:NO_x$ feed ratios [71].

Tronconi and co-workers, through an extensive study of the reactivity of NH_3–NO/NO_2 mixtures with different $NO:NO_2$ ratios over V–W–Ti–O SCR catalysts, proposed a novel mechanism for the fast SCR reaction that has been validated step by step by dedicated experiments [71–74].

The main features of this mechanism include:

1. Fast dimerization of NO_2 followed by reaction with water to give nitrous acid and nitric acid (this chemistry is well known from nitric acid plants):

$$2NO_2 \rightarrow N_2O_4 \tag{13.27}$$

$$N_2O_4 + H_2O \rightarrow HNO_2 + HNO_3 \tag{13.28}$$

2. In the presence of NH_3, nitrous acid forms ammonium nitrite, which readily decomposes to N_2:

$$HNO_2 + NH_3 \rightarrow [NH_4NO_2] \rightarrow N_2 + 2H_2O \tag{13.29}$$

3. Nitric acid is able to oxidize NO to NO_2, being reduced to nitrous acid:

$$HNO_3 + NO \rightarrow HNO_2 + NO_2 \tag{13.30}$$

4. In the presence of adsorbed NH_3, nitrous acid produces N_2 via decomposition of NH_4NO_2 [reaction (13.29)].
5. NH_4NO_3, formed by reaction of nitric acid and ammonia, can also oxidize NO to NO_2 and therefore it can participate in the same reaction pathway as described above.

The stoichiometry of the fast SCR reaction (Equation 13.24), is recovered once the overall reaction summarizing the chemistry of the NO_2–NH_3 system is coupled with that describing the addition of NO to this reacting system. The proposed reaction scheme accounts for the optimal equimolar $NO:NO_2$ feed ratio and also for the selectivity to all the observed products, namely N_2, NH_4NO_3, HNO_3 and N_2O (originated by decomposition of NH_4NO_3 at high temperature). Furthermore, it is in agreement with the observed kinetics of the fast SCR reaction, which at low temperature is limited by the reaction between HNO_3 and NO.

It is worth mentioning that the same chemistry is involved when surface nitrite and nitrate are considered instead of gas-phase nitrous and nitric acid. Indeed, surface nitrate has been suggested to take part in the reoxidation of the reduced catalyst sites, thus accounting for the higher rate of the fast SCR reaction compared with the standard SCR reaction [74].

The capability of NO to reduce nitrates, providing a pathway for the production of ammonium nitrite and thus of nitrogen, has also been demonstrated recently by Weitz and co-workers, mainly on the basis of IR data collected over BaNa-Y zeolite [75]; however, according to a parallel additional route NO would also react with NO_2 to form N_2O_3 and then nitrogen [reactions (13.25) and (13.26)], as already discussed. The oxidation of NO by surface nitrates over a Pt–Ba/Al_2O_3 catalyst has also been reported by Olsson et al. [76], whereas the formation of surface nitrates from NO_2 on bare Al_2O_3 has been reported by Apostolescu et al. [77] and previously observed in our laboratories also [78].

13.2.4.2 Unsteady-state Models of the Monolith NO/NO$_2$/NH$_3$ SCR Reactor

To describe the $NH_3 + NO/NO_2$ reaction system over a wide range of temperatures and $NO_2:NO_x$ feed ratios in addition to ammonia adsorption–desorption, ammonia oxidation and standard SCR reaction with the associated kinetics already discussed in Section 2.3.2, the following reactions and kinetics have been considered by Chatterjee and co-workers [79]:

- NO_2 disproportion: $2NO_2 + H_2O \rightarrow HNO_2 + HNO_3$

$$r_{amm} = k^\circ_{amm} \exp\left(\frac{-E_{amm}}{RT}\right) \left(\frac{P_{H_2O} C^2_{NO_2} - C_{HNO_3} C_{HNO_2}}{K_{EQamm}}\right) \tag{13.31}$$

- reaction of nitrous acid and ammonia: $HNO_2 + NH_3 \rightarrow [NH_4NO_2] \rightarrow N_2 + 2H_2O$

$$r_{nit} = k^\circ_{nit} \exp\left(\frac{-E_{nit}}{RT}\right) \theta_{NH_3} C_{HNO_2} \tag{13.32}$$

- adsorption–desorption equilibrium of ammonium nitrate:

$$r_{dec} = k_{adsnit} \theta_{NH_3} C_{HNO_3} - k^\circ_{dec} \exp\left(\frac{-E_{dec}}{RT}\right) \theta_{NH_4NO_3} \tag{13.33}$$

- reaction of HNO_3 and NO: $HNO_3 + NO \rightarrow HNO_2 + NO_2$

$$r_{FST} = k^\circ_{FST} \exp\left(\frac{-E_{FST}}{RT}\right) \gamma_{FST} \left(\frac{C_{NO} C_{HNO_3} - C_{NO_2} C_{HNO_2}}{K_{EQFST}}\right) \tag{13.34}$$

with $\gamma_{FST} = 1/[1 + K_{LH_2}\theta_{NH_3}/(1-\theta_{NH_3}-\theta_{NH_4NO_3})]$;

- NO$_2$ SCR: $6NO_2 + 8NH_3 \rightarrow 7N_2 + 12H_2O$

$$r_{NO_2r} = k°_{NO_2r}\exp\left(\frac{-E_{NO_2r}}{RT}\right)\theta_{NH_3}C_{NO_2} \qquad (13.35)$$

- formation of N$_2$O: $NH_4NO_3 \rightarrow N_2O + 2H_2O$

$$r_{N_2O} = k°_{N_2O}\exp\left(\frac{-E_{N_2O}}{RT}\right)\theta_{NH_3}C_{HNO_2} \qquad (13.36)$$

The rate parameters in Equations (13.31)–(13.36) were estimated by regression of transient microreactor experiments performed over commercial V_2O_5–WO_3/TiO_2 catalyst with medium–high V content following a sequential fitting strategy to minimize correlations in view of the large numbers of parameters required to account for the comprehensive reaction scheme.

The model of the honeycomb monolith SCR converter employed in the study is an extension of the heterogeneous dynamic 1D + 1D model of the single monolith channel already developed for standard SCR [62]. Briefly, the model includes the unsteady differential mass balances of six gaseous species (NH$_3$, NO, NO$_2$, N$_2$, N$_2$O and HNO$_3$) and of two adsorbed species (NH$_3^*$ and NH$_4$NO$_3^*$), while a pseudo-steady-state assumption is applied to HNO$_2$. Enthalpy balances for the gas and solid phases are included to account for the thermal effects. In addition, intra-phase diffusion limitations are accounted for by equations for diffusion–reaction of the gaseous reactants in the intra-porous monolith walls, with effective diffusivities evaluated from the catalyst morphological data according to a modified Wakao–Smith random pore model.

The model was validated against heavy duty and passenger car diesel engine test bench experiments. A good correlation was obtained between ESC and ETC experiments and simulation with 0 and 0.5% NO$_2$:NO ratios and a 'virtual oxidation' catalyst. The virtual oxidation catalyst model was realized by placing an oxidation catalyst model in front of the SCR catalyst.

Table 13.4 indicates that there is still a significant increase in the total NO$_x$ conversion compared with the simulations without NO$_2$ in the inlet feed. However, an appropriate design of the oxidation catalyst is necessary.

Additionally, the presence of NO$_2$ in the inlet feed allows a reduction in the catalyst volume by keeping the SCR efficiency at the same level.

Table 13.4 Total NO$_x$ conversion efficiencies (%) with $\alpha = 1$.

NO$_x$ conversion	ESC	ETC
0% NO$_2$	90.7	87.5
50% NO$_2$	94.0	94.9
DOC NO$_2$	93.6	91.8

The extended SCR model has been integrated with other catalyst and diesel particulate filter (DPF) models in the Daimler exhaust gas after-treatment systems simulation environment ExACT. The model was also extended to other SCR catalytic materials [79].

13.3
NO$_x$ Storage Reduction

13.3.1
NSR Technology

The NSR catalysts, also referred to as NO$_x$ adsorbers or lean NO$_x$ traps (LNTs), represent an alternative to urea/ammonia SCR catalysts for the removal of NO$_x$ under lean conditions.

In the NSR process, NO$_x$ are stored under lean conditions in the form of nitrites and nitrates and are reduced to nitrogen and water under rich conditions by UHCs, CO and H$_2$ present in the exhausts or by HCs injected into the exhaust gases. Accordingly, the engine operates under cyclic conditions, by alternating long lean periods with short rich excursions.

The NSR catalysts typically consist of an NO$_x$ storage component, such as an alkaline earth metal oxide, and of a noble metal that catalyzes the oxidation of NO$_x$, CO and UHCs and the reduction of stored NO$_x$ during the lean and rich phases, respectively. Catalysts made with a high surface area alumina support on which BaO and Pt are well dispersed represent typical materials for NSR applications. Among alkali and alkaline earth elements, Ba has been reported as the most effective element to store the NO$_x$ [80], but commercial catalyst formulations may also include potassium. Other catalyst components commonly added to the ternary Pt–BaO/Al$_2$O$_3$ system comprise TiO$_2$, which minimizes the adsorption of SO$_x$, together with Rh and ZrO$_2$ or CeO$_2$–ZrO$_2$, which are used to promote the formation of hydrogen through steam reforming and water gas shift reactions [4].

So far, the NSR technology has been commercialized for passenger cars equipped with lean burn gasoline engines in Japan, where low-sulfur gasoline has been available since the 1990s [4–6, 80]. Indeed, the major present limitation of NSR catalysts is associated with the fact that sulfates are more stable than nitrates and tend to poison the catalyst basic sites where NO$_x$ are adsorbed [81]. Accordingly, the catalyst must be periodically regenerated [82]; this is accomplished by decomposition of the sulfates at high temperature under reducing conditions to give SO$_2$ and H$_2$O. The durability of the NSR catalysts is not yet fully satisfactory: materials more stable at high temperature and more resistant to poisoning from sulfates along with less severe regeneration strategies of catalyst sulfates must be developed. Another major point is the possibility of running the engine at lambda 1 conditions without affecting performance, drivability and HC emissions. Taking into account that no large layout modifications are needed, NSR systems could be appealing for low segment vehicles. The use of the NSR technology in vehicles will be favored by the introduction of ultra-low-sulfur diesel fuel in the US market since October 2007 as per US Environmental Protection Agency (EPA) regulations (<15 ppm S) and by the decrease in the sulfur content of fuels in Europe due to new legislation requirements (<10 ppm S by 2009).

The NSR technology has also been proposed for stationary applications by EmeraChem under the trade name SCONOx [83]. The US EPA declared this technology the 'lowest achievable emission rate' (LAER) technology for NO_x control from stationary gas turbine installations. In contrast to the SCR process, SCONOx requires no ammonia and is able to remove CO and UHCs while simultaneously absorbing NO_x on a proprietary catalyst sorber. This sorber is periodically regenerated using a superheated steam–dilute hydrogen gas mixture which is produced on-site and in an automated 'on-demand' basis, using the same fuel as utilized by the turbine. The regeneration process results in the chemical reduction of the adsorbed NO_x compounds into nitrogen and water vapor. To accomplish the catalyst regeneration in an oxygen-free environment, the system is composed of several separate modules which are alternatively absorbing NO_x or operated in the regeneration mode.

The catalyst, used in the form of a ceramic honeycomb monolith, is constituted, as in mobile applications, by a noble metal and an absorber element, such as potassium, deposited on a γ-Al_2O_3 wash-coat layer. In the oxidation and absorption cycle, the SCONOx catalyst works by simultaneously oxidizing CO and UHCs to CO_2 and H_2O, while NO_x are captured on the adsorber compound. Catalyst regeneration is accomplished by passing a controlled mixture of regeneration gases across the surface of the catalyst in the absence of oxygen.

The SCONOx process has been used for the removal of NO_x in gas turbines in a few installations in the USA (32 MW Sunlaw Federal cogeneration facility, 5 MW Wyeth Biopharma plant, 15 MW University of California cogeneration facility in San Diego). However, the high cost of the SCONOx process, primarily due to the cost of the noble metal in the catalysts, represents a serious limitation on its extensive use.

The potential of NSR catalysts primarily in the removal of NO_x in vehicles has motivated in the last few years extensive investigations from both academia and the motor industry and numerous papers have been published dealing with fundamental and practical aspects of this technology [4–6, 80–82, 84–132]. In the following, the chemistry involved in the storage of NO_x and in the reduction of stored NO_x is addressed.

13.3.2
Storage of NO_x

13.3.2.1 Mechanistic Features
The mechanism involved in the storage of NO_x was extensively investigated in our laboratories over a model Pt–Ba/Al_2O_3 (1:20:100 w/w) system, the corresponding binary Ba/Al_2O_3 (20:100 w/w) and Pt/Al_2O_3 (1:100 w/w) samples and the bare γ-Al_2O_3 support for comparison purposes [78, 97–100]. The γ-alumina support was obtained by calcination at 700 °C of a commercial alumina material; Ba/Al_2O_3 and Pt/Al_2O_3 catalysts were prepared by the incipient wetness impregnation of the alumina

support with aqueous solutions of dinitrodiammine Pt and of barium acetate. The powder was dried overnight in air at 80 °C and then calcined at 500 °C for 5 h. Pt–Ba/Al_2O_3 samples with different Ba loadings (5–30% w/w) were prepared by impregnating the calcined Pt/Al_2O_3 (1:100) with different solutions of barium acetate [78, 97].

The adsorption of NO_x under lean conditions was studied by imposing a step change of NO and NO_2 feed concentrations in the presence and absence of excess oxygen over the reference catalysts in a fixed-bed flow microreactor operated at 350 °C and analyzing the transient response in the outlet concentrations of reactants and products [transient response method (TRM)]. The adsorption/desorption sequence was repeated several times in order to condition the catalytic systems fully: due to the regeneration procedure adopted (either reduction with 2000 ppm H_2 + He or TPD in flowing He), BaO was the most Ba-abundant species present on the catalyst surface. FT-IR spectroscopy was used as a complementary technique to investigate the nature of the stored NO_x species.

The adsorption of NO_2 was investigated first, because it was customarily believed that NO_2 is the intermediate in the adsorption of NO in the presence of oxygen [5, 94].

As shown in Figure 13.10, the storage of 1000 ppm NO_2 in He over Ba/Al_2O_3 (20:100 w/w) is accompanied by the evolution of NO in the gas phase which is observed with no dead time. The uptake of NO_2 is fairly effective since large amounts of NO_x are stored at the end of the pulse, near 9×10^{-4} mol g_{cat}^{-1}, and results in the formation of only nitrates [bands at 1320, 1420 and 1560 cm^{-1} in inset (b) in Figure 13.10].

The adsorbed nitrates are related to the Ba component as the surface of the alumina support was almost completely covered by Ba, as indicated by FT-IR data [98], which showed the disappearance of OH groups of the alumina support and in line with estimation of the Ba coverage [101]. In agreement with the literature [5, 87, 97], the nitrates are formed through NO_2 disproportionation with the following global stoichiometry (Equation 13.37):

$$BaO + 3NO_2 \rightarrow Ba(NO_3)_2 + NO\uparrow \quad (13.37)$$

which also accounts for the evolution of gaseous NO.

It is worth noting that the molar ratio of NO evolved to NO_2 consumed over Ba/Al_2O_3 is initially <3, as expected from reaction (13.37), which indicates that the NO_2 uptake at the beginning of the pulse does not obey the overall stoichiometry of the reaction. This has been explained in the literature, considering that the NO_2 disproportionation consists of consecutive elementary steps and that the first step is faster than the following ones, which account for the evolution of NO (Equations 13.38–13.40) [94]:

$$BaO + NO_2 \rightarrow BaONO_2 \quad (13.38)$$

$$BaONO_2 \rightarrow BaO_2 + NO \quad (13.39)$$

$$BaO_2 + 2NO_2 \rightarrow Ba(NO_3)_2 \quad (13.40)$$

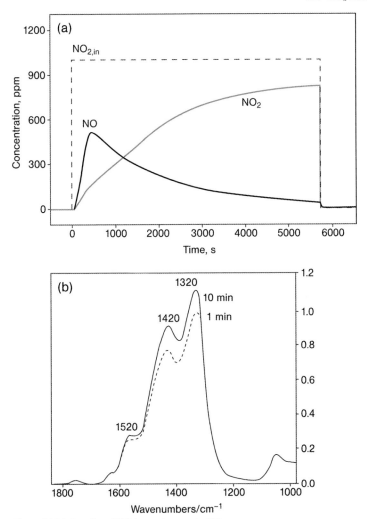

Figure 13.10 Results of TRM experiments in NO_2 (1000 ppm) + He at 350 °C over Ba/Al$_2$O$_3$ (20:100 w/w). Inset (b): FT-IR spectra recorded after 1 and 10 min exposure times to 5 mbar of NO_2 at 350 °C. Adapted from ref. [78].

Here the oxidation of BaO to BaO$_2$ by NO_2 is suggested. The disproportionation of NO_2 has also been described by a first step where nitrite and nitrate species are formed [BaO + 2NO$_2$ → Ba(NO$_2$)(NO$_3$)], followed by a slower second step where nitrites are oxidized to nitrates by NO_2 with release of NO in the gas phase [Ba(NO$_2$)(NO$_3$) + NO$_2$ → Ba(NO$_3$)$_2$ + NO↑] [133].

Similar results have been obtained over Pt–Ba/Al$_2$O$_3$ (1:20:100 w/w), but in this case the NO outlet concentration is related to NO_2 disproportionation and to NO_2 decomposition (Figure 13.11); in fact, O$_2$ formation was observed which originates

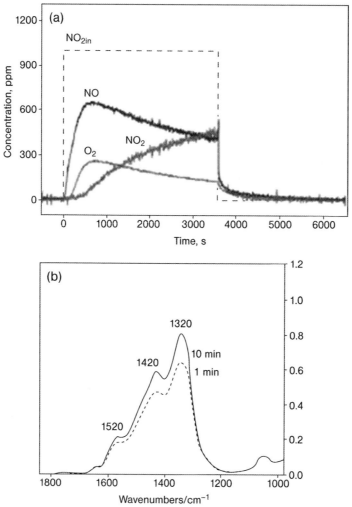

Figure 13.11 Results of TRM experiments in NO_2 (1000 ppm) + He at 350 °C over Ba/Al_2O_3 (20:100 w/w). Inset (b): FT-IR spectra recorded after 1 and 10 min exposure times to 5 mbar of NO_2 at 350 °C. Adapted from ref. [78].

from NO_2 decomposition at Pt sites. Also in this case IR spectra showed the formation of nitrate species only, already from the very beginning of the experiment.

Figure 13.12 shows the results obtained during adsorption of 1000 ppm NO in the presence of oxygen (3% v/v).

On Ba/γ-Al_2O_3, the NO adsorption proceeded as illustrated in Figure 13.12a, with small quantities of NO_x species adsorbed on the catalyst surface. The FT-IR spectra [inset (b) in Figure 13.12] showed the progressive formation of nitrite species (band at 1220 cm^{-1}) up to a 10 min exposure time, along with small amounts of nitrate

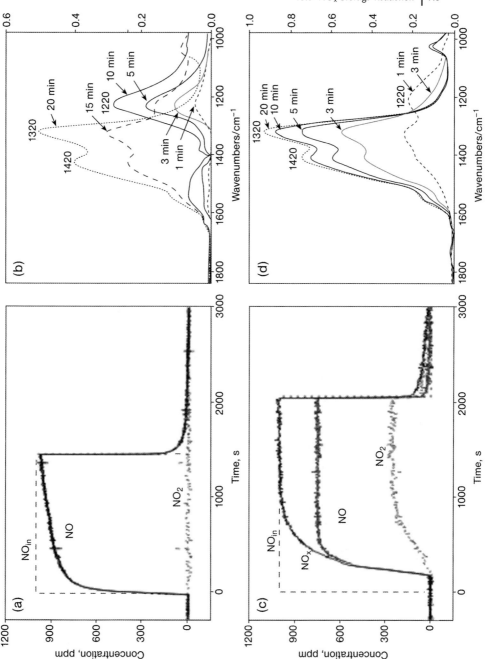

Figure 13.12 Storage of NO (1000 ppm) in 3% O_2 + He at 350 °C: TRM experiments over (a) Ba/Al$_2$O$_3$ (20:100 w/w); and (c) Pt–Ba/Al$_2$O$_3$ (1:20:100 w/w) at 350 °C; FT-IR spectra after 1, 3, 5, 10, 15, 20 min exposure times to NO–O$_2$ (1:4, p_{NO} = 5 mbar) over inset (b) Ba/Al$_2$O$_3$ (20:100 w/w) and inset (d) Pt–Ba/Al$_2$O$_3$ (1:20:100 w/w). Adapted from ref. [78].

species. At higher exposure times, the band due to nitrite species decreased in intensity and disappeared completely after 20 min. At the same time, bands characteristic of nitrate species (1420, 1320, 1030 cm^{-1} and shoulders at 1560 cm^{-1}) developed, so that after 20 min of exposure only nitrates were evident in the spectra.

Upon admission of 1000 ppm NO in the presence of 3% v/v oxygen over the Pt–Ba/γ-Al$_2$O$_3$ sample (Figure 13.12c), both the NO and NO$_2$ outlet concentrations show a significant delay. Then the NO concentration increases, followed by that of NO$_2$. NO$_2$ formation is ascribed to the oxidation of NO by O$_2$ according to the stoichiometry of reaction (13.41):

$$NO + \frac{1}{2}O_2 \rightarrow NO_2 \tag{13.41}$$

The breakthrough time observed in the NO$_x$ concentration profile indicates that during the initial part of the pulse the NO fed to the reactor is completely stored on the catalyst surface. As shown by FT-IR spectroscopy [inset (d) in Figure 13.12], the initial NO uptake occurs primarily in the form of nitrites, which are readily transformed into nitrates so that at the end of the NO pulse, at catalyst saturation, nitrates were the prevalent species. Notably, the rate of both nitrite formation and their oxidation to nitrates was higher on Pt–Ba/γ-Al$_2$O$_3$ than on Ba/γ-Al$_2$O$_3$, thus indicating a catalytic role of Pt.

FT-IR spectra collected under operating conditions [98] during NO adsorption in O$_2$ (3%) + He over Pt–Ba/γ-Al$_2$O$_3$ (1:20:100 w/w) confirmed that (i) nitrites represent the major NO$_x$ adsorbed species at the initial stage of adsorption before NO$_x$ breakthrough (observed after 4 min), (ii) nitrites are progressively transformed into nitrates during storage and (iii) a parallel route involving the direct formation of nitrates and/or adsorption of NO$_2$, although of minor importance, cannot be excluded.

All the previous data lead to the proposal of the reaction pathway shown in Figure 13.13 for the storage of NO under lean conditions over BaO in Pt–Ba/alumina catalysts.

NO is effectively stored through a stepwise oxidation at a Pt site followed by adsorption at a neighboring Ba site to form Ba nitrites; these species are then oxidized to nitrates. This pathway is referred to as the 'nitrite route'. The oxidation of nitrites to nitrates is catalyzed by Pt and probably involves NO$_2$ formed by NO oxidation on Pt

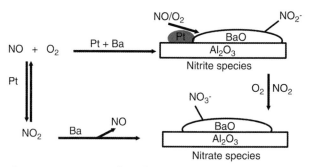

Figure 13.13 Reaction pathway for NO$_x$ adsorption under lean conditions over Pt–Ba/Al$_2$O$_3$.

and/or O_2 species. Accordingly, a cooperative effect between Pt–Ba neighboring couples might be relevant for this route.

NO is also oxidized to NO_2 over Pt in the presence of oxygen. NO_2 is adsorbed on the Ba sites to form Ba nitrates through the disproportionation reaction (13.37), which is accompanied by the evolution of NO and results in the formation of nitrates. This is referred to as the 'nitrate route'.

It is worth noting that the role of the alumina support, which showed a non-negligible NO_x adsorption capacity in the presence of NO_2 [78], is not considered in the scheme due to its almost complete coverage by the Ba component in the Pt–Ba/γ-Al_2O_3 (1:20:100 w/w) reference sample.

It has also been suggested that the presence of Pt–Ba couples (i.e. the existence of a Pt–Ba interaction) is relevant for the 'nitrite' route [78, 99, 122, 128].

To elucidate this point further, the effect of the Ba loading was investigated since it was expected that this would affect the number of Pt–Ba neighboring couples. Accordingly, catalysts with different Ba loadings, up to 30% w/w, were prepared and tested in the adsorption of NO_x.

The results obtained by applying a rectangular step feed of 1000 ppm NO in He + O_2 (3% v/v) showed that both the NO_x breakthrough and the storage capacity of the catalysts increase with the Ba content up to a maximum which is observed for the catalyst with a Ba loading of 23% w/w. Moreover, the increase in the Ba loading resulted in a strong increase in the percentages of Ba involved in the storage, namely from 4% for the sample with 5% w/w Ba up to $\sim$33% for the Pt–Ba/Al_2O_3 (1:23:100 w/w) catalyst. Notably, the estimation of the Ba coverage for the different samples showed that a value close to one was almost achieved for Ba contents of 16–20% w/w; this suggests that the maximum storage capacity was presented by the systems which are characterized by the formation of the monolayer.

The existence of Pt–Ba interactions has been confirmed by FT-IR analysis of CO chemisorption measurements. A band characteristic of CO linearly adsorbed on Pt sites is observed in all cases; however, the intensity of the band decreases on increasing the Ba loading (due to the decrease of the Pt dispersion) and shifts towards lower energy (from 2072 to 2049 cm^{-1}) according to the increase in the system basicity. Hence the data indicate a strong interaction between Pt and the basic oxygen anions of the Ba phase, thus suggesting that the exposed Pt sites and the Ba component are in close proximity [98].

Therefore, it has been suggested that on increasing the Ba loading up to around 20% w/w, the number of Pt–Ba neighboring couples is maximized and the storage of NO_x through the nitrite route is optimally favored.

13.3.2.2 Kinetics

The storage of NO_x on BaO has also been analyzed in the literature by adopting a detailed mechanistic kinetic model [76]. The model consists of 10 elementary reversible steps: adsorption of oxygen, adsorption of NO, adsorption of NO_2 and oxidation of NO to NO_2 at the Pt sites, adsorption of NO_2 at the BaO sites, release of NO into the gas phase and oxidation of BaO to BaO_2, formation of $BaONO_3$ and of

Ba(NO$_3$)$_2$, decomposition of BaO$_2$ to BaO and oxygen and the reversible spillover of NO$_2$ between Pt sites and BaO sites. Essentially the model assumes that the adsorption of NO$_x$ proceeds through the nitrate route and does not consider the nitrite route. Olsson et al. [76] estimated part of the rate parameters in their model from theoretical considerations, part were taken from the literature or calculated from thermodynamic constraints and part were estimated by fitting a set of experimental data.

The model of Olsson et al. was employed by Scotti et al. [103] to simulate the experimental results of NO adsorption over the physical mixture of Pt/Al$_2$O$_3$ (1:100 w/w) and Ba/Al$_2$O$_3$ (20:100 w/w) and over Pt–Ba/Al$_2$O$_3$ (1:20:100 w/w). The same parameters as estimated by Olsson et al. were used for the simulation, but Pt and Ba dispersions were fitted to experimental data.

The simulations showed that the model satisfactorily reproduces the results collected over the physical mixture upon NO step addition in the presence of oxygen. Both the oxidation of NO to NO$_2$ and the dead time for the breakthrough of NO$_x$ are reasonably accounted for by the model, as indeed expected according to the nitrate route. However, in the case of the Pt–Ba/alumina ternary sample, the dead time in the NO$_x$ breakthrough upon a step change of NO in the presence of oxygen is reasonably well predicted by the model but the oxidation of NO to NO$_2$ is markedly overestimated. If the parameter related to the Pt dispersion in the model is lowered to account for the correct NO to NO$_2$ oxidation level, the time delay in the NO$_x$ breakthrough is no longer predicted. Therefore, the model of Olsson et al. [76] is able to account for the NO$_x$ adsorption data over the mechanical mixture, but is not able to describe at the same time the oxidation of NO to NO$_2$ and the NO breakthrough in the Pt–Ba/alumina ternary sample. In fact, these two catalyst properties are strictly inter-related within the nitrate route, so that it appears that this route cannot account for the storage of NO in the presence of excess oxygen over the ternary catalyst.

A simplified kinetic model which includes both the nitrite and nitrate routes is currently under development in our laboratories [133]. The model is based on five reactions including the NO oxidation to NO$_2$ over Pt, the NO$_2$ disproportionation on Ba according to a two-step pathway, the formation of surface nitrites from NO + O$_2$ and the nitrate decomposition reaction. Preliminary results showed that this simplified kinetic model is able to account for both the NO oxidation to NO$_2$ and the NO$_x$ breakthrough over Pt–Ba/Al$_2$O$_3$ samples having different Ba loadings, provided that a rate expression which includes NO$_2$ inhibition is assumed to describe the oxidation of NO over Pt, as recently suggested [134]. However, it has been noted that, by increasing the kinetic constant of the NO oxidation reaction, the kinetic model satisfactorily describes the NO$_x$ storage even if the nitrite route has been omitted. Hence the relative importance of the nitrite and nitrate routes in NO$_x$ storage over BaO is still open.

13.3.2.3 Effect of CO$_2$

In order to verify if the NO$_x$ storage mechanism proposed for BaO is still operating in the presence of CO$_2$, that is, under conditions more representative of the real composition of the gas mixture in engine exhausts, the effect of CO_2 on the NO_x adsorption over the Pt–Ba/Al$_2$O$_3$ (1:20:100 w/w) catalyst and over the binary Ba/Al$_2$O$_3$ (20:100 w/w) sample was investigated. In the case of CO_2-containing feed streams, the Ba species involved in the storage is primarily BaCO$_3$ [97].

The results obtained upon 1000 ppm NO_2 adsorption in the presence of 0.3% v/v CO_2 on the ternary Pt–Ba/Al$_2$O$_3$ (1:20:100 w/w) catalyst at 350 °C are presented in Figure 13.14.

As in the absence of CO_2, the adsorption of NO_2 occurs with evolution of NO via reaction (13.37) and is accompanied by NO_2 decomposition on Pt sites, with evolution of NO and O_2 (Figure 13.14a). The release of CO_2 was observed in quantitative agreement with the following stoichiometry:

$$3NO_2 + BaCO_3 \rightarrow Ba(NO_3)_2 + NO\uparrow + CO_2\uparrow \tag{13.42}$$

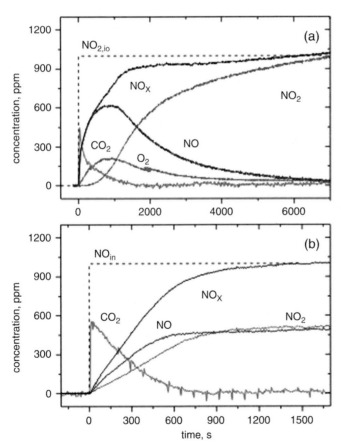

Figure 13.14 Storage of 1000 ppm NO_2 (a) and of 1000 ppm NO in He + O_2 (3%) + CO_2 (0.3%) (b) over Pt–Ba/Al$_2$O$_3$ (1:20:100 w/w) at 350 °C. Adapted from ref. [126].

which indicates that the catalyst surface is fully carbonated and that the NO_2 adsorption leads to the displacement of carbonates.

FT-IR spectra [126] showed that the presence of CO_2 does not significantly modify the spectral features of the NO_x surface species originating upon the adsorption of NO_2. In particular, both in the presence and in the absence of CO_2, nitrates are mainly formed at the catalyst surface (1410, 1320 and $1020\,cm^{-1}$ and $1550\,cm^{-1}$). Accordingly, it is concluded that the adsorption of NO_2–CO_2 mixtures strictly parallels that of NO_2 in the absence of CO_2, that is, the occurrence of the nitrate route is not greatly affected by the presence of CO_2.

A slightly different picture is obtained when the storage is carried out with 1000 ppm NO–3% v/v O_2 instead of NO_2. In fact, upon NO–O_2 admission in the presence of 0.3% v/v CO_2 at 350 °C over the reference Pt–Ba/Al_2O_3 sample (Figure 13.14b), the NO_x adsorption is slightly lower than in the absence of CO_2, but the time delay in the NO_x breakthrough in no longer observed (Figure 13.12c). Notably, also in this case the adsorption of NO_x was accompanied by the evolution of CO_2, in line with the stoichiometry of reaction (13.42), which confirms that only $BaCO_3$ sites are involved in the adsorption.

FT-IR data recorded upon NO–O_2 adsorption in the presence of CO_2 showed that nitrites are formed first over the carbonated surface. On increasing the time of contact, carbonates are partially displaced, while nitrites evolve to nitrate species [126]. Notably, for a given time of contact, the amount of surface nitrites is lower in the presence of CO_2, which suggests that the displacement of CO_2 from the carbonated Ba sites to give Ba nitrite is rate determining in the nitrite route, whereas the kinetics of the nitrate route are only marginally affected by the presence of CO_2. This may indicate that the nitrate route prevails over the nitrite route in the presence of CO_2, that is, under real operating conditions the nitrate route is the most important.

In agreement with these findings, the simplified kinetic model which has been used to describe the NO_x adsorption in the presence of CO_2 indicates that in the presence of CO_2 the data are well described by invoking the occurrence of the nitrate route alone.

13.3.3
Reduction of Stored NO_x

Figure 13.15 shows the results obtained upon stepwise addition of H_2 (2000 ppm) in He at 350 °C over Pt–Ba/γ-Al_2O_3 (1:20:100 w/w) after NO_x storage up to saturation at the same temperature.

H_2 is immediately and completely consumed and N_2 is the main product along with traces of NO; the reaction is very fast and is limited by the concentration of H_2. The concentration of nitrogen, which was constant at a level of 360 ppm until 500 s, progressively decreases to zero and the evolution of hydrogen in the gas phase and the formation of ammonia are observed. Accordingly, the reduction process is initially very selective to nitrogen (almost quantitative), then ammonia is formed in correspondence with the H_2 breakthrough. Ammonia is by far the most important

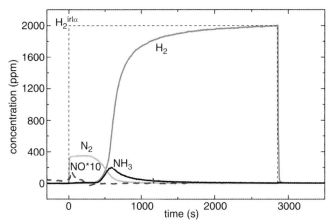

Figure 13.15 Reduction of stored NO_x with H_2 (2000 ppm) + He at 350 °C over Pt–Ba/Al_2O_3 (1:20:100 w/w) catalyst after storage at saturation at the same temperature.

by-product of the reduction of stored NO_x, as also reported by several other groups [5, 101, 124, 125]. It is worth noting that upon hydrogen admission, a very limited temperature increase of 3–5 °C was observed, so that the reduction was accomplished under nearly isothermal conditions.

The following main reactions can be invoked to explain the formation of N_2 and NH_3:

$$Ba(NO_2)_2 + 3H_2 \rightarrow BaO + N_2 + 3H_2O \tag{13.43}$$

$$Ba(NO_3)_2 + 5H_2 \rightarrow BaO + N_2 + 5H_2O \tag{13.44}$$

$$Ba(NO_2)_2 + 6H_2 \rightarrow BaO + 2NH_3 + 3H_2O \tag{13.45}$$

$$Ba(NO_3)_2 + 8H_2 \rightarrow BaO + 2NH_3 + 5H_2O \tag{13.46}$$

The N_2 concentration of 360 ppm in Figure 13.15 is close to that expected from reaction (13.45) (400 ppm) and is significantly lower than that expected from reaction (13.43) (660 ppm). This confirms that nitrates are the most abundant ad-species when the storage of NO_x is extended to saturation (Section 3.2).

13.3.3.1 Mechanism of the Reduction by H_2 of Stored NO_x

It is generally agreed that the reduction of stored NO_x over NSR catalysts implies first the release of NO_x from the catalyst surface in the gas phase, followed by the reduction of the released NO_x to N_2 or other products [5]. The reduction of the gaseous NO_x in a rich environment is thought to occur according to the TWC mechanism: it is suggested that NO is decomposed on reduced Pt sites [111] or that a direct reaction occurs between released NO_x species and the reductant molecules at the precious metal sites [112].

Based on this scheme, different proposals have been advanced to explain the NO_x release, that is, the nitrate decomposition. It has been suggested that the NO_x release is provoked by the heat generated upon the reducing switch (thermal release) [113], by the decrease in the gas-phase oxygen concentration that destabilizes the stored nitrates [114], by spillover and reduction of NO_2 at the reduced Pt sites or by the establishment of a net reducing environment which decreases the stability of nitrates [114–120].

In order to gain further insight into the mechanism governing the reduction of adsorbed NO_x species and to elucidate better the role of the different catalyst components, the reduction of stored NO_x was extensively investigated in our laboratories [117–119]. For this purpose, NO_x were stored under controlled conditions at 350 °C over different catalysts and were then removed not only by step addition of H_2 at the same constant temperature (Figure 13.15, already discussed) but also by thermal decomposition in He (TPD) or by heating in flowing 2000 ppm H_2 + He [temperature-programmed surface reaction (TPSR)]. In these last two experiments, the storage phase was followed by flowing in pure He for a prolonged period after adsorption and then by cooling in pure He to room temperature to allow for the desorption of weakly adsorbed NO_x species and to prevent readsorption of NO_x during cooling.

The results of TPD and TPSR experiments performed after storage of NO_x at 350 °C over Pt–Ba/Al_2O_3 (1:20:100 w/w) are presented in Figure 13.16.

In the TPD run, the onset of the desorption of stored NO_x is observed only above 350 °C, that is, above the NO_x adsorption temperature. Evolution of NO and O_2 was observed in this case, along with minor quantities of NO_2 [98, 78, 108, 106]. Complete desorption of NO_x was attained already slightly below 600 °C. Similar results were obtained when the adsorption was performed at 300 and 400 °C: the onset of desorption always occurred above the temperature of adsorption.

This behavior was confirmed over the binary Ba/γ-Al_2O_3 (20:100 w/w) sample. It is worth noting. however, that the complete decomposition of stored NO_x was achieved at lower temperatures over Pt–Ba/γ-Al_2O_3 than over Ba/γ-Al_2O_3. This indicates that Pt promotes the rate of nitrate decomposition [118], in line with several literature reports [5, 76, 78, 98, 114, 115, 123].

When the nitrates stored on Pt–Ba/Al_2O_3 (1:20:100 w/w) are heated in 2000 ppm H_2 + He instead of pure He (TPSR run in Figure 13.16), the reduction of stored NO_x is observed at temperatures as low as 140 °C: the reaction is very fast and shows complete consumption of H_2 at 170 °C with production of N_2 and also of significant amounts of NH_3. TPSR experiments were also performed after NO_x adsorption at different temperatures (300 and 400 °C) [117]: H_2 consumption was always observed near 140 °C.

The data indicate that the reduction of NO_x ad-species is initiated at temperatures well below that of thermal decomposition of nitrates; further, the temperature onset for the reaction is not affected by the adsorption temperature. This proves that the reduction of stored NO_x under near isothermal conditions occurs through a Pt-catalyzed chemical route, which is already active at low temperatures.

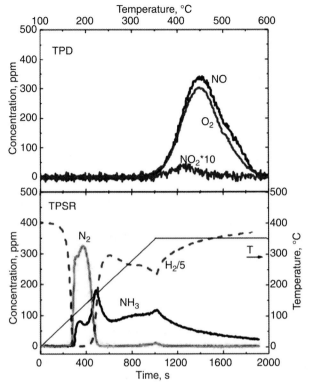

Figure 13.16 TPD in He and TPSR in H_2 (2000 ppm) + He after NO–O_2 adsorption at 350 °C over Pt–Ba/Al_2O_3 (1:20:100 w/w) catalyst. Adapted from ref. [119].

To clarify better the role of Pt in the reduction mechanism, a physical mixture of the binary Pt/γ-Al_2O_3 (1:100 w/w) and Ba/γ-Al_2O_3 (20:100 w/w) samples was prepared and tested [117]. Also in this case NO_x was stored at 350 °C upon adsorption of NO in the presence of excess O_2 [102], then the stability/reactivity of stored nitrates was investigated by means of TPD and H_2 TPSR. The TPD data showed that decomposition of adsorbed nitrates occurs at temperatures very close to that of adsorption; the presence of H_2 (TPSR run) does not decrease significantly the temperature threshold for nitrate decomposition (which is still observed near 350 °C), but instead provokes the reduction of the evolved NO_x to NH_3 and N_2. Hence the previously invoked Pt-catalyzed route is active only when Pt and Ba are dispersed over the same particle of the support.

Accordingly, the bulk of data collected over the different catalytic systems show that Pt catalyzes the reduction of stored NO_x already at low-temperatures (i.e. well below that of adsorption) and that the co-presence of Pt and Ba on the same support is required for the occurrence of this reaction. Notably, over Pt–Ba/Al_2O_3 the reduction of the stored NO_x leads primarily to the formation of N_2, but significant amounts of

NH_3 are also observed. In particular, when nitrates are reduced at constant temperature (Figure 13.15) ammonia is detected after N_2 evolution, whereas in the case of TPSR experiments (Figure 13.16), ammonia is observed in higher amounts and again mostly after N_2 evolution.

13.3.3.2 Identification of the Reaction Network During Reduction of Stored NO_x by H_2

To explain the formation of ammonia in the reduction of stored NO_x, a series of experiments were performed in which NO_x were stored on $Pt–Ba/Al_2O_3$ (1:20:100 w/w) at 350 °C and then reduced by H_2 at different temperatures in the range 150–350 °C [135].

When the reduction of stored NO_x is accomplished at 150 °C, the reaction shows a significant induction period (Figure 13.17). The decrease in the H_2 concentration is accompanied by the evolution of NH_3 and of minor amounts of N_2; however, a time delay is observed between product evolution and H_2 uptake. Therefore, the rate of reaction is low at this temperature and a critical high surface concentration of activated hydrogen species is needed for the reaction to occur. The regeneration of the catalyst is not complete, since only 80% of the stored NO_x could be reduced after prolonged treatment with H_2 at 150 °C. Also, the calculated overall N_2 selectivity is very poor, below 20%, and ammonia represents the main reaction product.

On increasing the reaction temperature (e.g. to 250 °C), the induction period observed at 150 °C disappears and N_2 formation is observed immediately upon admission of H_2 with no time delay, while NH_3 evolution is seen corresponding to

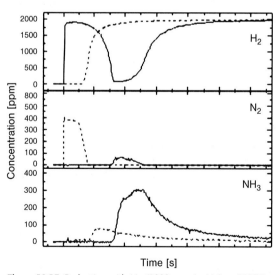

Figure 13.17 Reduction with H_2 (2000 ppm in He) at 150 °C (dashed lines) and at 250 °C (full lines) of NO_x stored after $NO–O_2$ adsorption at 350 °C over $Pt–Ba/\gamma-Al_2O_3$ (1:20:100 w/w) catalyst. Adapted from ref. [135].

the detection of unconverted H_2 (H_2 breakthrough). Notably, the increase in the reduction temperature favors N_2 formation at the expense of ammonia, so that at high temperatures (350 °C) N_2 is the most abundant reaction product.

The observed dependence of the N_2 selectivity on temperature may suggest that the reduction of stored nitrates by H_2 occurs via an in-series two-step pathway. The first step is fast even at low temperatures and is responsible for the consumption of hydrogen and for the formation of ammonia. The second step is slower and implies the reduction of residual nitrates with ammonia to form nitrogen; this reaction occurs to a significant extent only at higher temperatures.

To analyze these aspects better, H_2 TPSR and NH_3 TPSR experiments were performed in the presence of 1% v/v water over the Ba–Pt/Al_2O_3 (1:20:100 w/w) catalyst with stored NO_x after NO–O_2 adsorption at 350 °C [135]. The results are presented in Figures 13.18.

It appears that nitrates can be reduced by H_2 already at very low temperatures (from 60 °C), leading to the formation of ammonia and of minor amounts of N_2 (Figure 13.18a). The overall H_2 consumption is in line with the stoichiometry of the following reactions leading to the formation of NH_3 and N_2:

$$Ba(NO_3)_2 + 8H_2 \rightarrow 2NH_3 + BaO + 5H_2O \tag{13.47}$$

$$Ba(NO_3)_2 + 5H_2 \rightarrow N_2 + BaO + 5H_2O \tag{13.48}$$

Reaction (13.47) accounts for more than 95% of the overall H_2 consumption. Notably, the H_2 uptake in Figure 13.18a is seen before the evolution of the reaction products, which suggests that H_2 is first adsorbed and activated on the catalyst surface and then participates in the reduction of nitrates. However, a time delay in the detection of ammonia due to its slow desorption from the catalyst surface cannot be excluded.

The results of NH_3 TPSR experiment (Figure 13.18b) show that ammonia reduces the stored nitrates starting from 150 °C with formation of nitrogen according to the following overall stoichiometry:

$$3Ba(NO_3)_2 + 10NH_3 \rightarrow 8N_2 + 3BaO + 15H_2O \tag{13.49}$$

However, the reactivity of NH_3 is markedly lower than that of H_2, which is able to reduce stored nitrates at temperatures as low as 60 °C.

The decrease in the ammonia concentration at higher temperatures, above 350 °C, and the associated formation of N_2 and H_2 in a 1:3 ratio, on the other hand, are related to the decomposition of ammonia:

$$2NH_3 \rightarrow N_2 + 3H_2 \tag{13.50}$$

Accordingly, the data show that (i) the reduction of nitrates by H_2 is a fast reaction and results in the formation of ammonia and (ii) the reduction of nitrates by H_2 is slower because the onset of reaction is observed at higher temperature (150 vs 60 °C) and is highly selective to nitrogen.

It is worth noting that the reduction of stored nitrates by NH_3 is markedly enhanced in the presence of water because the onset of this reaction is observed at 60 °C in the

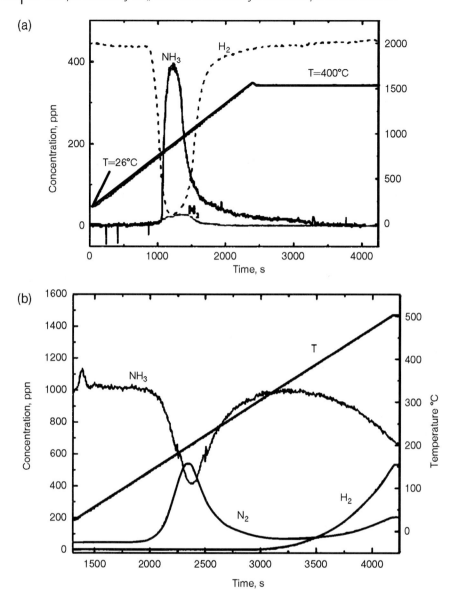

Figure 13.18 H_2 TPSR (2000 ppm + 1% v/v H_2O in He) (a) and NH_3 TPSR (1000 ppm + 1% v/v H_2O in He) (b) after NO–O_2 adsorption at 350 °C over Pt–Ba/γ-Al_2O_3 (1:20:100) catalyst. Adapted from ref. [135].

presence of water (Figure 13.18a) and at 150 °C in the absence of water (Figure 13.17). This may be explained considering that hydrogen spillover is favored by the presence of surface hydroxyl species. In contrast, the presence of water in the gas phase has a negligible effect, if any, on the reduction of nitrates by NH_3.

The results of TPSR experiments presented in Figure 13.18 were confirmed by isothermal TRM experiments with H_2 and NH_3 in the presence of water at different temperatures.

On the basis of these data, the following mechanism for NO_x reduction by hydrogen can be suggested. H_2, activated over the Pt sites according to the Pt-catalyzed pathway discussed previously, reduces the stored nitrates directly to ammonia or, more likely, induces the decomposition of nitrates to gaseous NO_x, which are then reduced by H_2 to NH_3 over the Pt sites [overall reaction (13.47)]. Once ammonia has been formed, it can react with adsorbed nitrates and this reaction is very selective towards nitrogen. It is worth noting that the reaction of ammonia with NO_x obeys the stoichiometry of reaction (13.49), which is different from that of the well-known NH_3–NO SCR reaction because it implies the participation of nitrates. Also, the NH_3 + nitrate reaction is catalyzed by the Pt component, since the reduction of nitrates (and also of nitrites) stored on Ba could not be accomplished by NH_3 over Ba/Al_2O_3 (20:100 w/w).

The in-series two-step pathway described above is able to account for the temporal evolution of the product which is observed during the reduction of stored NO_x, with nearly complete N_2 selectivity at the beginning of the reduction process and significant ammonia formation near the end of the regeneration step (see Figure 13.17).

As depicted in Figure 13.19, the occurrence of a fast reduction step of the adsorbed nitrates by H_2 to give ammonia together with the integral 'plug-flow' behavior of the laboratory microreactor implies the complete consumption of the reductant H_2 and the formation of an H_2 front traveling along the reactor axis. At a given time of the regeneration phase, different zones are present in the reactor: (i) a first zone, upstream of the H_2 front (zone I), where the Ba storage sites have been already restored through the reduction of nitrates by H_2 to give ammonia; (ii) the zone

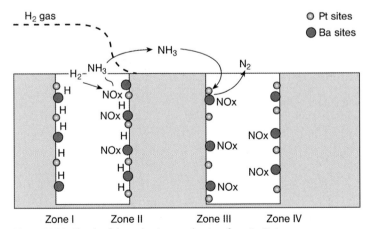

Figure 13.19 Sketch of the reduction mechanism for a Pt–Ba/γ-Al_2O_3 catalyst upon regeneration with H_2.

corresponding to the development of the H_2 front (zone II), where the concentration of hydrogen decreases from the inlet value to almost zero; here the formation of ammonia due to reduction of nitrates by H_2 (fast reaction) is accompanied by the subsequent reaction of ammonia with residual nitrates to give nitrogen (slower reaction) where the hydrogen concentration is sufficiently low; (iii) a zone immediately downstream of the H_2 front (zone III), where only the reaction of ammonia formed upstream with the stored nitrates to give N_2 takes place; (iv) the last zone (zone IV), where residual nitrates are still present and ammonia produced upstream (in zone I and II) has already been fully consumed (in zones II and III) . Accordingly, the regeneration of the trap proceeds both in the part of the reactor in which nitrates are reduced by H_2 to give ammonia (zone II) and in the zone in which the ammonia formed upstream reacts with nitrates leading to N_2 (zones II and III).

The in-series two-step pathway suggested above for nitrogen formation explains well the observed changes in the N_2 selectivity and in the efficiency of the reduction of the stored NO_x with temperature (Figure 13.17). In fact, when the temperature is low, reaction (13.47) is slow and the H_2 front is less steep and develops in a large part of the catalyst bed. In addition, due to the low temperature, ammonia can hardly react with the stored nitrates to form N_2 according to reaction (13.49). As a result, a limited efficiency of the reduction of the stored NO_x is observed together with a low nitrogen selectivity and the NH_3 breakthrough precedes that of H_2. On the other hand, at higher temperatures (e.g. 250–350 °C), the rate of reaction (13.47) is fairly high and the H_2 front is steep so that the efficiency in the reduction of the stored NO_x is almost complete. Further, ammonia readily reacts with nitrates left downstream of the H_2 front [reaction (13.49)] and this drives the selectivity to N_2. Accordingly, in this case N_2 is immediately observed at the reactor outlet and the NH_3 breakthrough is seen together or after the H_2 breakthrough, that is, when the H_2 front reaches the end of the trap and nitrates stored behind the H_2 front are depleted.

A similar pathway for the reduction of stored nitrates has been presented [136–138]. Here the importance of the integral behavior of the reactor is underlined but the occurrence of a regeneration front instead of a hydrogen front is suggested, whereas an in-series/parallel scheme is proposed and the stoichiometry of reaction (13.49) is not recognized.

13.4
Open Issues and Future Opportunities

Although the standard urea–ammonia SCR technology is nowadays well established and used world-wide for the control of NO_x from stationary sources, additional applications are predicted in large gas turbines, incinerators, stationary diesel engines and the cement and glass industries in view of the enforcement of stricter emission limits and of the high cost of alternative very efficient primary methods.

Concerning the application of the urea SCR technology to the abatement of NO_x emissions from vehicles, although the first commercialization for heavy-duty diesel engines is already in place, more fundamental and development work has to be done in order to consolidate and possibly extend its use to passenger cars. Advances in the mechanism of the fast SCR reaction and even more in the implementation of fast SCR in vehicles are very likely, particularly in view of the opportunity offered by an integrated approach to the after-treatment system where de-NO_x, DOC, DPF and primary measures such as CR, EGR and so on must be accommodated.

The advances in the knowledge of fast SCR and in its implementation may also offer a challenging opportunity for the use of urea–ammonia SCR at low temperatures in some stationary applications.

At present, the NSR technology is certainly less mature than ammonia standard SCR. However, NSR systems could be very appealing for low segment passenger cars because, in contrast to urea SCR, small layout modifications are needed. Still, NSR materials more stable at high temperature and more resistant to poisoning from sulfates together with less severe regeneration strategies of the NO_x trap must be developed. Extensive work is currently being devoted by the motor industry to improve the NSR process also in view of the large reduction in NO_x emissions from vehicles required by Euro VI emission limits in 2014. Some promising solutions to the above issues have been suggested recently in the patent literature but they need further fundamental and development work.

Finally, the advances in the identification of the pathway of the reduction of stored NO_x, where ammonia is suggested as the intermediate product in the formation of nitrogen, may favor the improvement of the combined NSR + SCR technology that has been proposed by several car manufacturers to make NO_x removal by NSR more effective and at the same time to limit the ammonia slip.

Acknowledgment

We thank Professor Isabella Nova for useful discussions.

References

1 Bosch, H. and Janssen, F. (1988) *Catalysis Today*, **2**, 369.
2 Forzatti, P. and Lietti, L. (1996) *Heterogenenous Chemistry Reviews*, **3**, 33.
3 Forzatti, P., Lietti, L. and Tronconi, E. (2003) *Encyclopedia of Catalysis*, Vol. **5**, Wiley-Interscience, New York, p. 298.
4 Matsumoto, S. (2000) *Cattech*, **4**, 102.
5 Epling, W.S., Campbell, I.E., Yezerets, A., Currier, N.W. and Park, J.E. (2004) *Catalysis Reviews: Science and Engineering*, **46**, 163.
6 Forzatti, P., Castoldi, L., Lietti, L., Nova, I. and Tronconi, E. (2007) *Studies in Surface Science and Catalysis*, **171**, 175.
7 Cristiani, C., Forzatti, P. and Busca, G. (1989) *Journal of Catalysis*, **116**, 586.
8 Ramis, G., Cristiani, C., Forzatti, P. and Busca, G. (1990) *Journal of Catalysis*, **124**, 574.
9 Lietti, L. and Forzatti, P. (1994) *Journal of Catalysis*, **147**, 241.
10 Alemany, L.J., Lietti, L., Ferlazzo, N., Forzatti, P., Busca, G., Giamello, E.

and Bregani, F. (1995) *Journal of Catalysis*, **155**, 117.
11 Lietti, L., Alemany, J.L., Forzatti, P., Busca, G., Ramis, G., Giamello, E. and Bregani, F. (1996) *Catalysis Today*, **29**, 143.
12 Paganini, M.C., Dell'Acqua, L., Giamello, E., Lietti, L., Forzatti, P. and Busca, G. (1997) *Journal of Catalysis*, **166**, 195.
13 Nova, I., Lietti, L., Casagrande, L., Dell'Acqua, L., Giamello, E. and Forzatti, P. (1998) *Applied Catalysis B–Environmental*, **17**, 245.
14 Lietti, L., Nova, I., Ramis, G., Dell'Acqua, L., Busca, G., Giamello, E., Forzatti, P. and Bregani, F. (1999) *Journal of Catalysis*, **187**, 419.
15 Casagrande, L., Lietti, L., Nova, I., Forzatti, P. and Baiker, A. (1999) *Applied Catalysis B–Environmental*, **22**, 63.
16 Mastikhin, V.M., Terskikh, V.V., Lapina, O.B., Filimonova, S.V., Seidl, M. and Knozinger, H. (1995) *Journal of Catalysis*, **156**, 1.
17 Went, G.T., Leu, L.J. and Bell, A.T. (1992) *Journal of Catalysis*, **134**, 479.
18 Wuurman, M.A., Wachs, I.E. and Hirt, A.M. (1991) *Journal of Physical Chemistry*, **95**, 9928.
19 Deo, G., Turek, A.M., Wachs, I.E., Machej, T., Haber, J., Das, N., Eckert, H. and Hirt, A.M. (1992) *Applied Catalysis A–General*, **91**, 27.
20 Marshneva, V.I., Slavinskaya, E.M., Kalinkina, O.V., Odegova, G.V., Moroz, E.M., Lavrova, G.V. and Salanov, A.N. (1995) *Journal of Catalysis*, **155**, 171.
21 Chen, J.P. and Yang, R.T. (1992) *Applied Catalysis A–General*, **80**, 135.
22 Smak, T.Z., Dumesic, J.A., Clausen, B.S., Tornquist, E. and Topsoe, N.-Y. (1992) *Journal of Catalysis*, **135**, 246.
23 Scharf, U., Schneider, M., Baiker, A. and Wokaun, A. (1994) *Journal of Catalysis*, **149**, 344. Scharf, U., Schneider, M., Baiker, A. and Wokaun, A. (1991) *Journal of Catalysis*, **11**, 57.
24 Rademacher, L., Borgamann, D., Hopfengartner, R., Wedler, G., Hums, E. and Spitznagel, G.W. (1992) *Journal of Surface and Interface Analysis*, **20**, 43.
25 Amiridis, M.D., Duevel, R.V. and Wachs, I.E. (1999) *Applied Catalysis B–Environmental*, **20**, 111.
26 Balling, L. and Hein, D. (1989) Proceedings of the Joint Symposium on Stationary NO_x Control, San Francisco, CA, 6–9 March 1989.
27 Tronconi, E., Forzatti, P., Gomez Martin, J.P. and Malloggi, S. (1992) *Chemical Engineering Science*, **47**, 2401.
28 Forzatti, P. (2000) *Catalysis Today*, **62**, 51; Forzatti, P. (2001) *Applied Catalysis A–General*, **222**, 221.
29 Ivanov, A.A. and Balzhinimaev, B.S. (1987) *Reaction Kinetics and Catalysis Letters*, **35**, 413.
30 Tronconi, E. and Forzatti, P. (1992) *AIChE Journal*, **38**, 201.
31 Svachula, J., Alemany, L.J., Ferlazzo, N., Forzatti, P., Tronconi, E. and Bregani, F. (1993) *Industrial and Engineering Chemistry Research*, **32**, 826.
32 Busca, G., Lietti, L., Ramis, G. and Berti, F. (1998) *Applied Catalysis B–Environmental*, **18**, 1.
33 Inomata, M., Miyamoto, A. and Murakami, Y. (1980) *Journal of Catalysis*, **62**, 140.
34 Janssen, F., van der Kerkhof, F., Bosch, H. and Ross, J. (1987) *Journal of Physical Chemistry*, **91**, 5931; Janssen, F., van der Kerkhof, F., Bosch, H. and Ross, J. (1987) **91** 6633.
35 Ramis, G., Busca, G., Forzatti, P. and Bregani, F. (1990) *Applied Catalysis*, **64**, 259.
36 Topsøe, N.Y., Topsøe, H. and Dumesic, J.H. (1995) *Journal of Catalysis*, **151**, 226; Topsøe, N.Y., Topsøe, H. and Dumesic, J.H. (1995) **151** 241.
37 Lietti, L., Nova, I., Camurri, S., Tronconi, E. and Forzatti, P. (1992) *AIChE Journal*, **43**, 2559.
38 Beeckman, J.W. and Hegedus, L.L. (1991) *Industrial and Engineering Chemistry Research*, **30**, 969.

39 Svachula, J., Ferlazzo, N., Forzatti, P., Tronconi, E. and Bregani, F. (1993) *Industrial and Engineering Chemistry Research*, **32**, 1053.
40 Buzanowski, M.A. and Yang, R.T. (1990) *Industrial and Engineering Chemistry Research*, **29**, 2074.
41 Lefers, J.B., Lodders, P. and Enoch, G.D. (1991) *Chemical Engineering and Technology*, **14**, 192.
42 Tronconi, E., Beretta, A., Elmi, A.S., Forzatti, P., Malloggi, S. and Baldacci, A. (1994) *Chemical Engineering Science*, **49**, 4277.
43 Koebel, M. and Elsener, M. (1998) *Industrial and Engineering Chemistry Research*, **37**, 327.
44 Beretta, A., Orsenigo, C., Ferlazzo, N., Tronconi, E. and Forzatti, P. (1998) *Industrial and Engineering Chemistry Research*, **37**, 2623.
45 Roduit, B., Bettoni, F., Baldyga, J., Wokaun, A. and Baiker, A. (1998) *AIChE Journal*, **44**, 2731.
46 Tronconi, E., Cavanna, A. and Forzatti, P. (1998) *Industrial and Engineering Chemistry Research*, **37**, 2341.
47 Tronconi, E., Orsenigo, C., Cavanna, A. and Forzatti, P. (1999) *Industrial and Engineering Chemistry Research*, **38**, 2593.
48 Borisova, E., Noskov, A.S. and Bobrova, L.N. (1997) *Catalysis Today*, **38**, 97.
49 Snyder, J.D. and Subramanian, B. (1998) *Chemical Engineering Science*, **53**, 727.
50 Khodayari, R. and Odenbrand, C.U.I. (1999) *Chemical Engineering Science*, **54**, 1775.
51 Andersson, S.L., Gabrielsson, P.L.T. and Odenbrand, C.U.I. (1994) *AIChE Journal*, **40**, 1911.
52 Lietti, L., Nova, I., Tronconi, E. and Forzatti, P. (2000) in *Reaction Engineering for Pollution Prevention* (eds M.A. Abrahamn and R.P. Hesketh), Elsevier, Amsterdam, p. 85.
53 Nova, I., Lietti, L., Tronconi, E. and Forzatti, P. (2000) in *Studies in Surface Science and Catalysis* (eds A. Corma, F.V. Melo, S. Mendioroz and J.L.G. Fierro), Elsevier, Amsterdam, p. 623.
54 Ramis, G., Busca, G., Forzatti, P., Cristiani, C. and Lietti, L. (1992) *Langmuir*, **8**, 1744.
55 Dumesic, J.A., Topsoe, N.-Y., Slabiak, T., Morsing, P., Tornquist, E. and Topsoe, H. (1993) in *New Frontiers in Catalysis* (eds L. Guczi, F. Solymosi and P. Tetenyi), Elsevier, Amsterdam, p. 1325.
56 Tronconi, E., Nova, I., Ciardelli, C., Chatterjee, D., Bandl-Konrad, B. and Burkhardt, T. (2005) *Catalysis Today*, **105**, 529; Nova, I., Ciardelli, C., Tronconi, E., Chatterjee, D. and Bandl-Konrad, B. (2006) *AIChE Journal*, **52**, 3222.
57 Nova, I., Lietti, L., Tronconi, E. and Forzatti, P. (2000) *Catalysis Today*, **60**, 73.
58 Nova, I., Lietti, L., Tronconi, E. and Forzatti, P. (2001) *Chemical Engineering Science*, **56**, 1229.
59 Winkler, C., Florchinger, P., Patil, M.D., Gieshoff, J., Spurk, P. and Pfeifer, M. (2003) *Society of Automotive Engineers (SAE) technical paper*, 2003-01-0845.
60 York, A.P.E., Watling, T.C., Cox, J.P., Jones, I.Z., Blakeman, P.G. and Ilkenhans, T. (2004) *Society of Automotive Engineers (SAE) technical paper*, 2004-01-0155.
61 Schar, C.M., Onder, C.H., Geering, H.P. and Elsner, M. (2004) *Society of Automotive Engineers (SAE) technical paper*, 2004-01-0153.
62 Chatterjee, D., Burkhardt, T., Bandl-Konrad, B., Braun, T., Tronconi, E., Nova, I. and Ciardelli, C. (2005) *Society of Automotive Engineers (SAE) technical paper*, 2005-01-0965.
63 Beekman, J.W. (1991) *Industrial and Engineering Chemistry Research*, **30**, 428.
64 Kato, A., Matsuda, S., Kamo, T., Nakajima, F., Kuroda, H. and Narita, T. (1981) *Journal of Physical Chemistry*, **85**, 4099.
65 Koebel, M., Elsener, M. and Madia, G. (2001) *Industrial and Engineering Chemistry Research*, **40**, 52.

66 Koebel, M., Madia, G., Raimondi, F. and Wokaum, A. (2002) *Journal of Catalysis*, **209**, 159.

67 Madia, G., Koebel, M., Elsener, M. and Wokaum, A. (2002) *Industrial and Engineering Chemistry Research*, **41**, 351.

68 Koebel, M., Madia, G. and Elsener, M. (2002) *Catalysis Today*, **73**, 239.

69 Sun, Q., Gao, Z., Wen, B. and Sachtler, W.M.H. (2002) *Catalysis Letters*, **78**.

70 Yeom, Y.H., Wem, B., Sachtler, W.M.H. and Weitz, E. (2004) *Journal of Physical Chemistry B*, **108**, 5386.

71 Ciardelli, C., Nova, I., Tronconi, E., Bandl-Konrad, B., Chatterjee, D., Weibel, M. and Krutzsch, B. (2006) *Applied Catalysis B–Environmental*, **70**, 80.

72 Ciardelli, C., Nova, I., Tronconi, E., Chatterjee, D. and Bandl-Konrad, B. (2004) *Chemical Communications*, 2718.

73 Nova, I., Ciardelli, C., Tronconi, E., Chatterjee, D. and Bandl-Konrad, B. (2006) *Catalysis Today*, **114**, 3.

74 Tronconi, E., Nova, I., Ciardelli, C., Chatterjee, D. and Weibel, M. (2007) *Journal of Catalysis*, **245**, 1.

75 Yeom, Y., Henao, J., Li, M., Sachtler, W.M.H. and Weitz, E. (2005) *Journal of Catalysis*, **231**, 181.

76 Olsson, L., Persson, H., Fridell, E., Skoglundh, M. and Andersson, B. (2001) *Journal of Physical Chemistry B*, **105**, 6895.

77 Apostolescu, N., Schroder, T. and Kureti, S. (2004) *Applied Catalysis B–Environmental*, **51**, 43.

78 Nova, I., Castoldi, L., Prinetto, F., Ghiotti, G., Lietti, L., Tronconi, E. and Forzatti, P. (2004) *Journal of Catalysis*, **222**, 377.

79 Chatterjee, D., Burkhardt, T., Weibel, M., Tronconi, E., Nova, I. and Ciardelli, C. (2006) *Society of Automotive Engineers (SAE) technical paper*, 2006-01-0468; Chatterjee, D., Burkhardt, T., Weibel, M., Tronconi, E., Nova, I. and Grossale, A. (2007) *Society of Automotive Engineers (SAE) technical paper*, 2007-01-1136; Grossale, A., Nova, I. and Tronconi, E. (2008) *Catalysis Today*, **136**, 18.

80 Miyoshi, M., Matsumoto, S., Kato, K., Tanaka, T., Harada, J., Takahashi, N., Yokota, K., Sugiura, M. and Kasahara, K. (1995) *Society of Automotive Engineers (SAE) technical paper*, 1995-01-0809.

81 Engstrom, P., Ambernstsson, A., Skoglundh, M., Fridell, F. and Smedler, G. (1999) *Applied Catalysis B: Environmental*, **22**, 241.

82 Gobel, U., Hohne, J., Lox, E.S., Muller, W., Okumura, A., Strehlau, W. and Hori, M. (1999) *Society of Automotive Engineers (SAE) technical paper*, 99FL-103.

83 Nova, I., Beretta, A., Groppi, G., Lietti, L., Tronconi, E. and Forzatti, P. (2006) in *Structured Catalyst and Reactors*, 2nd edn (eds A. Cybulski and J.A. Moulijn), CRC Press, Boca Raton, FL, p. 171.

84 Shinjoh, H., Takahashi, N., Yokota, K. and Sugiura, M. (1998) *Applied Catalysis B–Environmental*, **15**, 189.

85 Hodjati, S., Petit, C., Pichon, V. and Kiennemann, A. (2000) *Applied Catalysis B–Environmental*, **27**, 117.

86 Fridell, E., Skoglundh, M., Westerberg, B., Johansson, S. and Smedler, G. (1999) *Journal of Catalysis*, **183**, 196.

87 Fridell, E., Persson, H., Westerberg, B., Olsson, L. and Skoglundh, M. (2000) *Catalysis Letters*, **66**, 71.

88 Broqvist, P., Panas, I., Fridell, E. and Persson, H. (2002) *Journal of Physical Chemistry B*, **106**, 137.

89 Westerberg, B. and Fridell, E. (2001) *Journal of Molecular Catalysis A-Chemical*, **165**, 249.

90 Anderson, J.A., Bachiler-Baeza, B. and Fernandez-Garcia, M. (2003) *Physical Chemistry Chemical Physics*, **5**, 4418.

91 Hess, C. and Lunsford, J.H. (2002) *Journal of Physical Chemistry B*, **106**, 6358.

92 Hess, C. and Lunsford, J.H. (2003) *Journal of Physical Chemistry B*, **107**, 1982.

93 Huang, H.Y., Long, R.Q. and Yang, R.T. (2001) *Energy and Fuels*, **15**, 205.

94 Mahzoul, H., Brilhac, J.F. and Gilot, P. (1999) *Applied Catalysis B–Environmental*, **20**, 47.

95 Schmitz, P.J. and Baird, R.J. (2002) *Journal of Physical Chemistry B*, **106**, 4172.
96 Cant, N.V. and Patterson, M.J. (2002) *Catalysis Today*, **73**, 271.
97 Lietti, L., Forzatti, P., Nova, I. and Tronconi, E. (2001) *Journal of Catalysis*, **204**, 175.
98 Prinetto, F., Ghiotti, G., Nova, I., Lietti, L., Tronconi, E. and Forzatti, P. (2001) *Journal of Physical Chemistry*, **105**, 12732.
99 Prinetto, F., Ghiotti, G., Nova, I., Castoldi, L., Lietti, L. and Forzatti, P. (2003) *Physical Chemistry Chemical Physics*, **5**, 4428.
100 Nova, I., Castoldi, L., Prinetto, F., Del Santo, V., Lietti, L., Tronconi, E., Forzatti, P., Ghiotti, G., Psaro, R. and Recchia, S. (2004) *Topics in Catalysis*, **30/31**, 181.
101 Castoldi, L., Nova, I., Lietti, L., Tronconi, E. and Forzatti, P. (2004) *Catalysis Today*, **96**, 43.
102 Nova, I., Castoldi, L., Lietti, L., Tronconi, E. and Forzatti, P. (2005) *Society of Automotive Engineers (SAE) technical paper*, 2005-01-1085.
103 Scotti, A., Nova, I., Tronconi, E., Castoldi, L., Lietti, L. and Forzatti, P. (2004) *Industrial and Engineering Chemistry Research*, **43**, 4522.
104 Szailer, T., Kwak, J.H., Kim, D.H., Szanyi, J., Wang, C. and Peden, C.H.F. (2006) *Catalysis Today*, **114**, 86.
105 Kim, D.H., Chin, Y.H., Kwak, J.H., Szanyi, J. and Peden, C.H.F. (2005) *Catalysis Letters*, **105**, 259.
106 Szanyi, J., Kwak, J.H., Hanson, J., Wang, C., Szailer, T. and Peden, C.H.F. (2005) *Journal of Physical Chemistry B*, **109**, 7339.
107 Kabin, K.S., Khanna, P., Muncrief, R.I., Medhekar, V. and Harold, M.P. (2006) *Catalysis Today*, **114**, 72.
108 Piacentini, M., Maciejewski, M., Burgi, T. and Baiker, A. (2004) *Topics in Catalysis*, **30/31**, 71.
109 Fanson, P.T., Horton, M.R., Delgass, W.N. and Lauterbach, J. (2003) *Applied Catalysis B–Environmental*, **46**, 393.
110 Laurent, F., Pope, C.J., Mahzoul, H., Delfosse, I. and Gilot, P. (2003) *Chemical Engineering Science*, **58**, 1793.
111 Burch, R., Breen, J. and Meunier, F. (2002) *Applied Catalysis B–Environmental*, **39**, 283.
112 Maunula, T., Ahola, J. and Hamada, H. (2000) *Applied Catalysis B–Environmental*, **26**, 173.
113 Kabin, K.S., Muncrief, R.I. and Harold, M.P. (2004) *Catalysis Today*, **96**, 79.
114 Liu, Z. and Anderson, J.A. (2004) *Journal of Catalysis*, **224**, 18.
115 Cant, N.W. and Patterson, M.J. (2003) *Catalysis Letters*, **85**, 153.
116 Poulston, S. and Rajaram, R. (2003) *Catalysis Today*, **81**, 603.
117 Nova, I., Castoldi, L., Lietti, L., Tronconi, E. and Forzatti, P. (2006) *Society of Automotive Engineers (SAE) technical paper*, 2006-01-1368.
118 Nova, I., Lietti, L., Castoldi, C., Tronconi, E. and Forzatti, P. (2006) *Journal of Catalysis*, **239**, 244.
119 Forzatti, P., Castoldi, L., Nova, I., Lietti, L. and Tronconi, E. (2006) *Catalysis Today*, **117**, 316.
120 Szailer, T., Kwak, J.H., Kim, D.H., Hanson, J., Wang, C., Peden, C.H.F. and Szanyi, J. (2006) *Journal of Catalysis*, **239**, 51.
121 James, D., Fourre, E., Ishii, M. and Bowker, M. (2003) *Applied Catalysis B–Environmental*, **45**, 147.
122 Coronado, J.M. and Anderson, J.A. (1999) *Journal of Molecular Catalysis A–Chemical*, **138**, 83.
123 Sharma, M., Harold, M.P. and Balakotaiah, V. (2005) *Industrial and Engineering Chemistry Research*, **44**, 6264.
124 Lesage, T., Terrier, C., Bazin, P., Saussey, J. and Daturi, M. (2003) *Physical Chemistry Chemical Physics*, **5**, 4435.
125 Abdulhamid, H., Fridell, E. and Skoglundh, M. (2004) *Topics in Catalysis*, **30/31**, 161.
126 Frola, F., Prinetto, F., Ghiotti, G., Castoldi, L., Nova, I., Lietti, L. and Forzatti, P. (2007) *Catalysis Today*, **126**, 81.
127 Sveldberg, P., Jobson, E., Erkfeldt, S., Andersson, B., Larsson, M. and

Skoglundh, M. (2004) *Topics in Catalysis*, **30/31**, 199.

128 Nova, I., Castoldi, L., Lietti, L., Tronconi, E. and Forzatti, P. (2002) *Catalysis Today*, **75**, 431.

129 Theis, J.R., Jen, H.W., McCabe, R.W., Sharma, M., Balakotaiah, V. and Harold, M.P. (2006) *Society of Automotive Engineers (SAE) technical paper*, 2006-01-1067.

130 Epling, W.S., Campbell, G.C. and Parks, J.E. (2003) *Catalysis Letters*, **90**, 45.

131 Nova, I., Castoldi, L., Lietti, L., Tronconi, E. and Forzatti, P. (2007) *Topics in Catalysis*, **42–43**, 189.

132 Nova, I., Lietti, L. and Forzatti, P. (2008) *Catalysis Today*, **136**, 128.

133 Salfi, M. (2007) Master Thesis, Politecnico di Milano, Italy.

134 Mulla, S.S., Chen, N., Delgass, W.N., Epling, W.S. and Ribeiro, F.H. (2005) *Catalysis Letters*, **100**, 267.

135 Lietti, L., Nova, I. and Forzatti, P. (2008) *Journal of Catalysis*, **257**(2), 270.

136 Cumaranatunge, L., Mulla, S.S., Yezerets, A., Currier, N.W., Delgass, W.N. and Ribeiro, F.H. (2007) *Journal of Catalysis*, **246**, 29.

137 Pihl, J.A., Parks, J.E., Daw, C.S. and Root, T.W. (2006) *Society of Automotive Engineers (SAE) technical paper*, 2006-01-3441.

138 Choi, J.-S., Partridge, W.P. and Daw, C.S. (2007) *Applied Catalysis*, **77**, 145.

Index

12-tungstophosphoric acid (TPA)   335
30-cell DMFC stack   50, 57, 58, 59, 67
– steady-state V-I curves   58
– test results   59

a

absolute reaction rate   10
absorption curves   146
absorption-desorption isotherms   163
absorption process   130
acetaldehyde   22, 27
– formation   22
acetate-forming enzymes   272
acetic acid (AA)   23, 207–210, 216, 273
– activation   208
– deactivation mechanism   209
– formation   273
– steam reforming pathway   209
acid-base dissociation   256
acid-catalyzed reactions   330
activation energy   122
additional reaction pathway   208
ad hoc bacterial strain   283
adiabatic combustion temperatures   301, 364
adiabatic coefficient   112
ADP-Morgane membrane   30
adsorbate-surface interaction   181
adsorption isotherm   122
adsorption model   126
adsorption rate   122
advanced/emerging technologies   357–359
advanced systems integration   102–103
aerobic conditions tests   275
aerodynamic drag coefficients   91
aerodynamic factor   92, 95
aerosol propellant   204
Ag-based cathodes   38
agricultural policy   177

agricultural production   177
air cathode exhaust   53, 56
air feed stoichiometry   52, 53, 56, 64
Al_2O_3 catalysts   312, 415
alcohol oxidation   4, 11
– reaction mechanism   11
aldol condensation   214
– base-catalyzed   214
alkali metal boronates   154
– enthalpy diagram   154
alkaline anion-exchange membranes (AAEMs)   30
– poly(ethylene-co-tetrafluoroethylene)   30
– poly(hexafluoropropylene-co-tetrafluoroethylene)   30
alkaline derivatives   329
alkaline direct alcohol fuel cell   31
– tests   31
alkaline earth metal boronates   154
– enthalpy diagram   154
alkaline electrolyzers   251, 257
alkaline solutions   248
alkaline water electrolysis   266
alloy systems   163
aluminum hydride (AlH_3)   141, 160
ammonia   164, 401, 402, 405, 412
– adsorption-desorption kinetics   401, 402, 412
– decomposition catalysts   163
– precursor   395
– shut-off   405
– storage   163
– thermal desorption   164
anaerobic fermentation   273
anaerobic metabolism   272
analog-digital board   303
anion-exchange membrane (AEM)   16, 29, 31, 36

anisotropic lattice expansion 161
anode catalyst, layer 32, 49
anodic oxides 257
anti-reflective coating (ARC) 352
aqueous phase reforming process (APR) 185, 190–193, 214, 283
– advantage 191
– disadvantage 191
– reaction pathways 191, 211, 215
area velocity (AV) 401
Argonne national laboratory 297
Arrhenius equation 148
Arrhenius plots 147
atmospheric pollutants 174
– carbon monoxide 174
– fine particulate matter 174
– nitrogen oxides 174
– sulfur dioxide 174
atomic absorption spectrometry 273
ATR7B catalyst 312
– catalytic activity 312
Au nanoparticles 184
austenitic stainless steel 114
– AISI 304 114
– AISI 316 114
automotive industries 92
autothermal reforming (ATR) process 185, 189, 292, 294
– biodiesel processor 299
– catalyst formulations 294
– chemistry to engineering 294
– fuel processors 294
– reaction 295, 316
autothermal reforming (ATR) reactor 189, 287, 293, 298, 300, 303, 304, 305, 308, 316
– laboratory apparatus 303
– main parts 304
– operating conditions 306
– performance 309
– reactants preheating influence 307
– schematic drawing 304
– schematic representation 293
– setup 306, 307
– start-up phase 306
– temperature profile 312
autothermal reforming (ATR) technology 315
average biodiesel emissions 324
Avogadro's number 123
AZEP concept 372, 373
– flow diagram 372

b

Ba-abundant species 416
backbone polymer matrix 31
bacterial growth 281
bacterial strain(s) 275, 279
balance of system (BoS) 345, 359
– costs 360
– impact of 359
– prices 360
ball-milling catalyst-doping approach 146
base catalysts 326
batch reactor 277
batteries feasibility roadmap 95
bifunctional mechanism 25, 182, 208
bifunctional reaction pathway 213
bimetallic catalysts 218, 296
bimetallic systems 183, 210
binary hydrides 134
biodiesel 324–326, 328, 336
– advantages 324–326
– industry 279
biodiesel production 220, 326, 328, 336–341
– main features 328
– selective hydrogenation 336
bioethanol partial oxidation 203
biofuels 271
– alcohol 271
– biodiesel 271
– oil 271
biogas production 271, 275
bioglycerol dihydrogen 279–284
biomass 178, 185, 187, 222
– carbohydrates 210
– derived compounds 187
– derived feedstocks 185
– derived oxygenates 222
– production 177
– pyrolysis 186, 193
– resources 178, 213
bipolar plates 17, 19
– role 19
body-centered cubic (bcc) structures 160
– alloys 160
boil-off losses 119
boil-off hydrogen limits 120
borazane 163
– empirical formula 163
borohydride 156–157
– hydrolysis reaction 37
Bragg–William approximation 137
Brunauer, Emmett and Teller (BET) model 122
– surface area 126
bulk donor-acceptor heterojunction solar cells 358
Butler–Volmer equation 243, 244

c

cadmium indium selenide (CIS)   101
cadmium telluride (CdTe) systems   101
carbohydrates conversion pathway   210
carbon nanofibers (CNFs)   71–74, 76, 80, 84, 85
– catalytic performance   72, 76
– cyclic voltammograms   76
– dispersed mixture   74
– FE-SEM images   75
– FE-TEM images   75
– nanodispersion   85
– platelet (P-CNF)   73
– preparation   73
– selective synthesis   71
– single cell performances   77
– structural effects   74
– structure-controlled syntheses   72
– supported catalysts   76, 80
– surface areas   72
– thick herringbone (thick H-CNF)   73
– two-step gasification procedure   72
– very thin herringbone (very thin H-CNF)   73
carbon nanostructures   125
carbon nanotubes (CNTs)   72, 123
carbonaceous precursors   206
carbonaceous species   295
Carnot efficiency   100
Carnot's theorem   3, 6
catalyst activity   81
– dispersion effect   81–84
catalyst design   179–181
catalyst effects   147
catalyst formulation   297
catalyst impregnation   85
catalytic bed   304, 307, 310, 311, 313
catalytic burner   364
catalytic cycle   398
catalytic gasification   85
catalytic hydrogen generation   193
catalytic materials   375–387
– active catalyst layer   376
– PdO-based catalysts   377
– structured substrate   376
catalytic partial oxidation (CPO)   185, 188–189, 290
– advantages   188
– disadvantages   189
– methane   382
catalytic partial oxidation of methanol (CPOM)   194, 196
– catalysts   196
catalytic pilot burner processes   371
catalytic process(es)   180, 183, 292

catalytic reactor(s)   180, 247, 380, 395, 396
– annular reactor   380
– examples   380
– plate cell reactor   380
catalytic reforming reactions   178, 186
– mechanism   186
catalytic steam reforming (CSR)   185–188
– natural gas   187
catalytic steam reforming of methanol (CSRM)   194, 195
– drawback   195
– major pathways   194
catalytic system(s)   179, 200, 206, 427
catalyst precursor   147
catalyst preparation   83
cathode exhaust, stream   49, 51, 52, 64
cathode flooding   48
cathode performance   51
– air feed rate   51
– implications   51
– water balance   51
cathodic catalysts   38
cation-exchange membranes   31
cause-effect process   277
C–C bond(s)   190, 198, 200, 201, 209, 214, 218
– cleavage   209
– formation   182
cell design(s)   241–242
– advanced alkaline electrolysis   241
– conventional alkaline electrolysis   241
cell voltage   6, 18, 26, 27, 37, 61, 62
– definition   6
– limiting factors   6
– zone I   7
– zone II   7
– zone III   7
cellulosic biomass   178
ceria-zirconia-alumina matrix   202
ceria/zirconia-supported catalysts   201
cetane value   336, 337
charge density distribution   162
charge transfer coefficient   10
C–H bond cleavage   214
chemical acid-base equilibrium path   256
chemical hydrides   154
chemical oxide path   255
chemical potential   138
chlor-alkali industry   257
chloromethylated aromatic polymers   29
chlorosulfonic acid precursor   334
classic surface catalyst   147
clean hydrogen-rich synthesis gas (CHRISGAS)   205

Clostridium pasteurianum 275, 276
– active center 276
– active site 275
C–O bond(s) 190, 211, 215, 216
– cleavage 216, 218
CO-DRIFT spectra 385
CO shift reaction 293
coke formation 201, 302, 310
combined cycle gas turbine (CCGT) 175
combined heat and power (CHP) 175
combustion process 366
– kinetic regimes 366
commercial catalysts 396
commercial process 180
– catalyst selection 180
common rail (CR) fuel injection system 393
compact fuel processor 303
complex hydrides 141–154
complex metabolic food chain 271
complex methanation process 272
compressed hydrogen 112, 115, 120
– hydrogen density 120
– properties 112
– volumetric density 115
computational fluid dynamics (CFD) calculations 396
concentrator technology 357
conducting matrix, *see* electron-conducting polymer
conradson carbon residue (CCR) 339
– containing tubes 340
– evaluation 340
continuous fossil fuel 174
conventional chemisorptions techniques 379
conventional electrolyzer 246
conventional water electrolysis 241, 264
copper -based catalyst 195, 196, 197, 202
– Al_2O_3 system 201
– catalytic properties 195
– morphological characteristics 197
– nickel catalyst 73
– SiO_2 catalysts 338
– TiO_2 systems 341
– zinc-based catalysts 196
copper-based systems 337
copper gallium indium diselenide 101
copper hydrogenation catalysts 337
copper indium gallium selenide 354
copper indium selenide (CIS) 354
coprecipitation method 183, 184
– steps 184
corroding metals 257
counter-current exchanger 99
critical potential 258

critical value 404
crystalline silicon technology 348–353
– feedstock to wafers 348–349
– wafers to cells modules 349
crystallization process 356
current-potential curves 74
current densities 7, 8, 18, 26, 27, 245, 246, 253
cyclic voltammetry (CV) 74
Czochralski method 348

d

decomposition reaction 143
decomposition temperature 153
dehydration/hydrogenation reactions 214
dehydrogenation/isomerization reactions 216
dehydration reactions 192, 211
– acid-catalyzed 192
d-electron rule 251
demanding reaction 255
density-of-states curve 134
density functional theory (DFT) calculations 218
depolarization process 240
desorption experiments 147
desorption kinetics 160
Desulfomicrobium baculatum 276
diesel oxidation catalyst (DOC) 400
– vehicles layout 400
diesel particulate filter (DPF) 393
– models 414
differential scanning calorimetry (DSC) 149
diffusion coefficient 115
dihydrogen 282, 283
– catalytic production 282, 283
dimensionally stable anode (DSA) 251
dimethyl ether 203–205, 207
– advantage 204
– application 204
– hydration 205
– hydrolysis 205, 207
– partial oxidation 207
– reforming catalyst 206
– steam reforming 205
– steam reforming activity 206
– synthesis process 204
dioxygen 13
– electrocatalytic reduction 13
direct alcohol fuel cell (DAFC) 4, 11, 13, 27, 28
– electrocatalytic oxidation 11
– setup 26
– studies 26
– tests 26

direct borohydride fuel cell (DBFC)   36, 37
direct ethanol fuel cell (DEFC)   3, 4, 21
– challenge   3
– introduction   3
– principle   21
– schematic diagram   21
direct methanol fuel cell (DMFC)   4, 12, 32, 47, 48, 57, 71, 203
– anode catalysts   71
– cathode exhaust   48, 54, 55
– commercial applications   47
– current density   54, 55
– introduction   47
– noble metals   71
– performance   47
– power pack   57
– power system(s)   47, 57, 64, 68
– reactions   56
– schematic diagram   48
direct methanol fuel cell (DMFC) stack   12, 49, 52, 54, 56, 64, 65, 67, 68
– energy conversion efficiency   49
– performance parameters   65, 66, 68
– properties   65
dispersive interactions, *see* van der Waals interaction
disproportionation reaction   421
distributed energy system   288
doping agent   38
DRIFT spectroscopy   335
dry-gas reaction   144
dry oxides   257
dual reactor system   212, 213
– schematic representation   212
dynamic hydrogen electrode (DHE)   64

e

effective medium theory   135
– hydrogen absorption   135
efficient recovery systems   96
electric powertrain vehicle   92, 94
– hypotheses   92
– performance   92
electrical energy   239
electroactive alcohol fuels   16
electrocatalysis   244–245, 249, 259–260
– factors   252–255, 260–263
– targets   245
– theory   245–248, 249
electrochemical activation energy   10
electrochemical cell   4
electrochemical desorption   250
electrochemical kinetics   259
electrochemical oxidation   84

electrochemical processes   235, 239, 244
electrochemical reactions   6, 7, 9, 10, 19
– activation barrier   10
– electrocatalysis rate   9
electrochemical reactor   237
electrochemical system   237, 238
electrochemical technology   236
– advantage   236
electrochemical terms   240
electrochemical/non-electrochemical technology   243
electrode catalysts   17, 19
– anode   17
– cathode   17
electrode potential   6, 9, 37, 237
– anode   6
– cathode   6
electrode reaction   245
electrolytic cell   240, 244, 263
electrolytic method   235
electrolytic oxides   257
electrolytic water splitting   235
electron-conducting polymer   13
– polyaniline   13
– polypyrrole   13
electron-conducting substrate   12
– carbon powder   12
electron-deficient molecules   148
electronic factors   252
electro-osmotic drag coefficient   55
elemental cathode   251
elementary fuel cell   20, 32
elementary reaction processes   181
Eley–Rideal (ER) kinetic expression   398, 401
embedded catalyst(s)   183, 184
– design   185
– emerging strategies   183
– ideal structure   184
empirical models   133
end-of-pipe strategies   174
endothermic reaction   133, 189
energy-intensive techniques   351
energy balance   239
energy conversion efficiency   48
energy diagram   146
energy efficiency   6, 8, 9, 22
– definition   6, 22
energy generation   359
energy production   363, 364
– catalytic combustion   363, 364
energy systems   89, 173, 193
energy vector   235
engine injection systems   323
engine test-bench experiments   409

equilibrium term 240
ESR reactions 200, 201
esterification catalysts 328–336
ethanol oxidation 22
– reaction mechanisms 22
E-TEK Vulcan XC-72 catalyst 76, 78
ethylene glycol (EG) 32, 214, 217
– acid-catalyzed dehydration reactions 217
– aqueous phase reforming 214
– electro-oxidation 33–35
– polarization curves 34
– reaction pathways 215
– SPAIR spectra 35
EUCAR website 92
EU-funded studies 356
– MUSIC-FM 356
European emission standards 393–395
– Euro V emission standards 393
– Euro VI standards 393
European specification 339
European stationary cycle (ESC) 409
– experiments 413
European transient cycle (ETC) 409
– experiments 413
exhaust gas recirculation (EGR) 394
exhaust gases 399
exothermic reaction 81, 133, 189
extraction energy costs 174

f

face-centered cubic (fcc) lattice 109
– close-packed structure 123
faradaic efficiency 8, 9, 22, 26, 33
faradaic stoichiometry 48, 49, 52, 57, 60
Faraday constant 6, 236
Faraday's law 52
Fast selective catalytic reduction (SCR) mechanism 409
– features 411
– reaction 411, 433
– stoichiometry 411
fatty acid methyl esters (FAMEs) 323, 324, 340
feedback control system 407
fermentable biomass 273
fermentation processes 197
Fermi energy 134
Fe-Zn double metal cyanide (DMC) 335
fibrous nanocarbon materials 72
– carbon nanotubes (CNTs) 72
– CNFs 72
first-generation biofuels 197
Fischer–Tropsch processes 216
Fischer–Tropsch reactions 217

Fischer–Tropsch synthesis 191, 192, 220
five-cell stack 51, 52, 65, 67
– life test results 67
– test results 52
fixed-bed reactors 328
flex-fuel vehicles 4
fluidized-bed reactor 187, 351
fluidized-bed reformer 187
fluidized-bed system 187
fluorine-containing polymers 30
Food and Agriculture Organization of the United Nations 327
foreign metals, Cr 41
fossil fuels 173, 235, 236, 263, 287, 288, 289, 325
– advantages 289
– coal 173
– hydrogen production 287
– natural gas 173
– oil 173
– use 236
four-electron mechanism 13
Fourier transform infrared (FT-IR) spectroscopy 398, 416, 420
– analysis 421
– data 416, 424
– reflectance spectroscopy 23
– spectra 418, 419, 420, 424
free fatty acid (FFA) 327
fuel-air mixture 367
fuel cell 3–6, 14, 97, 196, 205, 213, 286, 296
– applications 3, 193
– catalysis 9
– catalysts 74
– catalytic combustion 373
– characteristics 28, 29
– different types 14–17
– hydrogen-rich 205
– kinetics 6
– performance 30
– performance characterization 74
– preparation 74
– reactions 9
– small-scale units 293
– stack characteristics 20
– system(s) 8, 177, 214
– technology status 15
– tests 26
– thermodynamics 5
– working principles 4–14
fuel electric mobility 103
– perspectives 103–104
fuel utilization efficiency 60

full electric mobility 89
– current grand challenges 89

g
gas bubbles 241
gas chromatography (GC) analysis 273
gas diffusion conditions 264
gaseous pollutants 175
gas hourly space velocity (GHSV) 298, 307
gasification catalyst 73
gasoline fuel processor 299
gas-phase combustion 370
gas-phase experiments 250
gas-phase reactions 366
gas turbines system (GTs) 363–370, 375, 376, 387
– applications 382
– catalytic combustion 375
– characteristics 365
– compressor 372
– design concepts 366
– fuel staging 367
– fully catalytic combustor 366
– lean catalytic combustion 364
– operations 365, 366
– partial catalytic hybrid combustor 367
– performance 366
– principles 364
– reference fuel 376
– system requirements 364
gas turbines system (GTs) combustors 365, 379–381, 387
– combustor catalysts 379
– design criteria 365
– operating conditions 365
geometric mean 258
Gibbs free energy 5, 21, 300
glass fibers 396
glass microspheres 117
glass spheres 117
– scanning electron micrograph 117
global civilization 174
global economy 174
global energy consumption 173
glycerol 219, 220, 222, 279, 281, 283, 327, 329
– aqueous phase reforming 222
– ATR 222
– bioconversion 274–275
– market 219
glycerol catalysts 220
– sulfated ZrO_2 220
gravimetric hydrogen density 111, 114, 116, 120, 126, 153, 158
green chemistry 210

greenhouse effect 174
greenhouse gases 236, 266, 288
– CO_2 236
– emissions 325

h
hazardous chemicals 176
health hazards 174
heat exchangers 305
– preheating 305
– water 305
heat-generating compound 165
– Fe_2O_3 165
– $LiAlH_4$ 165
– $NaBH_4$ 165
Hertz–Knudsen equation 121
heterogeneous catalysts 179, 190, 244
– acid catalysts 330–336
– basic catalysts 328–330
– use 328
heterogeneous catalytic systems 327
heterogeneous Esterfip-H based process 329
– advantage 329
– flow sheet 329
heterogeneous transesterification 328–336
heterojunction intrinsic thin layer (HIT) 353
– cells 353
– Sanyo modules 353
hexaaluminate (HA) catalysts 381–382
– metal-substituted 381
– materials 381
hexagonal plate-like crystallites 382
high cell packing density 62
high-efficiency cells 357
higher heating value (HHV) 92
high-frequency cell resistance 62, 63
high-performance liquid chromatography (HPLC) measurements 34
high-quality biodiesel 323
– catalysis 323
– production 323
high-resolution transmission electron microscopy (HRTEM) 385
high surface area graphite (HSAG) samples 126
high-temperature shift (HTS) reactors 299
hinder catalyst 365
Hofmann degradation reaction 31
homogeneous acid-catalyzed process 333
– flow sheet 333
homogeneous acid-catalyzed transesterification process 330
honeycomb catalyst 401
honeycomb monolith 376

– SCR converter   413
Hungate technique   275
hydrocarbon feedstock   292
hydrocarbon-selective catalytic reduction (HC-SCR)   394
hydrogen   109, 110, 118, 121, 124, 135, 137, 138, 143, 287
– adsorption isotherm   127
– based energy systems   288
– chemical potential   138
– consumes energy   112
– economy   236
– electrode   259
– evolution reaction   246
– fuel cells   92
– generation   157, 221
– nanotubes   123
– ortho-hydrogen   118
– oxygen fuel cell   5
– para-hydrogen   118
– physical properties   110
– potential   137
– primitive phase diagram   109, 110
– production costs   316
– production efficiency   199, 266
– production parameters   271
– production plants   313
– production process   201
– reaction exchange current density   11
– storage materials   109, 157
– storage methods   109, 111
– sustainable distributed production   287
– van der Waals interaction   121
– volumetric density   135
– wavefunction   118
hydrogenation process   337
hydrogenation reactor   213
hydrogen desorption   144, 152, 164
– reaction   150
– schematic mechanism   144
hydrogen evolution reaction   248–255
– cathodes materials   251
– electrocatalysis   249
– electrocatalysis factors   252–255
– electrocatalytic activity   253
– reaction mechanisms   248
– volcano curve   249, 252
hydrogen-fed fuel cells   16
– alkaline fuel cell   16
– direct methanol fuel cell   16
– molten carbonate fuel cell   16
– phosphoric acid fuel cell   16
– proton exchange membrane fuel cell   16
– solid alkaline membrane fuel cell   16

– solid oxide fuel cell (SOFC)   16
hydrogen on demand system   157
hydrogen oxidation reaction (HOR), rate constant   10
hydrogen-to-carbon ratio   193
hydrolysis reaction(s)   37, 156, 157
hydrous oxides, *see* anodic oxides
hypothetical desorption mechanism   151

i
impregnated catalysts   181
– promoters   181
– role of metal   181
– support   181
in-series two-step pathway   431, 432
in situ electrolytic separation   155
in situ IR reflectance spectroscopy   25, 34
– measurements   25
in situ Raman experiments   410
intensive energy process   292
intermetallic compound(s)   129, 133, 134
internal combustion engine (ICE)   92, 98
internal heat exchange monoliths   368
– examples   368
International Association for Hydrogen Energy   236
interstitial hydride   128–130, 132, 143
– formation   132
intrinsic chemical rate constant   398
intrinsic conversion efficiency   98
intrinsic kinetic constant   397
investment costs estimation   315
ion-exchange resins   333
– Amberlyst-15   333
– nafion   333
IR data   412
IRSA analytical method   273
IR spectra   418
isothermal plug flow reactor model   402

j
Jatropha curcas   342
Joule–Thompson cycle   118, 119
– expansion   118

k
Kilowatt-scale autothermal reforming (ATR)   298–299
– fuel processors   298–299
– reactor   303
kinetic considerations   239–240
kinetic constant   399
Klebsiella pneumoniae   279, 281
Klebsiella strain(s)   274, 279

– characterization 274
– strain K1 274, 279
– strain K2 274, 279
– strain K3 274, 279
– strain K4 274, 279
– use 274, 279
Kraft process 338
Kyoto protocol 222

l

Langmuir approach 402, 403
Langmuir isotherm 122
Langmuir kinetics 401
laser based method 194
lattice expansion 132
lattice gas 140
– chemical potential 140
– model 137–140
laves phase alloys 37
lean-feed scheme 60
lean fuel catalytic combustion 371
– advantages 371–372
Lennard–Jones potential 131
Lewis acid(s) 141, 335
lignocellulosic biomass 178
LiH-Si system 162, 163
Linde cycle, *see* Joule–Thompson cycle
liquefaction processes 178
liquid hydrogen 117–118
– liquefaction process 118
liquid hydrogen storage systems 121
liquid petroleum gas (LPG) fuels 204
– butane 204
– propane 204
liquid water flux 55
lithium tetrahydroboride 148
local band- structure model 136
logarithmic curve 7
low-bandgap semiconductor 358
lowest achievable emission rate (LAER) technology 415
low space velocity 310
low-temperature fuel cells 17, 32, 71
– DAFCs 17
– direct methanol fuel cell (DMFC) 71
– platinum-based catalysts 32
– polymer electrolyte membrane fuel cell (PEMFC) 17, 71
low-temperature shift (LTS) reactors 299

m

magnetoplumbite structures 381
matter organic non-glycerol (MONG) 329
mean field theory 137

membrane-electrode assembly (MEA) 17, 19, 26, 242
membrane fuel cell 29
– key component 29
metal-ammine complexes 163
metal-based catalysts 208
metal-carbon bonds 215
metal catalysts 181, 218
metal enzymes 272, 275
– role 275
metal honeycomb structure 298
metal hydride 128–141, 143, 161
– schematic enthalpy diagram 143
– short H-H-distance 161
metal lattice, hosts 130, 143
metal nanoparticles 184
metal precursor 181
metal-support interaction 183
methane partial oxidation 384
methane production 271
– process parameters 271
methanol/ethanol-fed fuel cells 16–17
methanol feed stoichiometry 60
methanol mass balance 55
methanol steam reforming component 206
methanol-to-oil ratios 331
methanol-to-substrate ratio 335
MgH_2 structure 162
– alloy formation 162
– destabilization 162
MH system 140
– coexistence curve 140
– spinodal line 140
microaerobic/anaerobic conditions tests 275
microcombustors 373–375, 387
– application 374
– based systems 373
– schematic drawing 375
microemulsion technique 183
microfabrication technologies 373
– development 373
microporous meta-organic framework 126
Miedema model 134
– schematic representation 134
millennium cell 157
MIT microengine project 373
modified redox (MR) rate law 405
monocrystalline silicon 348
monolayer surface area 123
– unsteady-state models 412–414
monolith SCR reactor 406, 410
– unsteady-state models 406
Monte Carlo program 123
muffin-tin orbital approach 161

multi-electron transfer reaction   13
multiple process technologies   288

n

NaBH$_4$ reaction   157
– energy diagram   157
nafion membranes   18, 28, 30, 47
– behavior   28
– N112   28
– N115   28, 30
– N117   28, 49
nafion solution   74
nanocrystalline oxide electrocatalysts   35
nanodispersion procedure   83
nanotunneled mesoporous H-CNF   78, 81
– FE-SEM photographs   79
– FE-TEM photographs   79
– preparation   73
– single cell performance   81
– structure   78
National Energy Policy   173
natural gas   288, 291, 292, 298, 313, 365
– catalytic ATR technology   316
– combustion   365
– economic scenario   313
– fuel processor   299
– pipelines   289
NDIR multiple analyzer   303
nichrome heating wire   165
nickel-based catalyst   182, 202, 211, 217, 291, 310, 341
– Al$_2$O$_3$-based catalysts   295
– MgO catalyst system   383
– platinum bimetallic catalysts   296
nickel-based systems   182
nickel-iron hydrogenases   276
nitrate routes, kinetic model   422
nitrite route pathway   420–422
– kinetic model   422
nitrogen oxides   367, 370, 393, 394, 397, 398, 400, 414–422, 425, 426, 431, 424
– adsorption reaction pathway   420, 421
– catalytic removal   393
– CO$_2$ effect   422
– emissions   323, 370, 400
– first-order rate constant   397
– formation   367
– kinetic constant   422
– kinetics   421
– mechanistic features   415–421
– oxidation reaction   422
– rate constant   398
– reduction   397, 398, 431
– reduction levels   393
– species   418, 425, 426
– storage component   414
– surface species   424
nitrogen oxides storage reduction (NSR)   394, 414–432
– catalysts   414, 415
– systems   414, 433
– technology   414–415, 433
noble metal catalysts, features   383
n-octanoic acid   334
non-catalytic processes   220, 292
non-petroleum renewable resource   176
non-thermal plasma (NTP)   394
nuclear power plants, advantage   175

o

ohmic dissipation term   241–242
ohmic drop(s)   6, 241, 243
oil-to-methanol ratio   335
oleic acid   336
open-circuit conditions   60
open circuit voltage   27, 28
ordinary differential equation (ODE) system   409
organic compounds, utilization   272
organic solar cells   347, 357
– utilization   271
organic waste energy   271
Organization for Economic Cooperation and Development (OECD)   89
ortho-hydrogen   118
over fire air   393
oxidation mechanism, step   11
oxidation reaction   10, 12, 23, 156
oxidative reforming process   198
oxide matrices   383
oxygen evolution reaction   255–264
– anodic oxides   256–257
– electrocatalysis   259–260
– intermittent electrolysis   263
– reaction mechanisms   255–256
– thermal oxides (DSA)   257–259
oxygen reduction reaction   41
– polarization curves   41
oxygenate reforming reactions   193
– acetic acid   207–210
– dimethyl ether   203–207
– ethanol   197
– ethylene glycol   214
– glycerol   219
– key examples   193
– methanol   193
– sugars   210
oxygenates reforming   173

– catalyst design   173
ozone-destroying chlorofluorocarbons (CFCs)   204

p
palladium catalysts   206, 379
palladium oxide (PdO)   368, 377, 378
– based catalysts   368, 377–381
– properties   377
– systems   376
– thermal decomposition   377, 378, 379
paramagnetic analyzer   303
partial catalytic hybrid combustor concept   367, 369
partial differential equations model   409
partial oxidation of ethanol (POE)   198
partial oxidation process   196, 292
particulate matter (PM) reduction   394
Pauling electronegativities   142
Pd-based electrocatalysts   35
petroleum-based feedstock   178
Photovoltaics   345
– current trends   345
– future vision   345
– module   355
– principles   345
– products   347
photovoltaic electricity   348, 361
photovoltaic market   346
– challenges   346–348
photovoltaic sector   347, 351, 353
– challenge   347
– roadmap   347–348
photovoltaic solar energy   345
photovoltaic systems, aspects   345, 359, 360
photovoltaic technologies   345, 355, 357, 361
– aspects   355–357
photovoltaics production   346
physisorption advantages   121, 128
pilot-scale reactor   387
pipeline systems   289
plant-derived fatty substances, oleic acid   336
plasma-enhanced chemical vapor deposition (PECVD)   352
platinum-based catalysts   19, 32, 42, 210
platinum-based electrodes   24
platinum-free materials   40
– cobalt macrocycles   40
– iron macrocycles   40
platinum-palladium catalysts   34
point of zero charge (pzc)   258
polybenzimidazole (PBI) membrane   32
– alkali-doped   32
polyepichlorhydrin homopolymer   31

polyethylene-type matrix   31
polymer degradation   30
– oxidative radical mechanism   30
polymer electrolyte membrane   53
polymer electrolyte membrane fuel cell (PEMFC)   71, 289
polymer electrolyte membranes, *see* Nafion membranes
polymer membrane   19
– Nafion   19
polymeric hydrides   141
polytetrafluoroethylene (PTFE)   120
portable electronic devices   3
– camcorders   3
– cell phones   3
– computers   3
portable micro fuel cells   211
portable power market   47
positional isomerization   337
potential catalytic materials   180
potential difference   243
potential efficiency   22
potential energy curve   130
power density   77, 82
power-energy needed vehicles   90
– basic formulation   90
power electronic, *see* Silicon Carbide technology
power electronics converters   102, 103
power generation system   288
power systems   89
POX reaction   292
POX systems   293
precious metal oxides   251
– IrO_2 251
– Rh_2O_3 251
– RuO_2 251
preferential oxidation (PROX) reactions   289
preheating effect   300
– reactor temperature profiles   308
preheating method   308
pressure-composition isotherms   132, 145, 160
pressure-swing adsorption   190, 289
– process   290
– unit   313
pressure vessel   113–114
– schematic representation   113
pressurized hydrogen   111–117
– density   113
process control techniques   357
proton exchange membrane fuel cell (PEMFC)   4, 18
– auxiliary/control equipment   19

- detailed scheme 20
- direct ethanol fuel cells 4
- schematic representation 18
- system 19, 20
proton exchange membrane (PEM) 4, 13, 18, 21
- applications 17
- fuel cell 193, 209
- principle 17
- role 18
pseudo-first-order kinetics 330
Pt-based binary catalysts 14
- Pt/Cr 14
- Pt/Ni 14
Pt-based electrocatalysts 27
Pt-catalyzed pathway 431
Pt/C catalyst 25, 38, 39
Pt/CeO$_2$ catalyst 296
Pyrex batch reactors 273
pyrolytic process 220

q

quantum confinement effect 358
quaternized polyether sulfone cardo polymers 30

r

Raman techniques 72
Raney Ni-based systems 218
Raney nickel catalysts 217, 218, 222
range extenders, fuel-based 96
rate parameters 408
reactants preheating effect 300
reaction enthalpy 159
reaction rate 19, 252
real-world application 66
redox mechanism 398
reduction mechanism 427, 431
reduction process 424
reduction reaction 10, 13
reforming catalyst 309
reforming efficiency 289, 298
reforming process 186, 192
reforming reaction 185, 188, 197
- aqueous phase reforming 185
- autothermal reforming 185, 290
- catalytic partial oxidation 185
- catalytic steam reforming 185
- partial oxidation (POX) 290
- process principles 185
- steam reforming (SR) 290
regeneration process 187, 415
regenerative methods 91
renewable biomass 177

renewable energy sources 16, 175, 176
- biomass 16
- hydroelectric power 16
- solar 16
- windmill 16
renewable energy technologies 176
renewable sources 4, 176, 204, 210
renewables-based routes 155
- electrolysis of water 155
- solar-generated electricity 155
reverse water gas shift reaction (r-WGSR) 205
Rh/Al$_2$O$_3$ catalysts 386
Rhodospirillum rubrum 283
ribbon technology 349, 352
rich catalytic lean (RCL) burn technology 370
- base configuration 370
rich combustion catalysts 382
- cobalt catalysts 382
- nickel catalysts 382
roll-to-roll technology 358
rotating ring disk experiments 39

s

Sabatier principle 247, 250
sacrificial reagents 194
Sanyo HIT modules 353
saturated calomel electrode (SCE) 238
scanning electron microscopy (SEM) 72
scanning tunneling microscopy (STM) 72
scarce fossil reSources 176
- future opportunities 432
- open issues 432
selective catalytic reduction (SCR) 394, 395, 400
- activity 398
- applications 399–400
- catalysts 395, 396, 397, 400, 405
- efficiency 413
- future 399
- reactor modeling 400–409
- steady-state modeling 400
- technology 400, 405, 432
- vehicles layout 400
self-sustaining reaction 165
serial hybrid electric vehicle 97
- general scheme 97
silica-based microspheres 115
silicon carbide technology 103
silicon feedstock 349, 351
- low-price 351
- process flow 349
- production 356
silicon nitride antireflective coating 350
silicon precursors 349

– trichlorosilane (SiHCl$_3$)   349
silicon solar cell   350
– principle   350
silicon wafer-based technology   347
single-cell tests   61
single-phase systems   139
single-process tools   356
single-walled carbon nanotube (SWNT)   125
SO$_2$ oxidation reaction   408
solar cell technologies   358
solar cells   349, 358
solar-grade silicon   351
solid-phase crystallization (SPC) technique   183
solar reactor   155
solid acid catalyst   213
– advantages   213
– silica-alumina   213
solid alkaline membrane fuel cell (SAMFC)   29–42
– development   29–32
solid metal lattice   137
solid polymer electrolyte (SPE) technology   237, 242
– electrolyzers   242
solid polymer fuel cell   199
solvent evaporation   238
– limitation   238
spectro-electrochemical study   26
spectroscopic techniques   180
stack anode exhaust stream   50
stack energy conversion efficiency   60
stainless-steel reactor   303
stand-alone process   292
standard electromotive force   6
standard hydrogen electrode (SHE)   5, 9, 238
standard SCR reacting system   409
standard SCR reaction   401
– unsteady-state Kinetics   401
start-up phase   306
static conversion devices   374
– thermoelectric systems   374
steam-to-carbon ratios   291, 293, 298
steam gasification   220
steam reforming of ethanol (SRE)   198
– advantage   198
– reaction pathways   199
steam reforming process   187, 208, 290, 294
steam reforming reaction   195, 294, 308
steam reforming technology   288
stoichiometric conditions   197
stoichiometric reactions   211

stored NOx reduction   424–432
– mechanism   425
– reaction network identification   428
strong metal-support interaction (SMSI)   254
sulfur-containing fuels   395
– coal   395
– oil   395
supercapacitors feasibility roadmap   95
surface oxide, formation   256
surface science techniques   181
surface-to-volume ratio   120, 374

t
Tafel equation   244
Tafel slope   248, 249, 250, 251, 253, 255, 259, 260, 261
temperature-programmed desorption (TPD) curves   402
temperature-programmed surface reaction (TPSR)   426
– experiments   426, 428, 431
tensile strength   114, 117
ternary system   129
thermal conductivity detector   303
thermal cracking process   220
thermal cycles   99
thermal decomposition (TPD)   426
thermal desorption   159
thermal desorption spectroscopy   149, 151
thermal expansion coefficients   350
thermal heat flux   99
thermal hydrogen desorption   150, 161
thermal oxides   260, 262
– electrocatalytic activity   260
– surface enrichment   262
thermal power   150
thermal reduction process   154
thermodynamic analysis   287, 299–302
thermodynamic evaluation   303
thermoelectric (TE) materials   98
thermoneutral potential   239
thin-film modules   354
thin-film photovoltaics   101
– cells   101
– module   354
thin-film technologies   353, 354
– advantages   355
– improvement requirements   354
three-way catalysts   174, 394
time on-stream testing   295
transesterification catalysts   334
transesterification pathway   330
transesterification reaction   327, 339
transient microreactor experiments   413

transient response method 403, 416
transition metal (s) 128, 129, 181, 249, 295
– binary hydrides 129
– Co 295
– Cu 295
– Fe 295
transmission electron microscopy (TEM) 72
tree producing non-edible oils 332
– types 332
trichlorosilane (SiHCl$_3$) 349
tungstated zirconia-alumina 334

u

ultra-low-sulfur diesel fuel 415
underpotential deposition (UPD) 33
University of New South Wales 352
University of Salerno 287
US Environmental Protection Agency (EPA) 369, 415
UV/Vis analysis 281

v

vacuum-insulated spherical dewars 119
Van't Hoff plots 136
Van der Waals interaction 121, 130
Vanadium oxide 396
vapor water flux 52
variable costs estimation 315
virent energy systems 210
virtual oxidation 413
– catalyst model 413
volatile fatty acids (VFAs) 275
volcano curve 11, 247, 249, 250, 255, 259
voltage efficiency 8
voltaic cells 263
volumetric hydrogen density 111, 114, 120, 158
volumetric storage density 115

w

wafer-based crystalline silicon 346
wafered silicon technology 350, 352
– aspect 350
wafering process 352
Wakao–Smith random pore model 401, 408
washcoated structured catalysts 297
waste heat rejection 64
water-methanol solution 194
water-soluble biomass-derived oxygenates 177
water-soluble ionic surfactant 184
water electrolysis 235, 237, 240, 265, 266
– state-of-the-art 265
water electrolyzers 255
water gas shift reaction (WGSR) 186, 194, 289, 290, 308
– promoter 218
water-to-ethanol molar ratio 202
water vapor fluxes 67
water vapor saturation level 55
well to wheel evaluations 92

x

X-ray diffraction (XRD) 75, 84, 147, 148
– pattern 153
– synchrotron 149
XONON combustor 368
– schematic configuration 368
XPS analysis 296

y

Young's modulus 114

z

zero-gap configuration 241
zinc cycle 154–156
– energy diagram 156
zirconia crystallization 195
Zn-O$_2$ separation technique 155